Developments in Block Copolymer Science and Technology

Developments in Block Copolymer Science and Technology

Edited by
Ian W. Hamley
Department of Chemistry, University of Leeds, UK

JOHN WILEY & SONS, LTD
Chichester • New York • Weinheim • Brisbane • Toronto • Singapore

Copyright © 2004   John Wiley & Sons Ltd, The Atrium, Southern Gate, Chichester,
West Sussex PO19 8SQ, England

Telephone   (+44) 1243 779777

Email (for orders and customer service enquiries): cs-books@wiley.co.uk
Visit our Home Page on www.wileyeurope.com or www.wiley.com

Reprinted June 2005

All Rights Reserved. No part of this publication may be reproduced, stored in a retrieval system or transmitted in any form or by any means, electronic, mechanical, photocopying, recording, scanning or otherwise, except under the terms of the Copyright, Designs and Patents Act 1988 or under the terms of a licence issued by the Copyright Licensing Agency Ltd, 90 Tottenham Court Road, London W1T 4LP, UK, without the permission in writing of the Publisher. Requests to the Publisher should be addressed to the Permissions Department, John Wiley & Sons Ltd, The Atrium, Southern Gate, Chichester, West Sussex PO19 8SQ, England, or emailed to permreq@wiley.co.uk, or faxed to (+44) 1243 770571.

This publication is designed to provide accurate and authoritative information in regard to the subject matter covered. It is sold on the understanding that the Publisher is not engaged in rendering professional services. If professional advice or other expert assistance is required, the services of a competent professional should be sought.

Other Wiley Editorial Offices

John Wiley & Sons Inc., 111 River Street, Hoboken, NJ 07030, USA

Jossey-Bass, 989 Market Street, San Francisco, CA 94103-1741, USA

Wiley-VCH Verlag GmbH, Boschstr. 12, D-69469 Weinheim, Germany

John Wiley & Sons Australia Ltd, 33 Park Road, Milton, Queensland 4064, Australia

John Wiley & Sons (Asia) Pte Ltd, 2 Clementi Loop #02-01, Jin Xing Distripark, Singapore 129809

John Wiley & Sons Canada Ltd, 22 Worcester Road, Etobicoke, Ontario, Canada M9W 1L1

Wiley also publishes its books in a variety of electronic formats. Some content that appears in print may not be available in electronic books.

Library of Congress Cataloging-in-Publication Data

Developments in block copolymer science and technology / edited by Ian W. Hamley.
    p. cm.
Includes bibliographical references.
  ISBN 0-470-84335-7 (Cloth : alk. paper)
 1. Block copolymers.   I. Hamley, Ian W.
  QD382.B5D492004
  547'.84—dc22

2003016092

British Library Cataloguing in Publication Data

A catalogue record for this book is available from the British Library

ISBN 10: 0-470-84335-7 (H/B)
ISBN 13: 978-0-470-84335-2 (H/B)

Typeset in 10/12 pt Times by Kolam Information Services Pvt. Ltd, Pondicherry, India.
Printed and bound in Great Britain by Antony Rowe Ltd, Chippenham, Wiltshire.
This book is printed on acid-free paper responsibly manufactured from sustainable forestry in which at least two trees are planted for each one used for paper production.

Contents

List of Contributors   vii

Preface   xi

1 Introduction to Block Copolymers   1
   Ian W. Hamley

2 Recent Developments in Synthesis of Model Block Copolymers using Ionic Polymerisation   31
   Kristoffer Almdal

3 Syntheses and Characterizations of Block Copolymers Prepared via Controlled Radical Polymerization Methods   71
   Pan Cai-yuan and Hong Chun-yan

4 Melt Behaviour of Block Copolymers   127
   Shinichi Sakurai, Shigeru Okamoto and Kazuo Sakurai

5 Phase Behavior of Block Copolymer Blends   159
   Richard J. Spontak and Nikunj P. Patel

6 Crystallization within Block Copolymer Mesophases   213
   Yueh-Lin Loo and Richard A. Register

7 Dynamical Microphase Modelling with Mesodyn   245
   JG.E.M. Fraaije, G.J.A. Sevink and A.V. Zvelindovsky

8 Self-consistent Field Theory of Block Copolymers   265
   An-Chang Shi

9 Lithography with Self-assembled Block Copolymer Microdomains   295
   Christopher Harrison, John A. Dagata and Douglas H. Adamson

10 Applications of Block Copolymer Surfactants   325
    Michael W. Edens and Robert H. Whitmarsh

11 The Development of Elastomers Based on Fully Hydrogenated Styrene–Diene Block Copolymers 341
Calvin P. Esneault, Stephen F. Hahn and Gregory F. Meyers

Index 363

List of Contributors

D. H. Adamson
Department of Physics, Princeton University, Princeton, NJ 08544, USA

K. Almdal
Danish Polymer Centre, Risø National Lab, DK-4000 Roskilde, Denmark

J. A. Dagata
Polymers Division, National Institute of Standards and Technology, Gaithersburg, MD 20899, USA

M. W. Edens
Dow Chemical Company, Research and Development, Freeport, TX 77541, USA

C. P. Esneault
Polymer Chemistry Discipline, Corporate Research and Development, The Dow Chemical Company, Midland, MI 48667, USA

J. G. E. M. Fraaije
Soft Condensed Matter Group, Leiden Institute of Chemistry, University of Leiden, NL-2300 RA Leiden, The Netherlands

S. Hahn
Polymer Chemistry Discipline, Corporate Research and Development, The Dow Chemical Company, Midland, MI 48667, USA

I. W. Hamley
Department of Chemistry, University of Leeds, Leeds, LS2 9JT, UK

C. Harrison
Schlumberger-Doll Research, Ridgefield, CT 06877, USA

C-Y. Hong
Department of Polymer Science and Engineering, University of Science and Technology of China, Hefei, Anhui 230026, China

Y-L. Loo
Department of Chemical Engineering, Princeton University, Princeton, NJ 08544, USA

G. F. Meyers
Polymer Chemistry Discipline, Corporate Research and Development, The Dow Chemical Company, Midland, MI 48667, USA

S. Okamoto
Department of Material Science and Engineering, Nagoya Institute of Technology, Nagoya 466–8555, Japan

C-Y. Pan
Dept of Polymer Science and Engineering, University of Science and Technology of China, Hefei, Anhui 230026, China

N. P. Patel
Department of Chemical Engineering, North Carolina State University, Raleigh, NC 27695, USA

R. A. Register
Department of Chemical Engineering, Princeton University, Princeton, NJ 08544, USA

K. Sakurai
Department of Chemical Processes and Environments, The University of Kitakyushu, Kitakyushu 808–0135, Japan

S. Sakurai
Department of Polymer Science and Engineering, Kyoto Institute of Technology, Kyoto 606–8585, Japan

G. J. A. Sevink
Soft Condensed Matter Group, Leiden Institute of Chemistry, University of Leiden, NL-2300 RA Leiden, The Netherlands

A-C. Shi
Department of Physics and Astronomy, McMaster University, Hamilton, Ont. L8S 4M1, Canada

R. J. Spontak
Department of Chemical Engineering and Department of Materials Science & Engineering, North Carolina State University, Raleigh, NC 27695, USA

List of Contributors

R. H. Whitmarsh
Dow Chemical Company, Research and Development, Freeport, TX 77541, USA

A. V. Zvelindovsky
Soft Condensed Matter Group, Leiden Institute of Chemistry, University of Leiden, NL-2300 RA Leiden, The Netherlands

Preface

Block copolymers are important materials in which the properties of distinct polymer chains are combined or "alloyed". A number of valuable books on block copolymers appeared in the 1980s and 1990s, in particular the two volumes "Developments in Block Copolymers" edited by Goodman [1,2] and my own monograph "The Physics of Block Copolymers" [3]. Recently, Hadjichristidis *et al.* [4] have provided an interesting overview of synthesis, together with physical properties. However, there have recently been significant advances in several aspects of the subject that have not been fully reviewed. For example, thin-film morphology characterization and nanoscience and technology applications are presently attracting a great deal of attention. There have also been major developments in computer modelling of phase behaviour and dynamics. New polymerization methods have been introduced that have led to the emergence of novel products and applications. At a more fundamental level, there has been substantial progress in understanding the crystallization process in block copolymers, and the mechanism of phase transformations in block copolymers in bulk phases. This volume is motivated by a desire to provide up-to-date reviews in these key topics. It is by no means exhaustive, but should be a useful introduction to the recent literature.

I wish to thank the contributors for providing the benefits of their considerable expertise in a timely and professional manner. I am also grateful to Jenny Cossham from Wiley for her help in the production of this volume. Finally, thanks to Valeria Castelletto for all her love, support and companionship.

Ian W. Hamley
Leeds, 2003

1. Goodman, I., Ed. *Developments in Block Copolymers – 1*, Applied Science, London, 1982.
2. Goodman, I., Ed. *Developments in Block Copolymers – 2*, Elsevier Applied Science, London, 1985.
3. Hamley, I. W. *The Physics of Block Copolymers*, Oxford University Press, Oxford, 1998.
4. Hadjichristidis, N., Pispas, S., Floudas, G. *Block Copolymers. Synthetic Strategies, Physical Properties and Applications*, Wiley, New York, 2003.

1 Introduction to Block Copolymers

I. W. HAMLEY
Department of Chemistry, University of Leeds, Leeds LS2 9JT, UK

1.1 INTRODUCTION

Block copolymers are useful in many applications where a number of different polymers are connected together to yield a material with hybrid properties. For example, thermoplastic elastomers are block copolymers containing a rubbery matrix (polybutadiene or polyisoprene) containing glassy hard domains (often polystyrene). The block copolymer, a kind of polymer alloy, behaves as a rubber at ambient conditions, but can be moulded at high temperatures due to the presence of the glassy domains that act as physical crosslinks. In solution, attachment of a water soluble polymer to an insoluble polymer leads to the formation of micelles in amphiphilic block copolymers. The presence of micelles leads to structural and flow characteristics of the polymer in solution that differ from either parent polymer.

A block copolymer molecule contains two or more polymer chains attached at their ends. Linear block copolymers comprise two or more polymer chains in sequence, whereas a starblock copolymer comprises more than two linear block copolymers attached at a common branch point. Polymers containing at least three homopolymers attached at a common branching point have been termed mixed arm block copolymers, although they can also be viewed as multigraft copolymers.

In the following, block copolymers prepared by controlled polymerization methods only are considered, primarily di- and tri-block copolymers (see Figure 1.1). Multiblock copolymers such as polyurethanes and poly (urethane-ureas) prepared by condensation polymerisation are not discussed. Whilst these materials do exhibit microphase separation, it is only short range in spatial extent due to the high polydispersity of the polymers.

A standard notation for block copolymers is becoming accepted, whereby X-*b*-Y denotes a diblock copolymer of polymer X and polymer Y. However, sometimes the *b* is replaced by the full term *block*, or alternatively is omitted, and the diblock is denoted X-Y.

A number of texts covering general aspects of block copolymer science and engineering appeared in the 1970s and 1980s and these are listed elsewhere [1]. More recently, specialised reviews have appeared on block copolymer melts and

---

Developments in Block Copolymer Science and Technology. Edited by I. W. Hamley
© 2004 John Wiley & Sons, Ltd.   ISBN: 0-470-84335-7

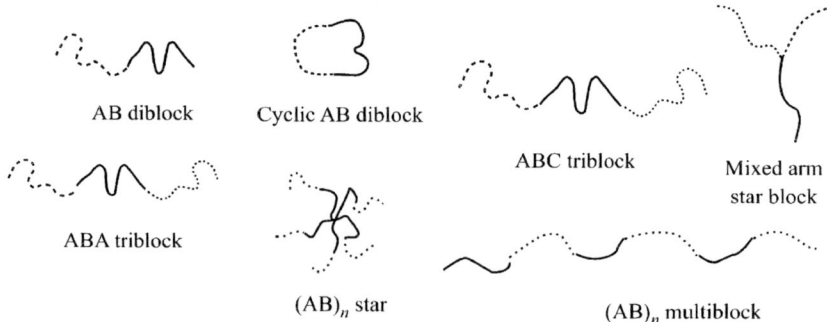

Figure 1.1 Block copolymer architectures.

block copolymer solutions, and these are cited in Sections 1.3 and 1.4 below. The burgeoning interest in block copolymers is illustrated by contributions covering various aspects of the subject in a review journal [2] and in a book [3].

Since the excellent review by Riess *et al.* [4] there have been many advances in the field of block copolymer science and engineering, including new synthesis methods, developments in the understanding of phase behaviour and the investigation of structure and dynamics in thin films. Many of these advances are likely to lead soon to novel applications.

1.2 SYNTHESIS

The main techniques for synthesis of block copolymers in research labs around the world are presently anionic polymerization and controlled radical polymerization methods. The older technique of anionic polymerization is still used widely in the industrial manufacture of block copolymers. Cationic polymerization may be used to polymerize monomers that cannot be polymerized anionically, although it is used for only a limited range of monomers. A summary of block copolymer synthesis techniques has been provided by Hillmyer [5].

1.2.1 ANIONIC POLYMERIZATION

Anionic polymerization is a well-established method for the synthesis of tailored block copolymers. The first anionic polymerizations of block copolymers were conducted as early as 1956 [6]. To prepare well-defined polymers, the technique is demanding, requiring high-purity starting reagents and the use of high-vacuum procedures to prevent accidental termination due to the presence of impurities. In the lab, it is possible to achieve polydispersities $M_w/M_n < 1.05$ via anionic polymerization. The method is also used industrially to prepare several important classes of block copolymers including SBS-type thermoplastic elastomers (S = polystyrene, B = polybutadiene) and polyoxyethylene-*b*-polyoxypropylene-*b*-polyoxyethylene Pluronic amphiphilic copolymers [3]. The principles of anionic polymerization are discussed in Chapter 2. There are a number of reviews that cover its application to block copolymers [7–11]. Recent advances have mainly been directed towards the synthesis of block copolymers with exotic architectures, such as mixed arm stars [12–14], H-shaped copolymers [12], ring-shaped (cyclic) block copolymers [15], etc. All of these require the careful choice of multifunctional initiators.

1.2.2 LIVING RADICAL POLYMERIZATION

Undoubtedly the main advance in block copolymer synthesis in the last decade has been the development of techniques of living radical polymerization (sometimes termed controlled radical polymerization). The principle of controlled radical polymerization methods is to establish a dynamic equilibrium between a small fraction of growing free radicals and a large majority of dormant species. Generated free radicals propagate and terminate as in conventional radical polymerization, although the presence of only a small fraction of radicals prevents premature termination. Among living polymerization methods, atom-transfer radical polymerization (ATRP) has been used most extensively to synthesize block copolymers. Here, the radicals are generated through a reversible redox process catalysed by a transition metal complex that undergoes a one-electron oxidation with the abstraction of a halogen atom from the dormant species. The ATRP method, and its application to the synthesis of block copolymers, has recently been reviewed [16].

ATRP has been used to prepare AB diblock, ABA triblock and most recently ABC triblock copolymers [17]. To date, the technique has been used to create block copolymers based on polystyrene and various polyacrylates [16]. However, it is possible to synthesize a so-called macroinitiator by other polymerization mechanisms (anionic, cationic, etc.), and use this in the ATRP of vinyl monomers. Examples, such as the anionic polymerization of PEO macroinitiators for ATRP synthesis of PEO/PS block copolymers, are discussed by Matyjaszewski and Xia [16].

1.2.3 OTHER METHODS

Sequential living cationic polymerization is primarily used to prepare block copolymers containing a vinyl ether block, or polyisobutylene [18–20]. It can also be coupled with other techniques [18,20]. However, the range of monomers that may be polymerized by this method is comparatively limited and consequently living cationic polymerization is only used in prescribed circumstances.

Ring-opening metathesis polymerization has also been exploited to build blocks from cyclic olefins, especially polynorbornene [5]. The development of ROMP for block copolymer synthesis has recently been facilitated by the introduction of functional-group-tolerant metathesis catalysts by Grubbs [21].

1.3 BLOCK COPOLYMER MELTS

The interest in the phase behaviour of block copolymer melts stems from microphase separation of polymers that leads to nanoscale-ordered morphologies. This subject has been reviewed extensively [1,22–24]. The identification of the structure of bicontinuous phases has only recently been confirmed, and this together with major advances in the theoretical understanding of block copolymers, means that the most up-to-date reviews should be consulted [1,24]. The dynamics of block copolymer melts, in particular rheological behaviour and studies of chain diffusion via light scattering and NMR techniques have also been the focus of several reviews [1,25,26].

The phase behaviour of block copolymer melts is, to a first approximation, represented in a morphology diagram in terms of χN and f [1]. Here f is the volume fraction of one block and χ is the Flory–Huggins interaction parameter, which is inversely proportional to temperature, which reflects the interaction energy between different segments. The configurational entropy contribution to the Gibbs energy is proportional to N, the degree of polymerization. When the product χN exceeds a critical value, $(\chi N)_{ODT}$ (ODT = order–disorder transition) the block copolymer microphase separates into a periodically ordered structure, with a lengthscale $\sim 5 - 500$ nm. The structure that is formed depends on the copolymer architecture and composition [1]. For diblock copolymers, a lamellar (lam) phase is observed for symmetric diblocks ($f = 0.5$), whereas more asymmetric diblocks form hexagonal-packed cylinder (hex) or body-centred cubic (bcc) spherical structures. A complex bicontinuous cubic gyroid (gyr) (spacegroup $Ia\bar{3}d$) phase has also been identified [27,28] for block copolymers between the lam and hex phases near the ODT, and a hexagonal-perforated layer (hpl) phase has been found to be metastable in this region [29–31]. A useful compilation is available of studies on the morphology of block copolymers of various chemistries [32].

The main techniques for investigating solid block copolymer microstructures are transmission electron microscopy (TEM) and small-angle X-ray or neutron scattering. TEM provides direct images of the structure, albeit over a small area of the sample. Usually samples are stained using the vapours from a solution of a heavy metal acid (OsO_4 or RuO_4) to increase the contrast for electrons between domains [33]. Small-angle scattering probes the structure over the whole sample volume, giving a diffraction pattern. The positions of the reflections in the diffraction pattern can be indexed to identify the symmetry of the phase [1,22]. The preparation method can have a dramatic influence on the apparent morphology, for example whether solvent casting or melt processing is performed. Numerous cases of mistaken identification of "equilibrium phases" have appeared in the literature, when the phase was simply an artifact. For instance, Lipic *et al.* [34] obtained different morphologies by varying the preparation conditions for a polyolefin diblock examined by them. In other cases, phases such as hexagonal perforated layers have been observed [29], which, although reproducible, have turned out to be only long-lived metastable phases, ultimately transforming to the equilibrium gyroid phase [30,31]. The ODT in block copolymers can be located via a number of methods – from discontinuities in the dynamic shear modulus [35–37] or small-angle scattering peak shape [38,39] or from calorimetry measurements [40].

To establish relationships between different block copolymer phase diagrams and also to facilitate comparison with theory, it is necessary to specify parameters in addition to χN and f. First, asymmetry of the conformation of the copolymer breaks the symmetry of the phase diagram about $f = 0.5$. For AB diblocks, conformational asymmetry is quantified using the "asymmetry parameter" $\varepsilon = (b_A^2/v_A)/(b_B^2/v_B)$ [41,42], where b_J is the segment length for block J and v_J is the segment volume. Composition fluctuations also modify the phase diagram, and this has been accounted for theoretically via the Ginzburg parameter $\bar{N} = Nb^6\rho^2$, where ρ is the number density of chains [43,44]. The extent of segregation of block copolymers depends on the magnitude of χN. For small χN, close to the order–disorder transition (up to $\chi N = 12$ for symmetric diblocks for which $\chi N_{\mathrm{ODT}} = 10.495$), the composition profile (density of either component) is approximately sinusoidal. This is termed the weak-segregation limit. At much larger values of $\chi N (\chi N > \sim 100)$, the components are strongly segregated and each domain is almost pure, with a narrow interphase between them. This is the strong-segregation limit.

The first theories for block copolymers were introduced for the strong-segregation limit (SSL) and the essential physical principles underlying phase behaviour in the SSL were established in the early 1970s [1]. Most notably, Helfand and coworkers [45–47] developed the self-consistent field (SCF) theory, this permitting the calculation of free energies, composition profiles and chain conformations. In the SCF theory, the external mean fields acting on a polymer chain are calculated self-consistently with the composition profile. The theory of Leibler [48] describes block copolymers in the weak-segregation limit. It employs a Landau–Ginzburg approach to analyse the free energy, which is expanded with reference to the average composition profile. The free-energy coefficients are computed within the random-phase approximation. Weak-segregation limit theory can be extended to allow for thermal-composition fluctuations. This changes the mean-field prediction of a second-order phase transition for a symmetric diblock copolymer to a first-order transition. Fredrickson and Helfand [43] studied this effect for block copolymers and showed that composition fluctuations, incorporated via the renormalization method of Brazovskii, lead to a "finite-size effect", where the phase diagram depends on $\bar{N}$. A powerful new method to solve the self-consistent field equations for block copolymers has been applied by Matsen and coworkers [49–52] to analyse the ordering of many types of block copolymer in bulk and in thin films. The strong- and weak-segregation limits are spanned, as well as the intermediate regime where the other methods do not apply. This implementation of SCF theory predicts phase diagrams, and other quantities such as domain spacings, in good agreement with experiment (see Figure 1.2) and represents an impressive state-of-the-art for modelling the ordering of soft materials. Accurate liquid-state theories have also been used to model block copolymer melts [53,54], although

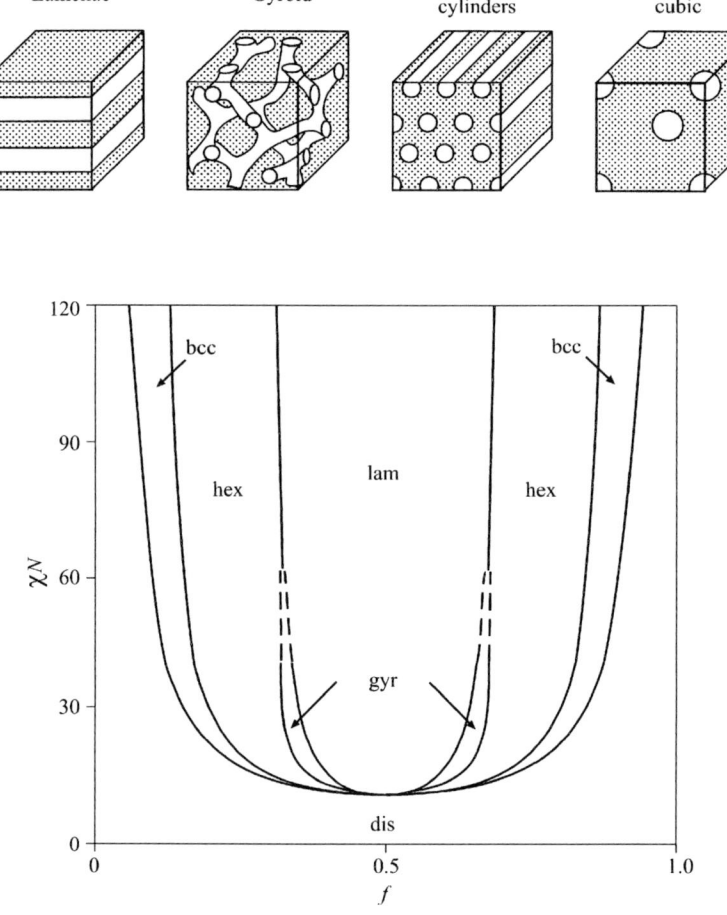

Figure 1.2 Phase diagram for a conformationally symmetric diblock copolymer, calculated using self-consistent mean field theory [49, 51], along with illustrations of the equilibrium morphologies. In the phase diagram, regions of stability of disordered (dis), lamellar (lam), gyroid (gyr), hexagonal (hex) and body-centred cubic (bcc) phases are indicated.

they are hard to implement and consequently the method is often, regrettably, overlooked [1]. Recently, a method has been developed to directly simulate field theories for polymers without introducing approximations such as mean-field approaches, perturbation expansions, etc. [55]. This technique holds much promise for examining the thermodynamics of block copolymers in the limit of low molecular weight where approximate methods such as mean-field theory or renormalization techniques break down.

A phase diagram computed using self-consistent mean field theory [49,51] is shown in Figure 1.2. This shows the generic sequence of phases accessed just below the order–disorder transition temperature for diblock copolymers of different compositions. The features of phase diagrams for particular systems are different in detail, but qualitatively they are similar, and well accounted for by SCF theory.

The phase behaviour of ABC triblocks is much richer [24] than two-component diblocks or triblocks, as expected because multiple interaction parameters (χ_{AB}, χ_{AC} and χ_{BC}) result from the presence of a distinct third block. Summaries of work on ABC triblock morphologies have appeared [1,56]. Because of the large number of possible morphologies, theorists are presently working to predict the phase behaviour of these copolymers using methods that do not require *a priori* knowledge of the space group symmetries of trial structures [57,58].

During processing, block copolymers are subjected to flow. For example, thermoplastic elastomers formed by polystyrene-*b*-polybutadiene-*b*-polystyrene (SBS) triblock copolymers, are moulded by extrusion. This leads to alignment of microphase-separated structures. This was investigated in the early 1970s by Keller and co-workers [22,59] who obtained transmission electron micrographs from highly oriented specimens of Kraton SBS copolymers following extrusion. Examples are included in Figure 1.3. Work on the effect of flow on block copolymer melts has been reviewed [1,25,60,61]. Due to the convenience and well-defined nature of the shear geometry, most model studies have exploited this type of flow. The application of shear leads to orientation of block copolymer microstructures at sufficiently high shear rates and/or strain amplitudes (in the case of oscillatory shear). Depending on shear conditions and temperature, different orientations of a morphology with respect to the shear plane can be accessed. This has been particularly well studied for the lamellar phase where so-called "parallel" (lamellar normal along shear gradient direction) and "perpendicular" (lamellar normal along the neutral direction) orientations have been observed [62]. Distinct orientation states of hexagonal and cubic phases have also been investigated, details being provided elsewhere [61]. The ability to generate distinct macroscopic orientation states of block copolymers by shear is important in future applications of block copolymers, where alignment will be important (reinforced composites, optoelectronic materials and separation media). Shear also influences thermodynamics, since the order–disorder transition shifts upwards on increasing shear rate because the ordered phase is stabilized under shear [63,64].

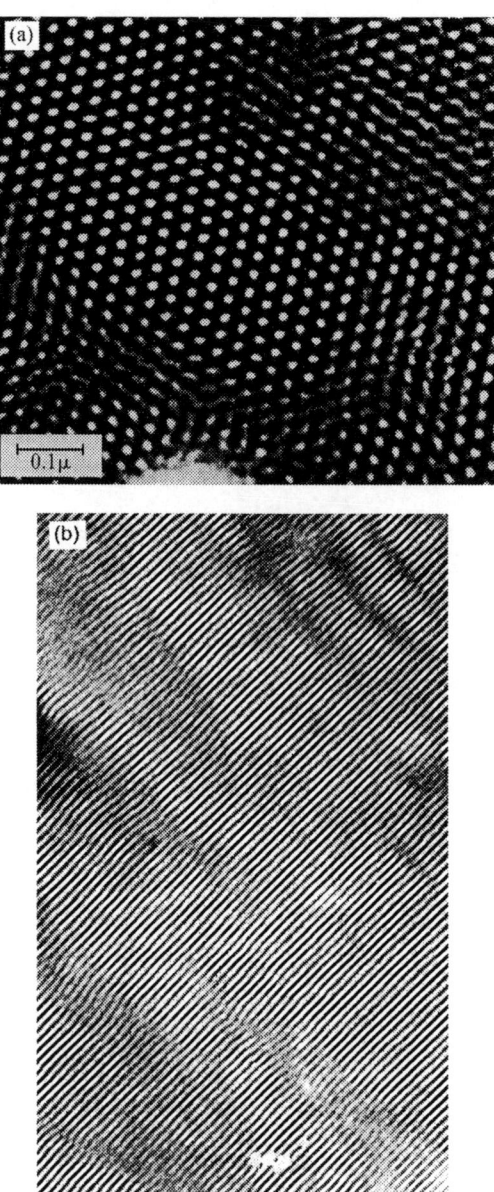

Figure 1.3 TEM micrographs from a hexagonal-packed cylinder structure subjected to flow during high-temperature extrusion. The sample was a PS-PB-PS tribock (Kraton D1102 [209]). (a) Perpendicular to the extrusion direction, (b) a parallel section.

The phase behaviour of rod–coil block copolymers is already known to be much richer than that of coil–coil block copolymers, because the rod block can orient into liquid-crystal structures [1]. The rod block may be analogous to a biomacromolecule, for example poly(benzyl glutamates) [65,66] and polypeptides [67] forming helical rod-like blocks have been incorporated in block copolymers. Possible applications of these materials arising from their biocompatibility are evident.

1.4 BLOCK COPOLYMER FILMS

Microphase separation by block copolymers in thin films has been investigated from several perspectives. First, the physics of self-assembly in confined soft materials can be studied using model block copolymer materials for which reliable mean-field statistical mechanical theories have been developed [68]. Second, interest has expanded due to potential exciting applications that exploit self-organization to fabricate high-density data-storage media [69], to lithographically pattern semiconductors with ultrasmall feature sizes [70,71] or to prepare ultrafine filters or membranes [72]. Research in this field is growing at a rapid pace, and the field has not been reviewed since 1998 [1,73], since when many new developments have occurred.

Block copolymer films can be prepared by the spin-coating technique, where drops of a solution of the polymer in a volatile organic solvent are deposited on a spinning solid substrate (often silicon wafers are used due to their uniform flatness). The polymer film spreads by centrifugal forces, and the volatile solvent is rapidly driven off. With care, the method can give films with a low surface roughness over areas of square millimetres. The film thickness can be controlled through the spin speed, the concentration of the block copolymer solution or the volatility of the solvent, which also influences the surface roughness [74]. Dip coating is another reliable method for fabricating uniform thin films [75]. Whatever the deposition technique, if the surface energy of the block copolymer is much greater than that of the substrate, dewetting will occur. The mechanism of dewetting has been investigated [76–78].

In thin films, the lamellae formed by symmetric block copolymers can orient either parallel or perpendicular to the substrate. A number of possible arrangements of the lamellae are possible, depending on the surface energies of the blocks and that of the substrate, and whether the film is confined at one or both surfaces. These are illustrated in Figure 1.4. In the case that a different block preferentially wets the interface with the substrate or air, wetting is asymmetric and a uniform film has a thickness $(n + \frac{1}{2})d$. If the initial film thickness is not equal to $(n + \frac{1}{2})d$, then islands or holes (quantized steps of height d) form to conserve volume [79]. As well as leading to distinct orientations, confinement of block copolymers can change the thermodynamics of ordering, in particular surface-induced ordering persists above the bulk order–disorder transition [80].

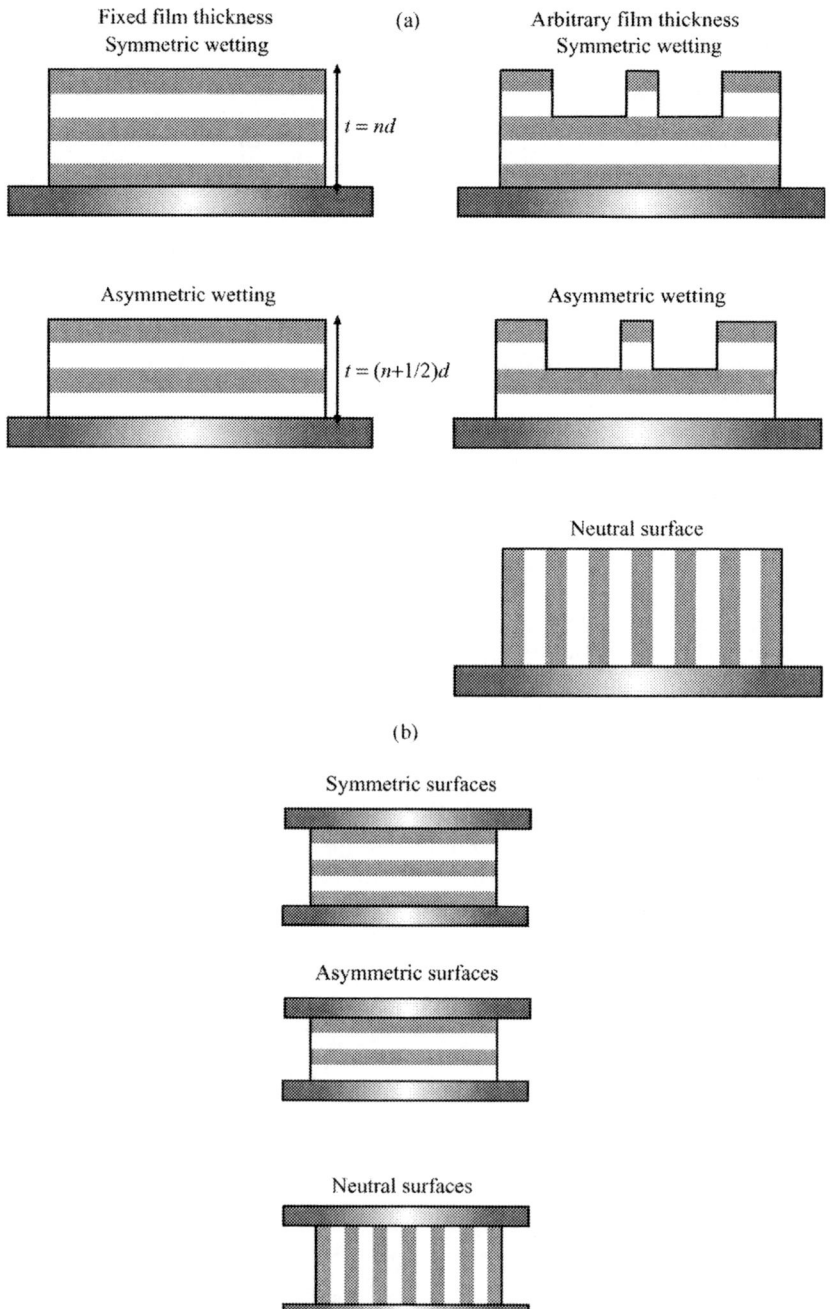

Figure 1.4 Possible configurations of lamellae in block copolymer films. (a) Confined at one surface. (b) Confined at both surfaces.

Introduction to Block Copolymers 11

Asymmetric block copolymers that form hexagonal or cubic-packed spherical morphologies in the bulk, form stripe or circular domain patterns in two dimensions, as illustrated in Figure 1.5. The stripe pattern results from cylinders lying parallel to the substrate, and a circular domain surface pattern occurs when cylinders are oriented perpendicular to the substrate, or for spheres at the surface. Bicontinuous structures cannot exist in two dimensions, therefore the gyroid phase is suppressed in thin films. More complex multiple stripe and multiple circular domain structures can be formed at the surface of ABC triblocks [81]. Nanostructures in block copolymer films can be oriented using electric fields (if the difference in dielectric permittivity is sufficient), which will be important in applications where parallel stripe [82] or perpendicular cylinder configurations [83] are desired.

The morphology of block copolymers on patterned substrates has attracted recent experimental [84,85] and theoretical [86–88] attention. It has been shown that block copolymer stripes are commensurate with striped substrates if the mismatch in the two lengthscales is not too large.

The surface morphology of block copolymer films can be investigated by atomic force microscopy. The ordering perpendicular to the substrate can be probed by secondary ion mass spectroscopy or specular neutron or X-ray reflectivity. Suitably etched or sectioned samples can be examined by transmission electron microscopy. Islands or holes can have dimensions of micrometers, and consequently may be observed using optical microscopy.

Theory for block copolymer films has largely focused on the ordering of lamellae as a function of film thickness. Many studies have used brush theories

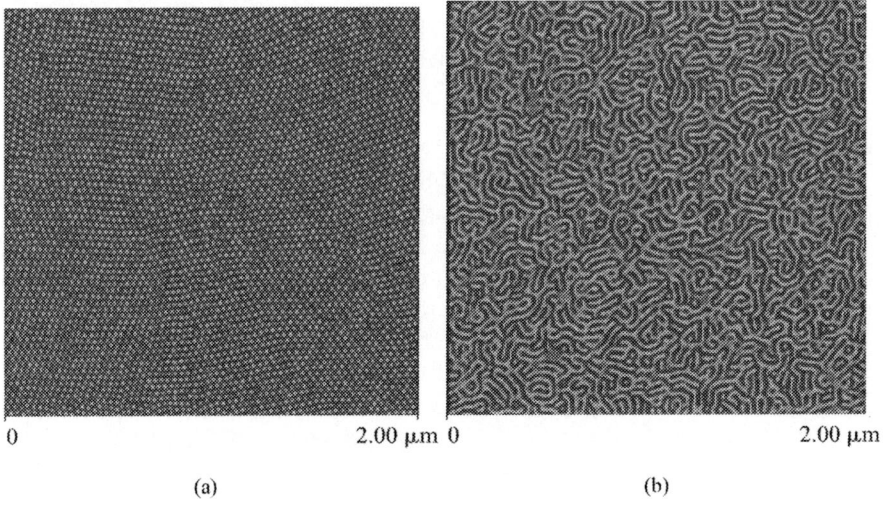

Figure 1.5 Hexagonal and stripe patterns observed via atomic force microscopy (Tapping Mode™). Phase contrast images of (a) polystyrene-*b*-poly(ethylene-co-butylene)-*b*-polystyrene Kraton G1657, (b) Kraton G1650 [210].

for block copolymers in the strong-segregation limit [89,90], although self-consistent field theory has also been employed [68,87,91]. Theory for weakly segregated block copolymers has been applied to analyse surface-induced order above and below the bulk order–disorder transition of a lamellar phase [92] and surface-induced layering in a hexagonal block copolymer film [93]. Computer simulations using the dynamic self-consistent mean-field method have predicted a range of "perforated lamellar" morphologies [94].

1.5 BLOCK COPOLYMERS IN SOLUTION

In a solvent, block copolymer phase behaviour is controlled by the interaction between the segments of the polymers and the solvent molecules as well as the interaction between the segments of the two blocks. If the solvent is unfavourable for one block this can lead to micelle formation in dilute solution. The phase behaviour of concentrated solutions can be mapped onto that of block copolymer melts [95]. Lamellar, hexagonal-packed cylinder, micellar cubic and bicontinuous cubic structures have all been observed (these are all lyotropic liquid-crystal phases, similar to those observed for nonionic surfactants). This is illustrated by representative phase diagrams for Pluronic triblocks in Figure 1.6.

The main classes of block copolymer examined in solution are those based on polyoxyethylene, which is water soluble and is the basis of most amphiphilic block copolymers, and styrenic block copolymers in organic solvents. Selected studies on these systems up to 1998 have been summarized [1]. Polyoxyethylene-based block copolymers include those of polyoxyethylene (E) with polyoxypropylene (P), especially EPE triblocks (commercial name: Pluronic or Synperonic), which are widely used commercially as surfactants in detergents and personal care products [96], and also in pharmaceutical applications, especially drug delivery [97–99]. A number of edited books on water-soluble polymers cover applications of block copolymers [100–105]. Related copolymers include those with a polyoxybutylene hydrophobic block [106,107]. Work on styrenic block copolymers in organic solvents has also been reviewed [1,108]. Block copolymers containing a polyelectrolyte chain have attracted attention from a number of research teams [109,110] (and references therein), copolymers containing a well-studied polyelectrolyte such as poly(styrene sulphonate) [111] or a polyacrylate [109] often being chosen.

Like surfactants, block copolymers form micelles above a critical concentration. The critical micelle concentration can be located by a variety of techniques [112], the most commonly used being surface tensiometry where the cmc is located as the point at which the surface tension becomes essentially independent of concentration. The primary methods to determine micelle size and shape are light scattering and small-angle X-ray and neutron scattering. The thermodynamic radius (from the thermodynamic volume, which is one eighth

Introduction to Block Copolymers

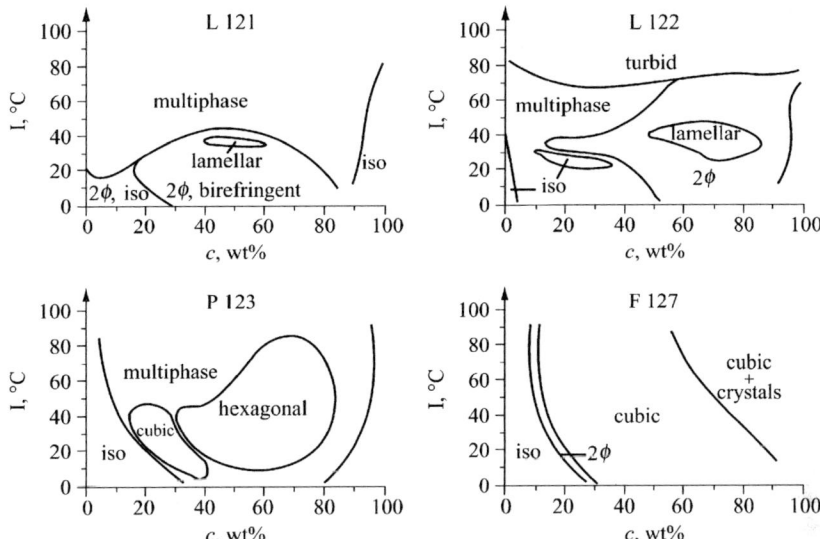

Figure 1.6 Phase diagrams in water of $E_m P_n E_m$ (E=polyoxyethylene, P=polyoxypropylene) Pluronics with $n = 69$ and $m = 4$ (Pluronic L121), $m = 11$ (Pluronic L122), $m = 20$ (Pluronic P123) and $m = 99$ (Pluronic F127). (Reproduced from G. Wanka et al. Macromolecules **27**, 4145 (1994). Copyright (1994) with permission from the American Chemical Society.)

of the excluded volume) of micelles can be obtained from static light scattering experiments by fitting the Debye function to the Carnahan–Starling equation for hard spheres [107]. This procedure can be used in place of Zimm plots when the angular dependence of the scattered intensity is weak, which is usually the case for block copolymer micelles, which are much smaller than the wavelength of light [107]. Static light scattering also provides the association number (from the micellar mass) and the second virial coefficient [1,107,113]. Dynamic light scattering provides the hydrodynamic radius from the mode corresponding to micellar diffusion obtained from the intensity distribution of relaxation times (often obtained from analysis of the intensity autocorrelation function using the program CONTIN (114)). The Stokes–Einstein equation can then be used to calculate the hydrodynamic radius from the diffusion coefficient [1,107]. Small-angle X-ray scattering or neutron scattering can be used to extract information on intra- and inter-micellar ordering [1]. Neutron scattering has the advantage compared to X-ray scattering that the contrast between different parts of the system (e.g. within the micelle or between the micelle and the solvent) can be varied by selective deuteration of solvent and/or one of the blocks. In dilute solution, only intramicellar structure contributes to the scattered intensity (the so-called form factor) and this can be modelled to provide information on micelle size and shape. The simplest model is that of a uniform hard sphere [115], although more sophisticated models are usually required for high-quality data fitting [115–118]. The intermicellar structure factor dominates at higher concentrations. It can be analysed using the hard sphere model [115,119,120] to give information on the micellar radius, and the micellar volume fraction. Where attractive interactions between micelles are significant, these also influence the structure factor and this can be modelled using the "sticky sphere" approximation [117].

A diverse range of theoretical approaches have been employed to analyse the structure of block copolymer micelles, and for micelle formation [1]. The first models were based on scaling relationships for polymer "brushes" and give predictions for the dependence of micelle dimensions on the size of the blocks, as well as the association number of the micelle. A "brush" theory by Leibler and coworkers [121] enables the calculation of the size and number of chains in a micelle and its free energy of formation. The fraction of copolymer chains aggregating into micelles can also be obtained. Self-consistent field theory was first applied to predict the cmc of a diblock in a homopolymer matrix, and then applied to block copolymers in solution. The lattice implementation of SCF theory has been applied by Linse and coworkers [122] to analyse the dimensions of micelles for specific (Pluronic) block copolymers.

In addition to applications as surfactants and in personal care products, block copolymer micelles have been extensively investigated as nanoparticles for solubilizing active agents for drug delivery [97,98,123,124], or as "nanoreactors" for the production of inorganic nanoparticles, e.g. of metals with potential applications in catalysis [125,126]. An alternative approach is to form vesicles (bilayers wrapped round into a spherical shell) [127,128]. These may be crosslinked or polymerized to form hollow-shell nanoparticles [129–131].

At higher concentrations, block copolymers in solution form a variety of lyotropic mesophases [1,132–135]. Due to fact that such phases possess a finite yield stress and so usually do not flow under their own weight, these are often termed gels. However, it must be emphasized that the gel properties result from the ordered microstructure rather than any crosslinks between polymer chains as in a conventional polymer gel. The symmetry of the ordered phase formed largely depends on the interfacial curvature, as for conventional amphiphiles [112], however, the phase behaviour can also be understood by mapping it onto that for block copolymer melts [95]. Shear can be used to orient block copolymer gels as for block copolymer melts. The effects of shear on lyotropic lamellar, hexagonal-packed cylindrical micellar and cubic micellar phases have all been investigated [132,136,137]. Large-amplitude oscillatory shear or high shear rate steady shear both lead to macroscopic orientation of the structures. In the case of cubic phases in particular the flow mechanisms are complex, as is the rheological behaviour with interesting nonlinear effects such as plateaus in the flow curve [138,139].

Theory for the phase behaviour of block copolymers in semidilute or concentrated solution is less advanced than that for melts or dilute solutions due to the complexity of interactions between polymer and solvent. The two main

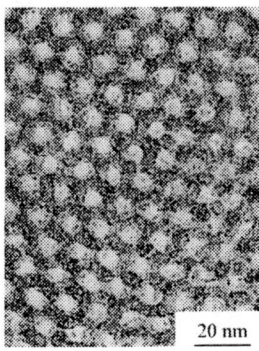

Figure 1.7 TEM image of calcined silica structure templated using an acidic solution of Pluronic poly(oxyethylene)-b-poly(oxypropylene)-b-poly(oxyethylene) triblock (Reproduced from D. Zhao et al. Science **279**, 548 (1998) Copyright (1998) with permission from the American Association for the Arrangement of Science.)

methods developed have been (a) SCF theory for density profiles and domain spacing scalings and (b) weak-segregation limit calculations of the shift in the order–disorder transition temperature with changing concentration. An overview of both approaches can be found elsewhere [1]. SCF theory calculations by Linse and coworkers [140,141] have produced phase diagrams for specific Pluronic copolymers in aqueous solution that are in remarkably good agreement with those observed experimentally. Simulations using the dynamic density functional theory (commercially available as the Mesodyn module of Cerius^2 from Accelerys) have also yielded surprisingly accurate predictions for the sequence of phases obtained on varying concetration [142].

Lyotropic block copolymer mesophases can be used to template inorganic materials such as silica [144, 212], this producing materials with a high internal surface area that could be useful in catalysis or separation technology. Figure 1.7 shows a transmission electron micrograph of hexagonal mesoporous silica, templated using a Pluronic block copolymer.

1.6 CRYSTALLIZATION IN BLOCK COPOLYMERS

In semicrystalline block copolymers, the presence of a noncrystalline block enables modification of the mechanical and structural properties compared to a crystalline homopolymer, through introduction of a rubbery or glassy component. Crystallization in homopolymers leads to an extended conformation, or to *kinetically controlled* chain folding. In block copolymers, on the other hand, *equilibrium* chain folding can occur, the equilibrium number of folds being controlled by the size of the second, noncrystallizable block. The structure of block copolymers following crystallization has been reviewed [1,145].

The most important crystallizable block copolymers are those containing polyethylene or poly(ethylene oxide) (PEO) (systematic name polyoxyethylene). Polyethylene (PE) in block copolymers is prepared by anionic polymerization of poly(1,4-butadiene) (1,4-PB) followed by hydrogenation, and has a melting point in the range 100–110 °C. This synthesis method leads to ethyl branches in the copolymer, with on average 2–3 branches per 100 repeats. These branches induce lengths for folded chains that are set by the branch density and not by the thermodynamics of crystallization. The melting temperature of PEO in block copolymers is generally lower than that of PEO homopolymer (melting temperature $T_m = 76$ °C for high molecular weight samples). In contrast to PE prepared by hydrogenation of 1,4-PB, there is usually no chain branching in PEO and the fold length depends on the crystallization procedure. Molecules with 1,2,3 ... folds can be obtained by varying the crystallization protocol (quench depth, annealing time, etc.). Crystallization has been investigated for other block co-polymers, in particular those containing poly(ε-caprolactone) (PCL) ($T_m = 57$ °C). The morphology in block copolymers where both blocks are crystallizable has also been investigated. It has been found that co-crystallization occurs in diblock copolymers, in contrast to blends of crystallizing homopolymers [146]. However, one block can influence the crystallization of another as shown by studies on polystyrene-b-polyethylene-b-poly(ε-caprolactone) ABC triblocks [147]. A suppression of the crystallization temperature of the poly(ε-caprolactone) block was noted when the polyethylene block crystals were annealed before crystallization of PCL at lower temperatures [147], this effect being termed "antinucleation".

It is now firmly established that confinement of crystalline stems has a profound influence on crystallization in block copolymers. Confinement can result from the presence of glassy domains or simply strong segregation between domains. In contrast, crystallization can overwhelm microphase separation when a sample is cooled from a weakly segregated or homogeneous melt [148–150]. The lamellar crystallites can then nucleate and grow heterogeneously to produce spherulites [148,151], whereas these are not observed when crystallization is confined to spheres or cylinders. Crystallization confined by glassy blocks leads to a drastic slowdown in crystallization kinetics and a reduction in the corresponding Avrami exponent [152,153]. Poly(ethylene) crystallites in a strongly segregated diblock have been observed to nucleate homogeneously within the PE spheres, leading to first-order kinetics, i.e. exponential growth in the degree of crystallinity [154]. Confined crystallization was first observed for a lamellar phase with glassy lamellae [155,156], and later in cylinders confined in a glassy matrix [157]. Crystallization of the polyethylene matrix in the inverse structure (i.e. a phase containing rubbery or glassy cylinders) occurs without disrupting the melt microstructure [158].

Chain folds can exist in equilibrium in block copolymers, in contrast to homopolymers, due to the finite cross sections of the blocks at the lamellar interface, which have to be matched if space is to be filled at normal densities.

The equilibrium fold diagram has been mapped out for poly(ethylene oxide)-based block copolymers in the melt [159] and in solution [160]. Nonequilibrium states of highly folded chains can also be trapped kinetically [160,161].

The orientation of crystalline stems in block copolymers depends on the morphology of the structure and the crystallization protocol. A parallel orientation of polyethylene stems with respect to a lamellar interface was reported for a series of polyethylene-*b*-polyethylethylene diblocks [162], and a similar orientation was later reported by Hamley *et al.* [155,156] for a series of PE-containing diblocks based on simultaneous SAXS/WAXS experiments, as shown in Figure 1.8. SAXS on aligned specimens gives the lamellar orientation, whereas WAXS provides information on unit cell orientation. Samples may be aligned in the melt, for example using large-amplitude oscillatory shear [155,163]. In constrast to these studies showing parallel stem orientation, Rangarajan *et al.* [148] proposed a perpendicular orientation of PE stems in a series of polyolefin diblocks investigated by them. Again using the combination of SAXS and WAXS, Quiram *et al.* [164] found that PE stems generally orient perpendicular to the cylinder axis, although tilted stems were observed when crystallization was confined by strong segregation or by a glassy matrix. These apparently conflicting observations of parallel and perpendicular stem orientations can be rationalised when it is recognised that in both orientations the *b* axis of the PE crystals is the fast growth direction – in the lamellar plane and along the cylinder axis, respectively. Recently, Zhu *et al.* [163] investigated the orientation of PE stems in a PS-b-PEO diblock forming a lamellar phase using SAXS and WAXS. Four regimes were identified: (i) A random stem orientation for a deep quench into liquid nitrogen, (ii) stems parallel to lamellae for a crystallization temperature $-50 \leq T_c \leq -10°C$, (iii) Stems inclined with respect to lamellae were observed for $-5 \leq T_c \leq -30°C$, (iv) Stems perpendicular to

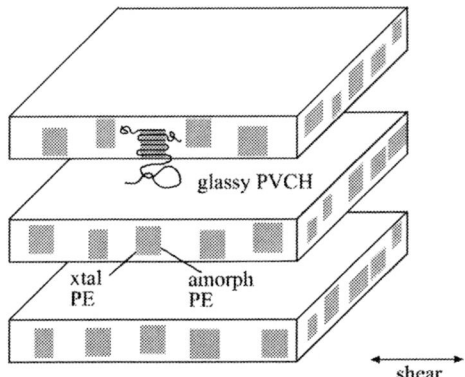

Figure 1.8 Model for confined crystallization in a lamellar phase formed by a polyethylene-*b*-poly(vinylcyclohexane) diblock (Reproduced from I. W. Hamley *et al.* Macromolecules **29**, 8835 (1996) Copyright (1996) with permission from the American Chemical Society.)

lamellae were observed for $T_c \geq 35°C$ [163]. For PEO cylinders formed in a PS-PEO diblock the parallel orientation of stems was not observed, although the states (i), (iii) and (iv) were confirmed [165]. These conclusions were supported by a separate study of the correlation lengths (apparent crystallite sizes) obtained from SAXS for different crystal orientations [166]. In this report it was also noted that it is the initial growth stage that determines the final crystal orientation in nanoconfined lamellae rather than the primary nucleation step. Crystal orientation and changes in lamellar thickness of a related diblock were examined in a companion paper, in which the change in the crystallization kinetics for confined and unconfined crystallization were deduced from Avrami plots of the degree of crystallinity [167].

Theories for semicrystalline block copolymers are able to provide predictions for the scaling of amorphous and crystal layer thickness with chain length [1,145]. A brush-type theory was developed by DiMarzio et al. [168] and a self-consistent field theory by Whitmore and Noolandi [169]. The latter approach predicts a scaling for the overall domain spacing $d \sim NN_a^{-5/12}$ (where N is the total degree of polymerization and N_a is that of the amorphous block) that is in good agreement with experimental results [170], as detailed elsewhere [1,145]. Approaches used for crystallization in homopolymers may be used to calculate the change in melting temperature due to finite crystal thickness (Thompson–Gibbs equation), lamellar crystal surface energies (Flory–Vrij theory), and growth rates (kinetic nucleation theory). Details can be obtained from [1].

The morphology of thin films of crystallized block copolymers can be probed most conveniently at the microscopic scale by atomic force microscopy (AFM), whereas spherulites can be observed optically. Crystallization in thin films of PE-b-PEO diblocks has recently been investigated by Reiter and coworkers [171,172]. For a diblock containing 45% PEO they observed, using AFM, parallel lamellae in the melt but lamellae oriented perpendicular to the substrate upon crystallization at a large undercooling [172]. This was ascribed to a kinetically trapped state of chain-folded PEO crystals. However, ultimately the morphology evolved into the equilibrium parallel one, which was also observed for three other diblocks with a higher PEO content [172]. Films of these copolymers were characterized by islands and holes at the surface due to an incommensurability between the film thickness and an integral number of lamellae, as discussed in Section 1.4. The island and hole structure was retained upon crystallization, although craters and cracks appeared in the lamellae. Within craters, terracing of lamellar steps was observed, from which the lamellar thickness could be extracted. Terracing of crystal lamellae oriented parallel to the substrate was also reported for a PEO-b-PBO diblock and a PEO-b-PBO-b-PEO triblock, probed via AFM [173]. In this work a comparison of the lamellar thickness was also made with the domain spacing obtained from SAXS and a model of tilted chains was proposed (fully extended for the diblock, once folded for the triblock). However, this is not in agreement with recent simultaneous SAXS/WAXS results that indicate PEO chains oriented perpendicular to lamellae in a PEO-*b*-PBO diblock [174].

1.7 BLENDS CONTAINING BLOCK COPOLYMERS

In blends of block copolymer with homopolymer, there is an interplay between macrophase separation (due to the presence of homopolymer) and microphase separation (of the block copolymer). Which effect predominates depends on the relative lengths of the polymers, and on the composition of the blend.

Macrophase separation can be detected by light scattering or via turbidity measurements of the cloud point since macrophase separation leads to structures with a length scale comparable to that of the wavelength of light. Regions of macrophase and microphase separation can also be distinguished by transmission electron microscopy or via small-angle scattering techniques. Microphase separation leads to a scattering peak at a finite wavenumber q, whereas macrophase separation is characterized by $q = 0$. The segregation of block copolymers to the interface between polymers in a blend can be determined in bulk from small-angle scattering experiments or TEM. In thin films, neutron reflectivity, forward recoil spectroscopy and nuclear reaction analysis have been used to obtain volume fraction profiles, which quantify the selective segregation of block copolymers to interfaces.

An important application of block copolymers is as compatibilizers of otherwise immiscible homopolymers. There are a number of useful reviews of work in this area [175–178]. The morphology of blends of polymers with block copolymer, and theories for this, have been reviewed [1]. The influence of added homopolymer on block copolymer structure has also been investigated, as have binary blends of block copolymers, and these systems are also considered in this section.

1.7.1 BLENDS OF BLOCK COPOLYMER WITH ONE HOMOPOLYMER

Block copolymers can solubilize homopolymers up to a certain amount, beyond which phase separation occurs. This ability to continuously swell block copolymer microstructures is the basis of a number of potential and actual applications in optoelectonics where the periodicity of the block copolymer structure is extended up to $0.1–1$ μm, which corresponds to wavelengths for reflection or guiding of light. The limit for macrophase separation in blends of block copolymer with homopolymer depends on the relative chain lengths, i.e. on $\alpha = N_{Ah}/N_{Ac}$, where N_{Ah} is the degree of polymerization of the homopolymer (A) and N_{Ac} is the degree of polymerisation of the same component of the copolymer. Work by the groups of Hashimoto [179] and Winey [180–182] has led to the identification of three regimes [1]. If $\alpha < 1$, the homopolymer tends to be selectively solubilized in the A domain of the microphase-separated block copolymer, and is weakly segregated towards the domain centre. If $\alpha \approx 1$, the homopolymer is still selectively solubilized in the A microdomains. However, it does not significantly swell the A block chains and tends to be more localized in the middle of the A microdomains. If $\alpha > 1$, macrophase separation occurs, with domains of microphase-separated copolymer in the homopolymer matrix. A transmission electron micrograph of the structure formed by a phase-separated lamellar diblock is shown in Figure 1.9.

Another important aspect of adding homopolymer to a block copolymer is the ability to change morphology (without synthesis of additional polymers). Furthermore, morphologies that are absent for neat diblocks such as bicontinuous cubic "double diamond" or hexagonal-perforated layer phases have been predicted in blends with homopolymers [183], although not yet observed. Transitions in morphology induced by addition of homopolymer are reviewed elsewhere [1], where a list of experimental studies on these systems can also be found.

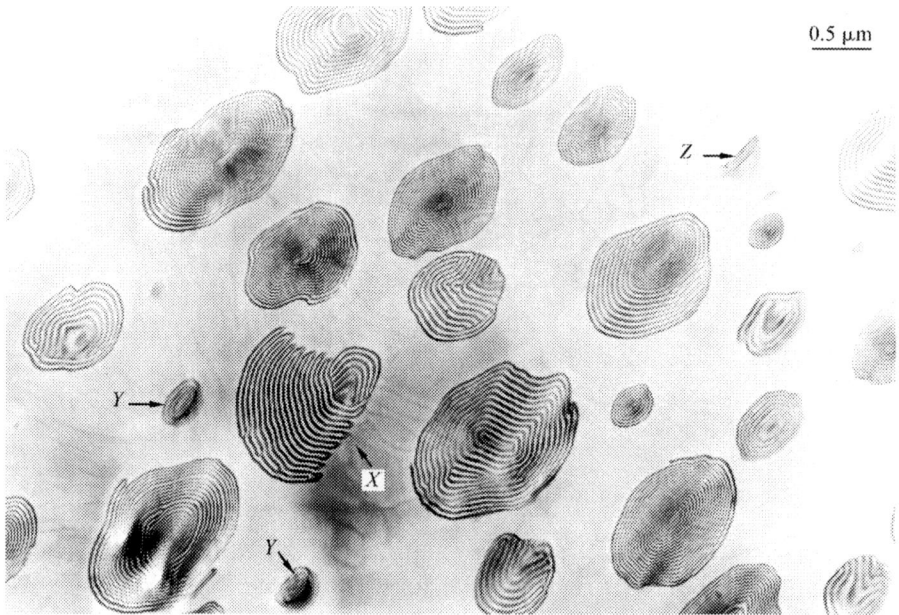

Figure 1.9 Electron micrograph showing macrophase separation of domains of microphase-separated polystyrene-*b*-polyisoprene block copolymer ($M_n = 100\,\text{kg}\,\text{mol}^{-1}$, $f_{PS} = 0.46$) in a PS homopolymer ($M_n = 580\,\text{kg}\,\text{mol}^{-1}$) matrix (Reproduced from S. Koizumi *et al*. Macromolecules **27**, 6532 (1994) Copyright (1994) with permission from the American Chemical Society.)

1.7.2 BLENDS OF BLOCK COPOLYMER WITH TWO HOMOPOLYMERS

The ability of block copolymers to act as compatibilizers is now established. However, a debate has occurred in the literature as to whether block copolymers are more effective compatibilizers than random copolymers. For example, it has been reported that polystyrene/poly(2-vinylpyridine) random copolymers act to compatibilize the parent homopolymers [184], but that random polystyrene/poly(methyl methacrylate) copolymers are much less effective than block copolymers [185]. The key appears to be the blockiness of the copolymer, which is much higher for the latter [186]. Theory suggests that compositional polydispersity is also important for effective compatibilization [186,187]. It leads to a greater gradation in composition across the interface, and consequently a lower configurational entropy of the homopolymers [187]. In practice, polymers are compatibilized during melt processing. Then kinetic quantities such as the rate of diffusion of the copolymers to the interface and the shear rate are important. Macosko and coworkers [188] have shown that the coalescence of polymer droplets is inhibited by diffusion of block copolymers. The molar mass must be low enough so that diffusion occurs rapidly but not too low to prevent entanglements at the interface. On the other hand, copolymers with a molar mass that is too high get stuck in micelles.

Block copolymers act as compatibilizers by reducing the interfacial tension between homopolymers. Recent work shows that block copolymers can reduce the interfacial tension between homopolymers to the extent that polymeric microemulsions can be formed where the copolymer forms a continuous film between spatially continuous homopolymer domains [189–191]. A TEM image of a microemulsion formed in a blend of two polyolefins and the corresponding symmetric diblock is shown in Figure 1.10. A bicontinuous microemulsion forms in the mixture composition range where mean-field theory predicts a Lifshitz point [192]. A Lifshitz point is defined as the point along the line of critical phase transitions at which macro- and microphase branches meet [1]. The observation of a microemulsion shows that mean-field theory breaks down due to the existence of thermal composition fluctuations. Although a theory for these composition fluctuations has not yet been developed, it has been shown that some properties of the microemulsion (elastic constants, composition profiles) can be modelled using an approach where the effective interaction between copolymer monolayers is computed [187,193,194]. Both SCF and SSL theories have been employed [194]. The effect of shear on polymeric microemulsions has recently been investigated, and it was shown that macrophase separation can be induced at sufficiently high shear rates [195]. The connection between microemulsions formed by block copolymers and those containing conventional amphiphilies (which can be used to stabilize oil/water mixtures) has been emphasized [190,196] due to the importance of this aspect of block copolymer phase behaviour to applications.

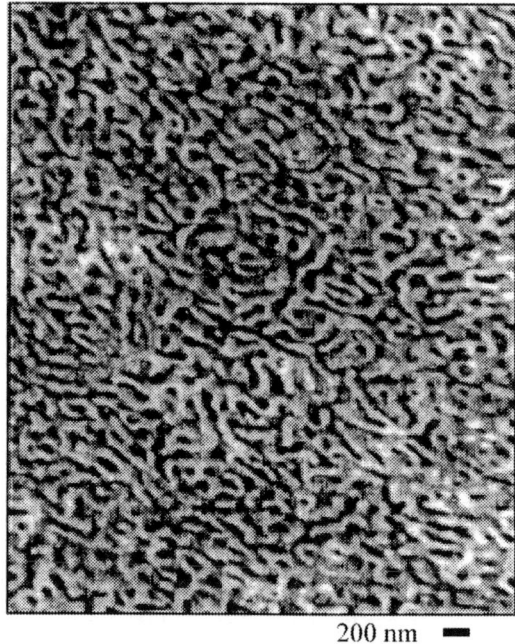

200 nm

Figure 1.10 Transmission electron micrograph image of a microemulsion formed in a ternary blend of polyethylene, poly(ethylene-propylene) and a symmetric diblock of these two polymers (Reproduced from M. A. Hillmyer *et al.* J. Phys. Chem. B **103**, 4814 (1999) Copyright (1999) with permission from the American Chemical Society.)

1.7.3 BLENDS OF BLOCK COPOLYMERS

Macro-versus micro-phase separation in blends of block copolymers has been investigated in particular for blends of polystyrene-*b*-polyisoprene diblock copolymers by Hashimoto and coworkers [197–201]. Writing the ratio of chain lengths as $\delta = N_1/N_2$, it was found that blends of lamellar diblocks are miscible for $\delta < 5$, whereas for $\delta > 5$, the mixtures are only partially miscible [197,200]. The same limiting value of δ was obtained by Matsen using self-consistent mean-field calculations [202]. The miscibility of pairs of asymmetric diblocks with the same [198] or complementary [198,199,203] compositions has also been investigated. By blending complementary diblocks (i.e. those with composition f and $1-f$), it is possible to induce a lamellar phase even for mixtures of asymmetric diblocks forming cylinder phases when pure [198,203]. Blends of diblocks with similar compositions and molecular weights can be used to map the phase diagram by interpolation in the composition range spanned [143]. By blending, the synthesis requirements to obtain a full phase diagram are reduced. The validity of this so-called "single-component" approximation has been tested using SCF theory. It was found that phase boundaries in the (f_1, f_2) plane, where f_1 and f_2 are the compositions of the two diblocks) map onto those of the corresponding pure diblock, at least if f_1 and f_2 do not differ too much [204,205]. In the case that either f_1 or f_2 becomes close to zero or unity, this approximation completely breaks down [205]. Thus, the one-component approximation is useful, although evidently the phase diagram of binary blends will contain biphasic regions.

Motivated by the possibility to prepare "exotic morphologies" exhibited by ABC triblocks just by blending diblocks, Frielinghaus et al. [206,207] have investigated phase diagrams of strongly interacting AB and BC diblocks where the common B block is polyisoprene and the other two blocks are polystyrene and poly(ethylene oxide). Although exotic phases were not found, regions of miscibility and immiscibility were mapped out. The phase diagrams obtained were in surprisingly good agreement with the predictions of a simple random-phase approximation calculation of the spinodals [208].

REFERENCES

1. I. W. Hamley, *The Physics of Block Copolymers* (Oxford University Press, Oxford, 1998].
2. P. Alexandridis and J. F. Holzwarth, *Curr. Opin. Colloid Interface Sci.* **5**, 312 (2000).
3. N. Hadijichristidis, S. Pispas and G. A. Floudas *Block Copolymers* (Wiley, New York, 2003).
4. G. Riess, G. Hurtrez, and P. Bahadur, in *Encyclopedia of Polymer Science and Engineering*, edited by H. F. Mark and J. I. Kroschwitz (Wiley, New York, 1985), Vol. 2, p. 324.
5. M. Hillmyer, *Curr. Opin. Solid State Mater. Sci.* **4**, 559 (1999).
6. M. Scwarc, M. Levy, and R. Milkovich, *J. Am. Chem. Soc.* **78**, 2656 (1956).
7. R. N. Young, R. P. Quirk, and L. J. Fetters, *Adv. Polym. Sci.* **56**, 1 (1985).
8. P. Rempp, E. Franta, and J. Herz, *Adv. Polym. Sci.* **86**, 147 (1989).
9. M. van Beylen, S. Bywater, G. Smets, M. Szwarc, and D. Worsfold, *Adv. Polym. Sci.* **86**, 89 (1989).
10. H. L. Hsieh and R. P. Quirk, *Anionic Polymerization: Principles and Practical Applications* (Marcel Dekker, New York, 1996).
11. N. Hadjichristidis, H. Iatrou, S. Pispas, and M. Pitsikalis, *J. Polym. Sci.A: Polym. Chem.* **38**, 3211 (2000).
12. M. Pitsikalis, S. Pispas, J. W. Mays, and N. Hadjichristidis, *Adv. Polym. Sci.* **135**, 1 (1998).
13. N. Hadjichristidis, *J. Polym. Sci.A: Polym. Chem.* **37**, 857 (1999).
14. N. Hadjichristidis, N. Pispas, M. Pitsikalis, H. Iatrou, and C. Vlahos, *Adv. Polym. Sci.* **142**, 71 (1999).
15. Z.-G. Yan, Z. Yang, C. Price, and C. Booth, *Makromol. Chem., Rapid Commun.* **14**, 725 (1993).
16. K. Matyjaszewski and J. Xia, *Chem. Rev.* **101**, 2921 (2001).
17. K. A. Davis and K. Matyjaszewski, *Macromolecules* **34**, 2101 (2001).
18. J. P. Kennedy and E. Maréchal, *Carbocationic Polymerization* (Wiley, New York, 1982).
19. J. P. Kennedy and B. Ivan, *Designed Polymers by Carbocationic Macromolecular Engineering: Theory and Practice* (Hanser, Munich, 1992).

20. M. Sawamoto and M. Kamigaito, in *New Methods of Polymer Synthesis*, edited by J. R. Ebdon and G. C. Eastmond (Blackie, London, 1995), Vol. 2.
21. S. T. Nguyen, L. K. Johnson, R. H. Grubbs, and J. W. Ziller, *J. Am. Chem. Soc.* **114**, 3975 (1992).
22. M. J. Folkes and A. Keller, in *The Physics of Glassy Polymers*, edited by R. N. Haward (Applied Science, London, 1973), p. 548.
23. F. S. Bates and G. H. Fredrickson, *Ann. Rev. Phys. Chem.* **41**, 525 (1990).
24. F. S. Bates and G. H. Fredrickson, *Physics Today* **52**, Feb issue, 32 (1999).
25. G. H. Fredrickson and F. S. Bates, *Annu. Rev. Mater. Sci.* **26**, 501 (1996).
26. R. H. Colby, *Curr. Opin. Colloid Interface Sci.* **1**, 454 (1996).
27. D. A. Hajduk, P. E. Harper, S. M. Gruner, C. C. Honeker, G. Kim, E. L. Thomas, and L. J. Fetters, *Macromolecules* **27**, 4063 (1994).
28. S. Förster, A. K. Khandpur, J. Zhao, F. S. Bates, I. W. Hamley, A. J. Ryan, and W. Bras, *Macromolecules* **27**, 6922 (1994).
29. I. W. Hamley, K. A. Koppi, J. H. Rosedale, F. S. Bates, K. Almdal, and K. Mortensen, *Macromolecules* **26**, 5959 (1993).
30. D. A. Hajduk, H. Takenouchi, M. A. Hillmyer, F. S. Bates, M. E. Vigild, and K. Almdal, *Macromolecules* **30**, 3788 (1997).
31. M. E. Vigild, K. Almdal, K. Mortensen, I. W. Hamley, J. P. A. Fairclough, and A. J. Ryan, *Macromolecules* **31**, 5702 (1998).
32. M. F. Schulz and F. S. Bates, in *Physical Properties of Polymers Handbook*, edited by J. E. Mark (American Institute of Physics, Woodbury, New York, 1996), p. 427.
33. K. Kato, *J. Electron Microsc.* (Japan) **14**, 220 (1965).
34. P. M. Lipic, F. S. Bates, and M. W. Matsen, *J. Polym. Sci. B: Polym. Phys.* **37**, 2229 (2001).
35. F. S. Bates, J. H. Rosedale, and G. H. Fredrickson, *J. Chem. Phys.* **92**, 6255 (1990).
36. J. H. Rosedale and F. S. Bates, *Macromolecules* **23**, 2329 (1990).
37. C. D. Han, D. M. Baek, J. K. Kim, T. Ogawa, N. Sakamoto, and T. Hashimoto, *Macromolecules* **28**, 5043 (1995).
38. S. M. Mai, J. P. A. Fairclough, I. W. Hamley, R. C. Denny, B. Liao, C. Booth, and A. J. Ryan, *Macromolecules* **29**, 6212 (1996).
39. N. Sakamoto and T. Hashimoto, *Macromolecules* **28**, 6825 (1995).
40. V. P. Voronov, V. M. Buleiko, V. E. Podneks, I. W. Hamley, J. P. A. Fairclough, A. J. Ryan, S.-M. Mai, B.-X. Liao, and C. Booth, *Macromolecules* **30**, 6674 (1997).
41. E. Helfand and A. M. Sapse, *J. Chem. Phys.* **62**, 1327 (1975).
42. F. S. Bates and G. H. Fredrickson, *Macromolecules* **27**, 1065 (1994).
43. G. H. Fredrickson and E. Helfand, *J. Chem. Phys.* **87**, 697 (1987).
44. G. H. Fredrickson and K. Binder, *J. Chem. Phys.* **91**, 7265 (1989).
45. E. Helfand, *Macromolecules* **8**, 552 (1975).
46. E. Helfand and Z. R. Wasserman, *Macromolecules* **9**, 879 (1976).
47. E. Helfand and Z. R. Wasserman, in *Developments in Block Copolymers 1*, edited by I. Goodman (Applied Science, London, 1982), p. 99.
48. L. Leibler, *Macromolecules* **13**, 1602 (1980).
49. M. W. Matsen and M. Schick, *Phys. Rev. Lett.* **72**, 2660 (1994).
50. M. W. Matsen and M. Schick, *Curr. Opin. Colloid Interface Sci.* **1**, 329 (1996).
51. M. W. Matsen and F. S. Bates, *Macromolecules* **29**, 1091 (1996).
52. M. W. Matsen, *J. Phys. Condens. Matter*, **14**, R21 (2001).
53. E. F. David and K. S. Schweizer, *J. Chem. Phys.* **100**, 7767 (1994).
54. E. F. David and K. S. Schweizer, *J. Chem. Phys.* **100**, 7784 (1994).
55. V. Ganesan and G. H. Fredrickson, *Europhys. Lett.* **55**, 814 (2001).
56. A. J. Ryan and I. W. Hamley, in *The Physics of Glassy Polymers*, edited by R. N. Haward and R. J. Young (Chapman and Hall, London, 1997).

57. F. Drolet and G. H. Fredrickson, *Phys. Rev. Lett.* **83**, 4317 (1999).
58. Y. Bohbot-Raviv and Z.-G. Wang, *Phys. Rev. Lett.* **85**, 3428 (2000).
59. A. Keller, E. Pedemonte, and F. M. Willmouth, *Nature* **225**, 538 (1970).
60. C. C. Honeker and E. L. Thomas, *Chem. Mater.* **8**, 1702 (1996).
61. I. W. Hamley, *J. Phys. Condens. Matter* **13**, R643 (2001).
62. K. A. Koppi, M. Tirrell, F. S. Bates, K. Almdal, and R. H. Colby, *J. Phys. France* II **2**, 1941 (1992).
63. K. A. Koppi, M. Tirrell, and F. S. Bates, *Phys. Rev. Lett.* **70**, 1449 (1993).
64. K. Almdal, K. Mortensen, K. A. Koppi, M. Tirrell, and F. S. Bates, *J. Phys. France* II **6**, 617 (1996).
65. A. Nakajima, T. Hayashi, K. Kugo, and K. Shinoda, *Macromolecules* **12**, 840 (1979).
66. A. Nakajima, K. Kugo, and T. Hayashi, *Macromolecules* **12**, 844 (1979).
67. J. J. L. M. Cornelissen, M. Fischer, N. A. J. M. Sommerdijk, and R. J. M. Nolte, *Science* **280**, 1427 (1998).
68. M. W. Matsen, *J. Chem. Phys.* **106**, 7781 (1997).
69. K. Liu, S. M. Baker, M. Tuominen, T. P. Russell, and I. K. Schuller, *Phys. Rev. B* **63**, 060403(R) (2001).
70. M. Park, C. Harrison, P. M. Chaikin, R. A. Register, and D. H. Adamson, *Science* **276**, 1401 (1997).
71. C. Harrison, M. Park, P. M. Chaikin, R. A. Register, and D. H. Adamson, *J. Vac. Sci. Technol. B* **16**, 544 (1998).
72. G. Widawski, M. Rawiso, and B. François, *Nature* **369**, 387 (1994).
73. M. W. Matsen, *Curr. Opin. Colloid Interface Sci.* **3**, 40 (1998).
74. K. E. Strawhecker, S. K. Kumar, J. F. Douglas, and A. Karim, *Macromolecules* **34**, 4669 (2001).
75. A. Böker, A. H. E. Müller, and G. Krausch, *Macromolecules* **34**, 7477 (2001).
76. I. W. Hamley, E. L. Hiscutt, Y.-W. Yang, and C. Booth, *J. Colloid Interface Sci.* **209**, 255 (1999).
77. R. Limary and P. F. Green, *Langmuir* **15**, 5617 (1999).
78. P. Müller-Buschbaum, J. S. Gutmann, and M. Stamm, *Phys. Chem., Chem. Phys.* **1**, 3857 (1999).
79. D. Ausserré, D. Chatenay, G. Coulon, and R. Collin, *J. Phys. France* **51**, 2571 (1990).
80. S. H. Anastasiadis, T. P. Russell, S. K. Satija, and C. F. Majkrzak, *Phys. Rev. Lett.* **62**, 1852 (1989).
81. N. Rehse, A. Knoll, M. Konrad, R. Magerle, and G. Krausch, *Phys. Rev. Lett.* **87**, 035505 (2001).
82. T. L. Morkved, M. Lu, A. M. Urbas, E. E. Ehrichs, H. M. Jaeger, P. Mansky, and T. P. Russell, *Science* **273**, 931 (1996).
83. T. Thurn-Albrecht, J. Schotter, G. A. Kästle, N. Emley, T. Shibauchi, L. Krusin-Elbaum, K. Guarini, C. T. Black, M. T. Tuominen, and T. P. Russell, *Science* **290**, 2126 (2000).
84. L. Rockford, Y. Liu, P. Mansky, T. P. Russell, M. Yoon, and S. G. J. Mochrie, *Phys. Rev. Lett.* **82**, 2602 (1999).
85. X. M. Yang, R. D. Peters, P. F. Nealey, H. H. Solak, and F. Cerrina, *Macromolecules* **33**, 9575 (2000).
86. G. G. Pereira and D. R. M. Williams, *Europhys. Lett.* **44**, 302 (1998).
87. D. Petera and M. Muthukumar, *J. Chem. Phys.* **109**, 5101 (1998).
88. Y. Tsori and D. Andelman, *Europhys. Lett.* **53**, 722 (2001).
89. M. S. Turner, *Phys. Rev. Lett.* **69**, 1788 (1992).
90. D. G. Walton, D. J. Kellogg, A. M. Mayes, P. Lambooy, and T. P. Russell, *Macromolecules* **27**, 6225 (1994).

91. G. T. Pickett and A. C. Balazs, *Macromolecules* **30**, 3097 (1997).
92. G. H. Fredrickson, *Macromolecules* **20**, 2535 (1987).
93. M. S. Turner, M. Rubinstein, and C. M. Marques, *Macromolecules* **27**, 4986 (1994).
94. H. P. Huinink, M. A. van Dijk, J. Brokken-Zijp, and G. J. A. Sevink, *Macromolecules* **34**, 5325 (2001).
95. K. J. Hanley, T. P. Lodge, and C.-I. Huang, *Macromolecules* **33**, 5918 (2000).
96. J. E. Glass, Ed., *Water-soluble Polymers: Beauty with Performance* (American Chemical Society, Washington, D.C., 1986), Vol. 213.
97. I. R. Schmolka, in *Polymers for Controlled Drug Delivery*, edited by P. J. Tarcha (CRC Press, Boston, 1991).
98. M. W. Edens, in *Nonionic Surfactants. Polyoxyalkylene Block Copolymers*, edited by V. N. Nace (Marcel Dekker, New York, 1996), Vol. 60, p. 185.
99. M. Malmsten, in *Amphiphilic Block Copolymers: Self-assembly and Applications*, edited by P. Alexandridis and B. Lindman (Elsevier, Amsterdam, 2000).
100. P. Molyneux, *Water-soluble Synthetic Polymers: Properties and Behavior* (CRC Press, Boca Raton, 1983).
101. S. W. Shalaby, C. L. McCormick, and G. B. Butler, Eds., *Water-soluble Polymers. Synthesis, Solution Properties and Applications* (American Chemical Society, Washington, D.C., 1991), Vol. 467.
102. J. E. Glass, Ed., *Hydrophilic Polymers: Performance with Environmental Acceptability* (American Chemical Society, Washington, D.C., 1996), Vol. 248.
103. V. N. Nace, Ed., *Nonionic Surfactants. Polyoxyalkylene Block Copolymers* (Marcel Dekker, New York, 1996), Vol. 60.
104. P. Alexandridis and B. Lindman, Eds, *Amphiphilic Block Copolymers: Self-assembly and Applications* (Elsevier, Amsterdam, 2000).
105. J. E. Glass, Ed., *Associative Polymers in Aqueous Media* (American Chemical Society, Washington, D.C., 2000), Vol. 765.
106. C. Booth, G.-E. Yu, and V. M. Nace, in *Amphiphilic Block Copolymers: Self-assembly and Applications*, edited by P. Alexandridis and B. Lindman (Elsevier, Amsterdam, 2000), p. 57.
107. C. Booth and D. Attwood, *Macromol. Rapid Comm.* **21**, 501 (2000).
108. Z. Tuzar and P. Kratochvil, in *Surface Colloid Science.*, edited by E. Matijevic (Plenum, New York, 1993), Vol. 15, p. 1.
109. L. Zhang, K. Khougaz, M. Moffitt, and A. Eisenberg, in *Amphiphilic Block Copolymers: Self-assembly and Applications*, edited by P. Alexandridis and B. Lindman (Elsevier, Amsterdam, 2000).
110. K. Szczubialka, K. Ishikawa, and Y. Morishima, *Langmuir* **16**, 2083 (2000).
111. P. Guenoun, M. Delsanti, D. Gazeau, J. W. Mays, D. C. Cook, M. Tirrell, and L. Auvray, *Eur. Phys. J. B* **1**, 77 (1998).
112. I. W. Hamley, *Introduction to Soft Matter* (John Wiley, Chichester, 2000).
113. Z. Tuzar and P. Kratochvil, in *Light Scattering – Principles and Development*, edited by W. Brown (Oxford University Press, Oxford, 1996), p. 327.
114. S. W. Provencher, *Makromol. Chem.* **180**, 201 (1979).
115. J. S. Pedersen, *Adv. Colloid Interface Sci.* **70**, 171 (1997).
116. J. S. Pedersen and M. C. Gerstenberg, *Macromolecules* **29**, 1363 (1996).
117. Y. Liu, S.-H. Chen, and J. S. Huang, *Macromolecules* **31**, 2236 (1998).
118. J. S. Pedersen, *Curr. Opin. Colloid Interface Sci.* **4**, 190 (1999).
119. N. W. Ashcroft and J. Lekner, *Phys. Rev.* **145**, 83 (1966).
120. K. Mortensen and J. S. Pedersen, *Macromolecules* **26**, 805 (1993).
121. L. Leibler, H. Orland, and J. C. Wheeler, *J. Chem. Phys.* **79**, 3550 (1983).
122. P. Linse, *Macromolecules* **26**, 4437 (1993).

Introduction to Block Copolymers

123. J. Zipfel, P. Lindner, M. Tsianou, P. Alexandridis, and W. Richtering, *Langmuir* **15**, 2599 (1999).
124. T. Riley, S. Stolnik, C. R. Heald, C. D. Xiong, M. C. Garnett, L. Illum, and S. S. Davis, *Langmuir* **17**, 3168 (2001).
125. M. V. Seregina, L. M. Bronstein, O. A. Platonova, D. M. Chernyshov, P. M. Valetsky, J. Hartmann, E. Wenz, and M. Antonietti, *Chem. Mater.* **9**, 923 (1997).
126. L. M. Bronstein, S. N. Sidorov, P. M. Valetsky, J. Hartmann, H. Cölfen, and M. Antonietti, *Langmuir* **15**, 6256 (1999).
127. L. Zhang, K. Yu, and A. Eisenberg, *Science* **272**, 1777 (1996).
128. L. Luo and A. Eisenberg, *J. Am. Chem. Soc.* **123**, 1012 (2001).
129. S. Stewart and G. Liu, *Chem. Mater.* **11**, 1048 (1999).
130. B. M. Discher, Y.-Y. Won, D. S. Ege, J. C.-M. Lee, F. S. Bates, D. E. Discher, and D. A. Hammer, *Science* **284**, 1143 (1999).
131. C. Nardin, T. Hirt, J. Leukel, and W. Meier, *Langmuir* **16**, 1035 (2000).
132. K. Mortensen, *Curr. Opin. Colloid Interface Sci.* **3**, 12 (1998).
133. P. Alexandridis, U. Olsson, and B. Lindman, *Langmuir* **14**, 2627 (1998).
134. P. Alexandridis and R. J. Spontak, *Curr. Opin. Colloid Interface Sci.* **4**, 130 (1999).
135. K. Mortensen, *Coll. Surf. A* **183–185**, 277 (2001).
136. I. W. Hamley, *Curr. Opin. Colloid Interface Sci.* **5**, 342 (2000).
137. I. W. Hamley, *Philos. Trans. R. Soc. Lond.* **359**, 1017 (2001).
138. E. Eiser, F. Molino, G. Porte, and O. Diat, *Phys. Rev. E* **61**, 6759 (2000).
139. P. Holmqvist, C. Daniel, I. Hamley, W. Mingvanish, and C. Booth, *Colloid Surf. A*, in press (2001).
140. J. Noolandi, A.-C. Shi, and P. Linse, *Macromolecules* **29**, 5907 (1996).
141. M. Svensson, P. Alexandridis, and P. Linse, *Macromolecules* **32**, 637 (1999).
142. B. A. C. van Vlimmeren, N. M. Maurits, A. V. Zvelinodvsky, G. J. A. Sevink, and J. G. E. M. Fraaije, *Macromolecules* **32**, 646 (1999).
143. J. Zhao, B. Majumdar, M. F. Schulz, F. S. Bates, K. Almdal, K. Mortensen, D. A. Hajduk, and S. M. Gruner, *Macromolecules* **29**, 1204 (1996).
144. P. Kipkemboi, A. Fogden, V. Alfredsson, and K. Flodström, *Langmuir* **17**, 5394 (2001).
145. I. W. Hamley, *Adv. Polym. Sci.* **148**, 113 (1999).
146. S. Nojima, M. Ono, and T. Ashida, *Polym. J.* **24**, 1271 (1992).
147. V. Balsamo, A. J. Müller, and R. Stadler, *Macromolecules* **31**, 7756 (1998).
148. P. Rangarajan, R. A. Register, and L. J. Fetters, *Macromolecules* **26**, 4640 (1993).
149. P. Rangarajan, R. A. Register, D. H. Adamson, L. J. Fetters, W. Bras, S. Naylor, and A. J. Ryan, *Macromolecules* **28**, 1422 (1995).
150. A. J. Ryan, I. W. Hamley, W. Bras, and F. S. Bates, *Macromolecules* **28**, 3860 (1995).
151. V. Balsamo, G. Gil, C. Urbina de Navarro, I. W. Hamley, F. von Gyldenfeldt, V. Abetz and E. Cañizales, *Macromolecules* **36**, 4515 (2003).
152. I. W. Hamley, J. P. A. Fairclough, F. S. Bates, and A. J. Ryan, *Polymer* **39**, 1429 (1998).
153. T. Shiomi, H. Tsukuda, H. Takeshita, K. Takenaka, and Y. Tezuka, *Polymer* **42**, 4997 (2001).
154. Y.-L. Loo, R. A. Register, and A. J. Ryan, *Phys. Rev. Lett.* **84**, 4120 (2000).
155. I. W. Hamley, J. P. A. Fairclough, N. J. Terrill, A. J. Ryan, P. M. Lipic, F. S. Bates, and E. Towns-Andrews, *Macromolecules* **29**, 8835 (1996).
156. I. W. Hamley, J. P. A. Fairclough, A. J. Ryan, F. S. Bates, and E. Towns-Andrews, *Polymer* **37**, 4425 (1996).
157. D. J. Quiram, R. A. Register, G. R. Marchand, and A. J. Ryan, *Macromolecules* **30**, 8338 (1997).

158. Y.-L. Loo, R. A. Register, and D. H. Adamson, *Macromolecules* **33**, 8361 (2000).
159. S.-M. Mai, J. P. A. Fairclough, K. Viras, P. A. Gorry, I. W. Hamley, A. J. Ryan, and C. Booth, *Macromolecules* **30**, 8392 (1997).
160. M. Gervais and B. Gallot, *Makromol. Chem.* **171**, 157 (1973).
161. A. J. Ryan, J. P. A. Fairclough, I. W. Hamley, S.-M. Mai, and C. Booth, *Macromolecules* **30**, 1723 (1997).
162. K. C. Douzinas and R. E. Cohen, *Macromolecules* **25**, 5030 (1992).
163. L. Zhu, S. Z. D. Cheng, B. H. Calhoun, Q. Ge, R. P. Quirk, E. L. Thomas, B. S. Hsiao, F. Yeh, and B. Lotz, *J. Am. Chem. Soc.* **122**, 5957 (2000).
164. D. J. Quiram, R. A. Register, and G. R. Marchand, *Macromolecules* **30**, 4551 (1997).
165. P. Huang, L. Zhu, S. Z. D. Cheng, Q. Ge, R. P. Quirk, E. L. Thomas, B. Lotz, B. S. Hsiao, L. Liu, and F. Yeh, *Macromolecules* **34**, 6649 (2001).
166. L. Zhu, B. H. Calhoun, Q. Ge, R. P. Quirk, S. Z. D. Cheng, E. L. Thomas, B. S. Hsiao, F. Yeh, L. Liu, and B. Lotz, *Macromolecules* **34**, 1244 (2001).
167. L. Zhu, S. Z. D. Cheng, B. H. Calhoun, Q. Ge, R. P. Quirk, E. L. Thomas, B. S. Hsiao, F. Yeh, and B. Lotz, *Polymer* **42**, 5829 (2001).
168. E. A. DiMarzio, C. M. Guttman, and J. D. Hoffman, *Macromolecules* **13**, 1194 (1980).
169. M. D. Whitmore and J. Noolandi, *Macromolecules* **21**, 1482 (1988).
170. S. Nojima, S. Yamamoto, and T. Ashida, *Polym. J.* **27**, 673 (1995).
171. G. Reiter, G. Castelein, P. Hoerner, G. Riess, A. Blumen, and J.-U. Sommer, *Phys. Rev. Lett.* **83**, 3844 (1999).
172. G. Reiter, G. Castelein, P. Hoerner, G. Riess, J.-U. Sommer, and G. Floudas, *Euro. Phys. J. E* **2**, 319 (2000).
173. I. W. Hamley, M. L. Wallwork, D. A. Smith, J. P. A. Fairclough, A. J. Ryan, S.-M. Mai, Y.-W. Yang, and C. Booth, *Polymer* **39**, 3321 (1998).
174. J. P. A. Fairclough, S.-M. Mai, M. W. Matsen, W. Bras, L. Messe, S. Turner, A. J. Gleeson, C. Booth, I. W. Hamley, and A. J. Ryan, *J. Chem. Phys.* **114**, 5425 (2001).
175. D. R. Paul and S. Newman, Eds., *Polymer Blends* (Academic, London, 1978).
176. M. J. Folkes and P. S. Hope, Eds., *Polymeric Blends and Alloys* (Blackie, London, 1993).
177. S. Datta and D. J. Lohse, *Polymeric Compatibilizers: Uses and Benefits in Polymer Blends* (Hanser, Munich, 1996).
178. L. A. Utracki, *Commercial Polymer Blends* (Chapman and Hall, London, 1998).
179. H. Hasegawa and T. Hashimoto, in *Comprehensive Polymer Science. Second Supplement*, edited by S. L. Aggarwal and S. Russo (Pergamon, London, 1996), p. 497.
180. K. I. Winey, E. L. Thomas, and L. J. Fetters, *Macromolecules* **24**, 6182 (1991).
181. K. I. Winey, *Mater. Res. Soc. Symp. Proc.* **248**, 365 (1992).
182. K. I. Winey, E. L. Thomas, and E. L. Fetters, *Macromolecules* **25**, 2645 (1992).
183. M. W. Matsen, *Phys. Rev. Lett.* **74**, 4225 (1995).
184. C.-A. Dai, B. J. Dair, K. H. Dai, C. K. Ober, E. J. Kramer, C.-Y. Hui, and L. W. Jelinksi, *Phys. Rev. Lett.* **73**, 2472 (1994).
185. M. Sikka, N. N. Pellegrini, E. A. Schmitt, and K. I. Winey, *Macromolecules* **30**, 445 (1997).
186. M. D. Dadmun, *Macromolecules* **33**, 9122 (2000).
187. R. B. Thompson and M. W. Matsen, *Phys. Rev. Lett.* **85**, 670 (2000).
188. C. W. Macosko, P. Guégan, A. K. Khandpur, A. Nakayama, P. Marechal, and T. Inoue, *Macromolecules* **29**, 5590 (1996).
189. F. S. Bates, W. W. Maurer, P. M. Lipic, M. A. Hillmyer, K. Almdal, K. Mortensen, G. H. Fredrickson, and T. P. Lodge, *Phys. Rev. Lett.* **79**, 849 (1997).

190. M. A. Hillmyer, W. W. Maurer, T. P. Lodge, F. S. Bates, and K. Almdal, *J. Phys. Chem. B* **103**, 4814 (1999).
191. J. H. Lee, N. P. Balsara, R. Krishnamoorti, H. S. Jeon, and B. Hammouda, *Macromolecules* **34**, 6557 (2001).
192. F. S. Bates, W. Maurer, T. P. Lodge, M. F. Schulz, M. W. Matsen, K. Almdal, and K. Mortensen, *Phys. Rev. Lett.* **75**, 4429 (1995).
193. M. W. Matsen, *J. Chem. Phys.* **110**, 4658 (1999).
194. R. B. Thompson and M. W. Matsen, *J. Chem. Phys.* **112**, 6863 (2000).
195. K. Krishnan, K. Almdal, W. Burghardt, T. P. Lodge, and F. S. Bates, *Phys. Rev. Lett.* **87**, 098301 (2001).
196. T. L. Morkved, P. Stepanek, K. Krishnan, F. S. Bates, and T. P. Lodge, *J. Chem. Phys.* **114**, 7247 (2001).
197. T. Hashimoto, K. Yamasaki, S. Koizumi, and H. Hasegawa, *Macromolecules* **26**, 2895 (1993).
198. S. Koizumi, H. Hasegawa, and T. Hashimoto, *Macromolecules* **27**, 4371 (1994).
199. D. Yamaguchi, M. Takenaka, H. Hasegawa, and T. Hashimoto, *Macromolecules* **34**, 1707 (2001).
200. D. Yamaguchi and T. Hashimoto, *Macromolecules* **34**, 6495 (2001).
201. D. Yamaguchi, H. Hasegawa, and T. Hashimoto, *Macromolecules* **34**, 6506 (2001).
202. M. W. Matsen, *J. Chem. Phys.* **103**, 3268 (1995).
203. A. D. Vilesov, G. Floudas, T. Pakula, E. Y. Melenevskaya, T. M. Birshtein, and Y. V. Lyatskaya, *Macromol. Chem. Phys.* **195**, 2317 (1994).
204. A.-C. Shi and J. Noolandi, *Macromolecules* **28**, 3103 (1995).
205. M. W. Matsen and F. S. Bates, *Macromolecules* **28**, 7298 (1995).
206. H. Frielinghaus, N. Hermsdorf, K. Almdal, K. Mortensen, L. Messé, L. Corvazier, J. P. A. Fairclough, A. J. Ryan, P. D. Olmsted, and I. W. Hamley, *Europhys. Lett.* **53**, 680 (2001).
207. H. Frielinghaus, N. Hermsdorf, R. Sigel, K. Almdal, K. Mortensen, I. W. Hamley, L. Messé, L. Corvazier, A. J. Ryan, D. van Dusschoten, M. Wilhelm, et al., *Macromolecules* **34**, 4907 (2001).
208. P. D. Olmsted and I. W. Hamley, *Europhys. Lett.* **45**, 83 (1999).
209. J. Dlugosz, A. Keller, and E. Pedemonte, *Kolloid Z.u.Z. Polymere* **242**, 1125 (1970).
210. S. Collins, T. Mykhaylyk, and I. W. Hamley, unpublished work (2002).
211. G. Wanka, H. Hoffmann, and W. Ulbricht, *Macromolecules* **27**, 4145 (1994).
212. D. Zhao, J. Feng, Q. Huo, N. Melosh, G. H. Fredrickson, B. F. Chmelka, and G. D. Stucky, *Science* **279**, 548 (1998).
213. S. Koizumi, H. Hasegawa, and T. Hashimoto, *Macromolecules* **27**, 6532 (1994).

2 Recent Developments in Synthesis of Model Block Copolymers Using Ionic Polymerisation

KRISTOFFER ALMDAL
The Danish Polymer Centre, Risø National Laboratory, DK-4000 Roskilde, Denmark

2.1 INTRODUCTION

Model polymers and block copolymers are most often synthesized because the target molecules are to be used for some specific purpose either in applications or as a custom-made tool that will allow the investigation and elucidation of some specific scientific problem. Thus, new synthetic methods are often developed and presented in conjunction with other results. However, in the present chapter the focus is on the science and craft of synthesis and many exciting results presented during the last few years in publications that are referenced in the present chapter for some synthetic detail are not discussed.

The present chapter reviews recent developments (work published in 1997 and later) in the synthesis of model block copolymers with a primary focus on ionic polymerisations. During this period controlled radical polymerisation techniques have attracted considerable interest and are emerging as a new method providing the synthesis of model polymers and copolymers. It is not the purpose of this chapter to cover this development since ionic methods still allow for better control of the polymers synthesised. Radical polymerization methods are the subject of Chapter 3. The question of how important differences in the widths of the molar mass distributions are has prompted the inclusion of a section on the MMD of model block copolymers.

Within the field of ionic polymerisation a number of specialized reviews have appeared. As an example the preparation of polymers bearing zwitterionic end groups has been reviewed [1]. In the field of polymer synthesis there is a strong tradition to use acronyms as a shorthand notation for chemical species. Apart from saving space this practice makes a text more readable for the well-informed reader. However, this benefit is achieved at the cost of the texts being unapproachable for the less-informed reader. This fact has prompted the inclusion in this chapter of an extensive list of acronyms that are used in the text and where the commonly used forms of abbreviation are maintained such that the list can serve as a small, admittedly incomplete, dictionary of acronyms in ionic polymer synthesis. The length of this list has also prompted the unusual practice of *not* defining an abbreviation at the first use in the text but only to specify it in the *list of abbreviations*. Before the actual review, sections on a description of standard polymerisation conditions and on MMDs of model block copolymers are given.

The effort to expand the scope of polymer synthesis can be grouped in many ways. Here, a section is devoted to work that is primarily interesting because new types of material are generated *Block copolymers by stepwise synthesis, new monomers, and post-polymerisation techniques*. Another angle is new ways of linking molecules together including macroinitiators, couplings, change of mechanism, and other architectural methods, which are described in the section *Methods for generating new block copolymer architectures*. Finally, substantial activity is found in the area of relatively polar monomers such as lactones, lactides, carboxyanhydrides and similar monomers, which is found in the section *Ring-opening polymerisation of lactones, lactides, carboxyanhydrides, and similar monomers*.

2.1.1 POLYMERISATION CONDITIONS

Controlled anionic polymerisation is applicable to a wide range of monomers and, in fact, a trend in current research is to expand the scope of the method especially towards new polymers, new architectures or an expanded range of molar masses. However, if one chooses to categorize according to the conditions under which the reaction is performed then a few sets of conditions describe the vast majority of experiments reported in the literature. A set of conditions is given by the solvent, an initiator, a temperature and a terminator. In some cases a chain end modifier is also involved. Thus a specification of the type /solvent/initiator/temperature/modifier/terminator/ will give the informed reader a fairly precise idea about how the polymerisation experiment was carried out. For example /THF/sBuLi/−78 °C//CH_3OH/ specifies conditions suitable for polymerising S and 2VP. The // between 78 °C and CH_3OH indicates that no chain-end modifier was used in this case. The chain-end modifier can either react with the chain end to modify the reactivity of the carbanion or it can act as a ligand to the counter ion. Here the effect of amine complexing agents on the polymerisation of dienes under apolar conditions is a classic example. For the block copolymerisation of S and MMA it is easily understood that ///// DPE // implies the addition of DPE between the polymerisation of the two blocks. In fact only two main types of conditions find wide application. The /THF/sBuLi/−78 °C//CH_3OH/ system used as an example is one of those and will be referred to as low-temperature polar conditions. Under low-temperature conditions the identity of the counterion is important but this is generally

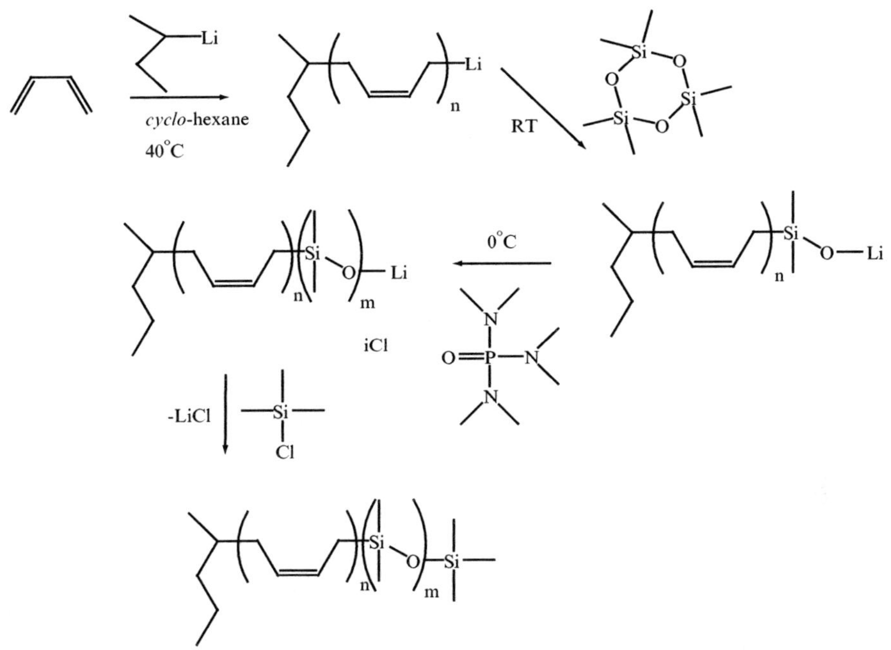

Figure 2.1 Synthesis of a PB-PDMS diblock. The PB-block is obtained under /CHX/sBuLi/ 40°C//D_3 conditions. The PDMS block is obtained under /CHX/PB-SiOLi/0 °C/HMPA/ TCMS conditions.

determined by the initiator species. The other main type is either /CHX/sBuLi/40 °C//CH_3OH/ or /Bz/sBuLi/RT//CH_3OH/, which in many instances will lead to almost identical polymers (dienes, S). This type of condition will be referred to as RT apolar conditions. In order to facilitate reading a typical reaction scheme is given in Figure 2.1. Structures of most of the monomers described in this review are given in Figures 2.2 and 2.3. The experimental techniques involved in anionic polymerisation have been reviewed recently both for the inert-gas techniques [2] and for the high-vacuum technique [3].

2.1.2 M_w/M_n CONSIDERATIONS FOR BLOCK COPOLYMERS

In the synthesis of block copolymers one is often in the situation for AB-diblock copolymers that M_w/M_n is known for the entire block copolymer and for the block synthesized first. The question arises what the value is for the B-block. Under some reasonable assumptions one can calculate this number. If the number chain length distributions for the two blocks are described by the continuous probability distributions $P_1(x)$ and $P_2(y)$, where $P_1(x)$ gives

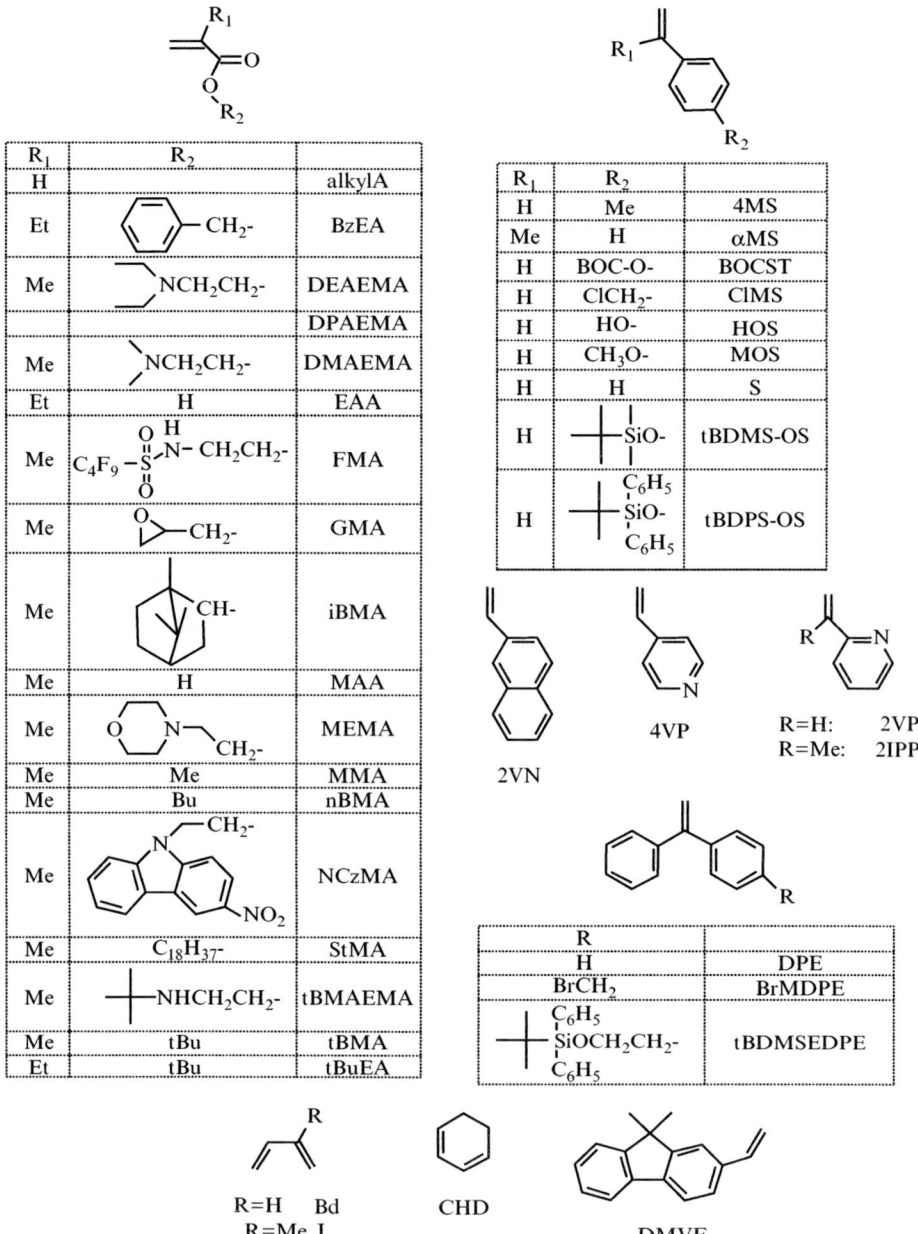

Figure 2.2 Selected monomers used for anionic polymerization.

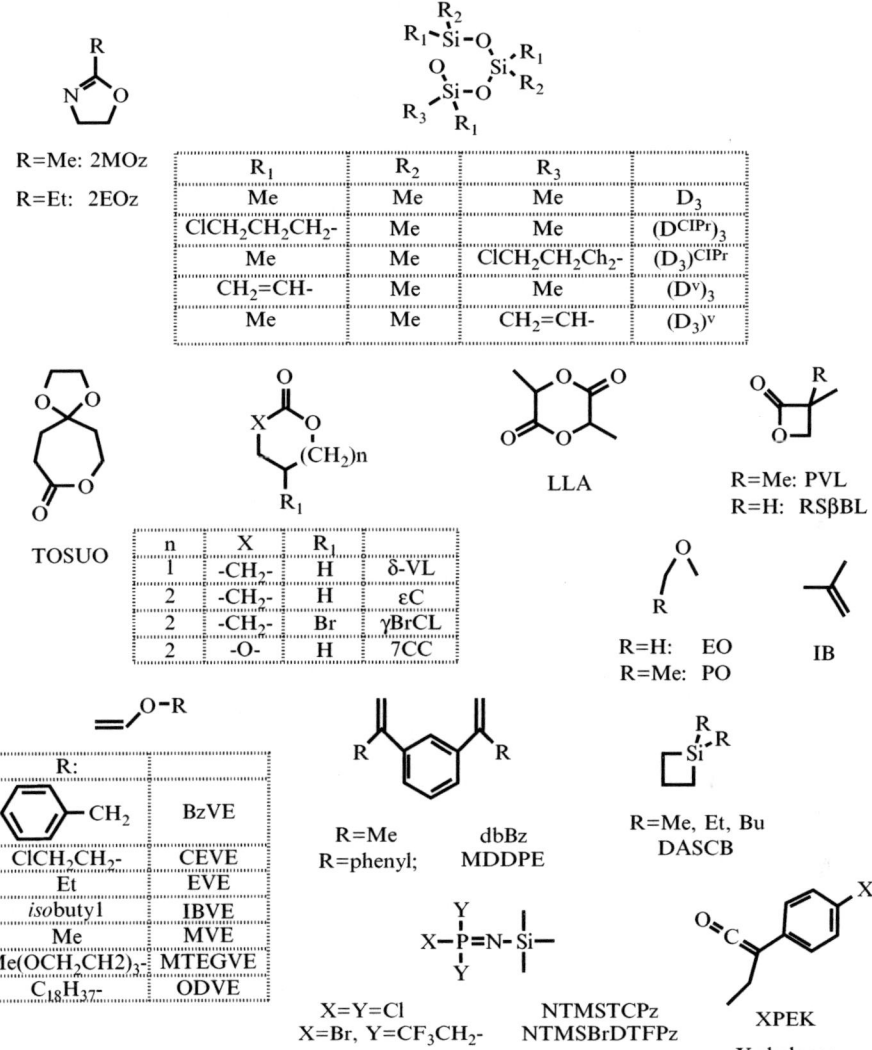

Figure 2.3 Selected monomers mostly used for ring-opening and cationic polymerization.

the probability that a given chain contains x A monomers and $P_2(y)$ the probability that a chain contains y B monomers, we want to be able to calculate the probability distribution that a given block copolymer has a certain length.

The relation between a number chain length distribution, $P(x)$, and the parameters normally used to describe molar mass distributions are given by:

$$M_n = \bar{x} M_0$$

$$M_w = \frac{\overline{x^2}}{\bar{x}} M_0 = \frac{\bar{x}^2 + \sigma^2}{\bar{x}} M_0 = M_n + \frac{\sigma^2}{\bar{x}} M_0$$

$$\frac{M_w}{M_n} = \frac{\overline{x^2}}{\bar{x}^2} = 1 + \frac{\sigma^2}{\bar{x}^2} = 1 + \frac{\sigma^2}{\left(\frac{M_n}{M_0}\right)^2} \tag{2.1}$$

$$U = \frac{M_w}{M_n} - 1 = \frac{\sigma^2}{\left(\frac{M_n}{M_0}\right)^2},$$

where $\bar{x}$ and σ^2 in Equation (2.1) are the mean and variance of $P(x)$, and M_0 is the molar mass of the monomer. A reasonable choice of probability distribution to describe the chain length of polymers and block copolymers is the Schultz–Zimm distribution, which is given by:

$$P(z) = \frac{k^k}{\Gamma(k)} z^{k-1} e^{-kz} \tag{2.2}$$

where $z = x/\bar{x}$ is a normalized degree of polymerisation and k is a parameter. The Schultz–Zimm distribution is an example of the Gamma distribution [4], $G(y)$ which has two free parameters, α and θ:

$$G(y) = \frac{y^{\alpha-1} e^{-y/\theta}}{\Gamma(\alpha) \theta^\alpha}. \tag{2.3}$$

For $G(y)$[4]:

$$\bar{y} = \alpha \theta$$
$$\sigma^2 = \alpha \theta^2. \tag{2.4}$$

In calculating MMDs for block copolymer it is more convenient to have molar mass, M, as the variable rather than z. Thus, with $M = zM_n$ and $dz = dM/M_n$:

$$P(M) = \frac{k^k}{\Gamma(k)} \left(\frac{M}{M_n}\right)^{k-1} \exp\left(-k \frac{M}{M_n}\right) \frac{1}{M_n}$$
$$= \frac{M^{k-1}}{\Gamma(k)} \left(\frac{M_n}{k}\right)^{-k} \exp\left(-\frac{M}{M_n/k}\right), \tag{2.5}$$

which is a Gamma distribution with parameters k and M_n/k. The parameters relevant to characterise the polymer are:

$$\overline{M} = \int_0^\infty M P(M) \mathrm{d}M = M_\mathrm{n}$$

$$\overline{M^2} = \int_0^\infty M^2 P(M) \mathrm{d}M = \frac{1}{\Gamma(k)} \left(\frac{M_\mathrm{n}}{k}\right)^{-k} \frac{\Gamma(k+2)}{(M_\mathrm{n}/k)^{k+2}} = M_\mathrm{n}^2 \left(1 + \frac{1}{k}\right)$$

$$\sigma^2 = \overline{M^2} - \overline{M}^2 = \frac{M_\mathrm{n}^2}{k} \qquad (2.6)$$

$$\frac{M_\mathrm{w}}{M_\mathrm{n}} = \frac{\overline{M^2}}{\overline{M}^2} = \left(1 + \frac{1}{k}\right)$$

$$U = k^{-1}.$$

Here, use is made of the definite integral $\int_0^\infty x^n \mathrm{e}^{-ax} \mathrm{d}x = \Gamma(n+1)/a^{n+1}$ for $n > -1; a > 0$. Note that the substitution $y = kz$ in Equation (2.5) yields a continuous version of the Poisson distribution, which is theoretically obtained for an ideal living polymerisation. On the other hand $k = 1$ yields:

$$P(M) = M_\mathrm{n} \exp\left(-\frac{M}{M_\mathrm{n}}\right), \qquad (2.7)$$

which is the most-probable distribution. The Schulz–Zimm distribution thus provides a continuous transition between two of the most important distributions in synthetic polymers. The MMD for an AB-diblock copolymer $P_{AB}(M)$ is thus a probability distribution describing the sum $M_A + M_B$ of two independent Schulz–Zimm distributed variables. This distribution is not easily analytically tractable in the general case. However, for the case where the two distributions have the same value of M_n/k the distribution is a Schultz–Zimm distribution and the variance of the combined distribution can be written:

$$\sigma_{AB}^2 = w^2 \sigma_A^2 + (1-w)^2 \sigma_B^2, \qquad (2.8)$$

where $w = M_{\mathrm{n},A}/(M_{\mathrm{n},A} + M_{\mathrm{n},B})$ is the mass fraction of A in the AB diblock copolymer. Thus:

$$U_{AB} = U_A w^2 + U_B (1-w)^2, \qquad (2.9)$$

and

$$U_B = \frac{U_{AB} - U_A w^2}{(1-w)^2} \qquad (2.10)$$

Computer simulations on a wide range of diblock copolymers, where it was assumed that the two blocks both follow a Schultz–Zimm distribution have shown that this expression is a very close approximation also in the general case where the M_n/k values for the two blocks are independent [5].

Another distribution function that is mostly ignored in the literature is the distribution in composition, $P(w)$, the probability that a given chain has the composition w. Following the treatment of chain lengths $P(w)$ is a probability distribution describing the fraction $M_A/(M_A + M_B)$ involving two independent Schulz–Zimm distributed variables. Again the general case is not analytically tractable but for the case where the two distributions have the same value of M_n/k, $P(w)$ is a Beta distribution with parameters k_1 and k_2. For the Beta distribution:

$$P(w) = \frac{\Gamma(k_1 + k_2)}{\Gamma(k_1)\Gamma(k_2)}(1-w)^{k_2-1}w^{k_1-1}$$

$$\overline{w} = \frac{k_1}{k_1 + k_2} \quad (2.11)$$

$$\sigma^2 = \frac{k_1 k_2}{(k_1 + k_2)^2(k_1 + k_2 + 1)}.$$

For a symmetric block copolymer where the two blocks have equal M_w/M_n the spread of the distribution $P(w)$ is found from

$$k = k_1 = k_2 = 1/U_A = 1/U_B$$

$$\sigma = \frac{1}{2}\sqrt{\frac{U_A}{2 + U_A}}. \quad (2.12)$$

Thus even for a modest $U_A = U_B = 0.1$ then $\sigma(P(w)) = 0.109$. $U_A = U_B = 0.01$ would be the theoretical (Poisson distribution) values for a 10.4/10.5 kg/mol PS-P2VP diblock copolymer which leads to $\sigma(P(w)) = 0.035$.

Figures 2.4 and 2.5 give $P(w)$ for a range of a representative set of U-values for the A and B block in a symmetric and asymmetric diblock copolymer. The data was obtained in computer simulations [5].

It is a consequence of Equation (2.10) that the M_w/M_n values for an AB-block copolymer should be smaller than the values normally observed for A and B homopolymers with molar mass comparable to the blocks provided the block copolymerisation reaction proceeds in a similar manner to the homopolymerisation. The vast majority of the M_w/M_n data presented in the literature is based on SEC measurements. In fact SEC is problematic for the characterization of very narrow MMDs. For homopolymers the axial dispersion phenomenon is the main problem, whereas for block copolymers it is also questionable to what extent true noninteracting conditions are accessible. A development has started towards the use of alternative techniques to SEC for the characterization of diblock copolymers. Apart from the popular MALDI-TOF mass spectroscopy various newer chromatographic techniques have been used. A series of PS samples prepared under as identical conditions as possible (/CHX/sBuLi/45 °C//CH$_3$OH/) were analysed by SEC and TGIC[6] and the measured M_w/M_n values compared with the Poisson distribution predictions.

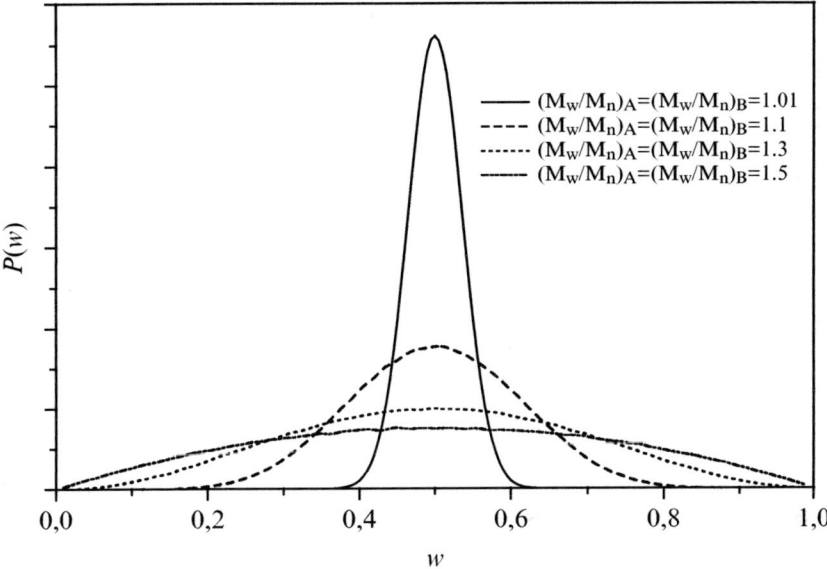

Figure 2.4 The probability distribution, $P(w)$ for a symmetric diblock copolymer with the M_w/M_n value given in the figure.

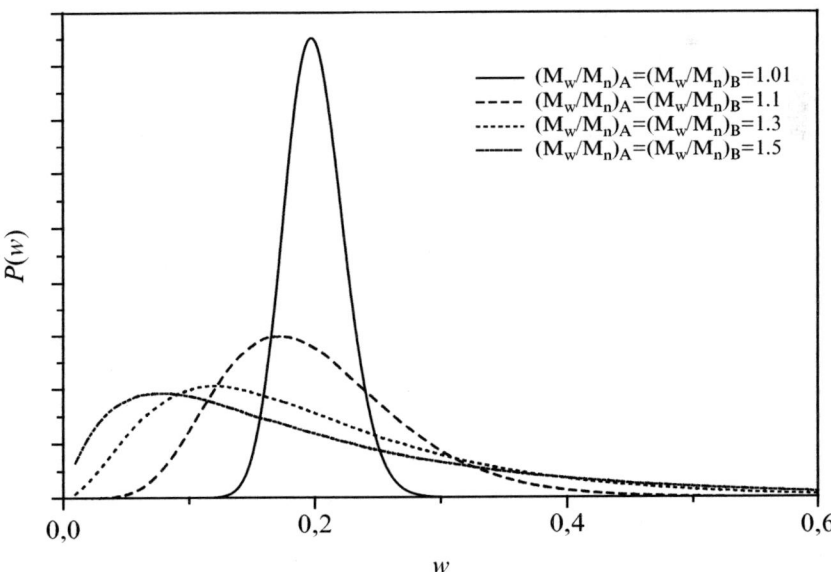

Figure 2.5 The probability distribution, $P(w)$ for an asymmetric ($w = 0.2$) diblock copolymer with the M_w/M_n value given in the figure.

It was confirmed that M_w/M_n is a decreasing function of the degree of polymerisation. Furthermore, the M_w/M_n value from SEC was considerably higher than that obtained by TGIC. Another technique, LCCC has the potential of measuring the MMD of one of the blocks in a block copolymer. However, the extent to which this is possible has been questioned [7].

The intrinsic width of the composition distribution even for block copolymers that have a very narrow MMD has been experimentally confirmed [8]. Two anionically prepared (/CHX/sBuLi/45 °C//CH$_3$OH/) PS-PI block copolymers were very carefully characterized. The highest molar mass block copolymer (34.8 wt% PI) showed (SEC): $M_w/M_n = 1.01$, $M_n = 24$ kg/mol. The length distribution of PS and PI was individually characterized by normal and reversed-phase liquid chromatography. The general idea is to find conditions under which one block is separated according to interaction chromatography (the interaction is molar mass dependent) whereas the other does not interact (that is ideal SEC conditions). In order to separate PI under interaction conditions, while PS does not interact (with the column packing material) the packing material must be less polar than PS and the mobile phase a good solvent for PS. This is reversed phase conditions (e.g. C$_{18}$-modified column packing material and CH$_3$CN/CH$_2$Cl$_2$ mixture as eluent). For separation of PS the situation is reversed (normal phase conditions: e.g. a diol bonded column packing and isooctane/THF mixture as the eluent). MALDI-TOF analysis on a 2.4 kg/mol block copolymer was used to confirm that it is possible to separate the polymers such that each fraction has a very narrow distribution of one of the blocks. The highest molar mass PS-PI was separated in 3 fractions with compositions ranging from 32.3 wt% to 39 wt% PI in excellent agreement with the above-mentioned spread of the composition. A drawback of this technique is that it is necessary to find an optimal solvent mixture in each case for which temperature variation alone can be used to fine tune the interaction chromatography.

2.2 BLOCK COPOLYMERS BY STEPWISE SYNTHESIS, NEW MONOMERS, AND POST POLYMERISATION TECHNIQUES

2.2.1 METHACRYLATE BLOCK COPOLYMERS

Stepwise synthesis is the predominant route to linear block copolymers and many publications deal with some aspects of this procedure. An example of how far this can be taken is the sequential block copolymerisation of 5 different monomers to obtain a S-I-2VP-tBMA-EO pentablock copolymer [9] ($M_n = 92$ kg/mol, $M_w/M_n = 1.04$).

A considerable effort goes into improving the procedures for preparing controlled methacrylates both in terms of utilising new monomers and in terms of obtaining higher molar mass. A number of studies were concerned with triblock and linear pentablock copolymers of S and MMA, tBMA, and MAA [10–13]. A difunctional initiator x was obtained from dPBz under either/ CHX/tBuLi/50 °C/// or /CHX/tBuLi/–20 °C/TEA// conditions. dPBz contains two double bonds that can react with an organolithium species to form a difunctional initiator. A whole class of difunctional initiator precursors are based on molecules containing two double bonds, which ideally are equally reactive towards organolithium species. The difunctional initiator is formed *in situ* by reaction with a monofunctional organolithium species. Bd was polymerised followed by S under /CHX/x/25 °C/Et$_2$O// conditions. The methacrylate was polymerised in 50% THF at -78 °C to obtain PtBMA-PBd-PtBMA, PMAA-PBd-PMAA, PtBMA-PS-PBd-PS-PtBMA, and PMAA-PS-PBd-PS-PMAA (typically $M_n = 80$ kg/mol; $M_w/M_n < 1.1$, 78% Bd and various tBMA/styrene ratios). The degree of 1,2-incorporation of the Bd units is $43 \pm 1\%$ under these conditions. The tBMA was hydrolysed to MAA with 2% methane sulfonic acid in 20/80 acetic acid/toluene for 1 h at 140 °C. iBMA can be used instead of tBMA in the polymerisation. The central Bd-block can be completely saturated using an Et$_3$Al/Co hexanoate complex as the hydrogenation catalyst. Furthermore, GMA can be incorporated in triblock copolymers in a similar manner to obtain a polymeric epoxide.

Methacrylate polymers have been reported that contain unusual functional groups (for polymers prepared by ionic polymerisation) including nitro, hydroxy, secondary amine, and cyanoazobenzene groups. The dipotassium salt of PMMA afforded the polymerisation of the nitro-containing NCzMA under low-temperature polar conditions provided a sufficient concentration of Et$_2$Zn (close to equimolar with respect to NCzMA) was present during the reaction. Addition of MMA to the dipotassium salt of P(NCzMA) yields a PMMA-P(NCzMA)-PMMA triblock copolymer ($M_n = 23$ kg/mol; $M_w/M_n = 1.32$)[14]. A triblock copolymer of S, protected hydroxystyrene (tBDMS-OS), and MMA was synthesised using low-temperature /THF/sBuLi//DPE// conditions ($M_w/M_n = 1.16$; $M_n = 22$ kg/mol). PS-PHOS-PMMA was obtained by hydrolysis. Clear signs of nonideal SEC for PS-PHOS-PMMA are seen [15]. The potential side-chain liquid-crystalline-forming group CPPHMA was homopolymerised and incorporated in block copolymers with S under low-temperature polar conditions (-110 °C for S and -40 °C for CPPHMA) in the presence of LiCl. No DPE was added to facilitate the crossover between the two monomers ($4.3 \leq M_n/(\text{kg/mol}) \leq 16.5$; $1.17 \leq M_w/M_n \leq 1.39$ for the diblocks and $9.3 \leq M_n/(\text{kg/mol}) \leq 80.1$; $1.17 \leq M_w/M_n \leq 1.38$ for the homopolymers) [16].

The synthesis of methacrylate-containing polymers has been developed to include new monomers. Stearyl side chains were obtained by sequential polymerisation of S and StMA under low-temperature polar conditions with DPE and LiCl present to yield PS-PStMA diblock copolymers ($30 \leq M_n/(\text{kg/mol}) \leq 464$; $1.04 \leq M_w/M_n \leq 1.18$) [17].

Amine functional side chains were also introduced in block copolymers. The potassium salt of a hydroxy-terminated PEO or C_nH_{2n+1}OH with $n = 12$

or 16 was used to polymerise a range of tertiary amine functional methacrylates including DMAEMA, DEAEMA, and MEMA, DPAEMA under /THF/potassium alcoholate/30–50 °C//CH$_3$OH/ conditions ($7 \leq M_n$/(kg/mol)≤ 14; $1.23 \leq M_w/M_n \leq 1.35$) [18,19]. Copolymers of S and DMAEMA or tBuMA were obtained under /THF/CmK/ – 78 °C/DPE/CH$_3$OH/ conditions ($M_n = 6$–134 kg/mol, $M_w/M_n = 1.02$–1.15 for PS-PtBuMA. No characterization data was presented for PS-PDMAEMA). The tBu groups can by hydrolysed in 5M HCl/Dioxane [20]. Secondary amine-containing homopolymers from the monomer tBAEMA ($M_n = 4$–50 kg/mol, $M_w/M_n = 1.05$–1.1) and block copolymers with S, MMA, and tBMA ($M_n = 13$–20 kg/mol, $M_w/M_n \leq 1.04$) were prepared at low-temperature polar conditions in the presence of a 10-fold excess of LiCl [21].

The polymerisation of oxazoline protected 4-vinyl-α-methylcinnamic acid is initiated under low-temperature polar conditions by organopotassium initiators. The living chain can initiate tBuMA quantitatively whereas S, 2VP, or I cannot be initiated quantitatively. On the other hand living chains of S, 2VP, or I initiated the polymerisation of the oxazoline protected 4-vinyl-α-methylcinnamic acid [22].

Block copolymers of PDMS and methacrylates are not accessible by sequential monomer addition. However, a methacrylate-terminated PDMS was reacted with EtMe$_2$SiH to form the macroinitiator silyl ketene acetal-terminated PDMS, which is a group transfer polymerisation initiator. TBABB catalysed the polymerisation in THF at RT of tBMA and TMSA ($9 \leq M_n$/(kg/mol) ≤ 29; $1.15 \leq M_w/M_n \leq 1.32$) [23].

An alternative initiator for MMA was reported. MMA was polymerised by the bulky carbanion in the NTPP, TPM salt at RT or above in THF to obtain PMMA ($1.3 \leq M_n$/(kg/mol) ≤ 24; $1.03 \leq M_w/M_n \leq 1.24$). The characterization data in this report is questionable since $U = 0.05$ is reported for a polymer with DP = 12. PMMA-PtBMA was prepared under similar conditions [24].

BzEA can be polymerised under low-temperature polar conditions with DPHLi as initiator ($2.4 \leq M_n$/(kg/mol) ≤ 20; $1.07 \leq M_w/M_n \leq 1.16$). The benzyl ester is readily cleaved and PBzEA can be converted to PEAA by treatment with TMSI in CHCl$_3$. The polymerisation of tBuEA did not lead to narrow MMD polymers under similar conditions [25]. PMMA-PalkylA-PMMA, where alkyl is n-butyl or isooctyl was obtained from PMMA-PtBA-PMMA by transalcoholation in pure alkyl alcohol with pTSA 10 mol% as catalyst. The transalcoholation results in more than 95% conversion. The precursor triblock ($75 \leq M_n$/(kg/mol) ≤ 290, $M_w/M_n \leq 1.07$) was synthesized at low-temperature polar conditions in the presence of LiCl [26–28].

A number of papers have discussed the polymerisation of silicon-containing 4-membered ring monomers, DASCB [29–33]. Typical polymerisation conditions are /THF, nHx/nBuLi/–48 °C//CH$_3$OH/. Block copolymers of methacrylates and S, HEMA, MMA, and tBMA were prepared with alkyl = Bu (M_n/(kg/mol) ≤ 17; $1.13 \leq M_w/M_n \leq 1.25$ for HEMA block copolymers and $1.08 \leq M_w/M_n \leq 1.18$ for the other comonomers). For alkyl = propyl, $M_n = 80$ kg/mol, and $M_w/M_n = 1.22$ was obtained.

Another exotic monomer type that has received some attention are the silicon- and phosphorous-strained ring-bridged ferrocene analogues that yield PFS and PFP as well as germanium and tin analogues [34–40]. The monomers can be prepared from dilithiated ferrocene to form a strained ring with Si or P bonded to both cyclopentadiene rings of the ferrocene. This monomer can be polymerised under /THF/nBuLi/RT/// conditions. For PFP DPs ≤ 100 with $M_w/M_n < 1.25$ are obtained. PSLi functions as an initiator as well, whereas PILi only give partial initiation ($M_n \sim 10$ kg/mol; $M_w/M_n = 1.1$). The functional initiator tBDMSPrLi allowed for the synthesis of block copolymers of PFS with DMAEMA. PFP-PFS and PFP-PDMS were likewise obtained ($M_n < 30$ kg/mol and $M_w/M_n \sim 1.1$).

A class of inorganic main-chain polymers that have attracted some attention for the possibility of controlled synthesis are the nitrogen-phosphorous alternating main-chain polyphosphazenes. For example, a series of phosphazene block copolymers with fluorine and fluorinated side chains was synthesised [41] by sequential cationic polymerisation of NTMSTCPz and substituted phosphoranimines initiated by PCl_5 in CH_2Cl_2 at RT followed by substitution of the chlorine with various groups ($7.5 \leq M_n/(kg/mol) \leq 23$; $1.06 \leq M_w/M_n \leq 1.35$). Di- and triblock copolymers of PEO and DTFPz were obtained from mono- or diamine-functional PEO by conversion to the corresponding phosphoranimine macroinitiator via treatment with NTMSBrDTFPz. The macroinitiator was used to cationically polymerise NTMSTCPz in CH_2Cl_2 with PCl_5 as initiator followed by substitution of the chloride by trifluorethoxy groups ($15 \leq M_n/(kg/mol) \leq 29$; $1.16 \leq M_w/M_n \leq 1.4$ for the products with $3.4 \leq M_n/(kg/mol) \leq 5.7$ for the precursor PEO) [42,43]. Likewise phosphazene siloxane block copolymers were obtained by hydrosilylation coupling of dihydrosilane terminated PDMS and allyl terminated PDTFPz in THF at 66 °C in the presence of a nonacidic Pt-based catalyst [44–46]. The functionally terminated DTFPz was obtained by reacting NTMSBrDTFPz with an appropriate amine (here allyl amine) in THF/TEA at -78 °C and using the product as an initiator for the cationic polymerisation of NTMSTCPz ($3 \leq M_n/(kg/mol) \leq 40$; $1.06 \leq M_w/M_n \leq 1.2$ for the functionally terminated DTFPz, similar values being obtained for the coupled products). In another route to the same block copolymer a dihydroxy-terminated PDMS was reacted with NTMSBrDTFPz and the product used as a macroinitiator to obtain PDTFPz-PDMS-PDTFPz triblock copolymers ($1.05 \leq M_n (kg/mol) \leq 1.13$; $5.6 \leq M_w/M_n \leq 16$).

2.2.2 FLUOROPOLYMERS AND BLOCK COPOLYMERS

Due to their chemical resistance and low surface energy properties, fluorinated polymers have attracted considerable recent interest. A couple of reviews describing various new routes to fluorinated polymers [47] by postpolymerisation chemistry and to fluorinated block copolymers [48] have appeared. In a number of studies fluorine-containing monomers have been polymerised directly. A perfluoroalkylsulfonamide-containing methacrylate block copolymer was obtained by sequential polymerisation of tBMA and FMA in /THF/ DPHLi/–78 °C/LiCl// ($3 \leq M_n$/(kg/mol) ≤ 90; $1.03 \leq M_w/M_n \leq 1.19$) [49]. Triblocks of silyl-protected HEMA, tBuMA and $C_4F_9C_2H_4$-metacrylate (all sequences of monomers possible) were synthesised under low-temperature polar conditions ($M_n = 20$ kg/mol overall, $M_w/M_n = 1.07$ in all cases) [50]. Diblock copolymers of MMA and side-chain fluorinated methacrylates ($C_xH_{2x-y+1}F_y$ with (x, y) = (5, 8); (8, 15); (8, 13))) were prepared under /Tol/ Al(BHT) Bu$_2^i$tBuLi/ 0 °C/// conditions. For the highest (x, y) values BTFB was used as cosolvent ($17 \leq M_n$/(kg/mol) ≤ 217; $1.1 \leq M_w/M_n \leq 1.5$). Fluoroalkyl methacrylates containing CF_3CH_2, $C_4F_9C_2H_4$, or $C_8F_{17}C_2H_4$ were obtained under /THF/DPHLi,TPMK,or TPMCs/t/LiCl// conditions ($M_n =$ 6–20 kg/mol, $M_w/M_n = 1.06–1.39$) [51]. Cationic polymerisation has also been utilized for the incorporation of fluorine in block copolymers. Block copolymerisations of the monomers $ROCH_2CH_2OCHCH_2$ where R is AcO, 2,2,3,3,3-pentafluoropropoxy, and n-butyl were performed under /CH_2Cl_2, Et_2O/HCl/-20 °C or -40 °C//CH_3OH, NH_3/ conditions ($M_n < 13$ kg/mol, $M_w/M_n = 1.05–1.20$) [52].

Another approach to fluorinated polymers is post-polymerisation methods. Partially fluorinated fatty acid ester side chains ($F(CF_2)_x(CH_2)_y$COO- with (x, y) = (6, 9); (8, 3); (8, 5); (8, 9); (10, 9) and ($H(CF_2)_x(CH_2)_y$COO- with (x, y) = (8, 10); (10, 10)) were introduced in a PS-hydroxylated PI diblock copolymer by reacting the corresponding acid chloride with the hydroxylated polymer [53]. The hydroxylated polymer was obtained by a hydroboration reaction on the PI block of a PS-PI diblock copolymer, where the PI block was a high vinyl type with a 40/60 ratio of 1,2 to 3,4 units ($100 \leq M_n$/(kg/mol) ≤ 126; $1.05 \leq M_w/M_n \leq 1.10$, $0.25 \leq f_{\text{fluorinated block}} \leq 1$ for the trifluoromethyl containing polymers and somewhat larger M_w/M_n values for the difluromethyl-containing polymers). Perfluorohexyl groups in the form of $F(CF_2)_6(CH_2)_2Si(CH_3)_2$ were introduced in a block copolymer by the hydrosilylation reaction between the fluoralkyl silane and the vinyl group in a PS-1,2-PBd diblock copolymer. The hydrosilylation reaction was catalysed by a nonacidic Pt based catalyst at 100–120 °C for 24 h (for $M_n \approx 60$ kg/mol no significant change was caused by the reaction in the M_w/M_n values, which were all less than 1.10) [54]. Perfluorohexyl side chains were introduced in PBd and PS-PBd diblock copolymer with varying microstructure by treatment of the parent polymer with perfluorohexyl iodide and triethylborane in nHx/ CFC113 mixtures under the controlled access of air [55,56]. The treatment was followed by catalytic hydrogenation to remove iodide and remaining unsaturation. Mechanistic studies indicate that the fastest addition mode involves two neighbouring pendant vinyl groups and perfluorohexyl substituted cyclopentane rings are generated ($M_n \approx 10$ kg/mol; $M_w/M_n \approx 1.03$ for the precursor polymers, which is not changed significantly by the reaction). The double bonds of PI and PBd were converted into difluorocyclopropane groups by treatment of the parent polymer with difluorocarbene generated by thermal decomposition of hexafluoropropylene oxide *in situ*. Crosslinking and chain-scission reactions can almost be avoided by running the reaction in the presence of a substantial amount of radical inhibitor (BHT) as evidenced by the small increase of M_w/M_n from 1.12 to 1.16 for a PI with $M_n = 63$ kg/mol [57,58].

Compared to the rather substantial activity involving the incorporation of fluorine in block copolymers very little effort has been spent on incorporation of the other halogens. In one example a diblock copolymer of chlorinated PBd and PS was obtained from PS-PBd diblock ($M_n \approx 50$ kg/mol) by chlorination (80% conversion) exclusively of the double bonds of PBd at $-50\,°C$ in CH_2Cl_2 under oxygen-free conditions without measurable chain scission, crosslinking, hydrogen substitution, or polystyrene chlorination reactions occurring [59].

2.2.3 HYDROCARBON POLYMERS

Hydrocarbon polymers are probably the most important class of polymers in terms of use and the field still attracts scientific interest in terms of finding new polymers or new methods for synthesising known structures in a better way. *Cyclo*-hexadiene is a diene-type monomer with special properties and the potential for generating a very stiff polymer in the case of a high level of 1,4-addition. CHD was polymerised under /Bz,nHx(60,40)/PSLi/–12 °C/TMEDA or DABCO/CH$_3$OH/ for long periods of time, 6 to 30 days, to achieve 85% 1,4 addition [60,61] ($22 \leq M_n/(\text{kg/mol}) \leq 37$; $1.01 \leq M_w/M_n \leq 1.07$ for PS-PCHDs containing ~33% PCHD). A very detailed investigation of the polymerisation behaviour of CHD showed that under all investigated conditions some level of side reactions was observed. The investigated conditions included low-temperature polar and apolar room-temperature conditions with and without amine additives in a small molar excess with respect to Li. DABCO was the most efficient of the additives tested, which included TMEDA and DME. In spite of the observed side reactions PS-PCHD diblock copolymers were obtained with a CHD content up to 81% ($6 \leq M_n/(\text{kg/mol}) \leq 44$; $1.04 \leq M_w/M_n \leq 1.07$). In another study CHD was polymerised under /CHX/nBuLi/40 °C/TMEDA// conditions to obtain a controllable polymerisation giving slightly more than 50% 1,2-addition of CHD provided the [TMEDA]/[Li] ratio was at least 1 ($M_w/M_n \approx 1.07$ and $M_n = 10$ kg/mol). Furthermore, a PCHD-PS-PCHD triblock copolymer was obtained ($M_w/M_n = 1.14$, $M_n = 64$ kg/mol) [62].

Polystyrene analogues where the phenyl group is replaced by naphthyl or fluorenyl have attracted recent interest [63–66]. 2VN can be polymerised under controlled /Tol/tBuLi/– 78 °C/THF/CH$_3$OH/ conditions in the presence of up to a 20-fold excess of THF with respect to Li. A purification step involving sublimation from LiAlH$_4$ is necessary to achieve controlled polymerisation ($0.5 \leq M_n/(\text{kg/mol}) \leq 21$; $1.04 \leq M_w/M_n \leq 1.11$). Initiation under similar conditions with KNaph yields the linear dipotassium salt of P2VN ($0.72 \leq M_n/(\text{kg/mol}) \leq 845$; $1.11 \leq M_w/M_n \leq 1.18$), which by reaction with DBX or BCMA under high dilution conditions was rendered cyclic ($0.79 \leq M_n/(\text{kg/mol}) \leq 11$; $1.11 \leq M_w/M_n \leq 1.12$). Macrocycles of PS and PDMVF containing a single 9,10-anthracenylidene group were prepared in a similar fashion with somewhat higher values of M_w/M_n. Cyclization of PS-P2VP in the presence of macrocyclic PS afforded a catenated block copolymer of PS and P2VP with $M_w/M_n = 1.3$.

A theme is the introduction of hydroxyl groups in PS either as phenolic or hydroxymethyl functional groups with the objective of either obtaining a hydrophilic polymer or providing a handle for further functionalisation. Homopolymers and block copolymers of S and protected hydroxymethyl styrene were obtained. The hydroxy group was in the form of a benzyl ether linkage to acetal-protected (carbohydrate) moieties in the *meta* position. Polymerisation was conducted under low-temperature polar conditions ($M_n \approx 10$ kg/mol; $M_w/M_n \approx 1.07$). Attempts to polymerise similar monomers with a para-benzylic ether linkage failed [67].

Block copolymers of hydrocarbons with hydrophilic polymers have also been studied. Hydroxy-terminated PS was prepared under room-temperature apolar conditions by adding a 3-fold excess of EO to the living PSLi solution. The PS hydroxylate was used to initiate the polymerisation of PO under a variety of conditions, including modification of counter ion, solvent and solvent mixture, and the addition of cation chelating species. The reaction product contained various amounts of PPO homopolymers presumably formed by chain-transfer reactions. Cs counter ions with pure THF as solvent gave the least formation of PPO and highest molar mass PPO block in the block copolymer ($M_w/M_n = 1.07$ for the block copolymer; $M_n = 12$ kg/mol for the PPO block) [68]. Apparently the chain-transfer reaction to monomer does not involve deactivation of the chain but rather proton extraction from the monomer, which generates an alcohol group on the chain. However, this alcohol can exchange protons with the other living chains and thereby continue to propagate. A block copolymer of the two protected hydroxystyrenes tBDMS-OS and tBDPS-OS was prepared under /CHX/sBuLi/RT/THF/CH$_3$OH/ conditions ($M_w/M_n = 1.16$; $M_n = 14$ kg/mol). The tBDMS group was removed by HCl and the phenolic group reprotected (BOC) through reaction with (BOC)$_2$O. The tBDPS was removed by reaction with tetrabutylammonium fluoride to produce a BOCST-HOS block copolymer [69]. Another hydrophilic substituted PS-poly(4-vinyl-α-methylcinnamic acid) was obtained from the oxazoline-protected monomer under /THF/DPHK/–78 °C//CH$_3$OH/ conditions ($M_n < = 63$ kg/mol, $M_w/M_n = 1.04$)[22].

Introduction of rigid rod-like moieties has also been attempted. PS-PPphen-PS was synthesized under /THF/x/−60 °C//CH$_3$OH/ conditions where the initiator x was obtained by treating α,ω-bromo-oligophenylene with sBuLi in THF at −60 °C ($M_w/M_n = 1.16$; $M_n = 5.2$ kg/mol). P2VP-PPhen-P2VP was similarly obtained with a 6-fold excess of LiCl with respect to living chain ends present in the solvent. The M_w/M_n was somewhat higher than for the PS-analogues [70]. DMAPLi was used to introduce dimethylamiono end functions in PS-PI diblock copolymers. An azobenzene linked to a phospholane was reacted with the dimethyl amino group to introduce a mesogenic and phosphorus quarternary ammonium zwitterionic end group [71].

Another activity is the linking of different hydrocarbon monomers in a new order. PS-1,4-PBd-3,4-PI and 1,4-PBd-PS-3,4-PI triblock copolymers were synthesized by polymerising either S followed by Bd or Bd under apolar room-temperature conditions followed by addition of a small amount of THF before the remaining monomers were polymerised ($91 \leq M_n/(\text{kg/mol}) \leq 161$; $M_w/M_n \approx 1.06$)[72].

2.2.4 SILOXANES

For the controlled polymerisation of siloxanes only a few publications have appeared but these constitute a couple of steps forward in the elucidation of methods to find conditions where controlled polymerisation can be performed. Block copolymers of S and D$_3$ were obtained with close to 100% conversion of D$_3$ under /Bz,THF/sBuLi/−20 °C//TMCS/ conditions ($M_n = 49$ kg/mol; $M_w/M_n = 1.03$ with 68% S in the PS-PDMS) [73]. Block copolymers of D$_3$ and diene monomers were obtained with close to 100% conversion under /CHX/sBuLi/0 °C/HMPA/TMCS/ conditions ($6 \leq M_n/(\text{kg/mol}) \leq 17$; $1.07 \leq M_w/M_n \leq 1.14$) [74]. In the preparation of block copolymers containing siloxanes it is a known problem that the crossover from the very reactive carbanion species to siloxane monomers is actually quite slow under apolar conditions. A possible solution to this problem has been found. Di- and triblock copolymers of PDMS and PS or PαMS were obtained by low-temperature polar polymerisation of S or αMS followed by reaction with EDS. The resulting Li silanolate polymerised D$_3$ in THF at 25 °C. Termination with TMCS or DMDCS yields diblock or triblock copolymers, respectively. The use of EDS affords a much faster crossover reaction than is normally obtained between S or αMS and D$_3$ ($27 \leq M_n/(\text{kg/mol}) \leq 190$; $1.07 \leq M_w/M_n \leq 1.28$) [75].

The controlled introduction of side-chain functionalities in siloxane has been studied. A 50/50 D$_3$ and (D^v)$_3$ block copolymer was obtained under /THF/nBuLi/RT//TMCS/ conditions ($M_w/M_n = 1.1$; $M_n = 17$ kg/mol) at ∼ 90% monomer conversion. (D$_3$)^v/D$_3$ was similarly obtained. The vinyl group in these polymers can be converted to a carboxymethylthioethyl group through a free-radical-mediated addition of mercaptoacetic acid [76]. Block copolymers of D_3 and $\left(D^{ClPr}\right)_3$ or $(D_3)^{ClPr}$ were synthesized in THF at 25 °C by initiation with a lithium silanolate. The reaction proceeded to $\sim 90\%$ conversion in 2–4 hs ($M_n \approx 10\,\text{kg/mol}$; $1.08 \leq M_w/M_n \leq 1.37$ for both diblock and homopolymers) [77].

2.2.5 CATIONIC POLYMERISATION METHODS

Cationic polymerisation is still very much a developing technique and many new methods are appearing. It is also clear that with the exception of vinyl ethers the polymerisation conditions need very careful optimisation in order to achieve controlled polymerisations. Since the vinyl ethers are the easiest to deal with, most of the publications appear in this field. Triblock copolymers of MeVE, EVE, and MTEGVE were synthesised by sequential cationic polymerisation in CH_2Cl_2 at -78 °C for MeVE and -20 °C for the two other monomers ($1.9 \leq M_n/(\text{kg/mol}) \leq 7.5$; $1.10 \leq M_w/M_n \leq 1.29$) [78]. MTEGVE-BzVE block copolymers were synthesized under /Tol, EtAc/iBuOEtAc/0 °C/$Et_{1.5}AlCl_{1.5}$/ $LiBH_4$/ conditions ($10 \leq M_n/(\text{kg/mol}) \leq 18$; $1.20 \leq M_w/M_n \leq 1.46$). The Bz ether was cleaved by heterogeneous catalytic hydrogenation [79]. A vinyl-ether-type monomer containing a protected glucosamine function (GluEV) was copolymerised with IBVE cationically by sequential monomer addition under //IBVE, TFA, $EtAlCl_2$/0 °C/dioxane/CH_3OH, NH_3/ conditions ($11 \leq M_n/(\text{kg/mol}) \leq 24$; $1.04 \leq M_w/M_n \leq 1.08$) [80]. Triblock copolymers PMVE-PODVE-PMVE were obtained by sequential cationic polymerisation with difunctional initiation under /Tol/TMePr, TMSI, ZnI_2/ -5 °C// alcohol, TEA/ conditions ($M_n = 7.2\,\text{kg/mol}$, $M_w/M_n = 1.2$ for the triblock) [81]. Copolymers of vinyl-ether-type monomers (CEVE and AcOVE) and cyclopentadiene were obtained by cationic polymerisation. Typical conditions are /CH_2Cl_2/ activated chloride, $SnCl_4$, Bu_4NCl/ -78 °C/// ($2.5 \leq M_n/(\text{kg/mol}) \leq 22$; $1.19 \leq M_w/M_n \leq 1.46$) [82]. PCEVE can be prepared by cationic polymerisation /Tol/x/-30 °C//t/ where the initiator x was HCl, HI, TMSI, or 3-(p-vinylbenzyl) propionaldehyde diethyl acetal/TMSI, with $ZnCl_2$ as a cocatalyst. Termination was done by CH_3OH/lutidine. Combined with the use of termination under high-dilution conditions linear, functional linear, and cyclic PCEVE is accessible. These molecules were used as a multifunctional terminated agent for PSLi (prepared under /Bz/BuLi/RT/TMEDA// conditions). Comb-like graft copolymers were characterized by $48 \leq M_n/(\text{kg/mol}) \leq 2680$; $1.01 \leq M_w/M_n \leq 1.27$ [83].

Cationic polymerisation has also been utilized to produce side-chain functional polymers. HOS and MOS can by cationically copolymerised either in a random or block fashion under /CH_3CN, CH_2Cl_2/BF_3OEt_2, water/0 °C/// conditions ($2.6 \leq M_n/(\text{kg/mol}) \leq 6.5$; $1.28 \leq M_w/M_n \leq 1.48$). The hydroxy group does not have to be protected for polymerisation under these conditions [84].

A precursor to the nitrogen analogue of PEO, poly(N-acylethyleneimine), was obtained with a hydrocarbon and a fluorocarbon end group from MOz by initiatiation with perFC$_{17}$C$_2$H$_4$OH triflic ester or C$_{16}$H$_{33}$OH triflic ester terminated with alkyl piperazine (alkyl = C$_n$H2_{n+1}, n = 6, 8, 10, 12, 14, 16, 18) [85].

2.2.6 MISCELLANEOUS POLYMERS AND CONDITIONS

A new route to a controlled synthesis of heavily aromatic substituted polyesters has been investigated. The method involves an alternating copolymerisation of an aldehyde and a ketene. PEK and MBA can be copolymerised under /THF/nBuLi/ − 20 °C/LiCl/CH$_3$OH/ conditions to obtain an alternating PEK/MBA copolymer (M_w/M_n = 1.15, M_n = 5.1 kg/mol). The alternating structure is documented by reductive cleavage of the polymer [86]. Living anionic polymerisation of PEK and XPEK in THF at 20 °C initiated by lithium 4-methoxyphenoxide affords a polyester containing halogenated phenyl side chains (4.2 ≤ M_n/(kg/mol) ≤ 69; 1.12 ≤ M_w/M_n ≤ 1.26 for XPEK). An alternating living anionic polymerisation of PEK and SiO-BA yield a polyester containing tBDMS-protected phenol group side chains with modest molar mass (≈ 5 kg/mol and M_w/M_n ≈ 1.2) [87].

Organolanthanides have been investigated as anionic polymerisation initiators for methacrylates and acrylates and a review on this subject has appeared [88]. In one example THF was polymerised cationically using a tri- or tetrafunctional initiator (e.g. 1,2,4,5-tetrakis(bromomethylbenzene)) with AgOTf as coinitiator and terminated by NaOOCCMe$_2$Br. The bromo- terminated chain was treated with SmI$_2$ and MMA polymerised in THF at −78 °C (7.2 ≤ M_n/(kg/mol) ≤ 16; 1.04 ≤ M_w/M_n ≤ 1.21) [89].

A new method for the polymerisation of EO with lithium as the counterion has been investigated. This is particularly important since EO-containing block copolymers will be more easily accessible. The use of the extremely strong neutral phosphazene base tBuP$_4$ as a chain-end modifier makes the anionic polymerisation of EO with lithium counter ions feasible. Block copolymers PBd-PEO and PI-PEO were prepared by mixing tBuP$_4$ and Bd or I monomer and THF and initiating by sBuLi at −110 °C followed by addition of EO at −40 °C. Polymerisation of EO at 40 °C lasted approximately 24 h (5 ≤ M_n/(kg/mol) ≤ 78; 1.02 ≤ M_w/M_n ≤ 1.06). The tBuP$_4$ is a sufficiently strong base to abstract weakly acidic protons such as from phenylpropionitrile to yield the salt tBuP$_4$H$^+$,(C$_6$H$_5$)(CH$_3$)(CN)C$^-$, which polymerises EO at 45 °C in THF to afford cyano-terminated PEO (M_n = 2.5 kg/mol; 1.04 ≤ M_w/M_n ≤ 1.09). The alcohol p-cresol, which creates a phenolic tBuP4H$^+$, salt also functions as an initiator for EO polymerisation [90,91].

In another new method for a known polymer it was shown that 2VP will polymerise under /Bz/PILi/6 °C/LiCl// conditions provided that the excess of LiCl is at least 4-fold and that the polymerisation time is kept very short

($140 \leq M_n/(\text{kg/mol}) \leq 224$; $M_w/M_n \approx 1.05$). If insufficient LiCl is present or the reaction times are too long (here more than a few minutes) some amount of coupled product is formed. The microstructure of the PI block is unaffected by the presence of LiCl during the polymerisation. The synthesized block copolymers were rich in PI, which makes the procedure somewhat insensitive to MMD broadening in the P2VP block. Furthermore, the rather high molar masses utilized here combined with short reaction time leads to the expectation of problems for synthesis at low molar mass [92].

2.3 METHODS FOR GENERATING NEW BLOCK COPOLYMER ARCHITECTURES

2.3.1 COUPLING REACTIONS

Coupling reactions can be employed for a number of reasons. For instance, it can be the only possible route to a block copolymer where sequential polymerisation for some reason is not possible. It is a drawback of coupling reactions in the preparation of model polymers that a fractionation or other purification step is normally necessary to obtain a pure block copolymer because it is very difficult for reasons of stoichiometric or purify (or both) to avoid the presence of unreacted precursor polymer in the crude product. One instance where sequential polymerisation is not possible is when the monomers involved are not subject to polymerisation with the same mechanism. A prime reason for performing couplings is the preparation of triblock copolymers, star polymers, or star block copolymers. The most widespread type of coupling agents are the chlorosilanes. This class of coupling agents in conjunction with the elaborate DPE chemistry has made the synthesis of very complicated architectures possible. Further reference to this work is given in the section on DPE.

PCHD-PS three arm stars were prepared by linking the corresponding diblock with TCMS. The polymerisations were carried out under /Bz/SBuLi/ RT/DABCO/chlorosilane/ conditions ($18 \leq M_n/(\text{kg/mol}) \leq 41$; $1.05 \leq M_w/M_n \leq 1.09$ for the isolated stars) [93]. A PSLi prepared under /Bz/nBuLi/0 °C/THF// conditions was used to polymerise D_3 in the presence of one equivalent 15-C-5 in Bz/THF (5 to 80% THF) at 50 °C. The reaction was terminated with TMCS or DMDCS to obtain either di- or tri-block copolymer ($11.7 \leq M_n/\text{(kg/mol)} \leq 27.8$; $1.09 \leq M_w/M_n \leq 1.20$) [94].

The hydrosilylation reaction has also found widespread use as a coupling reaction. α, ω-methacryloyl-silane-PDMS was obtain from D_3 under /THF/ LiMAOPDMS/0 °C//DMCS/ conditions. It was possible to obtain α, ω-allyl-x-PEO where x is carboxylate, hydroxy, or benzyl under /THF/potassium allyl alcoholate/RT//y/ conditions with y=succinic anhydride, CH_3OH, or benzylbromide as terminator. The α, ω-PDMS and α, ω-PEO were coupled through a hydrosilylation reaction in CH_2Cl_2 at 45 °C catalysed by PtDVScat and phenothiazine to produce ~30% block copolymer ($M_w/M_n = 1.1$; $M_n \approx$ 7 kg/mol) [95].

C_{60} was placed in the centre of a star block copolymer by the reaction between PSPILi and C_{60} in Bz at RT. Hexa-arm star block copolymers were obtained with a C_{60} core ($35 \leq M_n/(\text{kg/mol}) \leq 74$; $1.07 \leq M_w/M_n \leq 1.12$ for the star block) where the functionality of the core was characterized by measurement of the absolute molar mass of the star and the arms [96].

2.3.2 MACROINITIATORS AND TERMINATORS

A common strategy for producing block copolymers is to use a functionally terminated polymer as either an initiator or terminator in a new polymerisation reaction. Living PTHF$^+$, SbF$_6^-$ was coupled with living PtBMALi in THF ($M_n(\text{total}) = 4$ kg/mol, $M_w/M_n = 1.18$) [97]. Chloroethyl-terminated PMeVE was prepared cationically under /CH$_2$Cl$_2$/HCl-CEVE/$-78\,°$C/SnCl$_4$, Bu$_4$NCl/ CH$_3$OH, NH$_3$/ conditions. The chloroethyl-terminated PMeVE was used as a terminating agent for PSLi, which was prepared under room-temperature apolar conditions to obtain a PS-PMeVE diblock copolymer ($M_w/M_n = 1.04$, $M_n = 23$ kg/mol) [98]. THF was initiated by triflic acid to obtain a difunctional PTHF, which was terminated by monomethyl polyethyleneglycol to obtain PEO-PTHF-PEO (M_n between 10–20 kg/mol, $M_w/M_n = 1.1$–1.2 for the PTHF block) [99]. A chloromethylphenyl-terminated PDMS was synthesized by terminating living PDLi (obtained in Bz/THF 1/1 at $-20\,°$C) with CMPDMS. The macro-terminating agent was used to terminate living P2VP and living PtBMA to obtain PDMS-P2VP and PDMS-PtBMA diblock copolymers ($20 \leq M_n/(\text{kg/mol}) \leq 65$; $1.02 \leq M_w/M_n \leq 1.07$) [100]. A PI-PS-PMeVE copolymer was obtained through a coupling reaction of chloroethoxy terminal PMeVE with PI-PSLi [101]. Chloroethoxy terminal PMeVE has been prepared under /CH$_2$Cl$_2$/2-chlorethylvinyl ether HCl adduct/$-78\,°$C/SnCl$_4$ Bu$_4$NCl LiBH$_4$/ conditions and used to terminate the living PI-PSLi prepared under /Bz/sBuLi/50$\,°$C/// conditions ($M_w/M_n = 1.26$; $M_n = 32$ kg/mol for the coupled product - PI-PS precursors: $M_w/M_n = 1.04$; $M_n = 23$ kg/mol and PMeVE precursors: $M_w/M_n = 1.03$; $M_n = 11$ kg/mol).

More complicated architectures can also be prepared by such a strategy. A dumbbell-shaped polymer with five PI chains on the ends of a PS chain was obtained via difunctional initiation (the product of the reaction of MDDPE and sBuLi) of S. The living chain was reacted (using the titration technique[3]) with a hexafunctional chlorosilane followed by coupling with PILi. Monofunctional instead of difunctional initiation lead to PS(PI)$_5$ hexa-arm stars ($76 \leq M_{n,\text{PS-connector}}/(\text{kg/mol}) \leq 168$; $M_{n,\text{arm}} = (8 \pm 2)$ kg/mol, $1.02 \leq M_w/M_n \leq 1.09$ for both connector, arm, and final product). A similar procedure leads to PS-PI-P2VP-arm stars [102]. PCHD(PBd)$_{2\text{ or }3}$ stars were prepared by reacting PCHD with a tri- or tetra-chlorosilane (the excess/evaporation technique[3]) followed by coupling with PBdLi ($M_{n,\text{arm}} = 20 \pm 5$ kg/mol; $M_n \leq 1.09$ for both arm and final product). The M_w/M_n for PCHD is the highest due to a limited side reaction [103].

If one uses a side group rather than an end group as the terminating agent it is relatively easy to get a multifunctional terminator although not with an entirely controlled functionality. A 1,2-PBd-PS-1,2-PBd triblock copolymer was synthesized by sequential polymerisation under /Bz/sBuLi/RT/DIPIP// conditions of Bd and S followed by DMSCl$_2$ coupling. The pendant vinyl groups were 75% hydrosilylated (Pt-catalyst plus MSCl$_2$) to obtain a PS with an average of 32 chlorosilane groups at each end. The chlorosilane dumbbell was used as a terminating agent for PBdLi (typical molecular characteristics: $M_{\text{PS-connector}} \approx 11$ kg/mol, $1.2 \leq M_{n,\text{arm}}/(\text{kg/mol}) \leq 43$; $950 \leq M_{n,\text{total}}/(\text{kg/mol}) \leq 2850$; $M_w/M_n \approx 1.04$ for both connector arms and the final dumbbell product) [104].

Instead of synthesizing a macro-terminator it is also possible to use a macro-initiator. IB was polymerised under /CH$_2$Cl$_2$, DMA/2ClTMP/ $-78\,^\circ$C/BCl$_3$, TiCl$_4$, DtBP/DPE, CH$_3$OH/ conditions to obtain – depending on the conditions, – either diphenylvinyl- or diphenylmethoxy-terminated PIB. Both types of chain ends can be quantitatively metalated with any of the alkali metals Li, Na, K, or Cs. The obtained macroanions were used to polymerise tBuMA and MMA under low-temperature polar conditions ($13 \leq M_n/(\text{kg/mol}) \leq 39$; $1.05 \leq M_w/M_n \leq 1.13$) [105]. The method was extended to triblock and 3-arm star-blocks by starting from di- or trifunctional initiators for PIB. α,ω-chloromethylphenylethyl-terminated PDMS was obtained by hydrosilylation of styrene with hydride-terminated PDMS (Tol, H$_2$PtCl$_6$, $100\,^\circ$C) followed by chloromethylation with paraformaldehyde/TMCS/SnCl$_4$ at 0–$20\,^\circ$C. Utilizing the difunctional PDMS macroinitiator, S was polymerised cationically in CHCl$_3$ with SnCl$_4$ as catalyst ($M_n = 17.7$ kg/mol; $M_w/M_n = 1.46$) [94]. Polymerisation of 2VP or 4VP ($-78\,^\circ$C, THF) with a protected alcohol initiator under /THF/ELiPAA/$-78\,^\circ$C//CH$_3$OH/ conditions followed by deprotection and conversion to the lithium alcoholate afforded a macroinitiator for the polymerisation of D$_3$, which was used at $25\,^\circ$C in THF to yield block copolymers of D$_3$ and 2VP or 4VP ($1.3 \leq M_n/(\text{kg/mol}) \leq 1.16$; $9.5 \leq M_w/M_n \leq 167$) [106]. An alternative preparation of the macroinitiator involves the termination of the tBuLi-initiated 2VP or 4VP polymerisation with EBrPAA or 2IPP followed by capping with EO, leading to similar block copolymers as the protected initiator route.

An alternative to the preparation of polymer stars by the multifunctional coupling route is a multifunctional initiation route. In practice this is not a viable possibility under apolar conditions for anionic polymerisation. However, multifunctional initiation under different conditions and for different mechanisms has been explored. A couple of reviews on the subject have appeared recently [107,108]. A trifunctional analogue of MDDPE, TDPE was investigated for its potential as a precursor for a hydrocarbon soluble trifunctional anionic polymerisation initiator and coupling agent. TDPE reacts with sBuLi in

Bz but trifunctional initiation is only obtained in the presence of a 20-fold excess of THF with respect to Li [109].

Star blocks of S and IB were obtained by multifunctional initiation by calix[8]arene or hexa-epoxidized squalene ($M_{n,arm}$ = 21 kg/mol, M_w/M_n = 1.2 and $M_{n,arm}$ = 35–50 kg/mol, M_w/M_n = 1.18–1.26) [110,111]. Very careful control (optimisation) of the reactions conditions is needed to avoid side reactions. Another hexafunctional initiator for cationic polymerisation is hexa-(1-chloroethylphenylethyl benzene) [112].

A study on a combination of cationic polymerisation and ATRP for the synthesis of hetero-arm stars has appeared. However, as is seen from the M_w/M_n value the control of the reaction is less than optimal. A cationic polymer end-capped with a bifunctional unit possessing one site that is a cationic initiator and one that is an ATRP initiator was investigated. PTHF was obtained from acetylchloride and $AgClO_4$ in THF/CH_2Cl_2, at $-15\,°C$ and terminated difunctionally by 2-bromosuccinic anhydride (M_n = 4.7–5.2 kg/mol; M_w/M_n = 1.17–1.23). The acid group was converted to acid chloride and subsequently to acetyl (addition of $AgClO_4$ at $-30\,°C$) to initiate cationic polymerisation of 1,3-oxepane. The –CHBr group serves as an ATRP initiator for S using the CuBr/bipy system at 110 °C to produce PTHF-PEO-PS star block copolymers (M_n = 12–33 kg/mol, M_w/M_n = 1.47–1.50) [113].

A scheme for synthesizing heteroarm stars has been developed based on different end-functionalised polymers. The controlled introduction of 1 to 3 XMP groups proceeds via BCl_3 or TMCS/LiX transformation of a protected benzylic alcohol functionality (MMP or SMP). The MMP and SMP groups are introduced by an end-capping reaction of PSLi or PILi with MMP- or SMP-substituted bromoalkylbenzene or DPE. S and DPE end-functionalisation is obtained through reaction of PSLi and PILi with XPr para-substituted S or DPE. Among other possibilities AB_2C_2 heteroarm stars are obtained, where A, B, and C are S, I and αMS. $M_{n,arm}$ was typically 5–15 kg/mol and $1.01 \leq M_w/M_n \leq 1.10$. In a variation of the method a controlled number (4 or 8) of functional groups such as phenol, D-glucose, MMP and SMP were placed in the middle of a polymer chain [114–117].

A less-controlled variant of this method is the use of DVB star-linking followed by the addition of a new monomer. DVB was used to link PSLi, which initiates the polymerisation of 2VP. Starting from oligo-PSLi a PVP-PtBuA star was synthesized. As usual with the DVB technique only average arm number control is obtained [118].

Another method that does not provide very rigorous control of the product is in practice the grafting-from technique. The 4MS block of a block copolymer of 4MS and hydrogenated I or S was metalated by treatment with a superbase (an alkyllithium/potassium alcoholate mixture). The exact conditions have to be optimised for the particular block copolymer. The metalated block copolymer served as an initiator for the polymerisation of 4MS. Appreciable amounts of homopolymer of 4MS were formed along with the graft copolymer ($206 \leq M_n/(\text{kg/mol}) \leq 890$; $1.16 \leq M_w/M_n \leq 2.27$ for the copolymer) [119].

2.3.3 METHODS INVOLVING DPE AND DPE DERIVATIVES

Probably the most investigated and developed method for the synthesis of complicated star polymers is the combined use of DPE and DPE-X-DPE derivatives and chlorosilane couplings. Ingenious (and laborious) schemes allow the synthesis of star polymers and branched polymers where the length and chemistry of virtually any individual part of the molecule can be controlled to create heteroarm stars and other polymers with complicated architectures. These methods have been extensively reviewed [120–122]. One note of caution, which is important in the context of this method, is the following observation. 1-(polybutadienyl)-1,1-diphenyl-methyl-lithium (and other polymeric analogues) are unstable in 10% THF in CHX at RT with a half-life of approximately 15 days. The half-life of the isoprene analog is much higher. The corresponding half-lives with THF substituted by diethyl ether are much higher [123]. Probably the most spectacular example of what can be done with the DPE/chlorosilane chemistry is the synthesis of a variety of analogues of branched PE. The linking techniques employing chlorosilanes and DPE-X-DPE-based difunctional initiation under room-temperature apolar conditions was utilized to produce controlled architectures of 1,4-PBd such as pom-poms, stars, H-shaped molecules, and combs. All products were saturated by catalytic hydrogenation employing heterogeneous or homogenous catalysts [124].

A method was developed that gives access to 5-armed stars of the type AA'_2B_2 where A = S, B = αMS and AA' means arms of the same monomer but with different molar masses. A dimethoxymethyl-substituted DPE end-capped PS can be obtained by reacting a PSLi with dimethoxymethyl-DPE. The methoxymethyl function can be changed to chloromethyl. Thus PS terminated by bis(chloromethyl)DPE can be synthesized and reacted with a living P(St-αMS) with the anion in the middle to form a 5-arm star. DPE-terminated αMS can by obtained from termination of living αMS with a bromobutyl-substituted DPE. Finally the DPE-terminated polymer can be reacted with PSLi to obtain the macroanion needed in the final coupling reaction ($M_{\text{arm}} \sim 5$ kg/mol, $M_w/M_n = 1.05$ (after SEC purification) [125]. $(PS)_2$-PEO and $(PS)_2$-PtBMA 3-arm stars were obtained by reacting PSLi with a tBDMS-methyl substituted DPE. Somewhat surprisingly this gave both addition to the DPE double bond and substitution at the tBDMS-methyl site. The obtained PS macrocarbanion was used to initiate tBMA and EO (the reaction works with K^+ counterions as well) ($M_{n,\text{arm}} \approx 6$ kg/mol; $M_w/M_n \approx 1.2$ for both arms and final product) [126]. A DPE derivative containing a protected hydroxy function (tBDMSEDPE) was utilized to obtain 3-arm stars, e.g. a PS-PMMA copolymer was obtained under /THF/CmK/−78 °C/tBDMSEDPE/CH$_3$OH/

conditions. The hydroxy group was deprotected and used to initiate a PEO polymerisation. Three arm stars of PS-PEO-PεC, PS-PMMA-PEO, and PS-PEO-PLL were obtained in a similar manner ($9 \leq M_n/(\text{kg/mol}) \leq 260$; $1.16 \leq M_w/M_n \leq 1.47$) [127]. Bd initiated with sBuLi and polymerised in THF at $-15\,°\text{C}$ was terminated with BrMDPE to afford a DPE-terminated 1,2-PBd macromer. The macromonomer was used as the crossover reagent in a block copolymerisation of S and 2VP under low-temperature polar conditions to obtain S,1, 2Bd,2VP 3-arm stars ($86 \leq M_n/(\text{kg/mol}) \leq 288$; $1.01 \leq M_w/M_n \leq 1.06$) [128].

MDDPE was used as a coupling agent to produce a coupled living polymer with two active sites in the middle. These sites can, through reaction with a bromoalkyl substituted DPE in THF at low temperature, be converted into two DPE sites, thus generating a polymeric DPE-X-DPE species. The polymeric DPE-X-DPE can assume the role of the MDDPE in the previous reaction providing a scheme that allows preparation, in principle, of star polymers $A_2B_2C_2$. . . . The principle has been demonstrated by synthesizing a hexa-arm PS-star ($4.2 \leq M_{n,\text{arm}}/(\text{kg/mol}) \leq 10.2$, $M_w/M_n = 1.01-1.05$) and a PS, PMOS, PαMS star with similar molar mass characteristics [129].

The DPE scheme can also be used to introduce functional groups in the polymers. The functionalised DPE method was used to introduce dimethylamiono functions at the junction points of di- and triblock copolymers. The dimethylamiono functions were in turn converted to zwitterionic groups [130].

The DPE scheme can also be used to generate a controlled version of heteroarm stars produced by using DVB. A_2B_2 star polymers of I or Bd and MMA or nBMA were obtained by sequential sBuLi-initiated polymerisation of I in hexane at RT followed by solvent exchange to THF, chain-end modification by a compound of the DPE-X-DPE type and polymerisation of MMA at $-78\,°\text{C}$ in the presence of LiCl. ($38 \leq M_n/(\text{kg/mol}) \leq 70$; $1.01 \leq M_w/M_n \leq 1.06$) [131,132].

DPE also works for cationic polymerisation. IB was polymerised under /nHx,CH$_2$Cl$_2$(50/50)/TiCl$_4$/$-80\,°\text{C}$/TMP// conditions at $-80\,°\text{C}$ with 2ClTMP as initiator. The living PIB was coupled by DPE-X-DPE and MeVE was polymerised at $-80\,°\text{C}$ to obtain A_2B_2-type stars of IB and MeVE ($M_n = 25\,\text{kg/mol}$; $M_w/M_n = 1.12$) [133].

2.3.4 CYCLIZATION

Cyclic polymers can be synthesised through careful choice of the reaction conditions. The most common strategy is to employ highly dilute reaction conditions in order to favour intramolecular reactions over intermolecular reactions. A S–Bd–S living difunctional polymer obtained under apolar room-temperature conditions was cyclized using the titration method[3] employing a dichlorosilane as linker in slight excess. As a trick for the purification step the excess chlorosilane and monoreacted precursor polymer was reacted with a high molar mass living PBd. This procedure facilitates the following purification step [134].

A number of schemes have been developed that attempt to alleviate the restrictions on the creation of cyclic polymers set by the standard anionic polymerisation procedures. The polymerisation of S in Bz initiated by the protected aldehyde functional initiator 3LiPDEA proceeded in the presence of TMEDA. Termination by a slight excess of EO yielded a protected aldehyde-alcohol α, ω-terminated PS. After forming the potassium salt of this alcohol EO is polymerised in THF. Termination by ClMS yielded protected aldehyde-styrene α, ω-terminated PS-PEO. Cyclization under high dilution conditions was effected under (Lewis) acidic conditions ($1.6 \leq M_n/(\text{kg/mol}) \leq 7.1$; $1.03 \leq M_w/M_n \leq 1.12$ for both linear precursors and cyclic ones) [135]. An α, ω block terpolymer of S, I and MMA was obtained by initiating S by the amine-protected initiator TLiAD in Bz in the presence of TMEDA followed by polymerisation of I. MMA was polymerised by the macroinitiator under /THF//−78 °C/DPE/BTMB/ conditions to yield a protected acid functionality. The end functionalities can be hydrolysed to obtain primary amine and carboxylic acid, respectively. The α, ω amino acid can be cyclicized in good yield under less rigorous conditions than the previous polymerisation steps [136]. In a similar effort a PI-PS-PI halogen-terminated block copolymer was synthesised under low-temperature polar conditions and end functionalised through reaction with DPE and 1,3-dibromopropane. Cyclization was done through condensation at an interphase at relatively high concentration (2% w/v) [137].

2.4 RING-OPENING POLYMERISATION OF LACTONES, LACTIDES CABOXYANHYDRIDES AND SIMILAR MONOMERS

Ring-opening polymerisation is receiving much interest presently probably due to the range of monomers that can be polymerised with this procedure. A number of recent reviews have appeared on aspects of ROP including polymerisation of lactones, lactides, and cyclic esters [138], cyclic carbonates and block copolymers with ureas and PTHF [139], and the use of a wide range of macroinitiators in ROP [140].

2.4.1 MACROINITIATORS IN ROP, ε-CAPROLACTONE AND LACTIC ACID COPOLYMERS

Macroinitiators are the primary means to access block copolymers consisting of monomers suitable for ROP and monomers polymerisable with other techniques. This field is attracting much activity.

PBd(1,4))-PLA was prepared from PBd-OH AlEt$_3$. In a mechanistic study where the ratio $[\text{PBd-OH}]_0/[\text{AlEt}_3]_0$ was varied from 1 to 6 the equilibrium lactide concentration in toluene was measured as a function of temperature, and likewise the apparent rate constants. The reaction order in [AlEt$_3$] and lactide was determined. The rate as a function of temperature in toluene was also determined ($M_{n,\,\text{PLA-block}} = 11-20$ kg/mol; the time to 50% conversion $t_{0.5} = 0.2-17.8$ h for $t = 120-70\,°\text{C}$ with [PBd-OH]$_0$/[AlEt$_3$]$_0$=1, 1.5, 2, 3, 6; highest rate for 120 °C and [PBd-OH]$_0$/[AlEt$_3$]$_0$ = 2 [141]. PI-PLA was obtained in a similar manner ([PI-OH]$_0$/[AlEt$_3$]$_0$=1, 70 °C, Tol, $M_{n,\,\text{PLA-block}}$ =2−49 kg/mol, $M_w/M_n = 1.05-1.23$, reaction time 96–141 h, 40–94% conversion) [142]. PE-PLL was likewise obtained from PE-OH (PBd hydrogenated) with AlEt$_3$ present in 50% excess based on Et ($M_{n,\,\text{PE-PLLA}} = 61$ kg/mol, $M_w/M_n = 1.06$ with 80% conversion of L-lactide in 24 h, 70 °C, in toluene) [143]. A diblock of the closely related PVL was reported. The potassium salt of mono- or dicarboxylate-terminated PIB was obtained under /CH$_2$Cl$_2$, Hexane/initiator/ −78 °C/TiCl$_4$,DtBP/DPE,MTSP/ conditions where the initiator is 2ClTMP for monofunctional and tBuDiCumCl for difunctional initiation. The potassium α, ω-salt was used as a macroinitiator for the polymerisation of PVL under /THF/α, ω-salt/25 °C/18-C-6/acidic CH$_3$OH/ conditions ($5 \leq M_n/(\text{kg/mol}) \leq 58$; $1.07 \leq M_w/M_n \leq 1.77$) [144].

A block copolymer of PS and an oxazoline was obtained from an OTs-terminated PS, which was obtained by ethylene oxide termination of PSLi in THF at −78 °C followed by addition of tosylchloride. The OTs-terminated polymer served as a cationic macroinitiator for EOz (M_n per block 10–20 kg/mol, $M_w/M_n = 1.3-1.5$), which means that control of the EOz polymerisation is limited [145]. Another oxazoline block copolymer was obtained via the polymerisation of EO by a potassium difunctional initiator under /THF/KPDP/RT//MSCl/ conditions. The α, ω-acetal-methylsulfone-terminated PEO was used as a macroinitiator for the cationic polymerisation of 2MOz in nitromethane to yield acetal-terminated PEO-P2MOz, which was hydrolysed to removed the acetyl side chains to yield PEO-PEI ($M_n = 3.2$ kg/mol; $M_w/M_n = 1.04$ for the PEO; $5 \leq M_n/(\text{kg/mol}) \leq 10.5$; $1.35 \leq M_w/M_n \leq 1.56$ for the PEO-PEI) [146].

Polyaminoacid block copolymers have also been synthesized. An amino-terminated PEG macroinitiator polymerised the N-carboxyanhydrides of aminoacids (leucine: $M_w/M_n = 1.2-1.4$, $M_n = 6-20$ kg/mol; phenylalanine: $M_n = 6-15$ kg/mol, $M_w/M_n = 1.6-2.0$) [147]. An amino-terminated polyoxazoline (phenyl or methyl) functioned as a macroinitiator for N-carboxyanhydrides of phenylalanine ($M_n = 4.6$ kg/mol, $M_w/M_n < 1,2$) and an acetylated glucose derivative ($M_n = 4.4$ kg/mol, $M_w/M_n = 1.2$) [148].

A couple of studies concerned the polymerisation of εC by macroinitiators. In one case an α-ω OH-functional or OH-terminated PIB was obtained from cationic polymerisation of IB under /60:40 n-hexane:methylchloride/TiCl$_4$/ −80 °C/2,6-ditertbutylpyridine// conditions. A polymer with an group End-functional hydroxy was obtained by capping with 1,1-diphenylethylene followed by trimethylsiloxy-protected methyl isobutyrate to obtain a methoxycarbonyl terminus, which was reduced by LiAlH$_4$ ($M_n = 3-4$ kg/mol; $M_w/M_n = 1.05-1.09$). The –OH terminated PIB was used as initiator for εC with catalytic amounts of HCl in Et$_2$O as initiator, 25 °C; CH$_2$Cl$_2$ 24 h; (yield 87–100%, $M_{block} = 1-17$ kg/mol, $M_w/M_n = 1.02-1.18$) [149]. A butyrolactone polymer was used in a similar manner. PRSβBLεC was obtained by first polymerising RSβBL under /THF/sodium 3-hydroxybutanoate/RT/18-C-6/methyliodide/ conditions. εC was then polymerised under/Tol/HO-PRSβBL,Et$_3$Al/RT//HCl/ conditions, yielding $17 \leq M_n/(\text{kg/mol}) \leq 46$; $1.20 \leq M_w/M_n \leq 1.30$ [150]. Block copolymers of εC and VP were also reported. 2VP was polymerised under /THF/sBuLi/-50 ± 10 °C/// conditions to give P2VPLi ($1.1 \leq M_n/(\text{kg/mol}) \leq 3.7$; $1.10 \leq M_w/M_n \leq 1.24$) which was used as an initiator for εC. During the εC polymerisation the temperature was varied between -50 °C to RT. In some instances LiCl was present in the reaction mixture. The added εC block has $M_n = 0.1-7.8$ kg/mol, $1.33 \leq M_w/M_n \leq 2.43$ for the block copolymer. The level of control on the εC polymerisation is thus limited [151]. Finally, the preparation of Bd-εC block copolymers has been reported. Bd was polymerised under /Bz/sBuLi/30 °C//EO/ conditions to obtain PBd-EO-Li, which served as an initiator for the polymerisation of εC ($25 \leq T/(°C) \leq 70$) to obtain PBd-PεC ($3.7 \leq M_n/(\text{kg/mol}) \leq 15$ kg/mol $1.29 \leq M_w/M_n \leq 1.90$). In all cases a fraction of PεC homopolymer could be extracted from the product [152]. PS-PEO di- and triblock copolymers have been used as macroinitiators for εC. Three methods were compared: Thermal polymerisation in bulk at 180 °C for 30 h, anionic polymerisation in benzene with trace THF present at RT for 30 s with Na$^+$ counterion (titration, the PE-PEO is used as a macroinitiator) or K$^+$ counterion (sequential, the PS-PEO living polymer is not terminated). None of the methods give good control [153].

Block copolymers and star polymers from εC have been synthesized under different conditions. Block copolymers of εC and γBrCL were obtained under /Tol/Al(O^{i}Pr)$_3$/0 °C//HCl/ conditions. ($16 \leq M_n/(\text{kg/mol}) \leq 30$; $1.15 \leq M_w/M_n \leq 1.20$) [154].

Various methods for the synthesis of εC-stars were reported. C(CH$_2$OH)$_4$ and EtC(CH$_2$OH)$_3$ were used as initiators with various acids as promoters to polymerise εC in solution and bulk. The best control was obtained with weak acid in 5–15 times excess (fumaric acid, pKa $= 3.02$, 90 °C, 6–24 h, $M_{n,\text{arm}} = 1.3-5.7$ kg/mol, $M_w/M_n = 1.06-1.14$; conversion of εC $= 78-99\%$). Stronger acid gives broader distributions [155]. A 2,2-bis(hydroxymethyl) propionic-acid-based hyperbranched polymer served as initiator for εC to generate a star polymer with up to 40 arms (M_n/arm < 5000; $M_w/M_n = 1.4-2.6$) [156]. A glycerol core with catalytic amounts of SnOct yielded a 3-arm PεC ($M_{\text{arm}} = 1-7$ kg/mol, $M_w/M_n = 1.45-1.68$, 130 °C, 48 h) [157].

Homopolymers of TOSUO and block copolymer with PεC were prepared with Al(O^{i}Pr)$_3$ as initiator in Tol at 25 °C ($M_n \leq 10$ kg/mol; $M_w/M_n = 1.15-1.25$ was obtained for the homopolymers; $3.7 \leq M_n/(\text{kg/mol}) \leq 16$, $M_w/M_n = 1.20-1.35$ for the block copolymer) [158].

A number of reports on the polymerisation of LLA to form block copolymers or star polymers have appeared. Di- and tri-block copolymers of the seven-membered ring lactone 7CC and εC or δ-VL were obtained by sequential cationic polymerisation under /CH$_2$Cl$_2$/nBuOH, HCl, in Et$_2$O/25 °C/// conditions for εC and at –40 °C for δ-VL (6.6 ≤ M_n/(kg/mol) ≤ 9.5; 1.12 ≤ M_w/M_n ≤ 1.16) [159].

C(CH$_2$OH)$_4$ (a 4-functional alcohol) has been oligomerised with ethylene oxide. A dibutyl-tin complex was used as initiator for a L-lactide 4-arm star (M_{arm} = 1–19 kg/mol; M_w/M_n = 1.06–1.09; 85–98% yield; 60°C, chloroform, $[M]_0$ = 0.5 M, 7–290 h) [160]. Pentaerythritol and trimethylolpropane with stannous octoate as a promoter at 110 °C, polymerise 3-dimethyl-1,4-dioxolane-2,5-dione (to produce lactic acid-*alt*-glycolic acid polymer) (M_{arm} = 2.2–3.1 kg/mol, M_w/M_n = 1.15–1.20). The products initiate D,L-lactide with SnOct catalyst (M_{arm} = 2–8 kg/mol, M_w/M_n = 1.09–1.25; 130°C, 12 h) [161]. With a glycerol core and stannous octoate present in catalytic amounts at 130 °C for 6 h a 3-arm PLLA is obtained (M_{arm} = 0.7–1.3 kg/mol, M_w/M_n = 1.25–1.38) [162]. Initation using 1,6-hexanediol of L-lactide polymerisation with SnOct as co-initiator 0.005 mol% for 5 h at 130 °C followed by coupling with acid-chloride-terminated PEO yields a PEO-PLL block copolymer (M_n = 4.5–11 kg/mol, M_w/M_n = 1.24–1.42). Bidirectional chain growth in lactide polymerisation is obtained at 60 °C in chloroform with $[M]_0$ = 0.5 M for 6–160 h using a double bond-containing dihydroxy compound of dibutyl-tin as initiator (M_n = 5–105 kg/mol, M_w/M_n = 1.05–1.11; 40–86 yield) [163].

3-sec-butylmorpholine-2,5-dione was polymerised in bulk with mono-, di-, and tetra-hydroxy terminated PEO as initiator and equimolar SnOct at 140 °C as catalyst for 9 h with a 69–85% yield (M_n = 26 kg/mol, M_w/M_n = 2.09 for the homopolymer; For A$_2$B, AB, A$_2$BA$_2$copolymers, M_{arm} = 6–50 kg/mol, M_w/M_n = 1.28–2.47 for PEO block copolymer) [164].

2.5 CONCLUDING REMARKS

Methods for obtaining model polymers and block copolymer continue to be developed. The field is exceedingly diverse and it is difficult to present a review that gives due credit to all interesting results. Thus the present chapter does not claim to be complete in that respect. However, it is the hope of the author that it can serve as a source of information and inspiration for the development of new methods in controlled synthesis and characterization of polymers.

ACKNOWLEDGEMENTS

Financial support from the Danish Polymer Centre, a Research Centre at Risø and at the Danish Technical University sponsored by the Programme for

Development of Materials Technology is acknowledged. Fruitful discussions on molar mass distributions with Walther Batsberg Pedersen, Peter Sommer-Larsen and Frank S. Bates were instrumental in developing the section on MMDs.

REFERENCES

1. N. Hadjichristidis, S. Pispas, M. Pitsikalis, *Prog. Polym. Sci.*, **24**, 875–915 (1999).
2. S. Ndoni, C. M. Papadakis, F. S. Bates, K. Almdal, *Rev. Sci. Instrum.*, **66**, 1090–1095 (1995).
3. N. Hadjichristidis, H. Iatrou, S. Pispas, M. Pitsikalis, *J. Polym. Sci. Polym. Chem.*, **38**, 3211–3234 (2000).
4. http://mathworld.wolfram.com/GammaDistribution.html.
5. K. Almdal: Computer simulations performed in the process of writing the present review.
6. W. Lee, H. Lee, J. Cha, T. Chang, K. J. Hanley, T. P. Lodge, *Macromolecules*, **33**, 5111–5115 (2000).
7. W. Lee, D. Y. Cho, T. Y. Chang, K. J. Hanley, T. P. Lodge, *Macromolecules*, **34**, 2353–2358 (2001).
8. Soojin Park, Donghyun Cho, Jinsook Ryu, Kyoon Kwon, Wonmok Lee, Taihyun Chang, *Macromolecules*, **35**, 5974–5979 (2002).
9. N. Ekizoglou, N. Hadjichristidis, *J. Polym. Sci. Polym. Chem.*, **40**, 2166–2170 (2002).
10. P. Dubois, Y. S. Yu, P. Teyssie, R. Jerome, *Rubber Chem. Technol.*, **70**, 714–726 (1997).
11. J. M. Yu, P. Dubois, R. Jerome, *J. Polym. Sci. Polym. Chem.*, **35**, 3507–3515 (1997).
12. J. M. Yu, P. Dubois, R. Jerome, *Macromolecules*, **30**, 4984–4994 (1997).
13. J. M. Yu, Y. S. Yu, P. Dubois, P. Teyssie, R. Jerome, *Polymer*, **38**, 3091–3101 (1997).
14. Y. S. Cho, J. S. Lee, *Macromol. Rapid Commun.*, **22**, 638–642 (2001).
15. J. Wang, S. Kara, T. E. Long, T. C. Ward, *J. Polym. Sci. Polym. Chem.*, **38**, 3742–3750 (2000).
16. O. Lehmann, S. Förster, J. Springer, *Macromol. Rapid Commun.*, **21**, 133–135 (2000).
17. M. Pitsikalis, E. Siakali-Kioulafa, N. Hadjichristidis, *Macromolecules*, **33**, 5460–5469 (2000).
18. M. Vamvakaki, N. C. Billingham, S. P. Armes, *Macromolecules*, **32**, 2088–2090 (1999).
19. M. V. D. Banez, K. L. Robinson, M. Vamvakaki, S. F. Lascelles, S. P. Armes, *Polymer*, **41**, 8501–8511 (2000).
20. E. Bigdeli, *Iran J. Chem. Chem. Eng.-Int. Engl. Ed.*, **15**, 93–97 (1996).
21. K. Y. Baek, M. Kamigaito, M. Sawamoto, *J. Polym. Sci. Polym. Chem.*, **40**, 633–641 (2002).
22. T. Ishizone, J. Tsuchiya, A. Hirao, S. Nakahama, *Macromolecules*, **31**, 5598–5608 (1998).
23. K. T. Lim, S. E. Webber, K. P. Johnston, *Macromolecules*, **32**, 2811–2815 (1999).
24. D. K. Dimov, W. N. Warner, T. E. Hogen-Esch, S. Juengling, V. Warzelhan, *Macromol. Symp.*, **157**, 171–182 (2000).
25. J. Kim, D. A. Tirrell, *Macromolecules*, **32**, 945–948 (1999).
26. J. D. Tong, P. Leclere, C. Doneux, J. L. Bredas, R. Lazzaroni, R. Jerome, *Polymer*, **41**, 4617–4624 (2000).

27. J. D. Tong, G. Moineau, P. Leclere, J. L. Bredas, R. Lazzaroni, R. Jerome, *Macromolecules*, **33**, 470–479 (2000).
28. J. D. Tong, R. Jerome, *Polymer*, **41**, 2499–2510 (2000).
29. K. Matsumoto, U. Mizuno, H. Matsuoka, H. Yamaoka, *Macromolecules*, **35**, 555–565 (2002).
30. K. Matsumoto, C. Wahnes, E. Mouri, H. Matsuoka, H. Yamaoka, *J. Polym. Sci. Polym. Chem.*, **39**, 86–92 (2001).
31. K. Matsumoto, M. Deguchi, M. Nakano, H. Yamaoka, *J. Polym. Sci. Polym. Chem.*, **36**, 2699–2706 (1998).
32. R. Knischka, H. Frey, U. Rapp, F. J. Mayer-Posner, *Macromol. Rapid Commun.*, **19**, 455–459 (1998).
33. K. Matsumoto, K. Miyagawa, H. Matsuoka, H. Yamaoka, *Polym. J.*, **31**, 609–613 (1999).
34. L. Cao, M. A. Winnik, I. Manners, *J. Inorg. Organomet. Polym.*, **8**, 215–224 (1998).
35. X. S. Wang, M. A. Winnik, I. Manners, *Macromol. Rapid Commun.*, **23**, 210–213 (2002).
36. T. J. Peckham, J. A. Massey, C. H. Honeyman, I. Manners, *Macromolecules*, **32**, 2830–2837 (1999).
37. I. Manners, *Can. J. Chem.-Rev. Can. Chim.*, **76**, 371–381 (1998).
38. X. S. Wang, M. A. Winnik, I. Manners, *Macromol. Rapid Commun.*, **23**, 210–213 (2002).
39. L. Cao, I. Manners, M. A. Winnik, *Macromolecules*, **34**, 3353–3360 (2001).
40. T. J. Peckham, J. Massey, D. P. Gates, I. Manners, *Phosphorus Sulfur Silicon Relat. Elem.*, **146**, 217–220 (1999).
41. H. R. Allcock, S. D. Reeves, J. M. Nelson, I. Manners, *Macromolecules*, **33**, 3999–4007 (2000).
42. J. M. Nelson, A. P. Primrose, T. J. Hartle, H. R. Allcock, I. Manners, *Macromolecules*, **31**, 947–949 (1998).
43. H. R. Allcock, R. Prange, T. J. Hartle, *Macromolecules*, **34**, 5463–5470 (2001).
44. R. Prange, H. R. Allcock, *Macromolecules*, **32**, 6390–6392 (1999).
45. H. R. Allcock, J. M. Nelson, R. Prange, C. A. Crane, C. R. de Denus, *Macromolecules*, **32**, 5736–5743 (1999).
46. H. R. Allcock, R. Prange, *Macromolecules*, **34**, 6858–6865 (2001).
47. J. J. Reisinger, M. A. Hillmyer, *Prog. Polym. Sci.*, **27**, 971–1005 (2002).
48. M. Hillmyer, *Curr. Opin. Solid State Mater. Sci.*, **4**, 559–564 (1999).
49. K. Busse, J. Kressler, D. van Eck, S. Horing, *Macromolecules*, **35**, 178–184 (2002).
50. T. Ishizone, K. Sugiyama, Y. Sakano, H. Mori, A. Hirao, S. Nakahama, *Polym. J.*, **31**, 983–988 (1999).
51. T. M. Yong, W. P. Hems, J. L. M. vanNunen, A. B. Holmes, J. H. G. Steinke, P. L. Taylor, J. A. Segal, D. A. Griffin, *Chem. Commun.*, **1997**, 1811–1812.
52. K. Matsumoto, R. Nishimura, H. Mazaki, H. Matsuoka, H. Yamaoka, *J. Polym. Sci. Polym. Chem.*, **39**, 3751–3760 (2001).
53. J. G. Wang, G. P. Mao, C. K. Ober, E. J. Krammer, *Macromolecules*, **30**, 1906–1914 (1997).
54. J. D. J. S. Samuel, R. Dhamodharan, C. K. Ober, *J. Polym. Sci. Polym. Chem.*, **38**, 1179–1183 (2000).
55. Yu Ren, Timothy P. Lodge, Marc A. Hillmyer, *Macromolecules*, **34**, 4780–4787 (2001).
56. M. A. Hillmyer, T. P. Lodge, *J. Polym. Sci. Polym. Chem.*, **40**, 1–8 (2002).
57. Y. Ren, T. P. Lodge, M. A. Hillmyer, *J. Am. Chem. Soc.*, **120**, 6830–6831 (1998).
58. Y. Ren, T. P. Lodge, M. A. Hillmyer, *Macromolecules*, **33**, 866–876 (2000).
59. V. W. Stone, A. M. Jonas, R. Legras, P. Dubois, R. Jerome, *J. Polym. Sci. Polym. Chem.*, **37**, 233–244 (1999).

60. J. L. David, A. S. P. Gido, K. Hong, J. Zhou, J. W. Mays, N. B. Tan, *Macromolecules*, **32**, 3216–3226 (1999).
61. K. L. Hong, J. W. Mays, *Macromolecules*, **34**, 3540–3547 (2001).
62. I. Natori, K. Imaizumi, *Macromol. Symp.*, **157**, 143–150 (2000).
63. G. G. Nossarev, T. E. Hogen-Esch, *Macromolecules*, **35**, 1604–1610 (2002).
64. G. G. Nossarev, T. E. Hogen-Esch, *Macromolecules*, **34**, 6866–6870 (2001).
65. G. G. Nossarev, T. E. Hogen-Esch, *J. Polym. Sci. Polym. Chem.*, **39**, 3034–3041 (2001).
66. Y. D. Gan, D. H. Dong, T. E. Hogen-Esch, *Macromolecules*, **35**, 6799–6803 (2002).
67. A. Hirao, M. Hayashi, S. Loykulnant, *Macromol. Symp.*, **161**, 45–52 (2000).
68. R. P. Quirk, G. M. Lizarraga, *Macromol. Chem. Phys.*, **201**, 1395–1404 (2000).
69. H. Ito, A. Knebelkamp, S. B. Lundmark, C. V. Nguyen, W. D. Hinsberg, *J. Polym. Sci. Polym. Chem.*, **38**, 2415–2427 (2000).
70. C. Tsitsilianis, G. A. Voyiatzis, J. K. Kallitsis, *Macromol. Rapid Commun.*, **21**, 1130–1135 (2000).
71. S. Pispas, N. Hadjichristidis, *Macromolecules*, **33**, 6396–6401 (2000).
72. A. Avgeropoulos, S. Paraskeva, N. Hadjichristidis, E. L. Thomas, *Macromolecules*, **35**, 4030–4035 (2002).
73. V. Bellas, H. Iatrou, N. Hadjichristidis, *Macromolecules*, **33**, 6993–6997 (2000).
74. K. Almdal, M. A. Hillmyer, F. S. Bates, *Macromolecules*, **35**, 7685–7691 (2002).
75. J. Lee, T. E. Hogen-Esch, *Macromolecules*, **34**, 2095–2100 (2001).
76. M. Scibiorek, N. K. Gladkova, J. Chojnowski, *Polym. Bull.*, **44**, 377–384 (2000).
77. W. Fortuniak, J. Chojnowski, G. Sauvet, *Macromol. Chem. Phys.*, **202**, 2306–2313 (2001).
78. C. S. Patrickios, C. Forder, S. P. Armes, N. C. Billingham, *J. Polym. Sci. Polym. Chem.*, **35**, 1181–1195 (1997).
79. C. Forder, C. S. Patrickios, S. P. Armes, N. C. Billingham, *Macromolecules*, **30**, 5758–5762 (1997).
80. A K. Yamada, M. Minoda, T. Miyamoto, *Macromolecules*, **32**, 3553–3558 (1999).
81. E. J. Goethals, W. Reyntjens, X. C. Zhang, B. Verdonck, T. Loontjens, *Macromol. Symp.*, **157**, 93–99 (2000).
82. M. Ouchi, M. Kamigaito, M. Sawamoto, *J. Polym. Sci. Polym. Chem.*, **39**, 398–407 (2001).
83. M. Schappacher, C. Billaud, C. Paulo, A. Deffieux, *Macromol. Chem. Phys.*, **200**, 2377–2386 (1999).
84. K. Satoh, M. Kamigaito, M. Sawamoto, *Macromolecules*, **33**, 5830–5835 (2000).
85. R. Weberskirch, J. Preuschen, H. W. Spiess, O. Nuyken, *Macromol. Chem. Phys.*, **201**, 995–1007 (2000).
86. D. Nagai, H. Kuramoto, A. Sudo, F. Sanda, T. Endo, *Macromolecules*, **35**, 6149–6153 (2002).
87. A. Sudo, S. Uchino, T. Endo, *J. Polym. Sci. Polym. Chem.*, **39**, 2093–2102 (2001).
88. H. Yasuda, *J. Organomet. Chem.*, **647**, 128–138 (2002).
89. M. Narita, R. Nomura, I. Tomita, T. Endo, *Macromolecules*, **33**, 4979–4981 (2000).
90. H. Schlaad, H. Kukula, J. Rudloff, I. Below, *Macromolecules*, **34**, 4302–4304 (2001).
91. S. Forster, E. Kramer, *Macromolecules*, **32**, 2783–2785 (1999).
92. R. P. Quirk, S. Corona-Galvan, *Macromolecules*, **34**, 1192–1197 (2001).
93. K. L. Hong, Y. N. Wan, J. W. Mays, *Macromolecules*, **34**, 2482–2487 (2001).
94. D. Rosati, M. Perrin, P. Navard, V. Harabagiu, M. Pinteala, B. C. Simionescu, *Macromolecules*, **31**, 4301–4308 (1998).
95. O. Rheingans, N. Hugenberg, J. R. Harris, K. Fischer, M. Maskos, *Macromolecules*, **33**, 4780–4790 (2000).

96. D. Pantazis, S. Pispas, N. Hadjichristidis, *J. Polym. Sci. Polym. Chem.*, **39**, 2494–2507 (2001).
97. G. J. Wang, D. Y. Yan, *Chin. Sci. Bull.*, **45**, 1948–1953 (2000).
98. T. Hashimoto, H. Hasegawa, T. Hashimoto, H. Katayama, M. Kamigaito, M. Sawamoto, *Macromolecules*, **30**, 6819–6825 (1997).
99. I. C. De Witte, E. J. Goethals, *Polym. Adv. Technol.*, **10**, 287–292 (1999).
100. V. Bellas, H. Iatrou, E. N. Pitsinos, N. Hadjichristidis, *Macromolecules*, **34**, 5376–5378 (2001).
101. K. Yamauchi, H. Hasegawa, T. Hashimoto, N. Kohler, K. Knoll, *Polymer*, **43**, 3563–3570 (2002).
102. G. Velis, N. Hadjichristidis, *Macromolecules*, **32**, 534–536 (1999).
103. T. Tsoukatos, N. Hadjichristidis, *J. Polym. Sci. Polym. Chem.*, **40**, 2575–2582 (2002).
104. S. Houli, H. Iatrou, N. Hadjichristidis, D. Vlassopoulos, *Macromolecules*, **35**, 6592–6597 (2002).
105. J. Feldthusen, B. Ivan, A. H. E. Muller, *Macromolecules*, **31**, 578–585 (1998).
106. J. Lee, T. E. Hogen-Esch, *Macromolecules*, **34**, 2805–2811 (2001).
107. S. Jacob, J. P. Kennedy, *Adv. Polym. Sci.*, **146**, 1–38 (1999).
108. J. P. Kennedy, S. Jacob, *Accounts Chem. Res.*, **31**, 835–841 (1998).
109. R. P. Quirk, Y. S. Tsai, *Macromolecules*, **31**, 8016–8025 (1998).
110. J. E. Puskas, W. Pattern, P. M. Wetmore, V. Krukonis, *Rubber Chem. Technol.*, **72**, 559–568 (1999).
111. S. Jacob, I. Majoros, J. P. Kennedy, *Rubber Chem. Technol.*, **71**, 708–721 (1998).
112. E. Cloutet, J. L. Fillaut, D. Astruc, Y. Gnanou, *Macromolecules*, **31**, 6748–6755 (1998).
113. X. S. Feng, C. Y. Pan, *Macromolecules*, **35**, 2084–2089 (2002).
114. M. Hayashi, S. Nakahama, A. Hirao, *Macromolecules*, **32**, 1325–1331 (1999).
115. A. Hirao, Y. Tokuda, Y. Morifuji, M. Hayashi, *Macromol. Chem. Phys.*, **202**, 1606–1613 (2001).
116. M. Hayashi, A. Hirao, *Macromol. Chem. Phys.*, **202**, 1717–1726 (2001).
117. A. Hirao, M. Hayashi, *Macromolecules*, **32**, 6450–6460 (1999).
118. C. Tsitsilianis, D. Voulgaris, *Macromol. Chem. Phys.*, **198**, 997–1007 (1997).
119. M. Janata, L. Lochmann, J. Brus, P. Holler, Z. Tuzar, P. Kratochvil, B. Schmitt, W. Radke, A. H. E. Muller, *Macromolecules*, **30**, 7370–7374 (1997).
120. N. Hadjichristidis, *J. Polym. Sci. Pol. Chem.*, **37**, 857–871 (1999).
121. R. P. Quirk, T. Yoo, Y. Lee, J. Kim, B. Lee, *Adv. Polym. Sci.*, **153**, 67–162 (2000).
122. S. Sioula, Y. Tselikas, N. Hadjichristidis, *Macromol. Symp.*, **117**, 167–174 (1997).
123. C. Zune, P. Dubois, R. Jerome, T. Werkhoven, J. Lugtenburg, *Macromol. Chem. Phys.*, **200**, 460–467 (1999).
124. N. Hadjichristidis, M. Xenidou, H. Iatrou, M. Pitsikalis, Y. Poulos, A. Avgeropoulos, S. Sioula, S. Paraskeva, G. Velis, D. J. Lohse, D. N. Schulz, L. J. Fetters, P. J. Wright, R. A. Mendelson, C. A. Garcia-Franco, T. Sun, C. J. Ruff, **33**, *Macromolecules*, 2424–2436 (2000).
125. A. Hirao, A. Matsuo, K. Morifuji, Y. Tokuda, M. Hayashi, *Polym. Adv. Technol.*, **12**, 680–686 (2001).
126. O. Lambert, S. Reutenauer, G. Hurtrez, P. Dumas, *Macromol. Symp.*, **161**, 97–102 (2000).
127. O. Lambert, S. Reutenauer, G. Hurtrez, G. Riess, P. Dumas, *Polym. Bull.*, **40**, 143–149 (1998).
128. H. Huckstadt, A. Gopfert, V. Abetz, *Macromol. Chem. Phys.*, **201**, 296–307 (2000).
129. A. Hirao, M. Hayashi, T. Higashihara, *Macromol. Chem. Phys.*, **202**, 3165–3173 (2001).

130. S. Pispas, N. Hadjichristidis, *J. Polym. Sci. Polym. Chem.*, **38**, 3791–3801 (2000).
131. C. M. Fernyhough, R. N. Young, R. D. Tack, *Macromolecules*, **32**, 5760–5764 (1999).
132. C. M. Fernyhough, R. N. Young, *Macromol. Symp.*, **161**, 103–111 (2000).
133. Y. C. Bae, R. Faust, *Macromolecules*, **31**, 2480–2487 (1998).
134. H. Iatrou, N. Hadjichristidis, G. Meier, H. Frielinghaus, M. Monkenbusch, *Macromolecules*, **35**, 5426–5437 (2002).
135. S. Cramail, M. Schappacher, A. Deffieux, *Macromol. Chem. Phys.*, **201**, 2328–2335 (2000).
136. D. Pantazis, D. N. Schulz, N. Hadjichristidis, *J. Polym. Sci. Polym. Chem.*, **40**, 1476–1483 (2002).
137. K. Ishizu, A. Ichimura, *Polymer*, **39**, 6555–6558 (1998).
138. K. M. Stridsberg, M. Ryner, A. C. Albertsson, *Adv. Polym. Sci.*, **157**, 41–65 (2002).
139. Y. Heischkel, H. W. Schmidt, *Macromol. Chem. Phys.*, **199**, 869–880 (1998).
140. H. Keul, H. Hocker, *Macromol. Rapid Commun.*, **21**, 869–883 (2000).
141. Y. B. Wang, M. A. Hillmyer, *Macromolecules*, **33**, 7395–7403 (2000).
142. S. C. Schmidt, M. A. Hillmyer, *Macromolecules*, **32**, 4794–4801 (1999).
143. Y. B. Wang, M. A. Hillmyer, *J. Polym. Sci. Polym. Chem.*, **39**, 2755–2766 (2001).
144. Y. Kwon, R. Faust, C. X. Chen, E. L. Thomas, *Macromolecules*, **35**, 3348–3357 (2002).
145. S. Q. Xu, H. Y. Zhao, T. Tang, B. T. Huang, *Chin. J. Polym. Sci.*, **17**, 145–150 (1999).
146. Y. Akiyama, A. Harada, Y. Nagasaki, K. Kataoka, *Macromolecules*, **33**, 5841–5845 (2000).
147. M. L. Yuan, X. M. Deng, *Eur. Polym. J.*, **37**, 1907–1912 (2001).
148. K. Tsutsumiuchi, K. Aoi, M. Okada, *Macromolecules*, **30**, 4013–4017 (1997).
149. M. S. Kim, R. Faust, *Polym. Bull.*, **48**, 127–134 (2002).
150. P. Kurcok, P. Dubois, W. Sikorska, Z. Jedlinski, R. Jerome, *Macromolecules*, **30**, 5591–5595 (1997).
151. F. L. Duivenvoorde, J. J. G. S. van Es, C. F. van Nostrum, R. van der Linde, *Macromol. Chem. Phys.*, **201**, 656–661 (2000).
152. B. A. Rozenberg, Y. I. Estrin, G. A. Estrina, *Macromol. Symp.*, **153**, 197–208 (2000).
153. M. L. Arnal, V. Balsamo, F. Lopez-Carrasquero, J. Contreras, M. Carrillo, H. Schmalz, V. Abetz, E. Laredo, A. J. Muller, *Macromolecules*, **34**, 7973–7982 (2001).
154. C. Detrembleur, M. Mazza, O. Halleux, P. Lecomte, D. Mecerreyes, J. L. Hedrick, R. Jerome, *Macromolecules*, **33**, 14–18 (2000).
155. F. Sanda, H. Sanada, Y. Shibasaki, T. Endo, *Macromolecules*, **35**, 680–683 (2002).
156. M. Trollsas, C. J. Hawker, J. F. Remenar, J. L. Hedrick, M. Johansson, H. Ihre, A. Hult, *J. Polym. Sci. Polym. Chem.*, **36**, 2793–2798 (1998).
157. M. D. Lang, R. P. Wong, C. C. Chu, *J. Polym. Sci. Polym. Chem.*, **40**, 1127–1141 (2002).
158. D. Tian, P. Dubois, R. Jerome, *Macromolecules*, **30**, 1947–1954 (1997).
159. Y. Shibasaki, H. Sanada, M. Yokoi, F. Sanda, T. Endo, *Macromolecules*, **33**, 4316–4320 (2000).
160. A. Finne, A. C. Albertsson, *Biomacromolecules*, **3**, 684–690 (2002).
161. C. M. Dong, K. Y. Qiu, Z. W. Gu, X. D. Feng, *J. Polym. Sci. Polym. Chem.*, **40**, 409–415 (2002).
162. S. Y. Park, D. K. Han, S. C. Kim, *Macromolecules*, **34**, 8821–8824 (2001).
163. M. Ryner, A. Finne, A. Ch. Albertsson, H. R. Kricheldorf, *Macromolecules*, **34**, 7281–7287 (2001).
164. Y. K. Feng, D. Klee, H. Keul, H. Hocker, *Macromol. Biosci.*, **1**, 30–39 (2001).

LIST OF ABBREVIATIONS

15-C-5	1,4,7,10,13-pentaoxacyclopentadecane
18-C-6	1,4,7,10,13,16-hexaoxacyclooctadecane
2ClTMP	2-chloro-2,4,4-trimethylpentane
2EOz	2-ethyl oxazoline
2MOz	2-methyl-2-oxazoline
2IPP	2-isopropenylpyridine
2VN	2-vinylnaphthalene
2VP	2-vinylpyridine
3LiPDEA	3-lithiopropionaldehyde diethyl acetal
4VP	4-vinylpyridine
4MS	4-methylstyrene
7CC	1,3-dioxepan-2-one
αMS	α-methylstyrene
γBrCL	γ-bromo-ε-caprolactone
δ-VL	δ-valerolactone
εC	ε-caprolactone
AcO	acetoxy
AcOVE	2-acetoxy vinyl ether
AgOTf	silver triflate
Al(O^{i}Pr)$_3$	aluminum isopropoxide
Al(BHT)Bu$^i_2$	(2,6-di-*tert*-butyl-4-methylphenylphenoxydiisobutylaluminium
alkylA	alkylacrylate
BCMA	9,10-bis(chloromethyl)anthracene
Bd	butadiene
BF$_3$OEt$_2$	borontrifluoride etherate
bipy	2,2'-bipyridine
BOC	4-tert-butoxycarbonyl
(BOC)$_2$O	di-t-butyl dicarbonate
BOCST	4-tert-butoxycarbonyloxystyrene
BrMDPE	bromomethyl-DPE or 1-(4-bromo-methylphenyl)-1-phenylethylene
BTFB	1,3-bis(trifluormethyl)benzene
BTMB	4-bromo-1,1,1-trimethoxybutane
Bu	butyl
Bu$_4$NCl	tetrabutylamonium chlorid
Bz	Benzene
BzEA	benzyl 2-ethylacrylate
BzVE	benzyl vinyl ether
C(CH$_2$OH)$_4$	2,2-bis(hydroxymethyl)-1,3-propanediol pentaerythritol
CEVE	chloroethyl vinylether
CFC113	1,1,2-trichlorotrifluoroethane

CHCl$_3$	chloroform
CH$_2$Cl$_2$	methylenechloride
CHD	cyclohexadiene
CHX	cyclohexane
ClMS	chloromethylstyrene
CmK	cumylpotassium
CMPDMS	2-(chloromethylphenyl) ethyldimethylchlorosilane
CPPHMA	6-[4-(4-cyanophenylazo) phenoxy]hexyl methacrylate
Cs	Cesium
D$_3$	hexamethyl-*cyclo*-trisiloxane
(D^{ClPr})$_3$	2,4,6-tri(3-chloropropyl)-2,4,6-trimethylcyclotrisiloxane
(D$_3$)^{ClPr}	2-(3-chloropropyl)-2,4,4,6,6-pentamethylcyclotrisiloxane
(D^v)$_3$	2,4,6-trimethyl-2,4,6-trivinylcyclotrisiloxane,
(D$_3$)^v	2,4,4,6,6,-pentamethyl-2-vinylcyclotrisiloxane,
DABCO	1,4-diazabicyclo[2.2.2]octane]
DASCB	1,1-dialkylsilacyclobutane
DBX	1,4-bis(bromomethyl)benzene
DEAEMA	2-(diethylamino)ethyl methacrylate
DIPIP	dipiperidinoethane
DLLA	D-lactide (3 S)-cis-3,6-dimethyl-1,4-dioxolane-2,5-dion; C$_6$H$_8$O$_4$
DMA	*N,N*-dimethylacetamide
DMAEMA	2-(dimethylamino)ethyl methacrylate
DMDCS	dimethyldichlorosilane
DMAPLi	3-dimethylaminopropyllithium
DMCS	dimethylchlorosilane
DME	dimethoxyethane
DMS	dimethylsiloxane
DMSCl$_2$	dimethyldichlorosilane
DMSO	dimethylsulfoxide
DMVF	9,9-dimethyl-2-vinylfluorene
DP	degree of polymerisation
DPAEMA	2-(diisopropylamino)ethyl methacrylate
dPBz	di(isopropenyl)benzene
DPE	1,1-diphenylethylene
DPE-X-DPE	denotes the class of molecules containing 2 DPE units linked together, e.g. MDDPE
DPHLi	1,1-diphenylhexyllithium (BuLi DPE adduct)
DPHK	1,1-diphenylhexylpotassium
DtBP	2,6-di-*tert*-butylpyridine
DTFPz	di(trifluorethoxy)phosphazene
EAA	2-ethylacrylic acid
EBrPAA	ethyl-3-bromopropyl acetaldehyde acetal
EDS	2,2,5,5-tetramethyl-2,5-disila-1-oxacyclopentane

EI	ethyleneimine
ELiPAA	ethyl-3-lithiopropyl acetaldehyde acetal
EO	ethylene oxide
Et_2O	diethyl ether
EtAc	ethyl acetate
$EtAlCl_2$	ethylaluminumdichloride
Et_3Al	triethylaluminum
$EtC(CH_2OH)_3$	2-ethyl-2-(hydroxymethyl)-1,3-propanediol or trimethylolpropane
$EtMe_2SiH$	ethyl dimethylsilane
Et_2Zn	diethylzinc
EVE	ethyl vinyl ether
FMA	2-(N-methylperfluorobutane sulfonamido) ethyl methacrylate
f_x	volume fraction of x
GluEV	2-(3,4,6-tri-O-acetyl-2-deoxy-2-phthalimido-β-D-glucos-1-yl)-ethyl vinyl ether
GMA	glycidyl methacrylate
HEMA	2-hydroxyethyl methacrylate
HMPA	hexamethylphosphoric triamide
HOS	4-hydroxystyrene
I	isoprene
IB	*iso*-butylene
IBVE	*iso*-butylvinylether
iBuOEtAc	1-Isobutoxyethyl acetate
KNaph	potassium naphthalide
KPDP	potasssium 3,3-diethoxypropanolate
KtAmOx	potassium tert-amyloxide
iBMA	iso-bornylmethacrylate
LCCC	liquid chromatography at the critical condition
LiMAOPDMS	lithium methacryloxypropyl dimethylsilanolate
$LiALH_4$	lithiumaluminumhydride
LL	L-lactide
LLA	Lactide racemic 3,6-dimethyl-1,4-dioxolane-2,5-dion;
MAA	methacrylic acid
MALDI-TOF	matrix-assisted laser desorption ionisation time-of-flight mass spectroscopy
MBA	4-methoxybenzaldehyde
MDDPE	*meta*-doublediphenylethylene or 1,3-Bis(1-phenylethenyl (benzene)
MeVE	methylvinylether
MEMA	2-(N-morpholino)ethyl methacrylate
MMA	methylmethacrylate
MMD	molar mass distribution

MMP	methoxy methylphenyl
MOS	4-methoxystyrene
MSCl	methanesulfonyl chloride
$MSCl_2$	methyldichlorosilane
MTEGVE	methyl tri (ethylene glycol) vinyl ether
MTSP	1-methoxy-1-trimethylsiloxypropene
$NaOOCCMe_2Br$	sodium bromoisobutyrate
nBMA	n-butylmethacrylate
nBuOH	*n*-butylalcohol
nBuLi	*n*-butyllithium
NCzMA	2-(3-nitrocarbazolyl)ethyl methacrylate
nHx	n-hexane
NTMSBrDTFP	N-trimethylsilyl-bromo-di(trifluorethoxy)phosphoranimine $(Br(CF_3CH_2O)_2P = NSiMe_3)$
NTMSTCPz	N-trimethylsilyl-trichlorophospharanimine
NTPP	1-naphthyltriphenylphosphonium
ODVE	octadecylvinylether
O^iPr	*iso*-propyloxide (anion)
OTs	*para*-toluenesulfonate
P	poly
PBdLi	polybutadienyllithium
PB-SiOLi	lithium polybutadienesilanolate
PE	polyethylene
PEK	phenylethylketene
PFP	polyferrocenylphosphine
PFS:	polyferrocenylsilane
PILi	polyisoprenyllithium
Phen	phenylene
PI-PSLi	polyisoprene-polystyryllithium
PO	propylene oxide
PSLi	polystyryllithium
PtDVScat	platinum divinyltetramethyldisiloxane catalyst
pTSA	*para*-toluenesulfonic acid
PVL	pivalolactone (3-hydroxy-2,2-dimethylpropanoic acid lactone)
ROP	ring-opening polymerisation
RSβBL	(R,S)-β-butyrolactone
RT	room temperature
S	styrene
sBuli	*sec*-butyllithium
SEC	Size exclusion chromatography
SiO-BA	4-(tert-butyldimethylsilyloxy) benzaldehyde
SMP	(tert-butyldimethylsilyloxy) methylphenyl
SnOct	stannous octanoate

StMA	stearyl methacrylate
tBDMSPrLi	tertbutyldimethylsiloxy-1-propyllithium
TDPE	1,3,5-tris(1-phenylethenyl (benzene)
TEA	triethylamine
TBABB	tetrabutylammonium bibenzoate
tBAEMA	2-(*tert*-butylamino) ethylmetacrylate
tBDMS	tert-butyldimethylsilyloxy
tBDMS-OS	p-[(tertbutyldimethylsilyl)oxy] styrene)
tBDPS-OS	p-[(tertbutyldiphenylsilyl)oxy] styrene)
tBDMSEDPE	1-(4-(2-tert-Butyldimethylsiloxy) ethyl)phenyl-1-phenylethylene
tBMA	t-butylmethacrylate
tBu	*tert*-butyl
tBuDiCumCl	1,3-bis(1-chloro-1-methylethyl)benzene
tBuEA	tert-butyl 2-Ethylacrylate
tBuLi	*tert*-butyllithium
tBuP$_4$	1-tert-butyl-4,4,4-tris(dimethylamino)-2,2-bis-[tris(dimethylamino)-phoshoranylidenamino]-2λ^5,4λ^5-catenadi(phosphazene)].
TCMS	trichloromethylsilane
TEA	triethylamine
TFA	trifluoracetic acid
TGIC	temperature gradient interaction chromatography
THF	tetrahydrofuran
TLiAD	2,2,5,5-Tetramethyl-1-(3-lithiopropyl)-1-aza-2,5-disilacyclopentane
TMCS	trimethylchlorosilane
TMEDA	tetramethylethylenediamine
TMePr	1,1,3,3-tetramethoxy propane
TMP	2-chloro-2,4,4-trimethylpentane
TMSA	trimethylsilyl acrylate
TMSI	trimethylsilyliodide
TPM	triphenylmethylanion
TPMCs	triphenylmethyl cesium
TPMK	triphenylmethyl potassium
Tol	toluene
TOSUO	1,4,8-trioxa[4.6]spiro-undecanone (5-ethylene ketal εC)
Ts	tosyl
X	halogen
XMP	halomethylphenyl
XPEK	halophenylethylketene; (4-chlorophenyl)ethylketene and (4-bromophenyl)ethylketene
XPr	3-halopropyl

3 Syntheses and Characterizations of Block Copolymers Prepared via Controlled Radical Polymerization Methods

PAN CAI-YUAN, HONG CHUN-YAN
Department of Polymer Science and Engineering, University of Science and Technology of China, Hefei, Anhui, 230026, P. R. China

3.1 INTRODUCTION

Every year, new kinds of block copolymer are synthesized and their properties extensively studied. Due to their distinctive structures, block copolymers have useful and desirable properties. For example, inexpensive adhesive tape employs linear triblocks to achieve pressure-sensitive adhesion. The familiar polyurethane foams used in upholstery and bedding are composed of multi-block copolymers known as thermoplastic elastomers that combine high-temperature resistance and low-temperature flexibility. The addition of appropriate block copolymers into commodity plastics, such as polystyrene can enhance toughness, or modify the surface properties for applications as diverse as colloidal stabilization, medical implantation and microelectronic fabrication. Recent advances in theoretical investigations of the phase behavior of block copolymer materials enable the prediction of the morphology, domain size, interfacial width and interfacial area of many types of block copolymers [1–6]. These developments in applications and fundamental understanding of thermodynamics are driving forces to synthesize more interesting block copolymers. Advances in synthetic chemistry have created fresh opportunities for using judicious combinations of multiple blocks in novel molecular architecture to produce a seemingly unlimited number of exquisitely structured materials endowed with tailored mechanical, optical, electrical, ionic, barrier and other physical properties. This chapter primarily summarizes the preparation of block copolymers by controlled radical polymerization methods and their combination with other living polymerization methods. Information on living anionic polymerization methods can be found in chapter 2.

3.1.1 THE MOLECULAR ARCHITECTURE OF BLOCK COPOLYMERS

Block copolymers are prepared by joining two or more chemically distinct polymer blocks together. The constituent polymers are often thermodynamically incompatible. As shown in **Scheme 3.1**, although a nearly limitless number of molecular architectures based on two, three or more monomer types can be constructed, the block copolymers can be divided into two categories from the point of view of topology: linear and nonlinear. A linear AB diblock copolymer consists of a chain of monomers of type A attached at the end to a chain of type B monomers. Similarly, the chains of monomers A, B and C or more monomer types are joined together to form ABC triblock copolymer or multiblock copolymers. When three or more different polymer chains are connected at one point, one of the types of nonlinear block copolymers, a miktoarm star copolymer is formed. Although innovative developments in polymer chemistry have stimulated the creation of many useful types of block copolymers, practical difficulties in copolymer synthesis still remain. In most polymer syntheses, normal chemical kinetics results in a distribution of molecular weights, and in block copolymers, this will produce compositional heterogeneity. This can be avoided by adopting so-called "living" polymerization techniques.

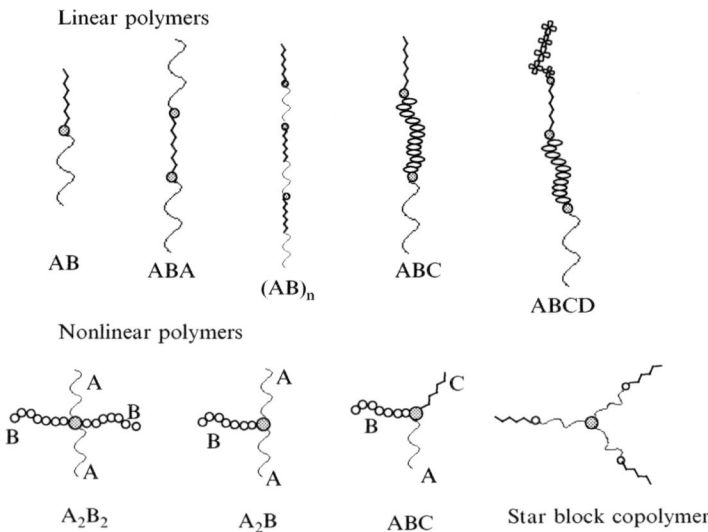

Scheme 3.1 Molecular architecture of block copolymers

3.1.2 CONTROLLED RADICAL POLYMERIZATION

Until recently, ionic polymerizations (cationic or anionic) were the only living techniques available that efficiently controlled the structure and architecture of vinyl polymers [7]. In contrast to ionic synthesis techniques where the growing species are mutually repulsive, radical polymerization suffers from bimolecular termination reactions such as radical recombination and disproportionation. Recent developments in controlled radical polymerization methods, such as atom-transfer radical polymerization (ATRP) [8–12], nitroxide-mediated living polymerization [13–16] and reversible addition-fragmentation transfer (RAFT) [17–19], have opened a new way to synthesize block copolymers. Although the mechanisms differ, the underlying principle for controlling radical polymerization is the same. The essential objective is to lower the instantaneous concentration of growing radical species by introducing an excess of covalent dormant species that exists in rapid equilibrium with the growth-active radical species. Such a dynamic and rapid equilibrium not only minimizes the probability of the radical bimolecular termination, but also gives an equal opportunity for all living (or dormant) chains to propagate via the frequent interconversion between the active and dormant species. These features lead to nearly uniform chain length. Here we should emphasize that although covalent dormant species reduce the termination reactions of the growing radicals, still around 10% of the growing radicals are terminated during the polymerization [20]. In some cases, it is thus important to separate the block copolymer product from homopolymer impurities.

3.2 DESIGN AND SYNTHESIS OF LINEAR BLOCK COPOLYMERS

3.2.1 SEQUENTIAL-CONTROLLED RADICAL POLYMERIZATIONS

Generally, block copolymers can be prepared in a traditional sequential fashion by the polymerization of one monomer, followed by a second monomer. This is also true for controlled radical polymerizations (see **Reaction 3.1**), which permit the synthesis of a wide variety of block copolymers, and may be more versatile

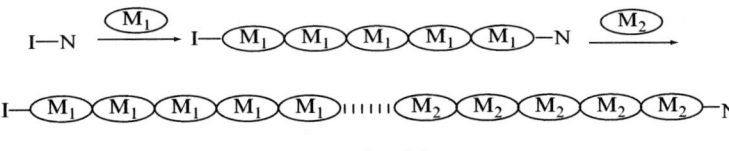

Reaction 3.1

than other living polymerization methods. Sequential-controlled radical polymerizations discussed here include (i) the addition of the second monomer into the polymerization system of the first monomer without any isolation, and (ii) the initiation of the polymerization of the second monomer using a macroinitiator obtained from controlled radical polymerization of the first monomer. The second method is used most widely.

3.2.1.1 Metal-Catalyzed-Controlled Polymerizations

A wide variety of block copolymers can be prepared by metal-catalyzed controlled radical polymerizations. Most of the block copolymers consist of methacrylates and/or acrylates and/or styrene (St), (LB-1 to LB-8 in **Scheme 3.2**) and can be synthesized both via sequential controlled radical polymerizations and via controlled radical polymerization using macroinitiators [21–31].

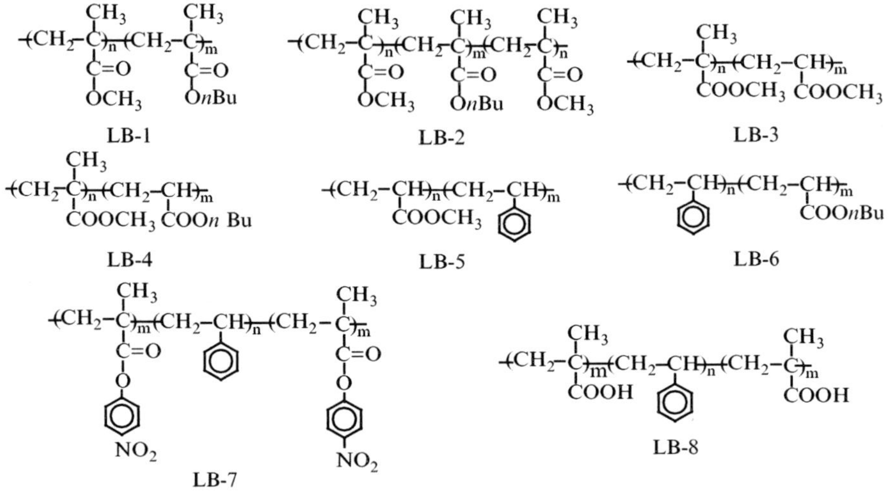

Scheme 3.2

Controlled radical polymerization of MMA followed by n-BuMA produces linear AB diblock copolymer LB-1 with a narrow molecular weight distribution (MWD, $M_w/M_n = 1.2$), which can be extended further into ABA triblock copolymer LB-2 with a similarly narrow MWD ($M_w/M_n = 1.2$) [21]. The block copolymers of MMA and MA, LB-3 and LB-4, were prepared using catalysts based on nickel, copper or iron complexes. Due to the higher activity of the carbon–halogen terminal bonds in poly(methacrylate)s than in poly(acrylate)s, the block copolymerization of MA using a PMMA macroinitiator can be achieved. The control of the polymerizations in the copper-based systems seems better than for the others [22–24].

Several papers reported the preparation of acrylate/St block copolymers, such as LB-5 and LB-6 using a copper-based complex. With such catalysts, the polymerization of both monomers can be controlled under common ATRP conditions [25–27]. Block copolymers can be prepared using a macroinitiator obtained from the ATRP of St or acrylate in the ATRP of second monomer.

The ATRP of St was performed using α, α'-dibromo-p-xylene (DBX) as initiator and CuBr/bipyridine (bpy) as catalyst and ligand. PSt with two terminal bromine groups was obtained and successfully used as macroinitiator in the polymerization of p-nitrophenyl methacrylate (NPMA). The ABA triblock copolymer LB-7 was obtained. Hydrolysis of LB-7 in acidic solution produced amphiphilic copolymer LB-8 [28,29].

The combination of reversible atom-transfer radical polymerization (RATRP) and ATRP, was successfully used to prepare block copolymers. For example, RATRP of MMA was carried out in CH_3CN/DMF, using AIBN/$CuCl_2$/bpy as an initiation system, leading to PMMA with MWD = 1.08. This was used in the successive ATRP of St to give diblock copolymer PMMA-b-PSt. No PMMA homopolymer was detected in the synthesis product [30].

As mentioned above, block copolymers can be prepared via sequential controlled radical polymerization, or via controlled radical polymerization using macroinitiators. The first method is the simple addition of a second monomer into the reaction medium after near-complete conversion of the first monomer. The second method involves the isolation and the purification of the first polymer, then using it as a macroinitiator. Although the first method is easy to carry out, the second block may produce a random copolymer because complete conversion in the controlled radical system is impossible, and loss of the terminal bromine at the end of polymer chain may occur after many steps of redox **Reaction 3.2**.

$$Cu(II) X_2 + \sim\!\sim\!\sim R\bullet \longrightarrow Cu(I)X + \sim\!\sim\!\sim RX \qquad (3.2)$$

Reaction 3.2

In order to avoid the troublesome separation of block copolymers from the homopolymer, the ATRP is generally stopped at lower conversion of the first monomer polymerization in order to ensure one bromine group at each end of the polymer chain.

Having discussed the principles of the two main routes to controlled radical polymerization of block copolymers, specific examples for tailored polymers are now outlined.

Fluorine-substituted polymers have excellent heat- and chemical-resistance properties, and low surface energy and low dielectric constant, etc. Another important application of these block copolymers is as amphiphilic emulsifiers in sc-CO_2 polymerizations, the fluorinated block being "CO_2-philic" and the conventional organic block being "CO_2-phobic" [31]. Thus perfluoroalkyl groups have also been incorporated into block copolymers with methacrylates, acrylates and St, which was prepared in sc-CO_2 or in the bulk [31]. For example, the first block was prepared from the bulk ATRP of St, MA or butyl acrylate (BA). After isolation, purification and characterization, the corresponding polymer was used as macroinitiator in the ATRP of fluorinated alkyl methacrylates, forming diblock copolymers with a hydrocarbon block and a fluorocarbon block. Using a difunctional initiator such as DBX, ABA triblock copolymers were obtained as shown in **Scheme 3.3**.

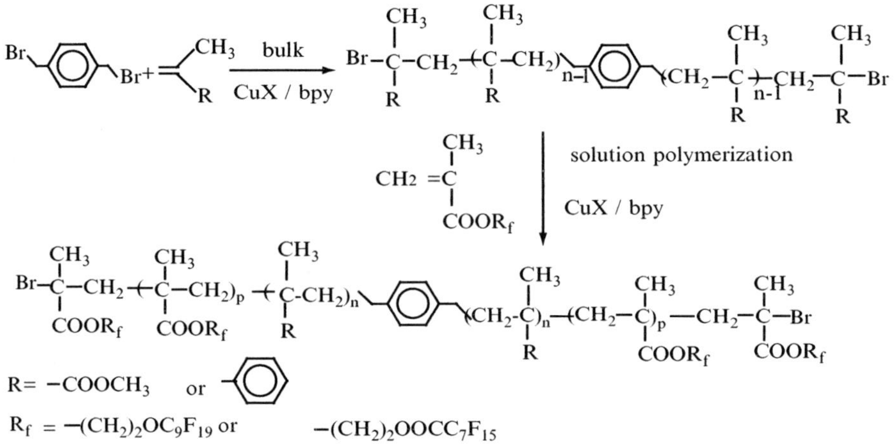

Scheme 3.3

Synthetic polymers with pendent carbohydrate moieties, which are referred to as glycopolymers, have potential applications in biomedical techniques [32]. Fukuda *et al.* reported the synthesis of diblock copolymer LB-9 by the sequential monomer addition technique. The first step was the polymerization of St in bulk with a CuBr/4,4'-di-n-heptyl-2,2'-bipyridine (2dHbpy) complex and 1-phenylethyl bromide as initiator at 110 °C. After more than 90 % conversion of St, a fresh feed of 3-O-methacrylol-1,2:5,6-di-O-isopropyldiene-D-glucofuranose with CuBr/2dHbpy complex dissolved in veratrole was added in vacuum to the precursor PSt-Br, and then polymerization was performed at 80 °C, forming diblock copolymer LB-9. When LB-9 was treated with formic acid, amphiphilic copolymer LB-10 was obtained [32].

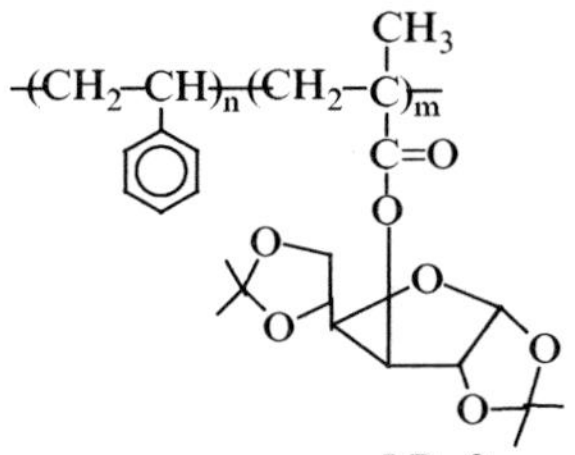

LB-9 and LB-10 structures

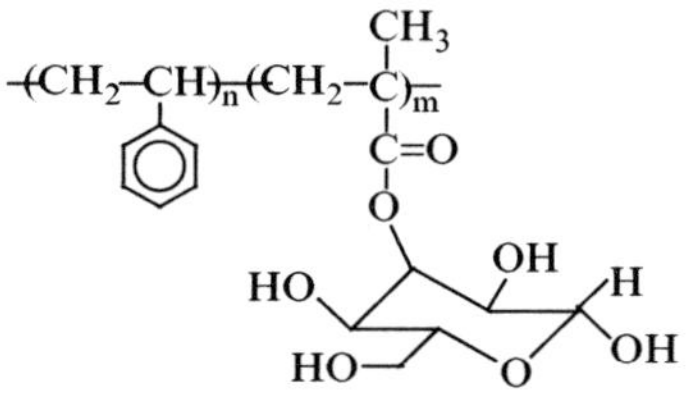

The monomer reactivity and the initiation efficiency of the macroinitiator must be considered in the synthesis of block copolymers. For example, in the ATRP synthesis of block copolymers from MMA and acrylates, the first step is the ATRP of MMA to give PMMA-Br, which generally provides good initiation efficiency. However, obtaining PMMA with close to 100% functionality proves difficult, because of side reactions that are more prevalent when Br is used as the end group than when Cl is the end group. Thus PMMA-Cl synthesized using TsCl/CuBr/dNbpy as a catalyst system, was used to initiate the ATRP of MA. After initiation, the end group of the growing polymer chain is predominantly the highly labile Br, thus a PMMA-*b*-PMA block copolymer with $M_w/M_n = 1.15$ was obtained [24]. When PMA-Cl is used as initiator in the ATRP of MMA with CuBr as catalyst, it is found that the initiation rate is too slow compared to the propagation rate of MMA, resulting in substantial amounts of PMA-Cl left over, and consequently a broad MWD of the block copolymer. Again "halide exchange" can be applied to solve this problem. For preparation of PMMA-*b*-PBA-*b*-PMMA triblock copolymers, Br-PBA-Br was used as macroinitiator in conjunction with CuCl in the ATRP of MMA to increase the rate of initiation relative to the rate of propagation [33].

Various amphiphilic block copolymers have been synthesized by sequential polymerizations of the protected forms of functional monomers, followed by deprotection [28,34–36]. One example is the synthesis of PMA-*b*-poly (acrylic acid) (PAA), as shown in **Scheme 3.4**. A diblock copolymer, PMA-*b*-poly (*tert*-butyl acrylate) (P*t*BA) was prepared by first polymerizing MA, followed by the growth of *t*BA from the PMA macroinitiator. This diblock was then treated in dichloromethane with anhydrous triflic acid (TFA) at room temperature. Selective cleavage of the *tert*-butyl ester group of P*t*BA block occurred to produce the amphiphilic block copolymer, PMA-b-PAA.

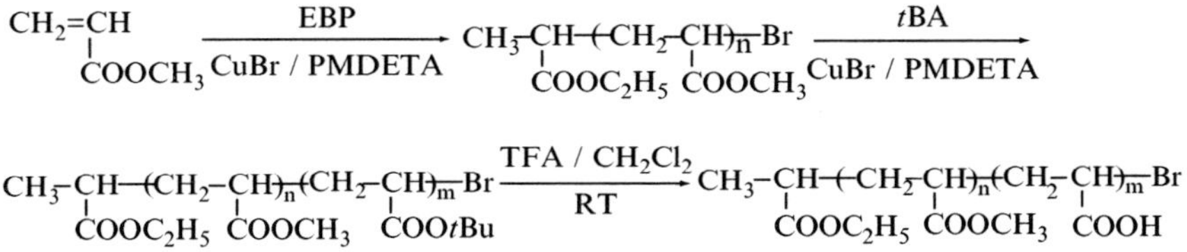

Scheme 3.4

ATRP cannot be used to polymerize some functional monomers, especially monomers having acidic groups, such as acrylic acid or methacrylic acid. The protection-deprotection technique is an important method to synthesize amphiphilic block copolymers containing such blocks. However, direct ATRP of some functional monomers has been developed to prepare amphiphilic copolymers [37–41]. Because the effects of different functional groups in the monomers on the polymerization are different, different conditions from those used for the polymerization of monomers without functional groups must be used. In the ATRP of hydroxyethyl methacrylate (HEMA), one key difference from MMA polymerizations is that the propagating radicals of HEMA are more stable and propagate more slowly. Unlike MMA, however, HEMA and the solvent used in the polymerization are very polar, which may dramatically affect the structure and function of the catalytic species, in the case of $Cu(bpy)_2X/Cu(bpy)_2X_2(X = Cl$ or Br), and may also lead to increased apparent rates of polymerization relative to that of MMA. Therefore, simply applying the same conditions used for MMA to the ATRP of HEMA does not result in a successful controlled radical polymerization. The solution ATRP of HEMA was achieved in 1-propanol/methyl ethyl ketone (30/70,v/v) at 60 °C, using PMMA-Cl as initiator in conjunction with a CuCl/bpy catalyst. Amphiphilic block copolymer, PMMA-*b*-PHEMA (LB-11) with MWD $= 1.17$ was successfully prepared [37,38].

Polymerization of 4-vinyl pyridine (4VP) poses a very challenging problem for ATRP, since both 4VP and P4VP are strong coordinating ligands that can compete for the binding of the metal catalysts in the system. Generally, the monomer is present in large excess over the ligand employed, which leads to the possibility of the formation of pyridine-coordinated metal complexes in the polymerization solution. The pyridine-coordinated copper complexes are not effective catalysts for ATRP. For example, the addition of 5 vol% pyridine in the solution polymerization of St catalyzed by CuBr complexed by 4,4'-di (5-nonyl)-2,2'-bipyridine significantly slowed the polymerization rate [39]. Therefore, a stronger binding agent, such as the tridentate N,N,N',N'',N''-pentamethyldiethylenetriamine (PMDETA) was used, faster polymerization rates being observed. In addition, the reaction of 1-phenylethyl bromide (PEBr) with 4-VP or/and P4VP must be considered. The polymerization of 4VP was carried out in 2-propanol using 1-phenylethyl chloride (PECl) as initiator and CuCl complexed with tris [2-(dimethylamino)ethyl] amine at 40 °C, the polymerization having living characteristics. The amphiphilic block copolymer, PMMA-*b*-P4VP (LB-12) with $M_{n,\,NMR} = 62\,500$, $M_w/M_n = 1.35$ was successfully prepared by solution ATRP of 4VP in DMF using PMMA-Cl ($M_{n,\,sec} = 7700$, $M_w/M_n = 1.06$) as macroinitiator in conjunction with CuCl/tris [2-(dimethylamino)ethyl] amine [40].

$$\text{-(CH}_2\text{-C}\overset{\overset{\displaystyle CH_3}{|}}{\underset{\underset{\displaystyle COOCH_3}{|}}{)_n}}\text{-(CH}_2\text{-C}\overset{\overset{\displaystyle CH_3}{|}}{\underset{\underset{\displaystyle COOCH_2CH_2OH}{|}}{)_m}}$$

LB-11

$$\text{-(CH}_2\text{-C}\overset{\overset{\displaystyle CH_3}{|}}{\underset{\underset{\displaystyle COOCH_3}{|}}{)_n}}\text{-(CH}_2\text{-CH)}_m\text{-Py}$$

LB-12

3.2.1.2 Reversible Addition-Fragmentation Chain Transfer (RAFT)

In 1998, Moad and coworkers reported a new controlled radical polymerization technique, RAFT, that offers exceptional versatility in providing polymers with predetermined molecular weight and narrow polydispersity (usually < 1.2, sometimes < 1.1) [17–19,42,43]. This is achieved by performing the polymerization in the presence of certain dithio compounds (e.g. RT1a-RT1d) which act as highly efficient reversible addition-fragmentation chain-transfer agents and give the polymerization living characteristics. The sequence of events for the polymerization can be illustrated as shown in **Scheme 3.5**. The propagating radicals initiated by the ordinary initiator react rapidly with the transfer agent [S=C(Z)S-R], resulting in a polymeric thiocarbonylthio compound [S=C(Z)S-P$_n$] in the early stage of polymerization. The S=C(Z)S-moiety is transferred between dormant and active chains, leading to the living character of the polymerization. Because the majority of chains in the target polymer have the S=C(Z)S-group, polymerization of a second monomer can be continued to give a block copolymer.

$$\text{Ph}-\underset{\underset{\displaystyle S-R}{||}}{\overset{\overset{\displaystyle S}{||}}{C}}$$

RT1a R=–CH$_2$Ph   RT1c R=–CH(CH$_3$)Ph
RT1b R=–C(CH$_3$)$_2$Ph   RT1d R=–C(CH$_3$)(CN)CH$_2$CH$_2$COOH

Chong and coworkers reported the preparation of AB type diblock copolymers by the RAFT process [44,45]. The results show the versatility and convenience of this process. Diblock copolymers PMMA-b-PMAA can be prepared directly from MAA monomer using PMMA-S-C(S)Ph as a macrotransfer agent.

In principle, one requirement for the formation of a narrow polydispersity AB block copolymer in a batch polymerization is that the first-formed polymeric thiocarbonyl transfer agent (S=C(Z)S-A) should have a high transfer constant in the subsequent polymerization step to give the B block. This requires that the leaving group of propagating radical A· is comparable to or better than that of the propagating radical B· under the conditions of the

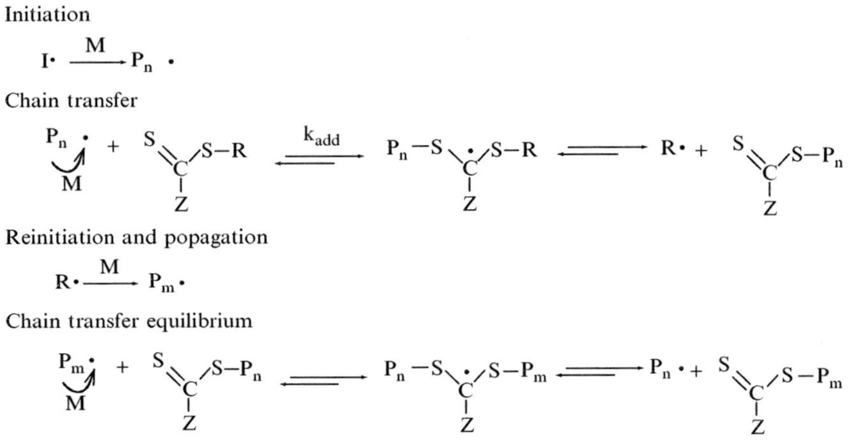

Initiation

$$I\cdot \xrightarrow{M} P_n\cdot$$

Chain transfer

$$P_n\cdot + \underset{Z}{\overset{S}{\underset{|}{C}}}\!\!=\!\!\underset{}{\overset{}{S\!-\!R}} \xrightleftharpoons{k_{add}} P_n\!-\!S\!-\!\underset{Z}{\overset{}{\underset{|}{C}}}\!-\!S\!-\!R \rightleftharpoons R\cdot + \underset{Z}{\overset{S}{\underset{|}{C}}}\!\!=\!\!S\!-\!P_n$$

Reinitiation and propagation

$$R\cdot \xrightarrow{M} P_m\cdot$$

Chain transfer equilibrium

$$P_m\cdot + \underset{Z}{\overset{S}{\underset{|}{C}}}\!\!=\!\!S\!-\!P_n \rightleftharpoons P_n\!-\!S\!-\!\underset{Z}{\overset{}{\underset{|}{C}}}\!-\!S\!-\!P_m \rightleftharpoons P_n\cdot + \underset{Z}{\overset{S}{\underset{|}{C}}}\!\!=\!\!S\!-\!P_m$$

Scheme 3.5

reaction (see **Scheme 3.6**). When A is a PMA or PSt chain, the transfer constants of $S=C(Z)$-A in the polymerization of MMA are too low due to the styryl – or acrylyl – propagating radicals being poor leaving groups relative to a methacryl propagating radical. This will cause the adduct radical in **Scheme 3.6** to return strongly to the starting materials, leading to a diblock copolymer with broad MWD and/or containing starting polymeric transfer agent. Therefore, when preparing a block copolymer, the polymerization of the monomer with the higher transfer constant should be carried out first. For example, macrotransfer agent PMMA-SC(Z)S is prepared first by the polymerization of MMA in the RAFT polymerization condition, and then used in the RAFT polymerization of the second monomer, such as St or MA. Diblock copolymers PMMA-*b*-PSt or PMMA-b-PMA with well-controlled MW and narrow MWD were then obtained.

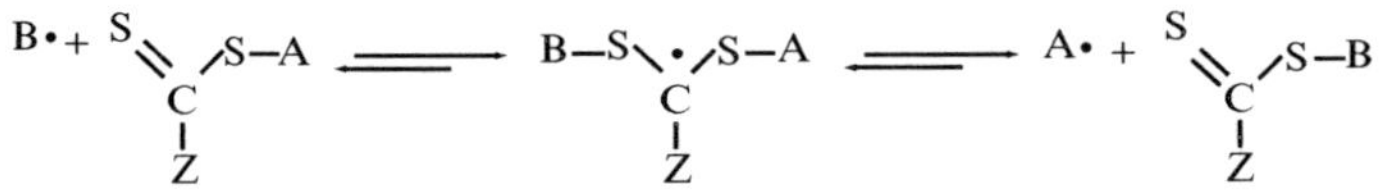

$$B\cdot + \underset{Z}{\overset{S}{\underset{|}{C}}}\!\!=\!\!S\!-\!A \rightleftharpoons B\!-\!S\!-\!\underset{Z}{\overset{}{\underset{|}{C}}}\!-\!S\!-\!A \rightleftharpoons A\cdot + \underset{Z}{\overset{S}{\underset{|}{C}}}\!\!=\!\!S\!-\!B$$

Scheme 3.6

In the RAFT process, the total number of chains formed will be equal to (or less than) the number of moles of dithio compound employed plus the number of moles of initiator-derived radicals generated during the course of the polymerization. In block copolymer synthesis, these additional initiator-derived chains are a source of homopolymer impurity. The level of impurities can be controlled by appropriate selection of the reaction conditions. To reduce impurities, it is desirable to use as low a concentration of initiator as practicable, and to choose solvents and initiators that give minimal chain transfer. Similar to conventional radical polymerization, the rate of RAFT polymerization is determined by the initiator concentration. In practice, it is usually not difficult to achieve block copolymers with no detectable homopolymer impurity (<5%), while still achieving an acceptable rate of polymerization. RAFT polymerization can be performed in bulk, in solution or in an emulsion [46–50]. A series of diblock copolymers, such as RB-2, RB-3 [47], RB-4 [46], RB-5 [48], RB-6 [49], RB-7 [50] and RB-8 [51] were prepared.

Poly(4-acetoxystyrene) with $M_n = 10\,000$, PDI = 1.12 was prepared by bulk RAFT polymerization of 4-acetoxystyrene at 90 °C using AIBN as initiator and α-acetic acid dithiobenzoate as chain transfer agent, and used as a macrotransfer agent in the block copolymerization of St with AIBN initiator after reprecipitation and the removal of residual monomer. The block copolymer, RB-2 was obtained [46]. The block copolymer can be hydrolyzed under mild basic conditions to give poly(4-hydroxy styrene)-b-PSt [46].

Double hydrophilic block copolymers, RB-3 and RB-4 have been prepared directly in aqueous media by using a dithioester-capped poly(4-styrene sulfate) or a dithioester-capped poly[(p-vinylbenzyl) trimethylammonium chloride] as the macrochain transfer agent in the successive RAFT polymerization of the second monomer [47]. The block copolymer, RB-5 was prepared using seeded emulsion polymerization via the RAFT mechanism. First, seeded particles consisting of PBA dormant chains were obtained by using active xanthate agent, [1-(O-ethylxanthyl)-ethyl]benzene, under bath and starved-feed conditions. These were then used in a second-stage emulsion polymerization of St under starved-feed and bath conditions to prepare polymer colloids of block PBA-*b*-PSt. Under starved-feed conditions, about 90% of the total polymer consists of blocks, whereas under batch conditions, only 70% is composed of block copolymers due to higher termination rates [48]. The copolymer RB-6 was prepared by a similar method [49]. The interesting block copolymer RB-7, is composed of poly(N-isopropylacrylamide) (NIPAM) with a lower critical solution temperature in an aqueous solution in the range of 32–34 °C, and poly[3-[N-(3-methacrylamidopropyl)-N,N'-dimethyl] amoniopropane sulfonate] (SPP), which possesses an upper critical solution temperature. It was prepared via the RAFT polymerization of SPP using dithioester-terminated PNIPAM as macrotransfer agent, with AIBN as initiator [50].

Li and coworkers [51] developed another strategy for the synthesis of block copolymers, prepared via one-pot polymerizations. For example, RAFT copolymerization of maleic anhydride (MAh) and St with a molar ratio of 1:9 was performed at 60 °C. Basically, the polymerization involves two stages, first, the copolymerization of St and MAh resulted in an alternating copolymer (stage 1). After the complete consumption of MAh, propagation reactions of St continued to produce the diblock copolymer RB-8.

$$-(CH_2-\underset{\underset{COOCH_2}{|}}{\overset{\overset{CH_3}{|}}{C}})_n-(CH_2-\underset{\underset{COOCH_2CH_2NMe_2}{|}}{\overset{\overset{CH_3}{|}}{C}})_m-(CH_2-\underset{\underset{COOt-Bu}{|}}{\overset{\overset{CH_3}{|}}{C}})_r-$$

RB-9

ABC triblock copolymers can be prepared by sequential RAFT polymerization. The procedure is simply to add a third monomer to a precursor AB diblock. For example, ABC triblock copolymer RB-9 was obtained by simply adding *t*BMA monomer into the polymerization system of a poly (benzyl methacrylate)-*b*-poly(N-dimethyl aminoethyl methacrylate) [44]. One approach to synthesize ABA triblock copolymers is to start with a difunctional transfer agent, so that two arms of polymer B are propagated simultaneously. The polymer obtained is then used in the next RAFT polymerization of the second monomer A as shown in **Scheme 3.7**. In this way, triblock copolymer, PMMA-*b*-PBMA-*b*-PMMA, RB-10 has been obtained [44].

The trithiocarbonates, such as RT-11 and RT-12 in **Scheme 3.8** are another type of difunctional transfer agent, and can be used in the RAFT polymerization of monomer A. One way to confirm the presence of an active trithiocarbonate group located in the center is to cleave the polymer chains at the trithiocarbonate function. This can be achieved in the presence of mild nucleophiles such as primary or secondary amines. Generally, the molecular weight of

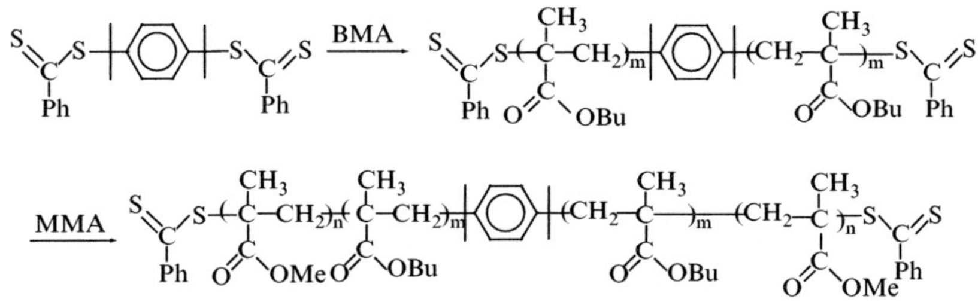

Scheme 3.7

the hydrolyzed polymers is half the molecular weight of the polymer before hydrolysis, with no appreciable change in polydispersity, indicating equal propagation rates in the two directions of trithiocarbonate group. The second monomer, B can be polymerized by RAFT in the presence of polymer A, and triblock copolymers are formed as shown in **Scheme 3.8**.

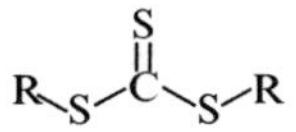

RT – 11 : R = – Ch_2 Ph
RT – 12 : R = – $CH(CH_3)Ph$

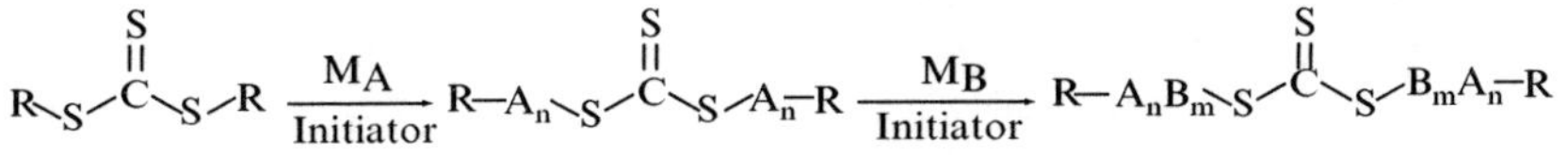

Scheme 3.8

A typical example is the synthesis of PSt-b-P*n*BA-b-PSt, which was obtained by successive RAFT polymerizations of *n*BA and St [45].

3.2.1.3 Stable Free-Radical Polymerization (SFRP)

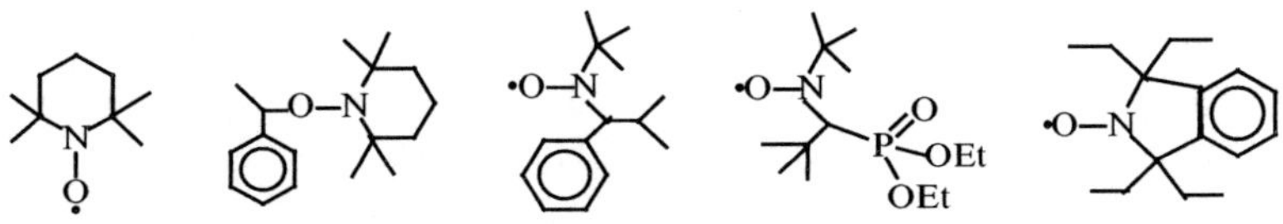

LN-1   LN-2   LN-3   LN-4   LN-5

SFRP can be performed via two approaches. The first is the addition of a stable radical, such as TEMPO (LN-1) into the conventional radical polymerization system [52]. At high temperature (e.g. 130 °C), the C–ON bond becomes unstable, releasing the nitroxide, which acts as a polymerization mediator, not as an inhibitor as at low temperature. The controlling process can be explained as shown in **Scheme 3.9**.

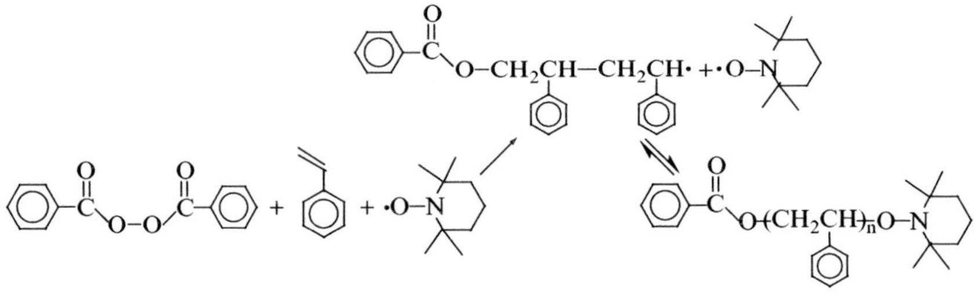

Scheme 3.9

The second method is to use unimolecular initiators, e.g. LN-2 in the polymerization system [53,54]. The C–O bond of LN-2 is expected to be thermolytically unstable and decompose on heating to give an initiating radical, i.e. the α-methyl benzyl radical as well as the mediating nitroxide radical. Following initiation the polymerization proceeds via propagation and reversible termination to give a polymer. However, there are a large number of problems with the use of TEMPO as a mediating radical. This includes the necessity to use high polymerization temperatures (125–145 °C), long reaction times (24–72 h) and an incompatibility with many important monomer families [7]. To overcome these deficiencies, it is apparent that changes in structure of the nitroxide are needed. The most significant breakthrough in the design of improved nitroxides was the use of alicyclic nitroxides, such as LN-3 to LN-5 [55–57]. The main difference in structure is the presence of a hydrogen atom on one of the α-carbons in contrast to the two quaternary α-carbons present in TEMPO. These nitroxides have subsequently been shown to be vastly superior to the original TEMPO derivatives, and can be used in the controlled radical polymerization of a wide variety of monomer families, such as acrylates, acrylamides, 1,3-dienes and acrylonitrile-based monomers. The universal nature of these initiators overcomes many of the limitations typically associated with nitroxide-mediated systems and leads to a level of versatility approaching ATRP- and RAFT-based systems. This advance benefits enormously the synthesis of a variety of block copolymers.

Block copolymers can be readily obtained by nitroxide-mediated polymerization of the first monomer to give the starting block, which is either isolated or used *in situ* [55]. The second monomer is then added, with or without the presence of a solvent to aid solubility, and on heating, the second block is grown. This is the same as ATRP and RAFT processes with one interesting feature that the first block can be characterized and stored before the controlled radical polymerization of the second monomer. Using this method, a variety of block copolymers, such as LNB-6 to LNB-9, were prepared [58–61].

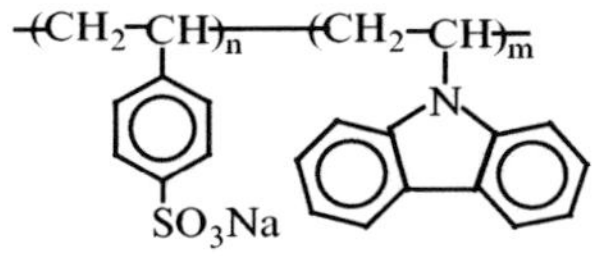

LNB-6

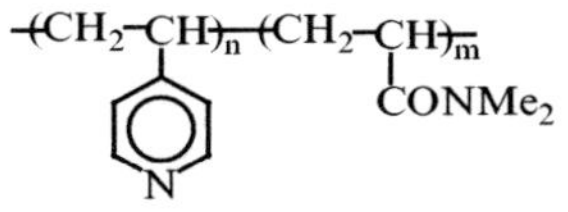

LNB-7

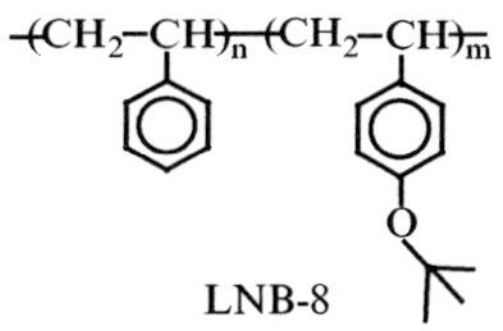

LNB-8

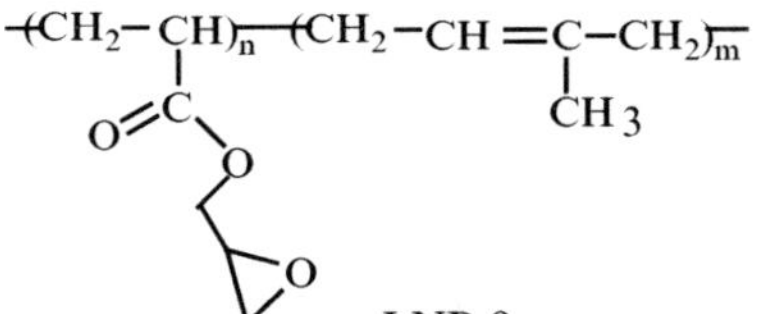

LNB-9

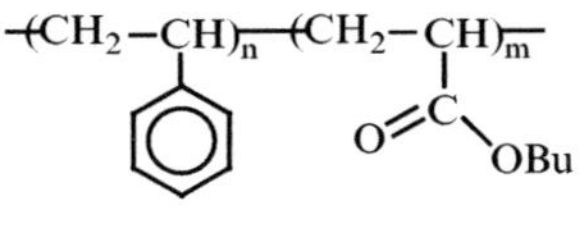

LNB-10

Similar to ATRP, nitroxide-mediated polymerizations suffer from a monomer-sequence issue when preparing specific block copolymers. In the synthesis of styrene-acrylate block copolymers, if a starting PSt macroinitiator is used to initiate the polymerization of nBA, a block copolymer, LNB-10 with a significant low molecular weight shoulder was observed [55]. This must be related to the relative rates of polymerization and initiation for St and nBA. However, when the starting PSt block is used as macroinitiator in the polymerization of isoprene, its initiating efficiency is extremely high, leading to a well-defined block copolymer, LNB-11, without homopolymer or low molecular weight polymer contamination. (**Scheme 3.10**).

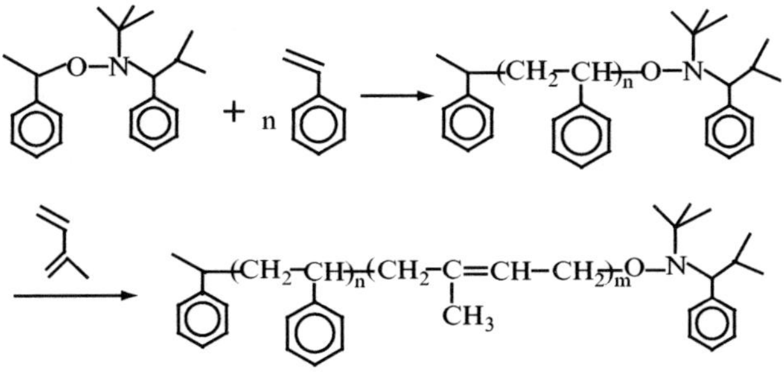

LNB-11

Scheme 3.10

The reverse synthetic strategy, that is the nitroxide-mediate polymerizations of acrylate to give the acrylate block followed by the polymerization of St, has been successfully used in the preparation of well-defined block copolymers with levels of control comparable to the ATRP procedure. For example, an alkoxyamine functionalized P*n*BA block has been used to polymerize St, resulting in a block copolymer, LNB-12 with $M_w/M_n = 1.06–1.19$, with no detectable amount of unreacted P*n*BA.

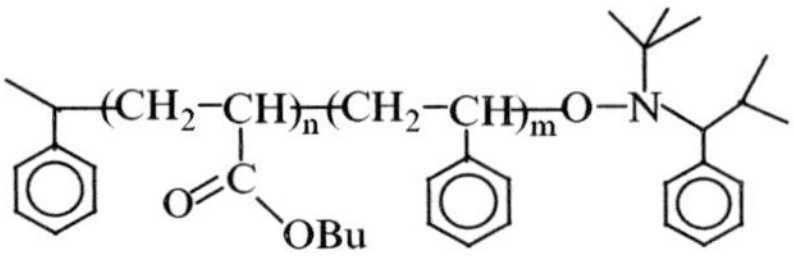

LNB-12

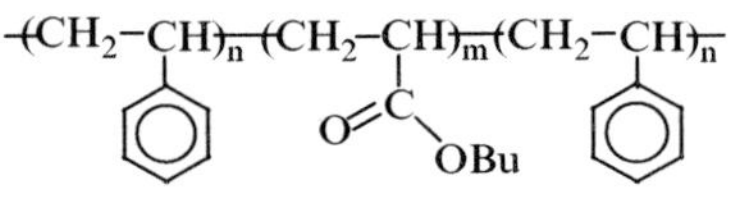

LNB-13

Recent papers reported the synthesis of well-defined ABA triblock copolymers LNB-13 by sequential polymerization of the two monomers. When using PSt-b-P*n*BA as macroinitiator, the polymerization conditions required must be carefully selected [62]. High blocking efficiency could be achieved while growing the third PSt block from PSt-b-P*n*BA, but the same result could not be obtained for the growth of the second P*n*BA block from PSt. Both the unfavorable kinetics of cross-addition and the incompatibility between the growing P*n*BA and the PSt precursor were found to affect the blocking process (blocking efficiency between 0.7–0.8). The method to solve this problem is adding *n*BA into the nitroxide-mediated radical polymerization system containing 10% of residual St because the presence of residual St will help to curb the rate of growth of the second block, giving in turn enough time for the first block to initiate the growth of the second block [62].

3.2.1.4 Photoinduced Controlled Radical Polymerization

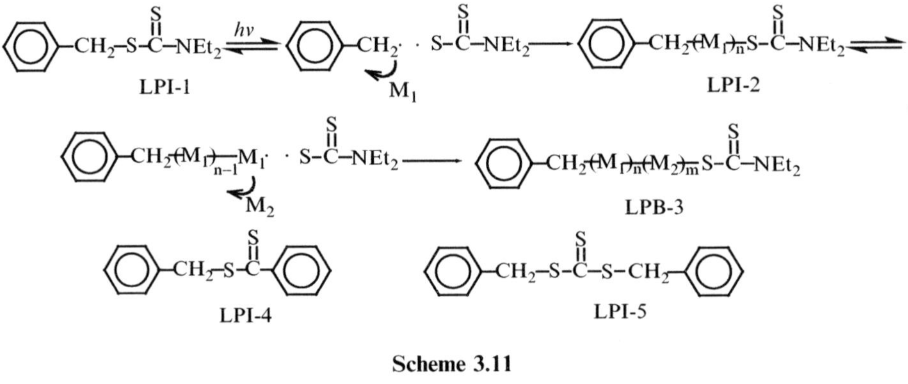

Scheme 3.11

The C–S bond in dithiocarbamates compounds is photosensitive, and can be homolytically split to form an active carbon radical and a less-reactive sulfur radical under UV irradiation [63]. For example, as shown in **Scheme 3.11**, the compound LPI-1 serves as an excellent photoiniferter because photochemical reaction of LPI-1 produces a benzyl radical similar to the propagating radical of PSt. When monofunctional end-reactive polymer LPI-2 was used in the photopolymerization of the second monomer, an AB-type block copolymer LPB-3 was obtained. However, block copolymers with narrow MWD are not obtained with dithiocarbamate initiators [64–66]. The probable reason is that the sulfur radical can initiate the polymerization [67] although the initiation is not so efficient. Increasing the stability of the sulfur radical should reduce its initiating ability. When the $-NR_2$' group is replaced with a phenyl group, the sulfur radical is more stable because of the resonance effect of the benzene ring. Therefore benzyl dithiobenzoate (LPI-4) and dibenzyl trithiocarbonate (DBTTC, LPI-5) have been used in the polymerizations of St, MA, MMA and other monomers, to give polymers with well-controlled molecular weight and narrow MWD [68–73]. As for other controlled radical polymerizations, the block copolymers can be prepared by sequential polymerizations of the monomers. For example, an ABA triblock copolymer, PSt-*b*-PMA-*b*-PSt, was prepared according to **Scheme 3.12** [69]. Under UV or γ-ray irradiation, DBTTC was excited, and the C–S bond was homolytically split into a benzyl radical and a stable benzyl trithiocarbonate radical. The former radical initiated the polymerization of St to give propagating radicals, which can be scavenged by the latter stable sulfur radicals, forming dormant chains. The fast equilibrium between dormant species and propagating radicals is established, which keeps the concentration of active radicals low, leading to control of MW and MWD. Since the photosensitive trithiocarbonate group is located in the middle of the

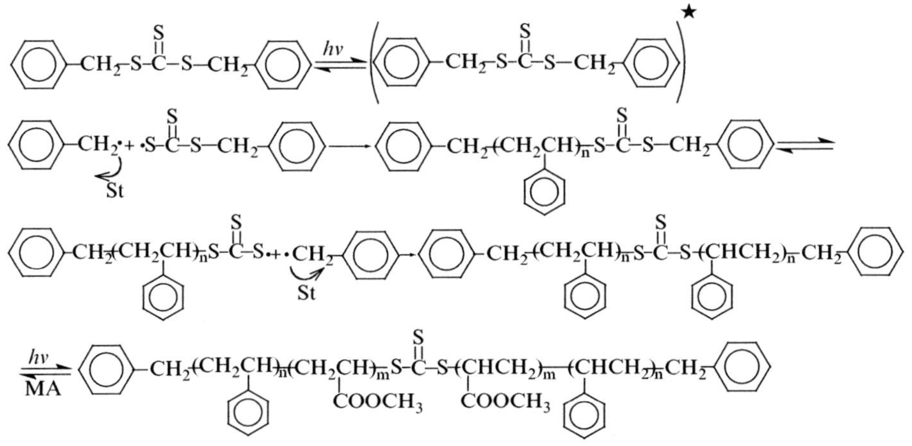

Scheme 3.12

PSt chain, the PSt with $M_w/M_n = 1.35$ obtained can be used as a macroinitiator in the polymerization of MA. A PSt-*b*-PMA-*b*-PSt triblock copolymer with $M_w/M_n = 1.18$ was obtained (**Scheme 3.12**).

This method can be applied to prepare block copolymers with more complicated structures. For example, in order to prepare comb-shaped copolymers with a handle as shown in **Scheme 3.13** [74], the following polymerization steps were adopted. The first step is to create a comb polymer by the homopolymerization of poly(tetrahydrofuran) acrylate (PTHFA) at room temperature using LPI-6 as initiator. Under ^{60}Co irradiation, LPI-6 was excited and homolytically decomposed to form an ethyl butanoate radical and a stable sulfur radical. The former radical initiated the polymerization of PTHFA. The propagation reactions are the same as that shown in **Scheme 3.12**. After isolation and purification, the PTHFA-SC(S)Ph polymer obtained was used as macroinitiator in the successive polymerization of MMA at room temperature under ^{60}Co irradiation, forming a PMMA handle block. A block copolymer LPB-8 with controlled branch chain length and number of grafts was obtained [74]. The PTHFA macromonomers were synthesized by the cationic ring-opening polymerization of tetrahydrofuran (THF) with acryloyl chloride/AgClO$_4$ as initiator. It should be noted that when other controlled radical polymerizations, such as ATRP, are attempted, the purification of poly PTHFA from macromonomer is very difficult, due to the low degree of polymerization.

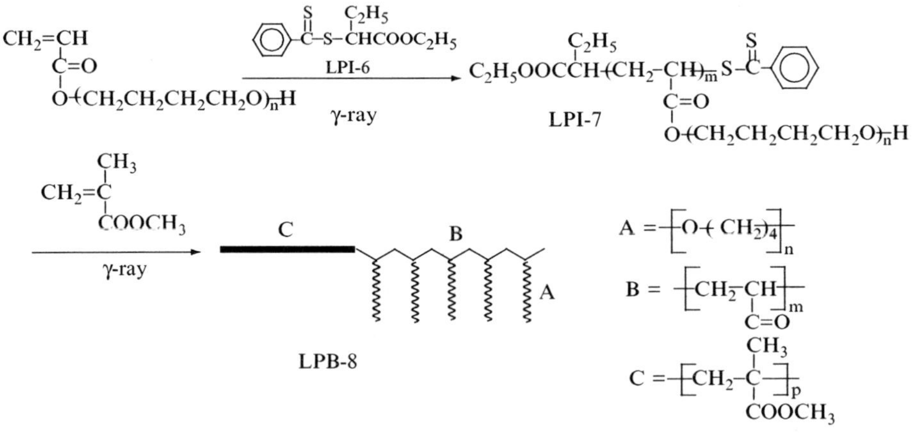

Scheme 3.13

3.2.2 MACROINITIATOR METHOD

As discussed above, macroinitiators can be obtained by isolation from metal-mediated, and nitroxide-mediated radical as well as RAFT polymerization systems. Here, we discuss only those macroinitiators prepared from functional polymers available commercially or obtained by other living polymerizations. By reacting functional polymers with functional initiators, macroinitiators are obtained. The preparation procedure can be outlined in **Scheme 3.14**.

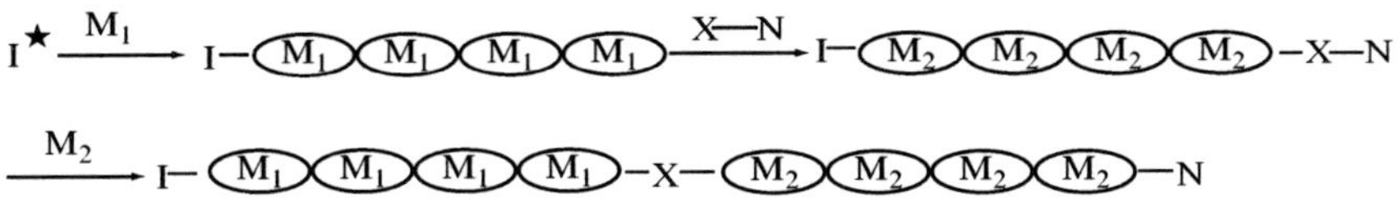

Scheme 3.14

3.2.2.1 α, ω–hydroxyl-terminated Polymers

Some commercially available α, ω–hydroxyl-terminated polyethers, polyesters and polybutadienes can react with 2-bromo- or 2-chloropropionic esters, to produce a number of macroinitiators. In order to avoid troublesome purification procedures, it is necessary to select an appropriate synthetic sequence to reduce the problems with separation from unreacted starting polymers. Poly (ethyleneoxide) (PEO) is of wide interest for a variety of applications, often covalently attached to other materials. Although PSt-*b*-PEO diblock copolymers with predictable block molecular weight and narrow MWD have been prepared by sequential anionic polymerizations of St and ethylene oxide (EO), the block copolymer PSt-*b*-PEO-*b*-PSt cannot be prepared directly by adding St into the anionic living polymerization of EO because the alkoxy anion used to polymerize EO is not basic enough to initiate the polymerization of St. By reacting hydroxy-terminated PEO with 2-bromo- or 2-chloropropionic acid, a macroinitiator can be obtained, which can be converted into the desired block copolymer by ATRP [75–77]. For example, macroinitiator LMI-2 was prepared according to **Scheme 3.15**. The esterification reaction of LMC-1 ($M_n = 2000$, $M_w/M_n = 1.05$) with 2-chloropropionyl chloride produced quantitatively LMI-2, which was used in the ATRP of St to give PEO-b-PSt block copolymer LMB-3 [76].

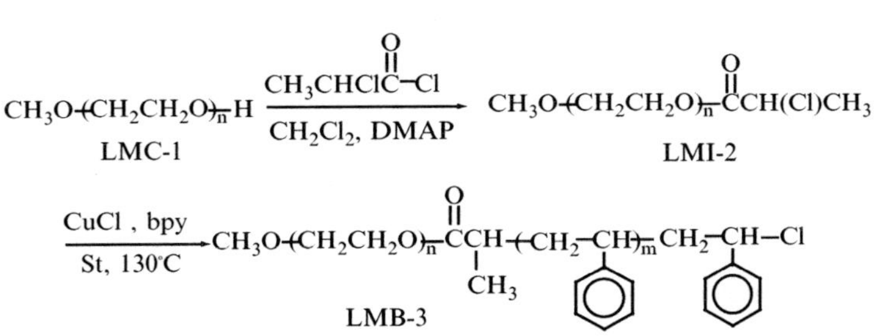

Scheme 3.15

Similarly ABA triblock copolymers, PSt-*b*-PEO-*b*-PSt have been prepared by the ATRP of St with the difunctional macroinitiator, α, ω–bis-(2-chloropropionyl) PEO in conjunction with CuCl/bpy. By reacting α, ω–hydroxyl-terminated poly(propylene oxide)s or polyesters or polybutadienes with chloroacetyl chloride in toluene, α, ω–dichloroacetyl polymers were obtained and used as macroinitiator in the ATRP of St. The series of triblock copolymers, LMB-4, LMB-5, LMB-6 were thus successfully prepared [77–80].

For nitroxide-mediated radical polymerizations and in the RAFT process, the same synthetic strategy as for ATRP can be used in the synthesis of AB and ABA block copolymers. The first step is coupling a functionalized alkoxyamine with a telechelic or monofunctional nonvinylic polymer to give a macroinitiator. This macroinitiator can be used in standard controlled free-radical polymerization procedures. This approach is best illustrated by the preparation of PEO-based block copolymers [81–84]. One example is the preparation of macroinitiator LMI-7 by the reaction of a monohydroxy-terminated PEO with sodium hydride followed by reaction with the chloromethyl-substituted alkoxy amine as shown in **Scheme 3.16**.

$$\text{ClCHCH}_2\text{(CH–CH}_2\text{)}_m\text{CH}_2\overset{\underset{\displaystyle\|}{O}}{C}\text{OCH}_2\text{CH}_2\text{O(CH}_2\overset{\underset{\displaystyle |}{CH_3}}{CH}\text{O)}_n\text{CH}_2\text{CH}_2\text{O}\overset{\underset{\displaystyle\|}{O}}{C}\text{CH}_2\text{(CH}_2\text{–CH)}_m\text{CH}_2\text{CHCl}$$

LMB-4

$$\text{ClCHCH}_2\text{(CH–CH}_2\text{)}_m\text{CH}_2\overset{\underset{\displaystyle\|}{O}}{C}\text{O(CH}_2\text{CH}_2\text{O}\overset{\underset{\displaystyle\|}{O}}{C}\text{–}\bigcirc\text{–}\overset{\underset{\displaystyle\|}{O}}{C}\text{O)}_n\text{CH}_2\text{CH}_2\text{O}\overset{\underset{\displaystyle\|}{O}}{C}\text{CH}_2\text{–CH}_2\text{–CH)}_m\text{CH}_2\text{CHCl}$$

LMB-5

$$\text{ClCHCH}_2\text{(CH–CH}_2\text{)}_m\text{CH}_2\overset{\underset{\displaystyle\|}{O}}{C}\text{O(CH}_2\text{–CH=CH–CH}_2\text{)}_n\text{–O}\overset{\underset{\displaystyle\|}{O}}{C}\text{CH}_2\text{(CH}_2\text{–CH)}_m\text{–CH}_2\text{CHCl}$$

LMB-6

The PEO-based macroinitiator (PDI = 1.05–1.10) LMI-7 can then be used to polymerize a variety of vinyl monomers, such as St, to give amphiphilic block copolymers of type LMB-8, which have accurately controlled molecular weight and very low polydispersity, 1.05–1.10 (**Scheme 3.16**) [84]. Compared to typical small molecule initiators, generally it is found that using a macroinitiator gives extremely low polydispersities because diffusion and reactivity of the macroinitiator are decreased, leading to the reduction in radical terminations and a more controlled polymerization. This phenomenon is also observed for the RAFT process. For example, the acid functional dithioester LMI-9 can be

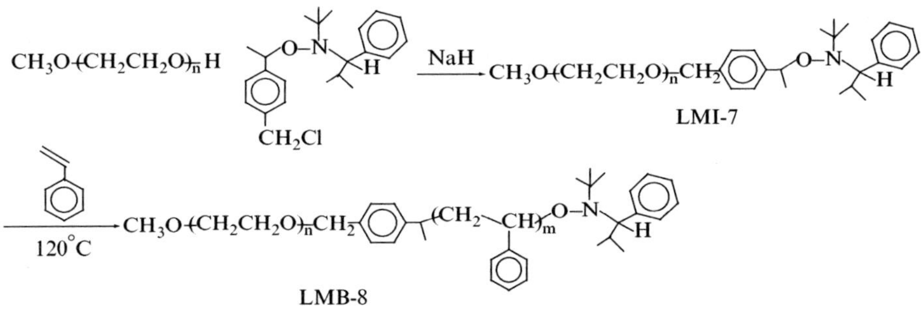

Scheme 3.16

coupled with PEO monomethyl ether with the aid of dicyclohexyl carbodiimide (DCC) to produce a polymeric dithioester LMI-10, which was then used in the block copolymerization of either St or benzyl methacrylate (BzMA) (**Scheme 3.17**). Block copolymer LMB-11 with $M_w/M_n = 1.10$ or LMB-12 with $M_w/M_n = 1.07$ were obtained, with no detectable PEO impurity in the copolymers [44].

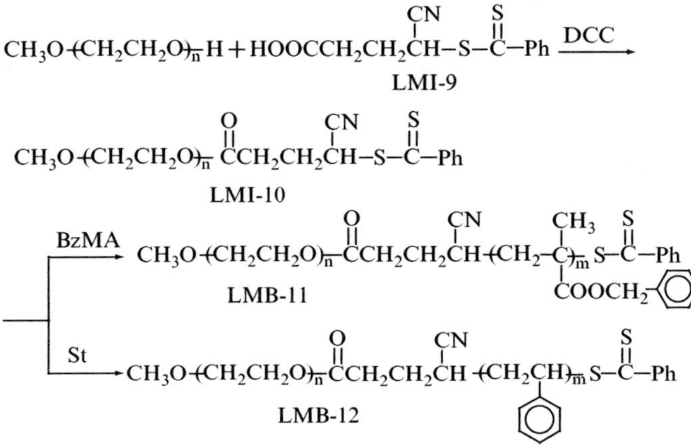

Scheme 3.17

The end hydroxyl group of PEO can be converted to an iniferter site as shown in **Scheme 3.18**. The reaction of methyl PEO with p-chloromethyl benzoyl chloride gave an intermediate LMI-13. Macroiniferter LMI-14 was obtained by subsequent reaction with $NaS-C(S)NEt_2$, and then was used in the UV polymerization of St to form diblock copolymers LMB-15 with $M_w/M_n = 1.2-1.3$ [85].

Block copolymers containing both a polyolefin block and a poly(St-co-MAh) block are useful as blend compatibilizers or as adhesion promoters

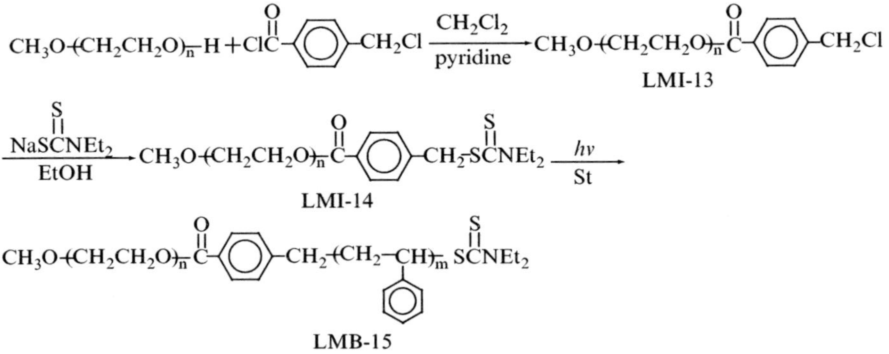

Scheme 3.18

for polyolefin coatings on polar substrates such as metals. The polyolefin block has been introduced in the polymerization in the form of a macromolecular transfer agent. A commercially available copolymer of ethylene and butylene containing one hydroxyl end group (Kraton L-T203) was treated with LMI-16 to yield a polyolefinic RAFT agent LMI-17. It was used as a macromolecular transfer agent in the RAFT polymerizations of St or St/MAh, to produce block copolymers, LMB-18 and LMB-19 with very narrow MWD (M_w/M_n = 1.12–1.20) [86] (**Scheme 3.19**).

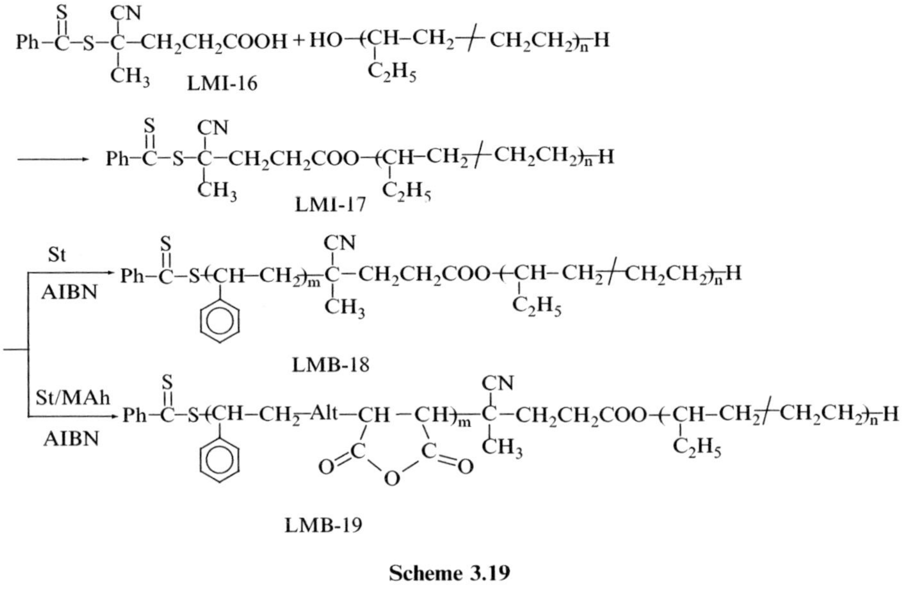

Scheme 3.19

3.2.2.2 Other Mono- and Difunctional Polymers

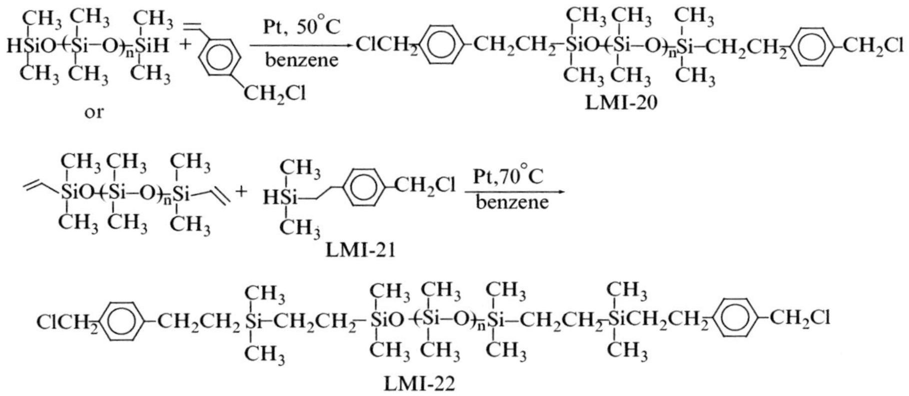

Scheme 3.20

As well as the hydroxy group, other end-functional groups can also be used in the preparation of macroinitiators. One example is the addition reaction of Si–H with a vinyl group. Poly(dimethylsiloxane)-b-PSt (PDMS-b-PSt) copolymers are of scientific and technological interest due to the combination of the rubbery, low T_g, PDMS with the glassy PSt. Hydrosilylation of hydride end-capped PDMS with 4-vinyl benzyl chloride yields a macroinitiator LMI-20. Alternatively vinyl-terminated PDMS and siloxane compound LMI-21 are reacted to form macroinitiator LMI-22. Both of these macroinitiators were used in the bulk or solution ATRP of St to give ABA triblock copolymers PSt-b-PDMS-b-PSt (see **Scheme 3.20**). The disadvantage of the method is that the molecular weight distribution of the block copolymers is broad due to the high polydispersity of the starting PDMS ($M_w/M_n > 2.0$) [87].

Using hydrosilylation reactions of vinyl groups with Si–H, nitroxide initiator sites can be introduced into the PDMS chains as shown in **Scheme 3.21** [88]. In this way, macroinitiator LMI-23 ($M_w/M_n = 1.20–1.22$) was used to synthesize AB diblock copolymer PDMS-b-PSt, LMB-24 ($M_w/M_n = 1.25–1.35$) by bulk stable free-radical polymerization of St.

Using the dechlorination reaction between dichloro-capped poly(methyl phenylsilylene) (PMPS) and p-(chlorodimethylsilylene propylene) benzyl chloride (LMC-25), the macroinitiator, LMI-26 was obtained, and used in the ATRP of St to give a block copolymer, LMB-27 as shown in **Scheme 3.22** [89].

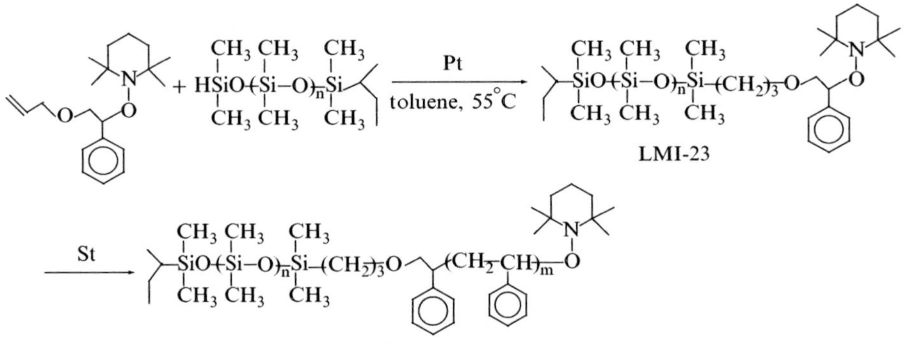

Scheme 3.21

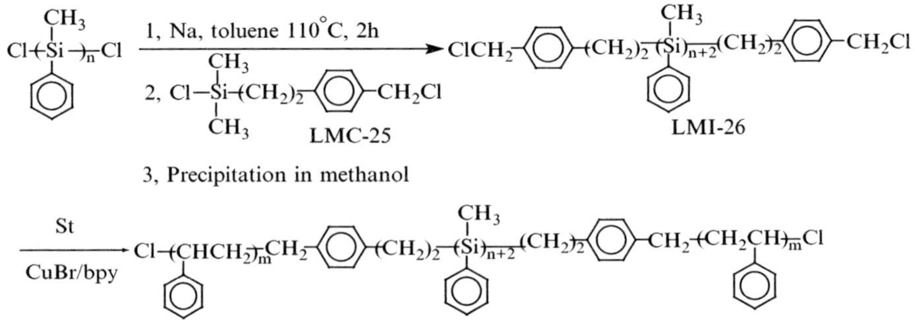

Scheme 3.22

3.2.3 COMBINATIONS WITH OTHER LIVING POLYMERIZATIONS

Since the number of monomers, and thus the resulting polymer structures, are limited by any of the specific living polymerization techniques, appropriate combination of different polymerization mechanisms can lead to a variety of new and useful polymeric materials. Therefore combinations of controlled radical polymerizations and other polymerizations applied to synthesize block copolymers have been developed. Generally, polymers with active sites, such as carbon-halogen or nitroxide or dithioester terminal groups, are synthesized by other living polymerizations, and the product is further used to initiate the controlled radical polymerization. In many cases, this method is essentially a variant of the macroinitiator method discussed above. However, in some cases, these kinds of macromolecules do not act as initiators, and may act as transfer agents. For example, an AB-type amphiphilic block copolymer, CLB-2 was prepared by RAFT polymerization of 2-(N-dimethylamino)ethyl methacrylate

(DMAEMA) using AIBN as initiator and PBzMA as a transfer agent (**Scheme 3.23**) [44]. Here, CLT-1 acts as a macromolecular transfer agent, not as a macroinitiator.

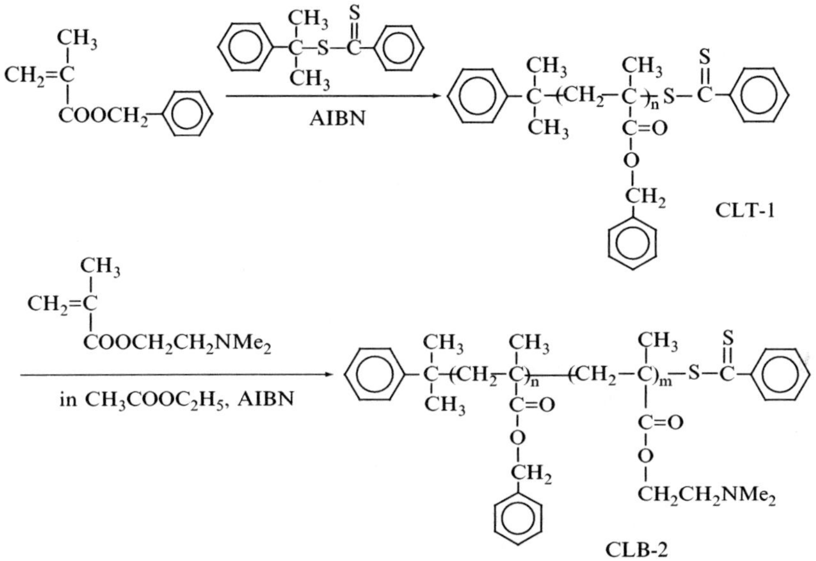

Scheme 3.23

The combination of controlled radical polymerizations with other living polymerizations can be realized by mechanism transformation. The transformation of an active chain end into another type of initiating site has been extensively used in the synthesis of block copolymers [90]. Well-defined block copolymers have been prepared by the transformation of initiating sites from living anionic to living cationic [91–94], from living cationic to living anionic [95–97], and from living coordination to living cationic polymerization [98]. Here we describe the recent advances in the synthesis of block copolymers via combination of controlled radical polymerizations with other living polymerizations.

3.2.3.1 Ionic Ring-opening Polymerization

Ionic ring-opening polymerizations of cyclic ether, acetal, ester and siloxane monomers give polymers with controlled molecular weights and well-defined terminal structure that are thus suitable for the synthesis of block copolymers when coupled with controlled radical polymerizations.

One synthetic strategy is transformation of controlled radical polymerizations into ring-opening polymerization without any modification of terminal groups. Generally, the terminal moiety of polymers obtained from ATRP is bromine or chlorine, which can be used as the initiator in a cationic ring-opening polymerization. For example, telechelic bromine-terminated PSt, Br-PSt-Br, CLI-3 can be prepared by the bulk ATRP of St using 1,2-bis(2'-bromobutyryloxy)ethane (BBrBE) as initiator in the presence of bpy and CuBr at 110 °C [99], followed by cationic ring-opening polymerization of THF in conjunction with silver perchlorate. The result is a PTHF-*b*-PSt-*b*-PTHF triblock copolymer, CLB-4, as shown in **Scheme 3.24** [99–101].

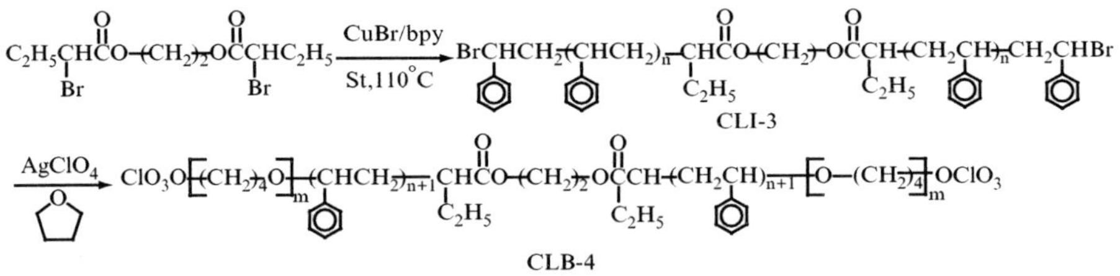

Scheme 3.24

In general, the initiation efficiency in this polymerization depends greatly on the initiation temperature, and model experiments have revealed that the β elimination is reduced from 63 to 33 to < 5%, when the initiation temperature decreases from room temperature to 0 °C to −78 °C. This conclusion, however, is based only on GPC analysis without any further evidence, such as high-resolution NMR spectra of the polymerization product [91,92,94]. Also, the decomposition of the terminal ester group induced by traces of acids as shown in **Scheme 3.25** should be considered [102]. Therefore, in order to get pure block copolymers, it is necessary to use highly purified $AgClO_4$, and carry out the polymerization at a low temperature [99].

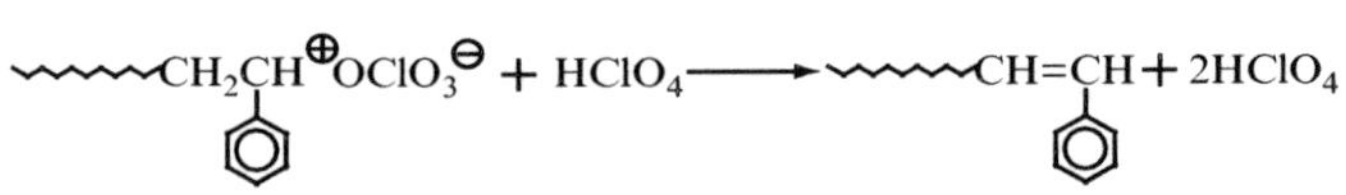

Scheme 3.25

For SFRP and RAFT polymerizations, directly using the polymers in the next cationic ring-opening polymerizations without any modification of the terminal group is impossible since the terminal groups are nitroxide and dithioester groups, respectively. Several transformation reaction steps are then required, and low transformation efficiency results.

Halogen-capped PSt and poly(p-methoxy styrene) (PMOSt) were modified with $Ph_2I^+PF_6^-$, and the resultant cationic species was subsequently used to initiate cationic ring-opening polymerization of cyclohexene oxide to produce CLB-5 and CLB-6, respectively [103].

Another synthetic strategy is the transformation of ring-opening polymerization to controlled radical polymerization. The transformation of the active site from one to the other can be performed in the polymerization system or after separation and purification. The active sites can be obtained by reacting functionalized polymer with appropriate reagents. Various block copolymers containing polyether (or polyester, or polysiloxane or polyacetal) blocks and vinyl polymer blocks have been prepared by this method.

Transformation of the cationic ring-opening polymerization of THF into ATRP of St, acrylates and methacrylates has been used to produce various block copolymers [104,105]. The ABA-type block copolymers CLB-7 to CLB-9 were prepared via termination of telechelic living PTHF with sodium 2-bromoisopropionate, followed by ATRP of St, or MA or MMA [104]. The transformation reaction was performed by adding sodium 2-bromoisopropionate into the cationic polymerization system of THF. Due to the existence of tertiary oxonium ions, the attack of 2-bromoisopropionate anion on the α-carbon of the oxonium ion produces 2-bromoisopropionate-terminated PTHF as shown in **Scheme 3.26** [104,105].

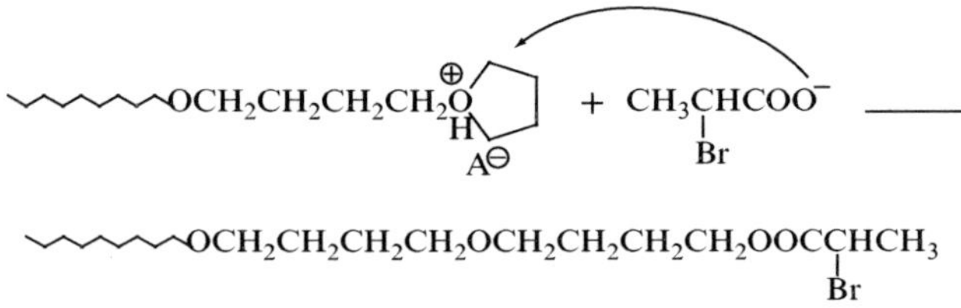

$$\text{\textasciitilde\textasciitilde\textasciitilde OCH}_2\text{CH}_2\text{CH}_2\text{CH}_2\overset{\oplus}{\text{O}} \underset{\text{A}^{\ominus}}{\overset{\text{H}}{\diagup}} + \text{CH}_3\underset{\text{Br}}{\text{CHCOO}^-} \longrightarrow$$

$$\text{\textasciitilde\textasciitilde\textasciitilde OCH}_2\text{CH}_2\text{CH}_2\text{CH}_2\text{OCH}_2\text{CH}_2\text{CH}_2\text{CH}_2\text{OOCCHCH}_3 \atop \text{Br}$$

Scheme 3.26

In order to obtain a well-defined block copolymer, the enhancement of transformation efficiency by selecting appropriate transformation reaction conditions is very important. The transformation efficiency for cationic ring-opening polymerizations of cycloacetals (e.g. trioxane, 1,3-dioxolane, and 1,3-dioxepane (DOP) etc), is lower than that for the cationic polymerization of THF because of the existence of tertiary oxonium and carbonium ions in the polymerization system [106]. In this case, use of a double-headed initiator may be the best method to prepare diblock copolymers such as P(St-b-DOP). It is well known that the cationic ring-opening polymerization of cyclic acetals is characterized by inter- and intra-molecular chain-transfer reactions, resulting in the formation of cyclic species. When DOP polymerization was initiated with triflic acid in the presence of ethylene glycol, the reaction proceeded mainly according to the active monomer (AM) mechanism [107]. By extending this technique to the synthesis of block copolymers, a double-headed initiator, such as 2-hydroxyethyl-2'-bromobutyrate (HEBrB) can be used in ATRP to give a hydroxy-terminated polymer. For example, PDOP-b-PSt diblock copolymers can be prepared by ATRP of St with HEBrB/CuBr/bpy as the initiator system, followed by the living cationic ring-opening polymerization of DOP with triflic acid as catalyst and hydroxy-terminated PSt as transfer agent, as shown in **Scheme 3.27** [108].

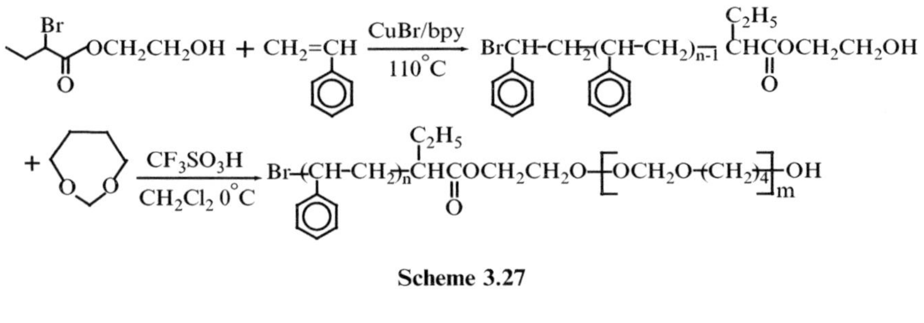

Scheme 3.27

The presence of a hydroxyl group in the initiator does not reduce the control over the living nature of the ATRP of vinyl monomers [109–113]. No loss of hydroxy group of initiator during the ATRP was found, indicating that each molecule of the polymer obtained contains one hydroxyl group [108].

Living ring-opening polymerization of ε-caprolactone (CL) with aluminum alkoxide or alkylaluminum can be combined with nickel-catalyzed living radical polymerization for the synthesis of linear and dendrimer-like star block copolymers such as CLB-10 to CLB-14 [109,110,114,115]. These block copolymers were first prepared via living radical polymerization with CBr_3CH_2OH, where the C–Br bond is a radical-initiating site and the hydroxyl group is used for the subsequent ring-opening polymerization. The two processes may also be reversed [109]. More interestingly, the two living polymerizations can be carried out simultaneously, because MMA and ε-caprolactone undergo parallel growth initiated by the $CBr_3CH_2OH/(PPh_3)_2NiBr_2/Al(O-i-Pr)_3$ system [110]. The aluminum compound might have a dual function, one as a catalyst for anionic ring-opening polymerization and the other as an additive enabling the Ni catalyst to facilitate the living radical process [116].

Block copolymers containing a poly(ε-caprolactone) block can also be prepared by a combination of living ring-opening polymerization with nitroxide-mediated free-radical polymerization using a hydroxy-substituted alkoxy-amine as a double-headed initiator, such as CLI-15 [109]. The primary hydroxy group was used as the initiating group for the ring-opening polymerization of caprolactone to give an alkoxyamine-terminated macroinitiator, which could then be used to initiate the controlled radical polymerization of St, yielding the well-defined block copolymer, CLB-10. In a similar vein, controlled radical polymerization has been combined with cationic ring-opening polymerization of oxazolines, or with anionic ring-opening polymerization using the same multifunctional initiator [117,118]. This trifunctional system has been shown to be highly effective, leading to well-defined block copolymers and can even be combined into a one-pot, one-step block copolymerization by simultaneous free-radical and either cationic ring-opening or anionic ring-opening procedures [118].

Another method of preparing block copolymers containing one or more polyester blocks is to combine free-radical ring-opening polymerization and controlled radical polymerization [119–121]. Controlled radical polymerizations of 2-methylene-4-phenyl-1,3-dioxolane (MPDO) and 5,6-benzo-2-methylene-1,3-dioxepane (BMDO) using ATRP have been investigated. Both show living characteristics, in particular constant concentration of propagating radicals and controlled molecular weight as well as narrow molecular weight distribution. BMDO undergoes complete ring-opening polymerization [122,123], but PMPDO polymerized via both addition and ring-opening mechanisms was obtained [124]. By using the SFRP method, it is possible to perform free-radical ring-opening polymerization of MPDO with quantitative ring-opening to produce a polyester with $M_w/M_n = 1.50$ [125].

Extending controlled radical polymerization methods to the block copolymerization of cyclic vinyl ethers using commercially available monomers is of great interest. Thus, macroinitiator, PSt-Br (or PMMA-Br, or PMA-Br) was prepared by ATRP, then used in the ATRP of cycloethers in chlorobenzene to produce diblock copolymers such as CLB-16 with $M_w/M_n = 1.18-1.40$ as shown in **Scheme 3.28** [119,121].

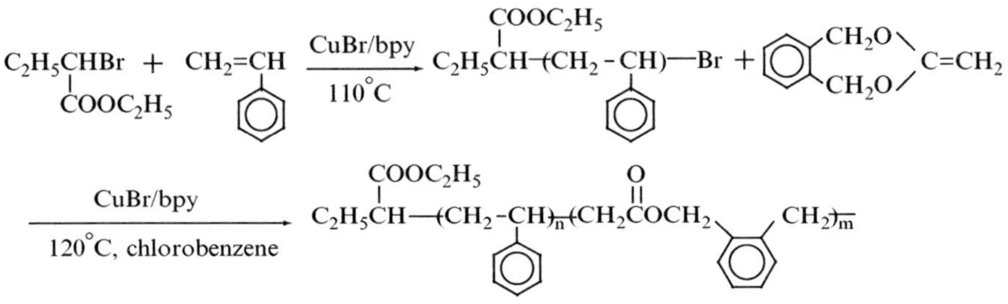

$$C_2H_5CHBr + CH_2=CH \xrightarrow[110°C]{CuBr/bpy} C_2H_5CH\text{--}(CH_2\text{-}CH)\text{--}Br + \text{benzene-}CH_2O\text{/}CH_2O\text{-}C=CH_2$$
$$\text{(with COOC}_2H_5\text{ groups)}$$

$$\xrightarrow[120°C, \text{chlorobenzene}]{CuBr/bpy} C_2H_5CH\text{--}(CH_2\text{-}CH)_n\text{--}(CH_2COCH_2\text{--benzene--}CH_2)_m$$

CLB-16

Scheme 3.28

3.2.3.2 Anionic Vinyl Polymerization

The carbanionic terminal groups in a living anionic polymerization can be transformed into carbon–halogen bonds suited for radical generation. Two approaches for the transformation mechanism have been proposed. First, ethylene oxide or one of its derivatives is used to quench the living anionic polymerization, resulting in hydroxyl-terminated polymers. After treatment with an appropriate reagent, a carbon–halogen terminal group is formed and used in a subsequent controlled radical polymerization [126–130]. One example is the preparation of block copolymers containing a PSt or poly(isopropylene) (PIP) block as shown in **Scheme 3.29**. The polystyrylithium (PSt$^-$Li$^+$) living anion, which was prepared by living anionic polymerization of St with BuLi as initiator, was converted into a bromine-terminated chain by reaction with styrene oxide [127], or ethylene oxide [128] followed by treatment with 2-bromoisobutyryl bromide. The same method can be adopted for the living anionic polymerization of isoprene [127]. Such macroinitiators can be employed in the controlled radical polymerizations of methacrylates, acrylates and styrene in the presence of copper catalysts to give block copolymers with narrow MWDs ($M_w/M_n = 1.1-1.2$) [127–129]. The reaction of the hydroxyl-terminated polymer with SOCl$_2$ produced a chlorinated end group that can be used, for example, in the subsequent polymerization of 2-vinylpyridine (2VP) to form PSt-*b*-P2VP copolymers [126].

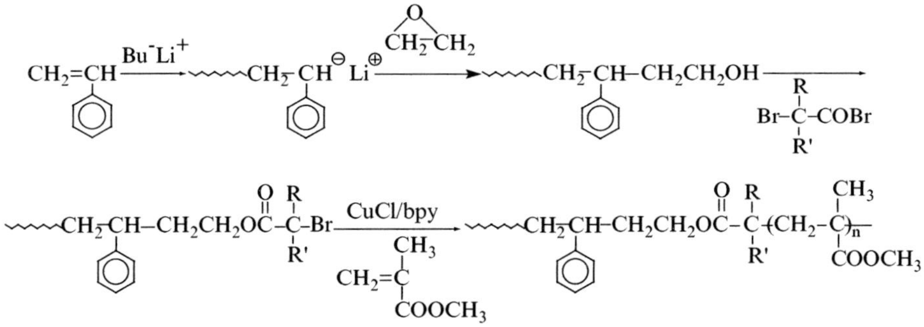

Scheme 3.29

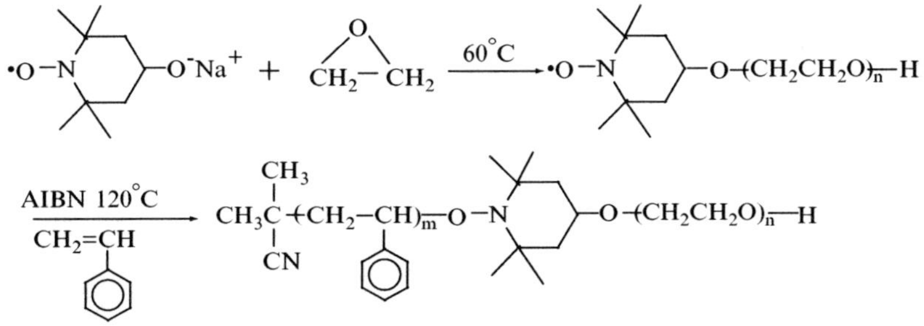

Scheme 3.30

The second approach for creating halogen-terminated polymers involves quenching living anionic polymerization by α-methylstyrene (αMSt), followed by addition of liquid bromine [130]. Similarly, poly(isoprene)-b-PSt (PI-b-PSt) block copolymers may be prepared by quenching the living anionic polymerization of isoprene with 1-(9-phenonthryl)-1-phenylethylene followed by addition of excess α,α'-dibromo-p-xylene, which leaves a C-Br terminal moiety effective for the copper-catalyzed radical polymerization of St [131].

There are fewer reports on the preparation of block copolymers via the combination of anionic polymerization with nitroxide-mediated syntheses [132,133]. As shown in **Scheme 3.30**, the reaction product of sodium with 4-hydroxyl-TEMPO initiated the anionic polymerization of ethylene oxide at 60 °C in THF solution. After treatment with methanol, TEMPO-terminated PEO was obtained, and then used in the nitroxide-mediated radical polymerization of St at 120 °C resulting in block copolymers of type CLB-17 [133]. Another method is the transformation of anionic polymerization into nitroxide-mediated radical polymerization. A poly(butadienyl)lithium solution in cyclohexane was prepared by living anionic polymerization with *sec*-butyl lithium as initiator. Into this solution, a 1.2-fold excess of CLI-18 was added. After reaction for 24 h, the TEMPO-terminated polymer CLI-19 was isolated and purified by precipitation to ensure the removal of any unreacted CLI-18. It was then used as a macroinitiator in the nitroxide-mediated radical polymerization to prepare block copolymers as shown in **Scheme 3.31** [132].

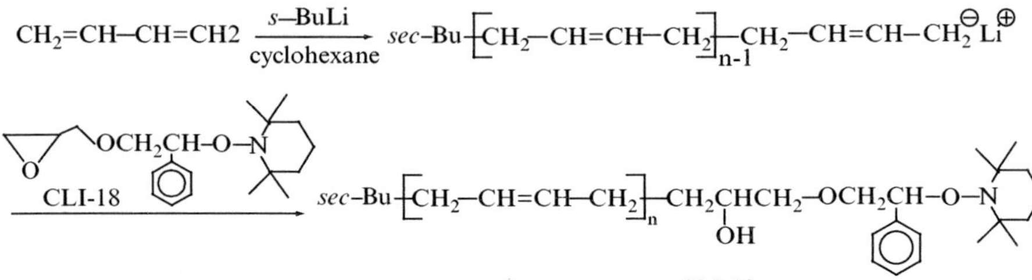

Scheme 3.31

3.2.3.3 Cationic Vinyl Polymerization

The living cationic polymerizations of some vinyl monomers with an initiating system based on alkyl halides and Lewis acids always leads to halogen-terminated polymers [134–136], which can be used as macroinitiators in subsequent ATRP without any modificaton [136–138]. For example, chloride-terminated polystyrenes, such as CLI-20, were obtained by living cationic polymerization of St using the 1-PhEtCl/SnCl$_4$ initiating system in the presence of tetrabutylammonium chloride at $-15\,°C$ in methylene chloride as shown in **Scheme 3.32**.

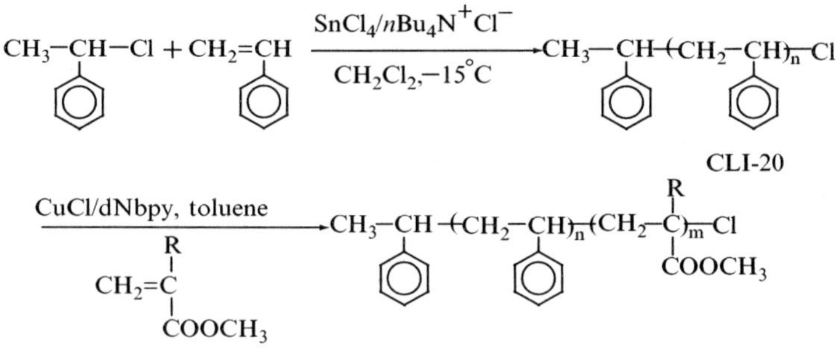

CLB-21 R=CH$_3^-$
CLB-22 R=H$^-$

Scheme 3.32

After isolation and purification, PSt-Cl was obtained and used to initiate homogeneous ATRP of MA or MMA in the presence of catalyst based on CuCl and 4,4'-di-5-nonyl-2,2'-bipyridine (dNbpy). Block copolymers CLB-21 and CLB-22 with $M_w/M_n = 1.10-1.57$ were obtained [137].

Adopting a similar synthetic technique, tethered PSt-*b*-polyacrylate brushes including PMMA, PMA and PDMAEMA terminal blocks were synthesized on flat silicate substrates [138, 139].

Chloride-capped poly(iso-butylene) (PIB) prepared via cationic polymerization was also used as macroinitiator for the copper-catalyzed radical polymerization of acrylates, methacrylates and St [140–143]. The C–Cl moiety at the end of the PIB chain cannot initiate living radical polymerization due to its lower activity for redox reactions, but it can be modified into an active form by inserting several units of St. Since the cross-reaction from the living PIB chain to St is a relatively rapid process, it is possible to add only a few St units to the PIB chain [144]. The resulting 1-chloro-1-phenylethyl end groups are potential initiating sites for ATRP of many vinyl monomers, leading to a variety of new block copolymers, such as CLB-23 (see **Scheme 3.33**), [140,142] which cannot be prepared by any direct polymerization techniques, because isobutylene can be polymerized only by cationic polymerization.

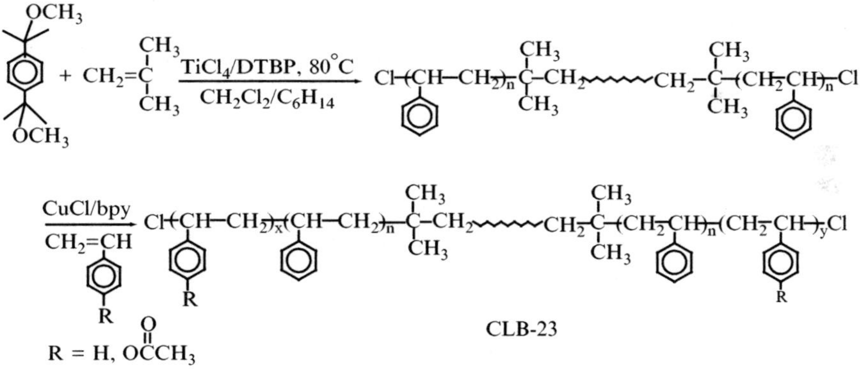

Scheme 3.33

The chloride end-function can also be transformed into a hydroxyl function, first by the quantitative conversion of the $-Cl$ to $-CH_2\text{-}CH = CH_2$, followed by oxidative conversion to give $-OH$. The $-OH$ termini were quantitatively esterified to the sought $-OC(O)C(CH_3)_2Br$ functions by the reaction of PIB-OH with 2-bromopropionyl or 2-bromoisobutyryl halide [143], and the macroinitiator obtained was used in ATRP for synthesizing block copolymers (see **Scheme 3.34**).

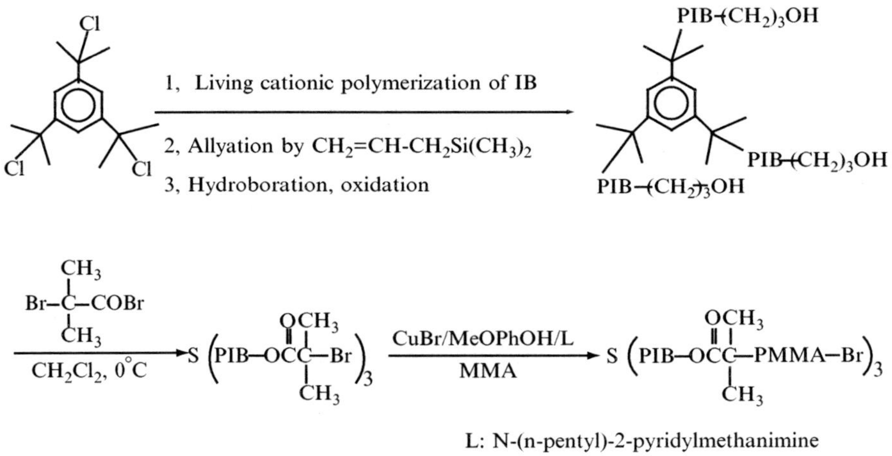

Scheme 3.34

3.2.3.4 Other Living Polymerizations

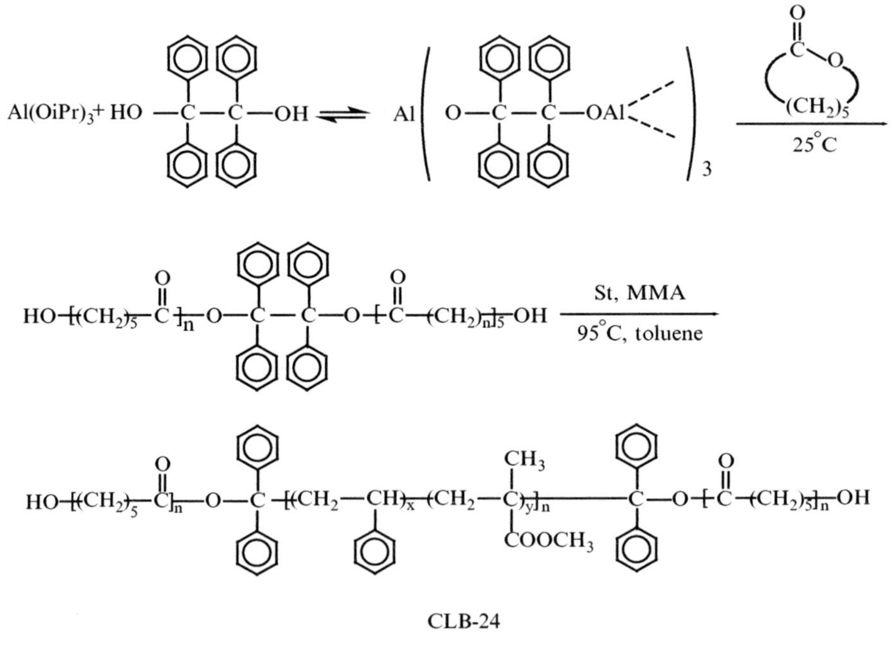

CLB-24

Scheme 3.35

As mentioned above, transformation polymerizations are efficient methods for the synthesis of block copolymers, which allow combinations of various polymerization mechanisms. Many monomers with different structures can be polymerized to yield block copolymers with novel properties. In addition to the transformation methods discussed above, other transformation polymerizations, such as the combination of coordination or photoinduced polymerization with controlled radical polymerizations, or the combination of two different controlled radical polymerizations have been reported [145–150]. One method involves the application of a bifunctional initiator, in which the two functional groups are stimulated either in sequence or concurrently to initiate the polymerization of two kinds of monomers via different polymerization mechanisms. It is well known that aluminum trialkoxides effectively initiate the polymerization of ε-caprolactone (CL) via coordination mechanisms [149, 150]. They can also be used as promoters for the polymerization of CL initiated with diols. Here, the active species are the reaction products of trialkoxides and diols as shown in **Scheme 3.35** [145]. In this example, after benzopinacole (BP) was treated with aluminium tri-isopropoxide, the CL solution in toluene was added, and the polymerization was carried out at 25 °C. Around 40% of the resulting PCL did not contain BP group, which does not affect the subsequent polymerization, and this was then used in the controlled radical polymerization of St and MMA at 95 °C. The resulting PCL-b-P(MMA/St)-b-PCL block copolymer structure, CLB-24 is presented in **Scheme 3.35** The PCL without the BP group can be removed with hot ethanol after the copolymerization of St and MMA.

Controlled radical polymerization has advantages over other living polymerization methods, including less sensitivity toward impurities present in the polymerization system and applicability to a wide range of monomers. The polymerization ability of the monomer with functional groups is related to the type of controlled radical polymerization. Block copolymers with desired structures can be prepared by the combination of two controlled radical polymerizations, such as combinations of photoinduced polymerization and ATRP [146], or ATRP and SFRP [147,148]. In most cases, the initiators carrying two different radical initiating sites were used in the preparation of block copolymers. In addition, the combination of conventional radical polymerization with controlled radical polymerization methods has been reported. For example, block copolymers were prepared by combining conventional polymerization with SFRP [151] and ATRP [152–154]. Photoinduced polymerization can be used in the transformation polymerization [155], although it has to be conducted at low temperature, usually room temperature, to prevent side reactions, leading to the formation of homopolymers. Combining UV-induced radical polymerization of MMA with ATRP of St produced a mixture of AB-type block copolymer and comb-like polymers. Since the termination mode of MMA polymerization is disproportionation, half of the PMMA formed by the photoinduced process should have a vinyl group at the end of the polymer chains. The macromonomers formed are thus capable of copolymerization, leading to the formation of mixtures of the block and comb-like polymers when ATRP is carried out in the subsequent step [146].

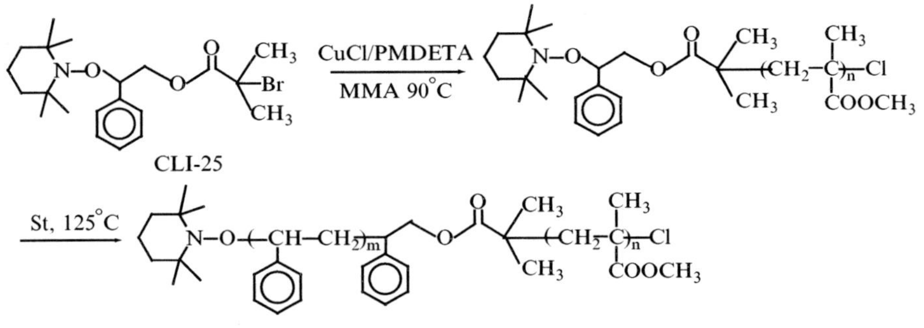

Scheme 3.36

Using novel asymmetric difunctional initiators containing TEMPO and 2-bromopropanoate or 2-bromo-2-methylpropanoate groups, block copolymers can be prepared via combination of ATRP and STRP [147,148]. For example, asymmetric difunctional initiator, CLI-25 was used in the ATRP of MMA with CuCl/N, N, N', N',N''-pentamethyldiethylene-triamine (PMDETA) as catalyst. The low initiator efficiency (i.e. 0.8) may be related to the side reactions that occur in the initiation step. Subsequently, the TEMPO-terminated PMMA was used as the macroinitiator in the nitroxide-mediated radical polymerization of St at 125 °C. A series of block copolymers, PMMA-*b*-PSt [147], P*t*BA-*b*-PSt [147] and P*t*BA-*b*-PMMA-*b*-PSt [148] have been obtained, an example of which is shown in **Scheme 3.36**.

3.3 NONLINEAR BLOCK COPOLYMERS

Distinct from linear block copolymers, such as AB diblocks, ABA triblocks, etc., nonlinear block copolymers, such as star block copolymers, miktoarm stars, umbrella polymers, etc., are formed by joining linear blocks at their center. Nonlinear block copolymers have attracted a great deal of industrial attention due to their potential applications as thermoplastic elastomers, tough plastics, compatibilizing agents for polymer blends, polymer micelles, etc. Academic interest arises, primarily, from the use of these materials as model copolymer systems where effects of thermodynamic incompatibility of the two (or more) components on the properties in bulk and solution can be probed. Also, segregation of the incompatible blocks on the molecular scale (5–100 nm) can produce astonishingly complex nanostructures. Compared to linear block copolymers, subtle variations in the composition or architecture of nonlinear block copolymers can lead to pronounced changes in morphology, as well as material properties, which is one of the driving forces for the preparation of new nonlinear block copolymers. Our focus here is on synthesis of well-defined block copolymers from controlled radical polymerizations. The same synthetic strategies for synthesis of linear block copolymers can be used to prepare nonlinear block polymers.

3.3.1 STAR-BLOCK COPOLYMERS

Star-block copolymers can be envisioned as star polymers where each arm is actually a diblock or a triblock copolymer. The presence of a central connecting point of the polymer chains is expected to bring about differences in the properties of the material compared to the linear diblock and triblock copolymers. Several synthetic approaches including sequential monomer addition and mechanism transformation have been used to synthesize star-block copolymers.

When traditional sequential monomer addition is employed in the preparation of star-block copolymers, the reaction is usually stopped before complete conversion of the first monomer to circumvent excessive terminations. The A block is usually isolated by precipitation, and used as macroinitiator for the polymerization of monomer B. If the second monomer is added before complete conversion of the first monomer, a gradient copolymer is formed with characteristics similar to those of the block [156]. The synthesis of gradient copolymers simplifies the process with some loss of control over the polymer structure.

For well-controlled arm number of the star polymers, an efficient approach is the use of multifunctional initiators [157,158]. For instance, the four-armed initiator, NLI-1, which was prepared by the condensation reaction of the hydroxy groups in $C(CH_2OCH_2CH_2CH_2OH)_4$ with α-bromoisobutyric acid, was used in the ATRP of (2,2-dimethyl-1,3-dioxolane-4-yl)methyl acrylate (DMDMA) with CuBr/bpy as catalyst. After isolation from the polymerization system, four-armed poly(DMDMA)s, such as NLI-2 with $M_w/M_n = 1.28-1.41$ were obtained, and used in the successive ATRP of MMA, giving star-block copolymers NLB-3. It is known that the cycloacetal ring is unstable in acidic conditions, so the hydrolysis of the block copolymer NLB-3 was accomplished in a 1 N HCl aqueous solution to give the amphiphilic star-block copolymer structure NLB-4 as shown in **Scheme 3.37** [159].

With a similar method, star-block copolymer NLB-6 was prepared from St and DMDMA using the hexafunctional initiator NLI-5. After hydrolysis of the star-block copolymer NLB-6, an amphiphilic block copolymer NLB-7 was formed [160]. NLB-9, was prepared from trifunctional initiator NLI-8, n-butyl methacrylate and 2-(N,N-dimethylamino)ethyl methacrylate [161]; NLB-11 from NLI-10, MMA, and n-butyl methacylate [162], NLB-14 and NLB-16 from multifunctional initiator NLI-12, t-butyl acrylate and MMA [163], and NLB-15 from 12-functional initiator NLB-13, tBuA and MMA.

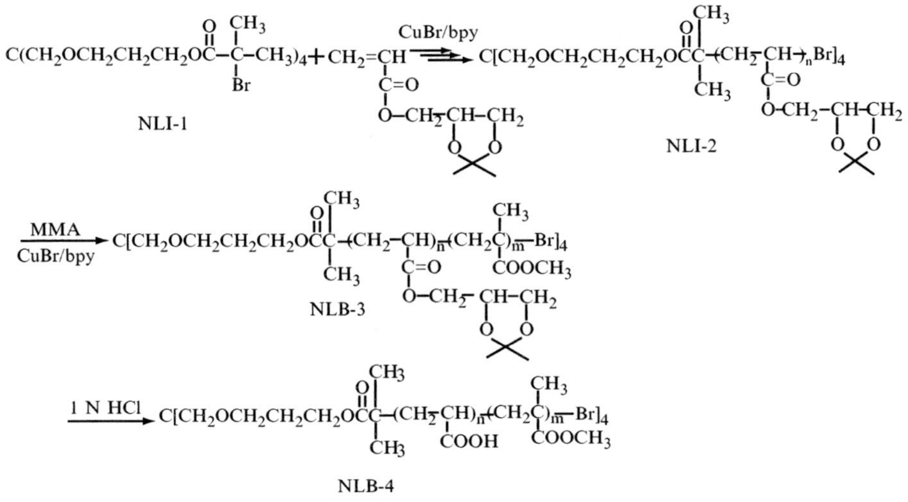

Scheme 3.37

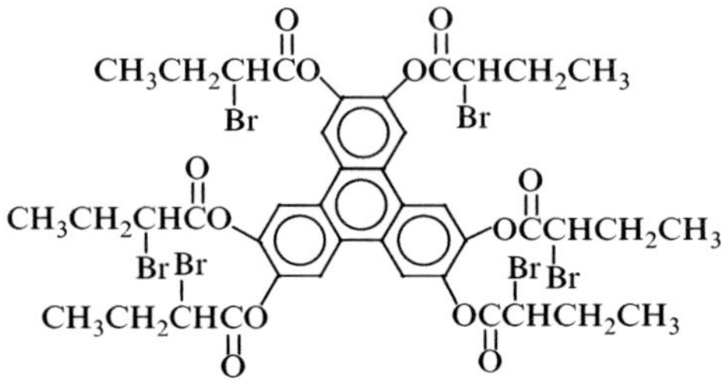

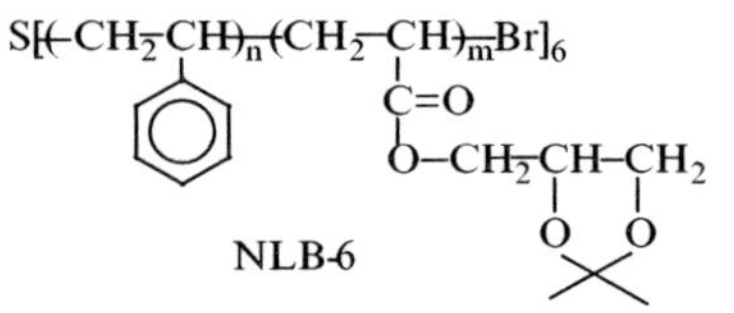

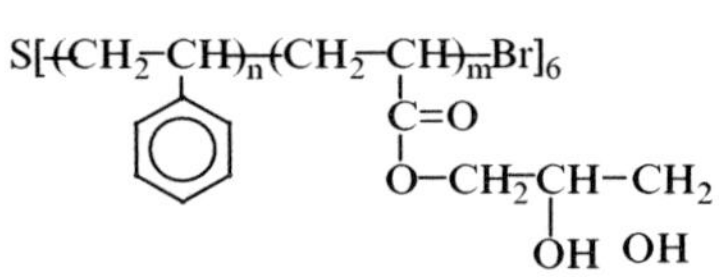

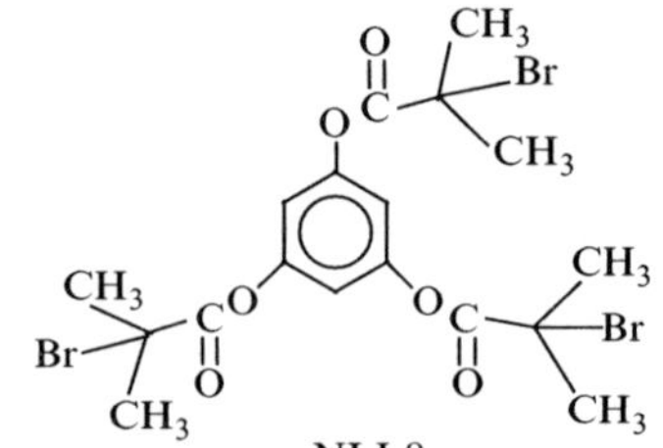

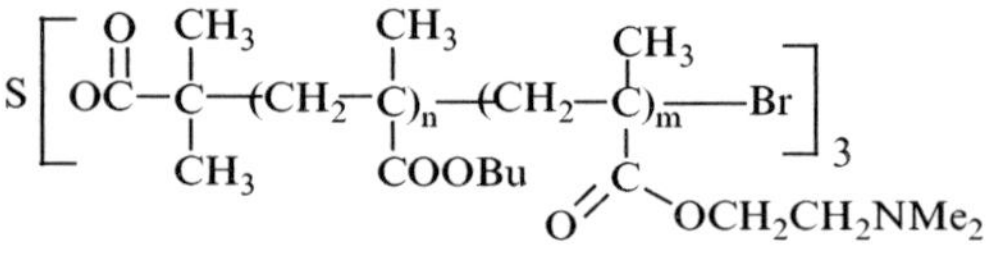

Syntheses and Characterizations of Block Copolymers

NLB-10

NLB-11

NLI-12

NLB-13

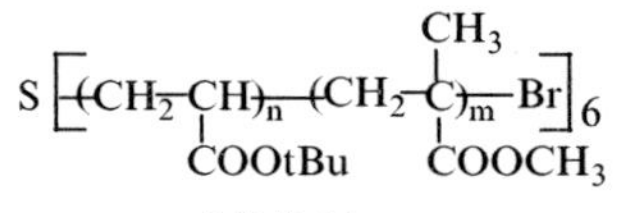

[Structures NLB-14, NLB-15, NLB-16 shown]

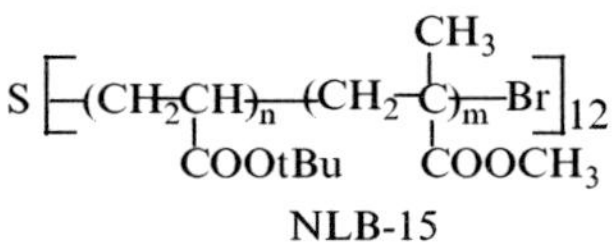

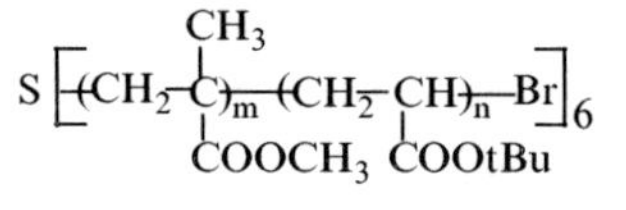

Star polymers with nBuMA-MMA block copolymer arms have been synthesized by the ruthenium-catalyzed sequential living radical polymerization of nBuMA and MMA [164], followed by a linking reaction with NLL-17 [165,166] as shown in **Scheme 3.38**.

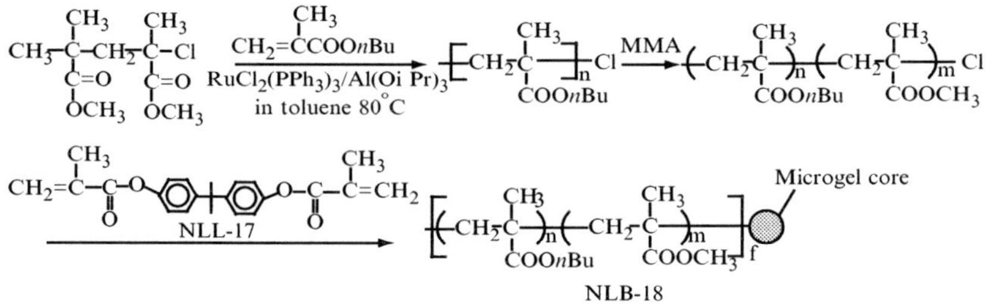

[Scheme 3.38 showing synthesis of NLB-18 via NLL-17 linking]

Scheme 3.38

The nBuMA was first polymerized using the $(MMA)_2Cl/RuCl_2(PPh_3)_3/$ Al(O-i-Pr)$_3$ system to give poly(nBuMA) with narrow MWD (M_w/M_n ~ 1.30). A fresh feed of MMA, equimolar to that of nBuMA, was then added into the unquenched polymerization system. Finally, a toluene solution of divinyl compound NLL-17 was added, and the linking reaction proceeded until the linking agent was reacted almost completely. The microgel particles with f arms (NLB-18) were formed due to the crosslinking reaction of divinyl compound NLL-17 [165].

This strategy for the synthesis of microgel particles with multiple arms is also suitable for nitroxide-mediated radical polymerization. One of the attractive features of this approach is that the starting linear chains can be isolated, characterized and stored before subsequent coupling. Additionally, a variety of different chains in terms of molecular weight, composition etc, can be copolymerized together to give heterogeneous star-block copolymers [167–170]. The procedures involve the preparation of alkoxyamine-terminated linear chains, and subsequent reaction of these dormant chains with crosslinking agents, such as divinylbenzene.

Another approach to synthesize star-block copolymers is the macroinitiator method. The first block with a terminal functional group is prepared by living anionic or cationic polymerizations or ionic ring-opening polymerization. After converting the terminal functional group into an appropriate controlled radical initiation site, the resulting polymer is used as the macroinitiator in the successive controlled radical polymerization. One example of this approach is the preparation of tetra-arm star-block copolymers, (PTHF-PSt)$_4$ and (PTHF-PSt-PMMA)$_4$. The tetra-arm PTHF polyol with controlled molecular weight and narrow MWD was prepared by cationic ring-opening polymerization of THF using tetra-acyl chloride/silver perchloride initiator at −15 °C [158,171], and then treated with α–bromoisobutylchloride in CH$_2$Cl$_2$ solution, macroinitiator NLI-19 being obtained. It can also be obtained by terminating the living cationic ring-opening polymerization with a methanol solution of sodium α–bromoisobutanoate, but the yield is low (~45%). Probably due to the high nucleophilicity of the perchloride group, an ester/ion equilibrium existed, and shifted towards the ester state during the polymerization. This may be the main reason for the low end-capping efficiency [158]. Macroinitiator NLI-19 was used in the bulk ATRP of St with CuBr/bpy as catalyst. Star-block copolymers PTHF-*b*-PSt-*b*-Br with structures NLB-20 were obtained [172]. The successive ATRP of MMA produced star-triblock copolymers, (PTHF-PSt-PMMA-Br)$_4$, as shown in **Scheme 3.39** [172]. Similarly, star-block copolymers with three polyisobutylene-*b*-PMMA arms, were obtained [143].

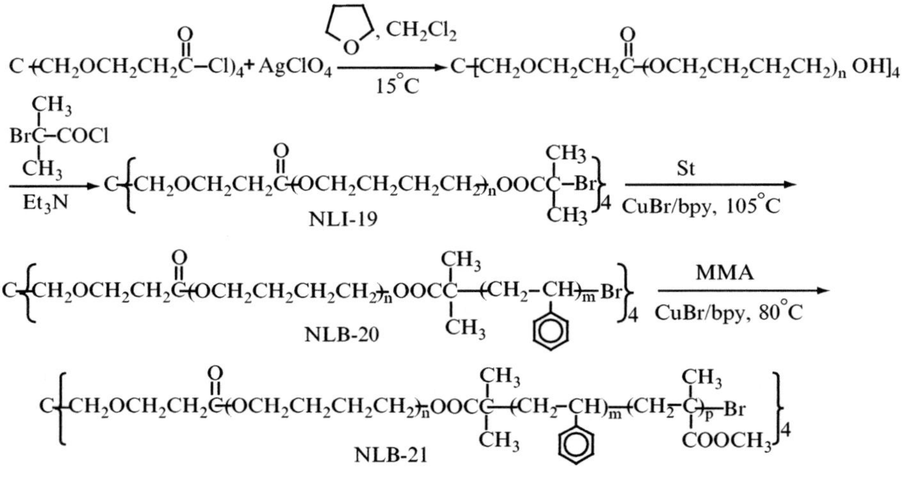

Scheme 3.39

3.3.2 MIKTOARM STAR COPOLYMERS

Miktoarm star copolymers are a special group of stars containing chemically different arms linked to the branch point. They differ from star block copolymers, where all the arms are chemically identical and consist of diblock or triblock copolymers of the A–B or A–B–A type. Although several methods have been developed for the synthesis of miktoarm stars, two general strategies, one a chlorosilane approach [173–176], another involving divinyl compounds [177–181], have been most extensively investigated. Both of them are based on living ionic, mainly anionic polymerizations. The disadvantages of the two methods are the high-vacuum techniques necessary, troublesome purification procedures, and difficulty in controlling the number of arms. In order to overcome these problems, mechanism transformation has been developed. Here we just discuss the application of mechanism transformations involving controlled radical synthesis of miktoarm star polymers.

Using an initiator with two or more different initiating sites, block copolymers can be prepared without any modification of initiating sites [182,183]. Several miktoarm star copolymers, A_2B_2 [184,185], and A_4B_4 [186], have been prepared by a combination of cationic ring-opening polymerization and ATRP with multiarm initiators. The cationic ring-opening polymerization of THF was performed at $-15\,°C$ with 3-{2,2-bis[2-bromo-2-(chlorocarbonyl) ethoxy] methyl-3-(2-chlorocarbonyl) ethoxy} propoxyl-2-bromopropanoyl chloride (BCPBC)/$AgClO_4$ as catalyst. The macroinitiator, $(CH_2OCH_2CHBr CO-PTHF)_4$ was obtained, and used in the ATRP of St with CuBr/bpy as catalyst. In this way A_4B_4 miktoarm star copolymer $(PTHF)_4(PSt)_4$ structure NLB-22, has been successfully prepared according to **Scheme 3.40** [186].

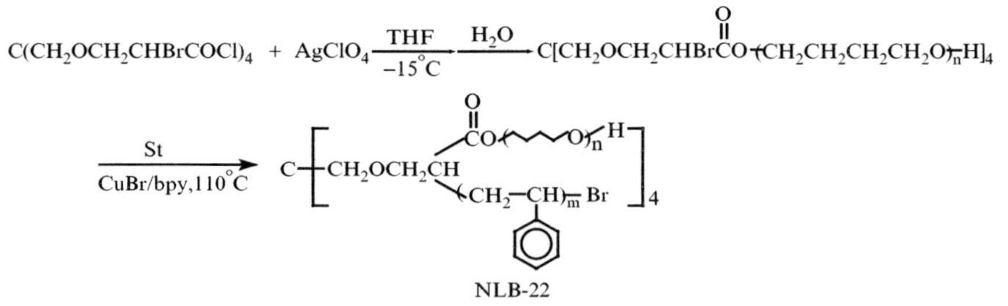

$$C(CH_2OCH_2CHBrCOCl)_4 + AgClO_4 \xrightarrow[-15\,°C]{THF} \xrightarrow{H_2O} C[CH_2OCH_2CHBrCO(CH_2CH_2CH_2CH_2O)_nH]_4$$

$$\xrightarrow[CuBr/bpy,110\,°C]{St} \quad C\left[CH_2OCH_2CH\begin{smallmatrix}CO(O)_nH\\CH_2-CH)_m Br\end{smallmatrix}\right]_4$$

NLB-22

Scheme 3.40

During the preparation of tetra-oxocarbonium species, no loss of bromine from the –CHBr group in the initiator was observed, probably due to the higher reactivity of –COCl with $AgClO_4$ relative to that of –CHBr. The molecular weight of each arm is almost the same and can be controlled through the initial feed radio of THF/initiator and the extent of conversion. The presence of four

PTHF blocks with four terminal hydroxyl groups does not affect the living character of the ATRP. Using a similar synthetic procedure, A_2B_2 miktoarm star copolymers, $(PSt)_2(PTHF)_2$ (NLB-23) and $(PSt)_2(PDOP)_2$ (NLB-24), were obtained [184,185].

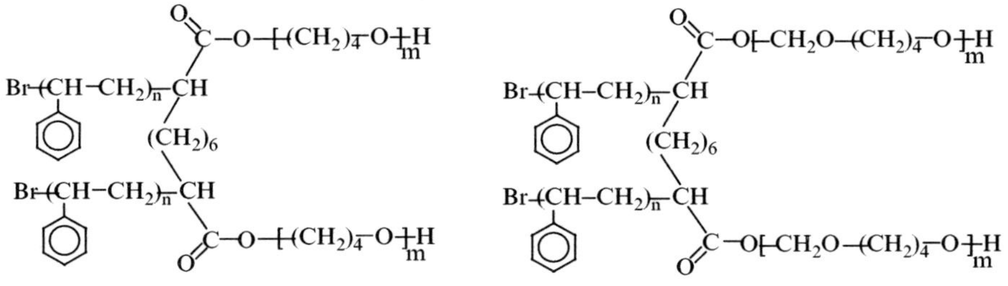

NLB-23 NLB-24

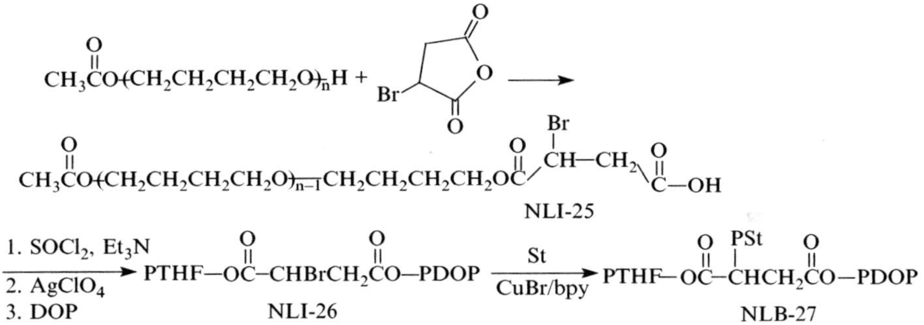

NLI-25

NLI-26 NLB-27

Scheme 3.41

Several methods for the synthesis of ABC miktoarm star copolymers have been reported based on living anionic or cationic polymerizations [187–191]. By combining cationic ring-opening polymerization with ATRP, ABC miktoarm star copolymers can be prepared according to **Scheme 3.41** [192]. PTHF with a terminal hydroxyl group, which was obtained from cationic ring-opening polymerization of THF with acetyl chloride/$AgClO_4$ initiator, reacted with 2-bromosuccinic anhydride (BSA) in benzene with a high conversion. The macroinitiator, NLI-25 with carboxylic acid and bromine groups, was used in the successive cationic ring-opening polymerization of 1,3-dioxepane (DOP) at $-30\,^{\circ}C$ with $AgClO_4$, after carboxylic acid was converted into acyl chloride. Importantly, the CHBr group of the macroinitiator was not lost during esterification and cationic polymerization. A diblock copolymer with a CHBr group in the center of the polymer chain was obtained. The ABC miktoarm star copolymer NLB-27 with well-controlled molecular weight and relatively narrow MWD was achieved by bulk copolymerization of St at 110 °C using macroinitiator NLI-26/CuBr/bpy as the initiation system [192].

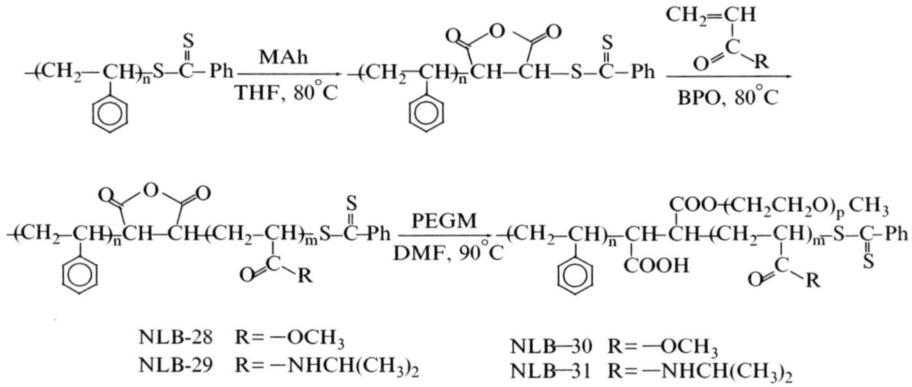

$$\text{+(CH}_2\text{-CH)}_n\text{S-C(=S)-Ph} \xrightarrow{\text{MAh, THF, 80°C}} \text{+(CH}_2\text{-CH)}_n\text{CH-CH-S-C(=S)-Ph} \xrightarrow{\text{CH}_2\text{=CH-C(=O)-O-R, BPO, 80°C}}$$

$$\text{+(CH}_2\text{-CH)}_n\text{CH-CH(CH}_2\text{-CH)}_m\text{S-C(=S)-Ph} \xrightarrow{\text{PEGM, DMF, 90°C}} \text{+(CH}_2\text{-CH)}_n\text{CH-CH(CH}_2\text{-CH)}_m\text{S-C(=S)-Ph}$$

with substituents:
COO-(CH$_2$CH$_2$O)$_p$-CH$_3$ and COOH

NLB-28 R= —OCH$_3$
NLB-29 R= —NHCH(CH$_3$)$_2$

NLB-30 R= —OCH$_3$
NLB-31 R= —NHCH(CH$_3$)$_2$

Scheme 3.42

Another method for synthesis of ABC miktoarm stars is based on nonhomopolymerizable monomers or linking agents, such as 1,1-diphenylethylene (DPE) and 1,1-bis(1-phenylethyl)benzene (DDPE), in anionic polymerizations [187–189]. Maleic anhydride (MAh) cannot undergo homopolymerization via the free-radical mechanism. Using ATRP, PSt with terminal bromine group reacted with MAh to give a PSt with terminal MAh-bromine groups [193]. This was used as a macroinitiator in the ATRP of a second monomer, although the technique was not successful due to the loss of some bromines in the reaction of PSt-Br with MAh [194]. Instead, dithio-terminated PSt was reacted with excess MAh, forming PSt with terminal anhydride and dithio groups. This was then used as a macromolecular transfer agent in the RAFT polymerization of MA or NIPAM with benzoyl peroxide as initiator. The polymers were isolated by precipitation in nonprotonic solvent, such as petroleum ether, in order to avoid opening of the anhydride ring. The anhydride group in the middle of diblock copolymers NLB-28 or NLB-29 reacted with one hydroxyl-terminated PEO. After the reaction mixture was washed with distilled water, ABC miktoarm star copolymers NLB-30 and NLB-31 with narrow MWD ($M_w/M_n = 1.08-1.12$) were obtained (see **Scheme 3.42**). The esterification efficiency was higher than 90%.

3.3.3 OTHER NONLINEAR BLOCK COPOLYMERS

Graft copolymers are composed of a main chain to which one or more side chains are connected through covalent bonds. The branches are usually randomly distributed along the backbone. Comb-like polymers are special graft copolymers in which many branches are connected to a polymer chain. They can be prepared by three general methods: graft onto, graft from and via macromonomers. Among them, the third method is more attractive for the synthesis of well-defined graft or comb-like polymers because the macromonomer can be thoroughly characterized before copolymerizing with another monomer. A variety of comb-like copolymers have been prepared by the homopolymerization of macromonomers with anionic, cationic and group-transfer polymerization [195–197]. The ATRP of vinyl ether-based macromonomers with a terminal methacryloyl group proceeded in a living fashion and gave products with fairly narrow polydispersities ($M_w/M_n \sim 1.2$) [198]. Since complete conversion of macromonomer is generally difficult, separation from precursor macromonomers is necessary. Another approach to the preparation of comb copolymers is the preparation of a copolymer with many initiating sites, followed by polymerization of the second monomer. For example, a mixture of St and p-chloromethylstyrene (CMS) can be polymerized under "living" free-radical conditions to give a well-defined linear copolymer NLI-33 with controlled molecular weight, and low polydispersity ($M_w/M_n = 1.10-1.25$). Reaction of NLI-33 with the sodium salt of the hydroxyl functionalized unimolecular initiator NLI-34 then gives the desired polymeric initiator NLB-35, which is a precursor to synthesize a variety of graft copolymers NLB-36 [199] as shown in **Scheme 3.43**.

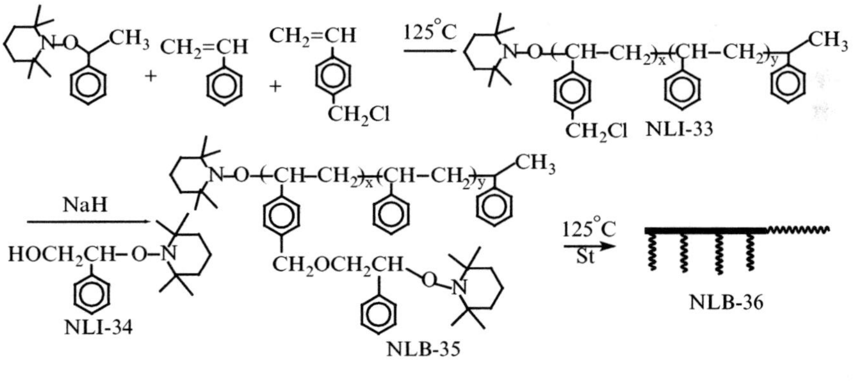

Scheme 3.43

The polymerization process is still radical in nature and radical–radical coupling reactions are decreased, but are not eliminated. In particular, for average grafting densities greater than six initiating sites per backbone, chain-chain radical coupling becomes apparent, and at densities greater than 15, it is a major process [199].

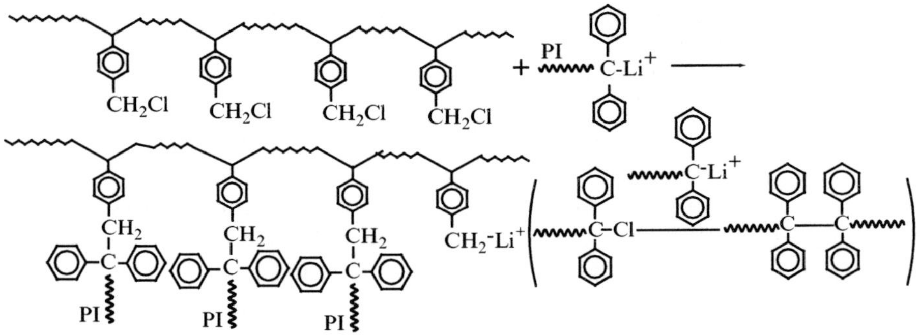

Scheme 3.44

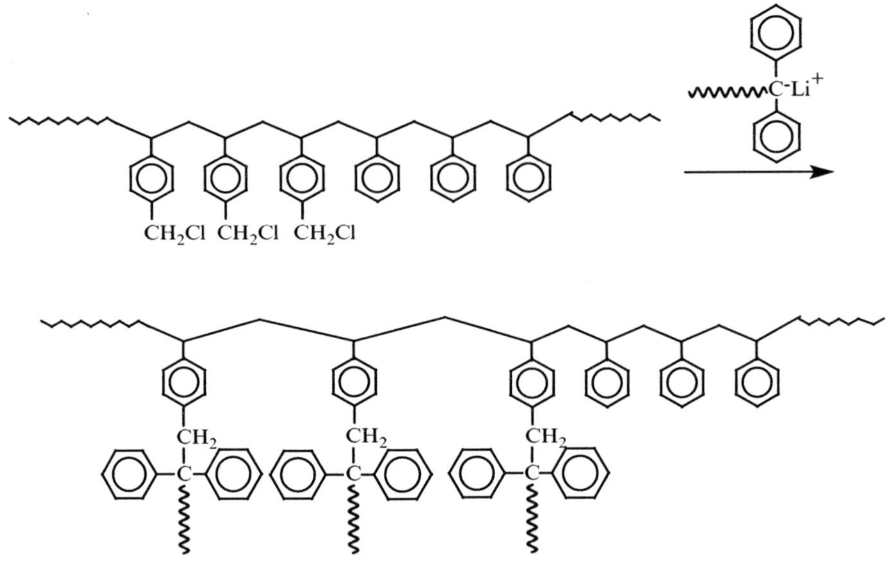

Scheme 3.45

PSt-*co*-PCMS copolymers were synthesized via the same procedure as shown in **Scheme 3.44**. The reactivity ratio of CMS: St for the LFRP is evidently the same as for conventional radical polymerization, and the chlorine atoms are randomly distributed along the PSt backbone. The coupling reaction of PSt-*co*-PCMS with an excess of about 50% DPE-terminated polyisoprenylithium, which was prepared by anionic polymerization of isoprene with BuLi as initiator, followed by termination upon addition of DPE into the polymerization system at $-20\,°C$ gave comb-like copolymer NLB-37 with a polydispersity of 1.07–1.13 (see **Scheme 3.44**) [200]. The grafting efficiency was near 100%, and the 4–5% of dimer formed could be isolated by precipitation. Using nitroxide-mediated radical polymerization, a linear diblock copolymer (PCMS-b-PSt) was synthesized, and used as the backbone. Then the living DPE-capped poly(isoprene) branches were linked to the chloromethyl groups of the diblock copolymer. A block-brush copolymer NLB-38, in which every monomeric unit of the PCMS block possesses one brush, was obtained (see **Scheme 3.45**) [200].

Unique dendritic-linear block copolymers have also been prepared by the coupling of functionalized initiators with dendritic macromolecules prepared by the convergent growth approach [201]. In this approach, the dendrimer can be attached either to the initiating fragment of the alkoxy amine or the mediating nitroxide, and the dendritic initiator is then used to initiate the growth of linear vinyl block under controlled radical conditions. As shown in **Scheme 3.46**, the coupling reaction of the dendrimer NLD-39, which contains a single bromomethyl group at its focal point, with the hydroxyl functionalized unimolecular initiator NLI-40 gives the dendritic initiator NLI-41. Hybrid dendritic–linear block copolymer, NLB-42 with well-controlled molecular weights and low polydispersities was then obtained by the nitroxide-mediated polymerization of NLI-41 with styrenic monomers or comonomer mixtures under living free-radical polymerization conditions [202,203].

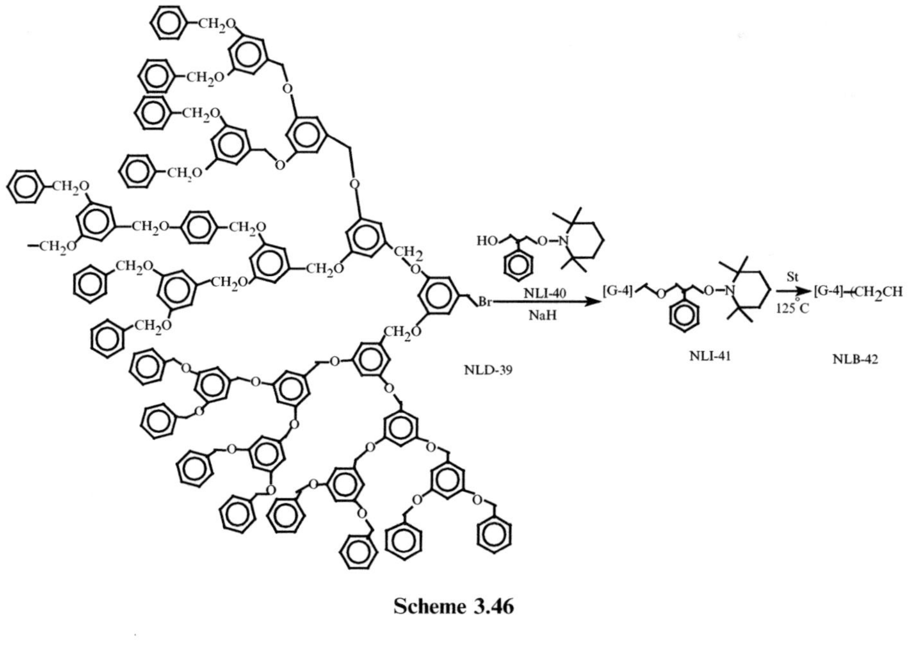

Scheme 3.46

Similar structures can also be prepared using ATRP chemistry and in this case the initiating group is simply a chloromethyl or bromomethyl species at a focal point [204]. Linear poly (acrylate) and poly (acrylic acid) blocks [205,206] are then connected to the dendrimer at its core. They were prepared by the copper-catalyzed living radical polymerizations of acrylates with dendrimer-type macroinitiators having a benzyl bromide at the focal point. After hydrolysis, amphiphilic block copolymers with a linear PAA hydrophilic block and a dendritic poly (benzyl ether) as hydrophobic block were obtained [205,206].

REFERENCES

1. Hamley, I. W., *The Physics of Block Copolymers*, Oxford U.P., Oxford, England (1998)
2. Bates F. S., Fredrickson G. H., *Phys. Today*, **52**: 32 (1999)
3. He X.-H., Liang H.-J., Pan C.-Y., *Phys. Rev. E*, **63**: 31804 (2001)
4. He X.-H., Huang L., Liang H.-J., Pan C.-Y., *J. Chem. Phys.*, **116**(23): 10508 (2002)
5. Stadler R., Auschra C., Beckmann J., Krappe U., Voigt-Martin I., Leibler L., *Macromolecules*, **28**: 3080 (1995)
6. Zheng W., Wang Z.-G., *Macromolecules*, **28**: 7215 (1995)
7. Hawker C. J., Bosman A. W., Harth E., *Chem. Rev.* **101**: 3661 (2001)
8. Wang J., Matyjaszewski K., *J. Am. Chem. Soc.* **117**: 5614 (1995); Wang J., Matyjaszewski K., *Macromolecules*, **28**: 7901 (1995)
9. Kato M., Kamigaito M., Sawamoto M., Higashimura T., *Macromolecules*, **28**: 1721 (1995); Endo T., Kato M., Kamigaito M., Sawamoto M., *Macromolecules*, **29**: 1070 (1996)
10. Patten T. E., Xia J.-H., Abernathy T., Matyjaszewski K., *Science*, **272**: 866 (1996); Xia J.-H, Matyjaszewski K., *Macromolecules*, **30**: 7697 (1997)
11. Percec V., Barboiu B., Neumann A., Ronda J. C., Zhao M.-Y., *Macromolecules*, **29**: 3665 (1996)
12. Pan C.-Y., Lou X.-D., Wang Y.-L., Wu C.-P., *Acta Polym. Sinica*, **3**: 311 (1998)
13. De León-Sáenz E., Morales G., Guerrero-Santos R., Gnanou Y., *Macromol. Chem. Phys.*, **201**: 74 (2000)
14. Yamada B., Nobukane Y., Miura Y., *Polym. Bull.*, **41**: 539 (1998)
15. Steenbock M., Klapper M., Mullen K., *Macromol. Chem. Phys.*, **199**: 763 (1998)
16. Puts R. D., Sogah, D. Y., *Macromolecules*, **29**: 3323 (1996)
17. Chiefari J., Chong Y. K, Ercole F., Krstina J., Jeffery J., Le T. P. T., Mayadunne R. T. A, Meijs G. F., Moad C. L., Moad G., Rizzardo E., Thang S.H, *Macromolecules*, **31**: 5559 (1998)
18. Le T. P., Moad G., Rizzardo E., Thang S. H., PCT Int. Appl. WO9801478A1 980115
19. Chiefari J., Mayadunne R. T. A., Moad G., Rizzardo E., Thang S. H., *PCT Int. Appl.* WO9931144A1 990624
20. Matyjaszewski K. "*Controlled Radical Polymerization*" in *ACS Symp. Ser.* **685**, Matyjaszewski K. Ed., American Chemical Society, Washington DC, Chapter 1 (1998)
21. Kotani Y., Kato M., Kamigaito M., Sawamoto M., *Macromolecules*, **29**: 6979 (1996)
22. Zhu S., Yan D., *J. Polym. Sci. Part A: Polym. Chem.* **38**: 4308 (2000)

23. Uegaki H., Kotani Y., Kamigaito M., Sawamoto M., *Macromolecules*, **31**: 6756 (1998)
24. Shipp D. A., Wang J.-L., Matyjaszewski K., *Macromolecules*, **31**: 8005 (1998)
25. Davis K. A., Matyjaszewski K., *Macromolecules*, **33**: 4039 (2000); Matyjaszewski K., Coca S., Gaynor S. G., Wei M., Woodworth B. E., *Macromolecules*, **31**: 5967 (1998)
26. Cassebras M., Pascual S., Polton A., Tardi M., Vairon J.-P., *Macromol. Rapid Commun.* **20**: 261 (1999)
27. Davis K. A., Charleux B., Matyjaszewski K., *J. Polym. Sci. Part A: Polym. Chem.* **38**: 2274 (2000)
28. Liu Y., Wang L.-X., Pan C.-Y., *Macromolecules*, **32**: 8301 (1999)
29. Yuan J.-Y., Wei G.-Y., Wang Y.-M., Pan C.-Y., *Acta Polym. Sinica*, **5**: 625 (2001)
30. Liu B., Hu C.-P., *Eur. Polym. J.* **37**: 2025 (2001)
31. Xia J., Johnson T., Gaynor S. G., Matyjaszewski K., DeSimone J., *Macromolecules*, **32**: 4802 (1999); Zhang Z., Ying S.-K., Shi Z., *Polymer*, **40**: 5439 (1999)
32. Ohno K., Tsujii Y., Fukuda T., *J. Polym. Sci. Part A: Polym. Chem.* **36**: 2473 (1998); Li Z.-C., Liang Y.-Z., Chen G.-Q., Li F.-M., *Macromol. Rapid Commun*, **21**: 375 (2000)
33. Matyjaszewski K., Shipp D. A., McMurtry G. P., Gaynor S. G., Pakula T., *J. Polym. Sci. Part A: Polym. Chem.* **38**: 2023 (2000)
34. Mühlebach A., Gaynor S. G., Matyjaszewski K., *Macromolecules*, **31**: 6046 (1998)
35. Ma Q.-G., Wooley K. L, *J. Polym. Sci. Part A: Polym. Chem.* **38**: 4805 (2000)
36. Pan C.-Y., Tao L., Wu D.-C., *J. Polym. Sci. Part A: Polym. Chem.* **39**: 3062 (2000)
37. Beers K. L., Boo S., Gaynor S. G., Matyjaszewski K., *Macromolecules*, **32**: 5772 (1999)
38. Wang X.-S., Lao N., Ying S.-K., *Polymer*, **40**: 4157 (1999)
39. Matyjaszewski K., Patten T. E., Xia J., *J. Am. Chem. Soc*, **119**: 674 (1997)
40. Xia J., Zhang X., Matyjaszewski K., *Macromolecules*, **32**: 3531 (1999)
41. Yamamoto S., Tsujii Y., Fukuda T., *Macromolecules*, **33**: 5995 (2000)
42. Hawthorne D. G., Moad G., Rizzardo E., Thang S. H., *Macromolecules*, **32**(16): 5457 (1999)
43. Rizzardo E., Chiefari J., Chong B. Y. K., Ercole F., Krstina J., Jeffery J., Le T. P. T., Mayadunne R. T. A., Meijs G. F., Moad C. L, Moad G., Thang S. H., *Macromol Symp.* **143**: 291 (1999)
44. Chong B. Y. K., Le T. P. T, Moad G., Rizzardo E., Thang S. H., *Macromolecules*, **32**: 2071 (1999)
45. Mayadunne R. T. A., Rizzardo E., Chiefari J., Kristina J., Moad G., Postma A., Thang S. H., *Macromolecules*, **33**: 243 (2000)
46. Kanagasabapathy S., Sudalai A., Benicewicz B. C., *Macromol. Rapid Commun.*, **22**: 1076 (2001)
47. Mitsukami Y., Donovan M. S., Lowe A. B., McCormick C. L., *Macromolecules*, **34**: 2248 (2001)
48. Monteiro M. J., Sjöberg M., Vlist J., Göttgens C. M., *J. Polym. Sci. Part A: Polym. Chem*, **38**: 4206 (2000)
49. Monteiro M. J., de Barbeyra C. L., *Macromolecules*, **34**: 4416 (2001)
50. Arotçaréna M., Heise B., Ishaya S., Laschewsky A., *J. Am. Chem. Soc.*, **124**(14): 3787 (2002)
51. Zhu M.-Q., Wei L.-H., Li M., Jiang L., Du F.-S., Li Z.-C., Li F.-M., *Chem. Commun.*, 365 (2001)
52. Georges, M. K., Veregin R. P. N., Kazmaier P. M., Hamer G. K., *Macromolecules*, **26**: 2987 (1993)
53. Hawker C. J., *J. Am. Chem. Soc.* **116**: 1185 (1994)

54. Hawker C. J., Barclay G. G., Orellana A., Dao J., Devonport W., *Macromolecules*, **29**: 5245 (1996)
55. Benoit D., Chaplinski V., Braslau R., Hawker C. J., *J. Am. Chem. Soc*, **121**: 3904 (1999)
56. Benoit D., Grimaldi S., Robins S., Finet J. P., Tordo P., Gnanou Y., *J. Am. Chem. Soc*, **122**: 5929 (2000)
57. Chong Y. K., Ercole F., Moad G., Rizzardo E., Thang S. H., Anderson A. G., *Macromolecules*, **32**: 6895 (1999)
58. Nowakowska M., Zapotoczny S., Karewicz A., *Polymer*, **42**: 1817 (2001)
59. Fischer A., Brembilla A., Lochon P., *Polymer*, **42**: 1441 (2001)
60. Ohno K., Ejaz M., Fukuda T., Miyamoto T., Shimizu Y., *Macromol. Chem. Phys*, **199**: 281 (1998)
61. Grubbs R. B., Dean J. M., Broz M. E., Bates F. S., *Macromolecules*, **33**: 9522 (2000)
62. Robin S., Gnanou Y., *Polym. Prep.* **40**(2): 387 (1999); Robin S., Gnanou Y., *Macromol. Symp.* **165**: 43 (2001)
63. Otsu T., Yoshida M., Kuriyama A., *Polym. Bull.*, **7**: 45 (1982)
64. Otsu T., Kuriyama A., *J. Macromol. Sci. Chem*, **A21**: 961 (1984)
65. Otsu T., Kuriyama A., *Polym. Bull.*, **11**: 135 (1985)
66. Ostu T., Kuriyama A., *Polym. J.*, **17**: 97 (1985)
67. Sebenik A., *Prog. Polym. Sci*, **23**: 876 (1998)
68. Bai R.-K., You Y.-Z., Pan C.-Y., *Macromol. Rapid Commum.*, **22**(5): 315 (2001)
69. You Y.-Z., Bai R.-K., Pan C.-Y., *Macromol Chem. Phys.* **202**(9): 1980 (2002)
70. Bai R.-K., You Y.-Z., Zhong P., Pan C.-Y., *Macromol. Chem. Phys.* **202**(9): 1970 (2002)
71. Hong C.-Y., You Y.-Z., Bai R.-K., Pan C.-Y., *J. Polym. Sci. Part A: Polym. Chem.* **39**(22): 3934 (2001)
72. You Y.-Z., Hong C.-Y., Pan C.-Y., *Eur. Polym. J.* **38**: 1279 (2002)
73. You Y.-Z., Hong C.-Y., Pan C.-Y., *Macromol. Chem. Phys.* **203**: 477 (2002)
74. He T., Zou Y.-F., Pan C.-Y., *J. Polym. Sci. Part A: Polym. Chem.* **40**: 3367 (2002)
75. Jankova K., Chen X., Kops J., Batsberg W., *Macromolecules*, **31**: 538 (1998)
76. Jankova K., Truelsen J. H., Chen X., Kops J., Batsberg W., *Polym. Bull.* **42**: 153 (1999)
77. Wang X.-S., Luo N., Ying S.-K., Liu Q., *Eur. Polym. J.* **36**: 149 (2000)
78. Gaynor S. G., Edelman S. Z., Matyjaszewski K., *Polym. Prep.* **38**(1): 703 (1997)
79. Gaynor S. G., Matyjaszewski K., *Macromolecules* **30**: 441 (1997)
80. Wang X.-S., Luo N., Ying S. K., *China Synth. Rubba. Ind.*, **20**(2): 115 (1997)
81. Chen X., Gao B., Kops J., Batsberg W., *Polymer* **39**: 911 (1998)
82. Wang Y., Chen S., Huang J., *Macromolecules*, **32**: 2480 (1999)
83. Wang Y., Huang J., *Macromolecules*, **31**: 4058 (1998)
84. Bosman A. W., Fréchet J. M. J., Hawker C. J. *Polym. Mater. Sci. Eng.* **84**: 376 (2001)
85. Nakayama Y., Miyamura M., Hirano Y., Goto K., Matsuda T., *Biomaterials* **20**: 963 (1999)
86. Brouwer D. H., Schellekens M. A. J., Klumperman B., Monteiro M. J., German A. L., *J. Polym. Sci. Part A: Polym. Chem.* **38**(19): 3596 (2000)
87. Nakagawa Y., Miller P. J., Matyjaszewski K., *Polymer*, **39**: 5163 (1998)
88. Morgan A. M., Pollack S. K., Beshah K., *Macromolecules*, **35**: 4238 (2002)
89. Lutsen L, Cordina G. P. G., Jones R. G., Schue F., *Eur. Polym. J.* **34**: 1829 (1998)
90. Stewart M. J., in "*New Methods of Polymer Synthesis*"; Ebdon J. R. Ed.; Blackie, London, p. 107 (1991)
91. Burgess F. J., Cunliffe A. V., Richards D. H., Sherrington D. C., *J. Polym. Sci. Polym. LeH Ed.* **14**: 471 (1976)

92. Burgess F. J., Cunliffe A. V., MacCallum J. R., Richards D. H., *Polymer*, **18**: 719 (1977)
93. Hallensleben M., *Macromol. Chem.*, **178**: 2125 (1977)
94. Burgess F. J., Cunliffe A. V., Dawkins J. V., Richards D. H., *Polymer* **18**: 733 (1977)
95. Lin Q., Konas M., Dacis R. M., Riffle J. S., *J. Polym. Sci. Part A: Polym. Chem.* **31**: 1709 (1993)
96. Kennedy J. P., Price J. L., Koshimura K., *Macromolecules*, **24**: 6567 (1991)
97. Feldthusen J., Ivan B., Muller A. H. E., *Macromolecules*, **31**: 578 (1998)
98. Doi Y., Watanabe Y., Ueki S., Soga K., *Macromol. Chem. Rapid Commun.*, **4**: 533 (1983)
99. Xu Y.-J., Pan C.-Y., *J. Polym. Sci. Part A: Polym. Chem.* **38**: 337 (2000)
100. Xu Y.-J., Pan C.-Y., *J. Polym. Sci. Part A: Polym. Chem.* **37**: 3391 (1999)
101. Matyjaszewski K., Penczek S., Franta E., *Polymer*, **20**: 1184 (1979)
102. Szwarc M., *Macromolecules*, **28**: 7309 (1995)
103. Düz A. B., Yagci Y., *Eur. Polym. J.*, **35**: 2031 (1999)
104. Kajiwara A., Matyjaszewski K., *Macromolecules*, **31**: 3489 (1998)
105. Xu Y.-J., Pan C.-Y., *Macromolecules*, **33**: 4750 (2000)
106. Szmanski R., Kubisa P., Penczek S., *Macromolecules*, **16**: 1000 (1983)
107. Liu Y., Wang H.-B., Pan C.-Y., *Macromol. Chem. Phys*, **198**: 2613 (1997)
108. Xu Y.-J., Pan C.-Y., Tao L., *J. Polym. Sci. Part A: Polym. Chem.* **38**: 436 (2000)
109. Hawker C. J., Hedrick J. L., Malmström E. E., Trollsås M., Mecerreyes D., Moineau G., Dubois Ph., Jérome R., *Macromolecules*, **31**: 213 (1998)
110. Mecerreyes D., Moineau G., Dubois Ph., Jerome R., Hedrick J. L., Hawker C. J., Malmström E. E., Trollsås M., *Angew Chem. Int. Ed.* **37**: 1274 (1998)
111. Haddleton D. M., Waterson C., Derrick P. J., Jasieczek C. B., Shooter A., *J. Chem. Soc. Chem. Commun.* **7**: 683 (1997)
112. Coca S., Jasieczek C. B., Beers K. L., Matyjaszewski K., *J. Polym. Sci. Part A: Polym. Chem.* **36**: 1417 (1998)
113. Percec V., Kim H. J., Barboiu B., *Macromolecules*, **30**: 8526 (1997)
114. Hedrick J. L., Trollsås M., Hawker C. J., Atthoff B., Claesson H., Heise A., Miller R. D., Mecerreyes D., Jérôme R., Dubois Ph., *Macromolecules*, **31**: 8691 (1998)
115. Zhang Q., Remsen E. E., Wooley K. L., *J. Am. Chem. Soc.*, **122**: 3642 (2000)
116. Uegaki H., Kotani Y., Kamigaito M., Sawamoto M., *Macromolecules*, **30**: 2249 (1997)
117. Puts R. D., Sogah D. Y., *Macromolecules*, **30**: 7050 (1997)
118. Weimer M. W., Scherman O. A., Sogah D. Y., *Macromolecules*, **31**: 8425 (1998)
119. Yuan J.-Y., Pan C.-Y., *Chinese J. Polym. Sci.*, **20**(2): 171 (2002)
120. Yuan J.-Y., Pan C.-Y., *Eur. Polym. J.* **38**: 2069 (2002)
121. Yuan J.-Y., Pan C.-Y., *Eur. Polym. J.* **38**: 1565 (2002)
122. Yuan J.-Y., Zou Y.-F., Pan C.-Y., *Chem. J. Chinese Universities*, **21**: 1494 (2002)
123. Yuan J.-Y., Pan C.-Y., Tang B.-Z., *Macromolecules*, **34**: 211 (2001)
124. Pan C.-Y., Lou X.-D., *Macromol. Chem. Phys.*, **201**: 1115 (2000)
125. Wei Y., Connors E. J., Jia X., Wang C., *J. Polym. Sci. Part A: Polym. Chem.* **36**: 761 (1998)
126. Ramakrishnan A., Dhamodharan R., *J. Macromol. Sci., Pure. Appl. Chem.* **A37**: 621 (2000)
127. Acar M. H., Matyjaszewski K., *Macromol. Chem. Phys.* **200**: 1094 (1999)
128. Liu B., Liu F., Luo N., Ying S.-K., Liu Q., *Chinese J. Polym. Sci.* **18**: 39 (2000)
129. Liu B., Liu F., Luo N., Ying S.-K., Liu Q., *Chem. J. Chinese. University*, **21**: 484 (2000)
130. Liu F., Liu B., Luo N., Ying S.-K., *Chem. Res. Chinese University*, **16**: 72 (2000)

131. Tong J.-D., Ni S., Winnik M. A., *Macromolecules* **33**: 1482 (2000)
132. Kobatake S., Harwood H. J., Quirk R. P., Priddy D. B., *Macromolecules* **30**: 4238 (1997)
133. Hua F.-J., Yang Y.-L., *Polymer*, **42**(4): 1361 (2001)
134. Kennedy J. P., Ivan B., *Designed Polymers by Carbocationic Macromolecular Engineering: Theory and Practice*; Hanser, Munich, Germany, (1992)
135. Storey R. F., Baugh D. W., Choate K. R., *Polymer*, **40**: 3083 (1999)
136. Lu J., Liang H., Zhang R.-F., Li B.-E., *Polymer*, **42**: 4549 (2001)
137. Coca S., Matyjaszewski K., *Macromolecules* **30**: 2808 (1997)
138. Zhao B., Brittain W. J., *Macromolecules* **33**: 8813 (2000)
139. Zhao B., Brittain W. J., *J. Am. Chem. Soc.* **121**: 3557 (1999)
140. Coca S., Matyjaszewski K., *J. Polym. Sci. Part A: Polym. Chem.* **35**: 3595 (1997)
141. Jankova K., Kops J., Chen X., Gao B., Batsberg W., *Polym. Bull.* **41**: 639 (1998)
142. Chen X., Ivan B., Kops J., Batsberg W., *Macromol. Rapid. Commun.* **19**: 585 (1998)
143. Keszler B., Fenyvesi G. Y., Kennedy J. P., *J. Polym. Sci. Part A: Polym. Chem.* **38**: 706 (2000)
144. Ivan B., Chen X., Kops J., Batsberg W., *Macromol. Rapid. Commun.* **19**: 15 (1998)
145. Guo Z.-R., Wan D.-C., Huang J.-L., *Macromol. Rapid. Commun.* **22**: 367 (2001)
146. Erel I., Cianga I., Serhatli E., Yagci Y., *Eur. Polym. J.* **38**: 1409 (2002)
147. Tunca U., Erdogan T., Hizal G., *J. Polym. Sci. Part A: Polym. Chem.* **40**: 2025 (2002)
148. Tunca U., Karliga B., Ertekin S., Ugur A. L., Sirkecioglu O., Hizal G., *Polymer* **42**: 8489 (2001)
149. Chong Y. K., Le T. P. T., Moad G., Rizzardo E., Thang S. H., *Macromolecules* **32**: 2071 (1999)
150. Moad G., Chiefari J., Chong Y. K., Kristina J., Mayadunne R. T. A., Postma A., Rizzardo E., Thang S. H., *Polym. Int.* **49**: 993 (2000)
151. Yildirim T. G., Hepuzer Y., Hizal G., Yagci Y., *Polymer* **40**: 3885 (1999)
152. Moineau C., Minet M., Teyssie P., Jerome R., *Macromolecules*, **32**: 8277 (1999)
153. Zhang Z., Ying S.-K., Shi Z., *Polymer* **40**: 1341 (1999)
154. Sedjo R. A., Mirous B. K., Brittain W. J., *Macromolecules* **33**: 1492 (2000)
155. Li I. Q., Howell B. A., Dineen M. T., Kastl P. E., Lyons J. W., Meunier D. M., Smith P. B., Priddy D. B., *Macromolecules* **30**: 5195 (1997)
156. Matyjaszewski K., *J. Polym. Sci. Part A: Polym. Chem.* **34**: 1785 (1999)
157. Xu Y.-J., Liu Y., Pan C.-Y., *J. Polym. Sci. Part A: Polym. Chem.* **37**: 2347 (1999)
158. Xu Y.-J., Liu Y., Pan C.-Y., *J. Polym. Sci. Part A: Polym. Chem.* **37**: 3391 (1999)
159. Pan C.-Y., Tao L., Wu D.-C., *J. Polym. Sci. Part A: Polym. Chem.* **39**: 3062 (2001)
160. Feng X.-S., Pan C.-Y., Wang J., *Macromol. Chem. Phys.* **202**: 3403 (2001)
161. Narrainen A. P., Pascual S., Haddleton D. M., *J. Polym. Sci. Part A: Polym. Chem.* **40**: 439 (2002)
162. Ohno K., Wong B., Haddelton D. M., *J. Polym. Sci. Part A: Polym. Chem.* **39**: 2206 (2001)
163. Heise A., Hedrick J. L., Trollsas M., Hilborn J. G., Frank C. W., Miller R. D., *Polym. Prep.* **40**(1): 452 (1999)
164. Kotani Y., Kato M., Kamigaito M., Sawamoto M., *Macromolecules* **29**: 6979 (1996)
165. Baek K. Y., Kamigaito M., Sawamoto M., *J. Polym. Sci. Part A: Polym. Chem.* **40**: 633 (2002)
166. Baek K.-Y., Kamigaito M., Sawamoto M., *Macromolecules*, **34**: 215 (2001)
167. Benoit D., Harth E., Hawker C. J., Helms B., *Polym. Prep.* **41**(1): 42 (2000)
168. Bosman A. W., Heumann A., Klaerner G. G., Benoit D. C., Fréchet J. M. J., Hawker C. J., *J. Am. Chem. Soc.* **123**: 6461 (2001)

169. Pasquale A. J., Long T. E., *J. Polym. Sci. Part A: Polym. Chem.* **39**: 216 (2001)
170. Tsoukatos T., Pispas S., Hadjichristidis N., *J. Polym. Sci. Part A: Polym. Chem.* **39**: 320 (2001)
171. Xu Y.-J., Pan C.-Y., *Chem. J. Chin. Univ.* **21**: 140 (2000)
172. May J. W., *Polym. Bull.* **23**: 247 (1990)
173. Sioula S., Tselikas Y., Hadjichristidis N., *Macromolecules* **30**: 1518 (1997)
174. Xie H., Xie J., *Makromol. Chem.* **188**: 2543 (1987)
175. Iatrou H., Hadjichristidis N., *Macromolecules* **26**: 2479 (1993)
176. Allgaier J., Young R. N., Efstratiadis K., Hadjichristidis N., *Macromolecules* **29**: 17 (1996)
177. Höcker H., Latterman G. J., *J. Polym. Sci. Symp.* **54**: 361 (1976)
178. Quirk R. P., Lee B., Schock L. E., *Makromol. Chem. Makromol. Symp.* **53**: 201 (1992)
179. Quirk R. P., Yoo T., Lee B., *J. Macromol. Sci. Pure Appl. Chem. A*, **A31**(8): 911 (1994)
180. Fernyhough C. M., Young R. N., Tack R. D., *Macromolecules* **32**: 5760 (1999)
181. Wright S. J., Young R. N., Croucher T. G., *Polym. Int.* **33**: 123 (1994)
182. Xu Y.-J., Pan C.-Y., *J. Polym. Sci. Part A: Polym. Chem.* **38**: 337 (2000)
183. Xu Y.-J., Pan C.-Y., Tao L., *J. Polym. Sci. Part A: Polym. Chem.* **38**: 436 (2000)
184. Guo Y.-M., Pan C.-Y., *Polymer* **42**(7): 2863 (2001)
185. Guo Y.-M., Xu J., Pan C.-Y., *J. Polym. Sci. Part A: Polym. Chem.* **39**(3): 437 (2001)
186. Guo Y.-M., Pan C.-Y., Wang J., *J. Polym. Sci. Part A: Polym. Chem.* **39**(13): 2134 (2001)
187. Fujimoto T., Zhang H., Kazama T., Isono Y., Hasegawa H., Hashimoto T., *Polymer* **33**: 2208 (1992)
188. Hückstädt H., Göpfort A., Abetz V., *Macromol. Chem. Phys.* **201**: 296 (2000)
189. Lambert O., Dumas P., Hurtrez G., Riess G., *Macromol. Rapid. Commun.* **18**: 343 (1997)
190. Lambert O., Reutenauer S., Hurtrez G., Riess G., Dumas P., *Polym. Bull., (Berlin)* **40**: 143 (1998)
191. Lu Z.-J., Chen S., Huang J.-L., *Macromol. Rapid. Commun.* **20**: 394 (1999)
192. Feng X.-S., Pan C.-Y., *Macromolecules* **35**: 2084 (2002)
193. Koulouri E. G., Kallistis J. K., *Macromolecules* **32**: 6242 (1999)
194. Feng X.-S., Pan C.-Y., *Macromolecules* **35**: 4888 (2002)
195. Rempp P., Lutz P., Masson P., Chaunont P., Franta E., *Macromol. Chem. Suppl.* **13**: 47 (1985)
196. Asami R., Takaki M., Moriyama Y., *Polym. Bull.* **16**: 125 (1986)
197. Kataoka S., Sueoka M., Moriyama Y., *J. Polym. Sci. Part A: Polym. Chem.* **31**: 2513 (1993)
198. Yamada K., Miyazaki M., Oho K., Fukuda T., Minoda M., *Macromolecules* **32**: 290 (1999)
199. Grubbs R. B., Hawker C. J., Dao J., Fréchet J. M. J., *Angew. Chem. Int. Ed. Engl.* **36**: 270 (1997)
200. Tsoukatos T., Pispas S., Hdajichristidis N., *Macromolecules* **33**: 9504 (2000)
201. Hawker C. J., Fréchet J. M. J., *J. Am. Chem. Soc.* **112**: 7638 (1990)
202. Emrick T., Hayes W., Fréchet J. M. J., *J. Polym. Sci. Part A: Polym. Chem.* **37**: 3748 (1999)
203. Leduc M. R., Hawker C. J., Dao J., Fréchet J. M. J., *J. Am. Chem. Soc.* **118**: 11111 (1996)
204. Leduc M. R., Hayes W., Fréchet J. M. J., *J. Polym. Sci. Part A: Polym. Chem.* **36**: 1 (1998)

205. Zhu L., Tong X., Li M., Wang E., *J. Polym. Sci. Part A: Polym. Chem.* **38**: 4282 (2000)
206. Zhu L.-Y., Tong X.-F., Li M.-Z., Wang E.-J., *Acta Chimica Sinica* **58**(6): 609 (2000)

ABBREVIATIONS

AIBN	2,2'-azobisisobutyronitrile
ATRP	atom transfer radical polymerization
BA	butyl acrylate
BBrBE	1,2-bis (2'-bromobutyryloxy) ethane
BCPBC	3-{2,2-bis[2-bromo-2-(chlorocarbonyl)ethoxy]methyl-3-(2-chlorocarbonyl) ethoxy}propoxyl-2-bromopropanoyl chloride
BMDO	5,6-benzo-2-methylene-1,3-dioxepane
BP	benzopinacol
Bpy	bipyridine
BSA	2-bromosuccinic anhydride
BzMA	benzyl methacrylate
CL	caprolactone
CMS	*p*-chloromethylstyrene
DBTTC	dibenzyl trithiocarbonate
DBX	α, α'-dibromo-p-xylene
DCC	dicyclohexyl carbodiimide
DDPE	1,1-bis (1-phenylethyl) benzene
DMA	dimethylacrylamide
DMAEMA	A2-(N,N'-dimethylamino) ethyl methacrylate
DMAP	4-(dimethylamino) pyridine
DMDMA	(2,2-dimethyl-1,3-dioxolane-4-yl) methyl acrylate
DMF	dimethyl formamide
DOP	1,3-dioxepane
DPE	1,1-diphenylethylene
EA	ethyl acrylate
EBP	ethyl 2-bromopropanoate
2dHbpy	4,4'-di-*n*-heptyl-2,2'-bipyridine
HEBrB	2-hydroxyethyl 2'-bromobutyrate
MA	methyl acrylate
MAh	maleic anhydride
MMA	methyl methacrylate
MPDO	2-methylene-4-phenyl-1, 3-dioxolane
αMSt	α-methylstyrene
MWD	molecular weight distribution
DNbpy	4,4'-di-5-nonyl-2,2'-bipyridine
NIPAM	N-isopropyl acrylamide

NPMA	p-nitrophenyl methacrylate
PAA	poly (acrylic acid)
PBMA	poly (n-butyl methacrylate)
PDI	polydispersity index
PDMS	poly (dimethylsiloxane)
PEBr	1-phenylethyl bromide
PECl	1-phenylethyl chloride
PEO	poly(ethylene oxide)
PHEMA	poly(hydroxyethyl methacrylate)
PIB	poly (isobutylene)
PIP	poly (isopropylene)
PMAA	poly (methacrylic acid)
PMDETA	N,N,N',N'',N''-pentamethyldiethylenetriamine
PMOSt	poly (p-methoxy styrene)
PMPS	poly(methylphenylsilylene)
PSt	polystyrene
PtBA	poly (*tert*-butyl acrylate)
P4VP	poly (4-vinyl pyridine)
RAFT	reversible addition fragmentation transfer
RATRP	reversible atom-transfer radical polymerization
SFRP	stable free-radical polymerization
SPP	3-[N-(3-methacrylamidopropyl)-N,N'-dimethyl] ammoniopropane sulfonate
TEMPO	2,2,6,6-tetramethylpiperidinyloxy
TFA	triflic acid
THF	tetrahydrofuran

4 Melt Behaviour of Block Copolymers

SHINICHI SAKURAI
Department of Polymer Science & Engineering
Kyoto Institute of Technology, Matsugasaki, Sakyo-ku, Kyoto 606-8585, Japan

SHIGERU OKAMOTO
Department of Material Science & Engineering
Nagoya Institute of Technology, Gokiso-cho, Showa-ku, Nagoya 466-8555, Japan

KAZUO SAKURAI
Department of Chemical Processes & Environments, The University of Kitakyushu
1-1, Hibikino, Wakamatsu-ku, Kitakyushu 808-0135, Japan

4.1 INTRODUCTION

Because of the potential abilities of block copolymers, their use and applications are becoming widespread. It follows that a huge number of studies have recently been published from both the academic and the technological points of view. Of course, there is no way to cover thoroughly all fields of block copolymer research, so in this chapter we will focus on the cutting edge of the melt phase behaviour of block copolymers. Although books and some comprehensive reviews are available [1–17], subsequently many discoveries have been made and promising implications have been noted in the last decade. Therefore, recent advances in the understanding of quiescent melt-phase behaviour, thermally induced order-to-order transitions (OOTs), structure evolution across the disorder-to-order transition, and crystallization behaviour of semicrystalline block copolymers are reviewed in this chapter.

4.2 AN OVERVIEW OF RECENT PROGRESS IN UNDERSTANDING MELT BEHAVIOUR OF BLOCK COPOLYMERS

The melt-phase behaviour of block copolymers has been theoretically examined by many researchers using the self-consistent mean-field theory (SCFT), cell dynamics simulations, and so forth. Especially, it is noteworthy that Matsen and Bates [10] have constructed the phase diagram for an A-B diblock copolymer with a stable phase of close-packed spheres (CPS) in addition to body-centred-cubic (bcc) packed spheres, hexagonally close packed (hex) cylinders, double gyroid (DG), and alternating lamellae (lam). Recall that Leibler [18] showed composition-dependent phase boundaries between different types of microphase regions in his theoretical phase diagram in the context of the weak segregation theory, but the CPS phase has not been predicted. Experimentally, only the bcc packing is observed for bulk sphere-forming block copolymers and an equilibrium CPS phase has not been identified. Note, however, that fcc (face-centred-cubic) packed spheres have been well studied for block copolymer micelles in solution [19–21] and interesting results were obtained: Micelles with relatively short corona chains form a fcc structure, whereas micelles with long corona chains adopt bcc packing [20]. Thomas et al. [22] already gave a possible explanation in the context of the Wigner–Seitz cell (which defines the territory of corona chains) model for the stability of the bcc packing in bulk, to which long corona chains are, of course, relevant. Very recently, thermoreversible, epitaxial fcc ↔ bcc transitions in block copolymer solutions have been reported [23]. Spheres with fcc packing have also been observed in block copolymer/homopolymer blends in the bulk state [24]. Note that the block copolymers form lamellar microdomains before blending with the homopolymers. Upon blending, spherical microdomains are formed with short corona chains, which may enable the spheres to arrange in an fcc lattice even in the bulk state. Even for neat block copolymers in bulk, the fcc packing of spheres has been found very recently [25]. Here, one-dimensional planar flow was imposed. Such an external field may be relevant to the formation of the fcc structure. The existence of the fcc or CPS phase should be subjected to further critical examination. Unlike other types of morphologies, the order–disorder transition (ODT) in sphere-forming block copolymers has a unique feature. Namely, the spherical micelles do not dissolve at the ODT, although the packing is randomized (undergoing the transition to the disordered micelle phase). Far above the ODT temperature, demicellization occurs for the case of the UCST (upper critical solution temperature)-type block copolymers [26–35]. The theoretical treatment of the ODT in sphere-forming block copolymers [34,35] is considerably less advanced than that for OOTs or ODT from nonspherical phases and is thus required. As for understanding the mechanism of OOTs, which include bcc↔hex, hex↔DG (or hex↔lam), and DG↔lam transitions, this has progressed and advances in theoretical and experimental studies will be extensively reviewed later in this chapter. Theoreticians have detailed the calculation of the phase diagram for more complicated architectures [36–38]. The second simplest architecture is an ABA triblock copolymer [36]. The phase diagram was calculated by SCFT and it was shown that the triblock copolymers remain ordered to higher temperature as compared to the case of diblock copolymers for which the total chain lengths (contour lengths) are half those of the triblock copolymers. Furthermore, the phase diagram is not symmetric with respect to the composition $f = 0.5$, even though equal lengths for A and B segments was considered. The phase diagram was tested experimentally using the small-angle X-ray scattering (SAXS) technique by Mai et al. [39]. They also confirmed that the domain spacing is slightly larger (by $\sim 10\%$) than that for diblocks [39], which showed good agreement with the theoretical prediction by Matsen and Thompson [36]. The theoretical approach was extended to A'BA" triblock copolymers with different length of outer block chains A' and A" comprising the same chemical structure of the repeat unit (A type) [37]. As well as this kind of architectural asymmetry, the asymmetry in the segment size of A and B has been also corrected for [36,37]. Ordering in A_2B_2-type four-arm starblock copolymers was also considered [38]. For experimental studies, model copolymers having well-defined architectures are required. Advances in synthetic techniques provide an opportunity to prepare of model architectures, such as I_5S (I: isoprene, S: styrene) miktoarm starblock copolymers [40]. The influence of packing constraints has been examined, and discontinuous *"chevron tilt grain boundaries"* have been reported. More complicated architectures have only recently been reported, for example by Beyer et al. [41] for graft copolymers with regularly spaced, tetrafunctional branch points. It is already well known that ABC-type triblock copolymers exhibit rich morphological behaviour [42]. More interesting phenomena have been found for the melt-phase behaviour of blends of lamellar triblock copolymers, including the formation of superlattices such as a noncentrosymmetric lamellar structure via blending of AB-, ABA-, and/or ABC-type block copolymers. For details, refer to recent comprehensive reviews [13,16].

Phase behaviour in external fields is another important aspect and extensive studies have been conducted. Since these are not the scope of this chapter, we just provide a brief overview here. The effect of shear on lamellar orientation has been studied by computer simulations (cell dynamics simulations) and insights into defect formation have been obtained [43–45]. A series of experimental studies have also been conducted on the shear orientation of lamellae [14,46]. Recently, interesting results were obtained for lamellae consisting of pentablock copolymers [47,48]. As for the orientation of hex cylinders, besides the utilization of shear flow [14,49], Thomas and coworkers [50,51] have obtained near-single crystals of hex cylinders using the roll-casting method, where a flow field is imposed concurrently with solvent evaporation. The oriented cylinders were further studied under perpendicular deformation [50,51]. It has also been shown that application of a flow field can induce an OOT and further that it is possible to obtain oriented hex cylinders via the OOT from the bcc structure [52–54] or from the DG structure [55]. To orient DG structures the roll-casting method is also effective [56]. It has been reported that stretching can be utilized as an effective external field to induce lamellar orientation in crosslinked lamellar block copolymers [57,58]. Panyukov and Rubinstein [59] have proposed theoretically the following mechanism of lamellar orientation. When a crosslinked block copolymer network is anisotropically deformed, frozen random elastic forces disrupt anisotropically the long-range order of lamellar microdomains. As a result, preferential orientation is induced with respect to the stretching direction. The lamellar orientation was confirmed experimentally [60]. This is a confirmation of the symmetry breaking of an isotropic lamellar orientation induced by an anisotropic deformation of the block-copolymer network. For a discussion of other interesting phenomena and significant results on the structure and flow behaviour of block copolymers, a comprehensive review is available [14]. It can be mentioned that electric fields have been utilized as a very efficient technique to align lamellar or cylinder structures [61–67].

Crystallization of a semicrystalline block copolymer quenched from the melt will also be briefly reviewed. Chu and Hsiao [68] comprehensively reviewed recent developments in SAXS where they discussed simultaneous measurements with other techniques. Among recently developed techniques, we will focus on simultaneous SAXS/WAXS (wide-angle X-ray scattering) and/or Hv-SALS (depolarized small-angle light scattering) measurements [69,70] because these are powerful techniques to study crystallization and spherulitic higher-order hierarchical structures in semicrystalline block polymers [71,72]. Current developments will also be reviewed later in the subsection on semicrystalline block polymers.

4.3 MELT-PHASE BEHAVIOUR OF AMORPHOUS BLOCK COPOLYMERS

4.3.1 2-DIMENSIONAL SMALL-ANGLE X-RAY SCATTERING (2D-SAXS) TECHNIQUE FOR PRECISE ANALYSES OF MICRODOMAIN STRUCTURES

Block copolymers are materials widely used as thermoplastic elastomers [6,73–75]. To quantify the relationship between structure and mechanical properties, structural deformation upon uniaxial stretching of block copolymers has been intensively studied using the 2-dimensional SAXS technique [73,75]. The advantage of 2D-SAXS is exemplified for instance by the evolution of new peaks due to the breaking of the diffraction extinction rule through the change in the space group symmetry upon deformation of the DG structure [75]. The 2D-SAXS technique was also applied to studies of herringbone structures [58,76–78], which result from yielding and successive microfracture of alternating glassy/rubbery lamellae [76]. Further discussion on mechanical properties and fracture of the microdomain structures is beyond the scope of this chapter. For more details, refer elsewhere [6,11,12,14].

4.3.2 MACROMOLECULAR ORIGIN OF COMPLEX MICRODOMAIN STRUCTURES

The characteristics of block copolymer chains strongly govern the melt-phase behaviour and the detailed microdomain structure. In this regard, the microdomain structure reflects the state of the chain dimensions in the block copolymer. Deviations from Gaussian conformations due to segregation between immiscible block chains and nonuniformity of the deviation are observed. Advances in theory are summarized from the viewpoint of a standard Gaussian model in a recent comprehensive review by Matsen [17]. Nonuniformity of the chain conformations in the Wigner–Seitz unit cell results from the constraint to maintain uniform interfacial curvature, which is generally observed in a diblock copolymer system. Minimizing the nonuniformity of the chain conformations in the Wigner–Seitz unit cell also accounts for the favoring of the body-centered cubic arrangement over the close packing of spherical microdomains in three dimensions [22]. The stability of interfaces of the complex phases such as gyroid, perforated lamella, and diamond structures has been discussed with respect to the nonuniformity (distribution) of the curvature of such interfaces [17]. Based on SCFT calculations, it was shown that the gyroid structure has a minimum distribution of the curvature among these three, supporting the recent theoretical result that the gyroid structure is thermodynamically stable, whereas the other complex phases are not [79–81]. Furthermore, the finite nonuniformity of the curvature of the gyroid interface rather suggests its metastability in the strong-segregation limit [17]. Note here that the curvature distribution of the cylinder interface is infinitesimally narrow.

4.3.3 ORDER-TO-ORDER TRANSITION

The dynamical aspects of melt-phase behaviour are also governed by interfacial reorganization. It has been recognized theoretically that interfacial fluctuations or corrugations [80–90] play an important role in order-to-order transitions and set the kinetic pathway (or intermediate state) [91–96]. Examples are shown in Figure 4.1 for hex-to-bcc, hex-to-lam, and bcc-to-hex transitions. These are the results obtained using a theory of anisotropic fluctuations [97] introduced to study the stability of ordered phases [80,81]. Experimental evidence for interfacial fluctuations has been presented for the hex-to-bcc transition by 2D-SAXS for well-aligned hex cylinders [86,87]. The presence of interfacial fluctuation was confirmed by transmission electron microscopy (TEM) observations performed by the same authors.

SAXS has been widely used for the quantitative analysis of OOTs. To check so-called epitaxial relationships [89,95,96,98,99] at an OOT, the change in the position of the first-order diffraction peak should be monitored. As for the transition mechanism, the temporal change in the peak area for the respective ordered phase is analyzed following Avrami-type kinetics [100,101]. Namely, the temporal change in the peak area, $A(t)$, is expressed by an exponential decay function with the annealing time to the power n, as follows:

$$A(t) = (A_0 - A_\infty)\exp\{(-t/\tau)^n\} + A_\infty. \quad (4.1)$$

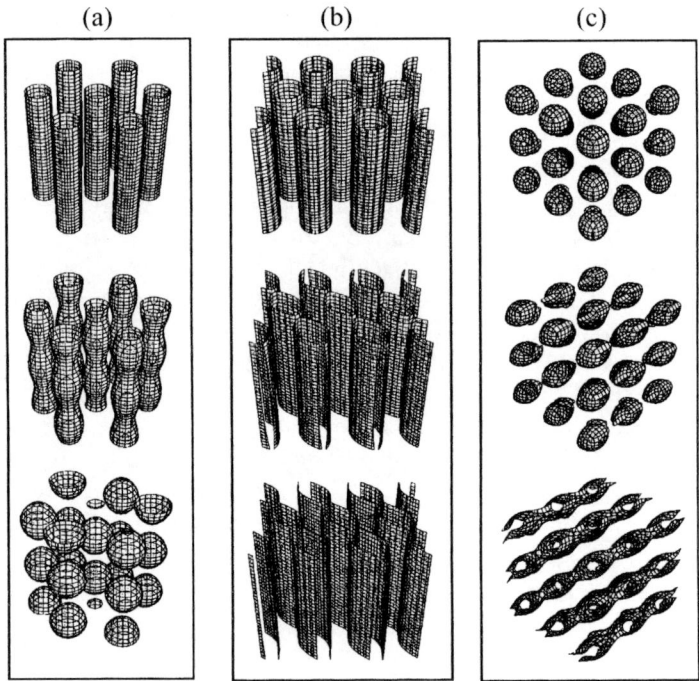

Figure 4.1 Examples of interfacial fluctuations or corrugations at an OOT (a) hex-to-bcc, (b) hex-to-lam, and (c) bcc-to-hex transitions. (Reproduced from M. Laradji *et al.* (1997) Macromolecules, 30: 3242–3255, Copyright (1997) with permission from the American Chemical Society.)

The quantity τ denotes the relaxation time, and A_0 and A_∞ stand for the peak area at $t = 0$ and ∞, respectively. More generally, the time-dependent change in the volume fraction of one phase, $\phi(t)$, is examined.

$$\phi(t) = (1-\phi_\infty)\exp\{(-t/\tau)^n\} + \phi_\infty. \qquad (4.2)$$

From the SAXS results, $\phi(t)$ can be evaluated simply from the fractional peak area of the respective phase, A_1 and A_2, as follows:

$$\phi(t) = k_1 A_1/(k_1 A_1 + k_2 A_2) \qquad (4.3)$$

where k_1 and k_2 are constants to convert the peak area to the volume of the respective phase. Avrami-type analyses have been used to analyse the kinetics of many types of OOT [95,96,102,103]. Jeong *et al.* [90] applied this method to the lam-to-hex transition and obtained interesting results for a SEBS (styrene-ethylene butylene-styrene) triblock copolymer. As shown in Figure 4.2, the Avrami exponent n evaluated from the SAXS results (open circles) exhibits a discrete change as a function of the annealing temperature. Note here that this transition is not a thermodynamically equilibrium one. Namely, the lamellar structure is

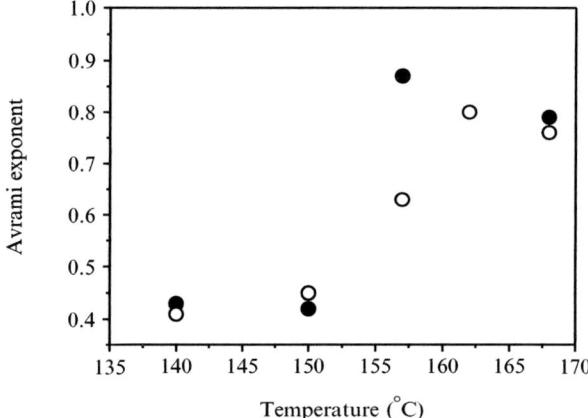

Figure 4.2 Avrami exponent n evaluated from the SAXS results (open circles) and from rheological results (closed circles) for the lam-to-hex transition in a SEBS triblock copolymer [90].

kinetically trapped in the as-cast film due to vitrification of one phase (PS phase in this particular case) when the sample is cast from a solution in a selective solvent. At lower temperatures, the value of n is about 0.4, while it is approximately 0.8 at higher temperatures. There is a threshold around 155 °C, and the mechanism of the transition differs above and below this temperature. At lower temperatures, the growth of the hex phase is suggested to be one-dimensional according to the Avrami dynamics of crystallization with diffusion control [101]. On the other hand, two-dimensional growth of the hex phase is suggested for higher annealing temperatures. These mechanisms were confirmed by TEM [90]. In-plane modulations and correlations of the in-plane modulation of lamellae may be relevant to one- and two-dimensional growth of the hex phase, respectively. It is of great interest that these mechanisms change discontinuously around 155 °C. Note that the results obtained from rheology (Avrami-type temporal change of the storage shear modulus G') included in Figure 4.2 (closed circles) agree very well with the results from the SAXS measurements.

The OOTs involving a more complex phase, DG, have also been studied theoretically. An epitaxial transition from hex to DG and a reverse transition are illustrated in Figure 4.3. These are along a low-energy pathway connecting the local minima of the hex and DG phases, revealed by studying the topography of the Landau free-energy surface [99]. Experimentally detecting intermediate structures is a future challenge. Very recently, Wang and Lodge [95,96] found the hexagonally perforated lamellar (HPL) phase as an intermediate state for the transition from hex to DG for a SI (styrene-isoprene) diblock copolymer solution in DBP (dibutyl phthalate). Note here that the HPL phase is theoretically known to be a nonequilibrium structure [79–81]. They also observed a direct transition from hex to DG for a shallow quench. The results are

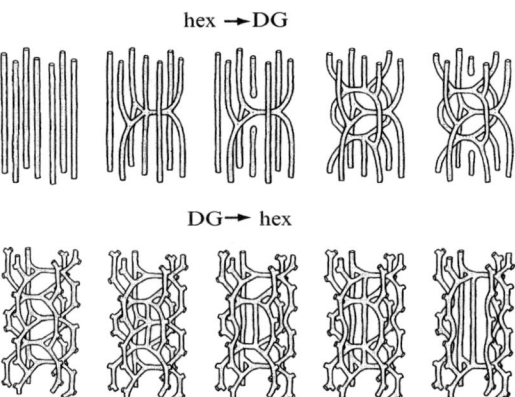

Figure 4.3 Schematic illustration of the epitaxial transition from hex to DG and the reverse transition. These are along a low-energy pathway connecting the local minima of the hex and DG phases, revealed by studying the topography of the Landau free-energy surface. (Reproduced from M. W. Matsen (1998) Phys. Rev. Lett. 80: 4470–4473, Copyright (1998) with permission from the American Physical Society.)

summarized in Figure 4.4(a). To explain such quench-depth-dependent transition pathways, a qualitative analysis has been presented. As shown in Figure 4.4(b), the free-energy curves of the respective phases are very complicated around the threshold temperature ($\sim 68\,^\circ$C). Note that the equilibrium morphology is in the sequence lam $\rightarrow$ DG $\rightarrow$ hex with increasing temperature and the HPL phase is always metastable. The free energy is in the order hex > HPL > DG for temperatures just below 68 $^\circ$C, which may account for the formation of the metastable HPL phase as an intermediate structure. On the other hand, it is in the order HPL > hex > DG at higher temperatures, giving rise to the direct transition of hex to DG. This interesting scenario should be subjected to further experimental examination.

It is also noted in Figure 4.3 that the modulation of cylinders is involved in the transition to the DG structure. Such modulations discussed above are all undulation (corrugation) modes. Wang *et al.* [104] pointed out the significance of another modulation, which is the amplitude-modulation mode. Those two modes are illustrated in Figure 4.5 and look very similar to zigzag and Eckhaus instability modes when a periodic ordered structure is brought into a nonequilibrium state by distortion of the period (by compression or extension) [43,105,106]. These instabilities are transient upon compression or extension of the periodic structure and the randomization of orientation or polygrain structure follows. In this regard, the herringbone structure as discussed above is an example of a transient instability, which is the only case at present for which the pattern dynamics (from herringbone to well-oriented lamellae) has been studied [58]. Pattern dynamics of block copolymer microdomains deserve further investigation in relation to the field of nonlinear science [105,106].

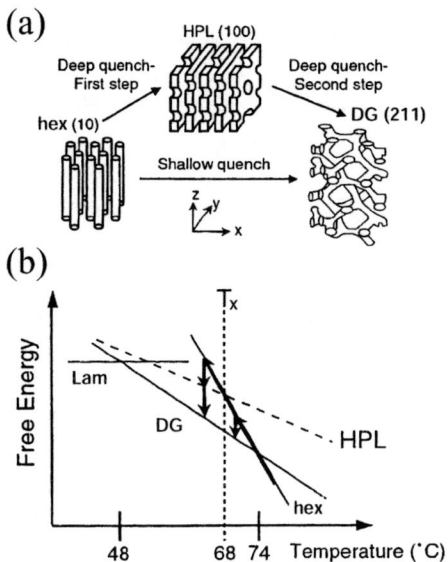

Figure 4.4 (a) Schematic illustration for direct and indirect transition from hex to DG, depending on the quench depth. (b) Free-energy curves of the respective phases. From a low-temperature region the equilibrium morphology changes in the order lam → DG → hex with an increase of temperature and the HPL phase is always metastable. (Reproduced from C.-Y. Wang and T. P. Lodge (2002) Macromolecules **35**: 6997–7006, Copyright (2002) with permission from the American Chemical Society.)

In the case of the hex-bcc OOT, a memory of the orientation of hex cylinders has been reported for a SI diblock copolymer melt [88]. As shown in Figure 4.6, the prominent POM (polarizing optical microscopic) image almost recovers upon the transition from well-oriented hex to bcc that was then followed by the successive reversing transition from bcc to hex. Since cylindrical particles exhibit form birefringence [107], the prominent POM image is ascribed to a grain structure [108–111] of oriented cylinders extending from 10 μm to 0.3 mm. Note here that spherical microdomains have no form birefringence due to their isotropic shape. Therefore, the disappearance of the prominent POM image upon heating to 118.6 °C clearly indicates the hex-to-bcc transition. As pointed out by Lee *et al.* [112], randomization of orientation of hex cylinders should result from cycling through transitions via the bcc phase, because the twinned bcc is generally obtained from the well-oriented hex and then the coalescence of spheres can take place equivalently in the seven degenerate [111] directions. The reason for the memory effect observed in Figure 4.6 originates in slow diffusion of the block copolymer molecules. Namely, the time scale required for the randomization of the hex orientation is quite long compared to the duration of annealing at 118.6 °C. To support this scheme, the idea of deterministic spherical coalescence to recover the original cylinder is required. The other

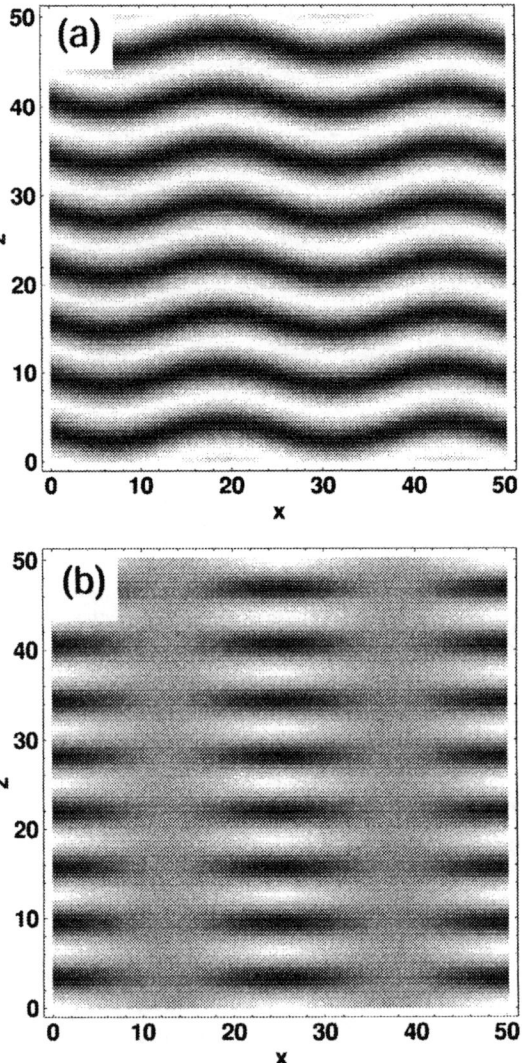

Figure 4.5 Illustration of two kinds of modulation modes (undulation and amplitude-modulation mode), based on theoretical considerations. (Reproduced from S. Qi and Z.-G. Wang (1999) J. Chem. Phys. 111: 10681–10688, Copyright (1999) with permission from the American Institute of Physics.)

equivalent [111] directions are necessarily excluded upon coalescence. The illustrations presented in Figure 4.7 explain this anomaly, where the long lifetime of poles on the spherical microdomains in the direction parallel to the original cylinder axis is indicated in Figure 4.7(a) and deterministic spherical coalescence thus results as shown in Figure 4.7(b). In Figure 4.7(a), the dots

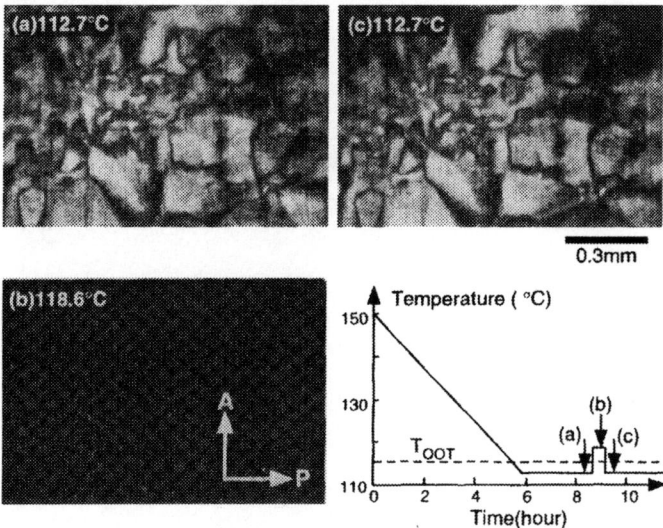

Figure 4.6 Change in POM images across the transition from well-oriented hex to bcc that was then followed by the reverse transition from bcc to hex. Since cylindrical particles exhibit form birefringence, the prominent POM image is ascribed to a grain structure of oriented cylinders spreading from 10 μm to 0.3 mm. Note here that spherical microdomains have no form birefringence due to their isotropic shape. (Reproduced from K. Kimishima *et al.* (2000) *Macromolecules* **33**: 968–977, Copyright (2000) with permission from the American Chemical Society.)

specify the on-interface distribution of chemical junctions between A and B block chains in A-B diblock molecules. When cylinder undulation sets in, a lowering of the number density of the chemical junctions is accompanied around a neck due to the outgoing diffusion of the block copolymer chains. The neck points evolve into poles during the pinching-off of a cylinder so that the poles are uniaxially arranged along the original cylinder axis even after a sphere forms. The process from step 4 to 5 in Figure 4.7(a) involves annihilation of poles, which is of course attained by diffusion of block copolymer molecules. Therefore, this process may be rate-determining and the poles will be long lived. Before the annihilation of poles is completed, they play a role as most-favoured transition channels for the reverse transition to cylinders as schematically explained in the lower panel of Figure 4.7(b), which in turn induce the deterministic coalescence of spheres.

4.3.4 STRUCTURAL EVOLUTION FROM THE DISORDERED STATE

Very little has been known until recently about the mechanism of microdomain structure evolution from the disordered state. Hashimoto and coworkers

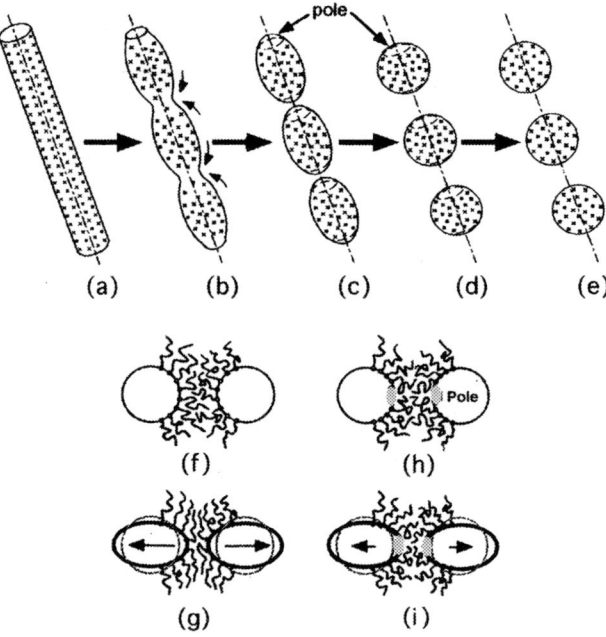

Figure 4.7 Schematic representation for the deterministic spherical coalescence to recover the original cylinder. Parts (a)-(e) show the process from cylinders to spheres via undulating cylinders. The long lifetime of poles on the spherical microdomains in the direction parallel to the original cylinder axis is relevant. Here, the dots specify the on-interface distribution of chemical junctions between A and B block chains in A-B diblock molecules. The sketches in (f)-(i) explain the role of poles in the deterministic coalescence of spheres. (Reproduced from K. Kimishima et al. (2000) Macromolecules 33:968–977, Copyright (2000) with permission from the American Chemical Society.)

[15,113–119] highlighted this problem using lamellar-forming block copolymers and presented interesting results, summarized in Figure 4.8. Panel (a) shows the 1D-SAXS profiles and the inset highlights those for samples in the disordered state, undergoing microphase separation, and in the ordered state, respectively, at 417.5, 415.6, and 413.7 K [117]. Here, the scattering intensity $I(q)$ is shown in a semilog plot as a function of the magnitude of the scattering vector, q, which is defined by

$$q = (4\pi/\lambda)\sin(\theta/2), \qquad (4.4)$$

where λ and θ are the wavelength of the X-rays and the scattering angle, respectively. For the sample undergoing microphase separation, the SAXS profile seems to be a weighted sum of those for the ordered and disordered samples, implying coexistence of an embryonic lamellar grain embedded in the disordered matrix and further suggesting that the evolution of the microdomains occurs via nucleation and growth. This conjecture is confirmed by the

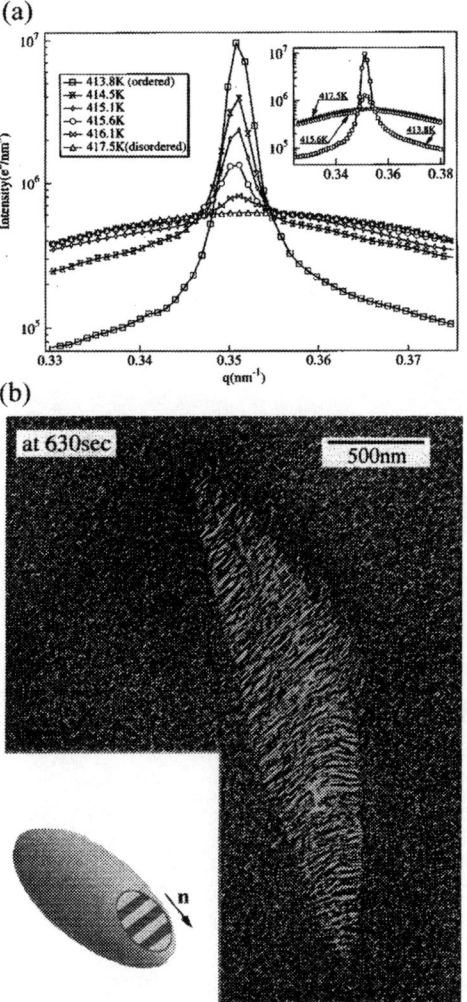

Figure 4.8 1D-SAXS profiles for a SI diblock copolymer upon microphase separation from the disordered state. The inset highlights the ones for samples in the disordered state, undergoing microphase separation, and in the ordered state, respectively, at 417.5, 415.6, and 413.8 K (Reproduced from T. Hashimoto (2001) Macromol. Symp. **174**: 69–83, Copyright (2001) with permission from Wiley–VCH). (b) TEM photograph showing an embryonic lamellar grain embedded in the disordered matrix for the sample undergoing microphase separation. (Reproduced from N. Sakamoto and T. Hashimoto (1998) Macromolecules **31**: 3815–3823, Copyright (1998) with permission from the American Chemical Society.)

TEM photograph [116] shown in part (b). The more interesting finding here is the anisotropic shape of the embryonic lamellar grain in which all the lamellae are oriented perpendicular to the interface between an ordered grain and disordered matrix (OD interface). The same authors performed computer simulations using a phenomenological model equation for the anisotropic nucleation process, taking into account the anisotropy of the interfacial free energy of the OD interface [115]. The result supports the TEM observation, which is in turn explained by anisotropic nucleation. The structure obtained by prolonged annealing is shown in Figure 4.9(a) [116]. Here, one can see a remaining memory of the anisotropic shape of embryos, although the volume

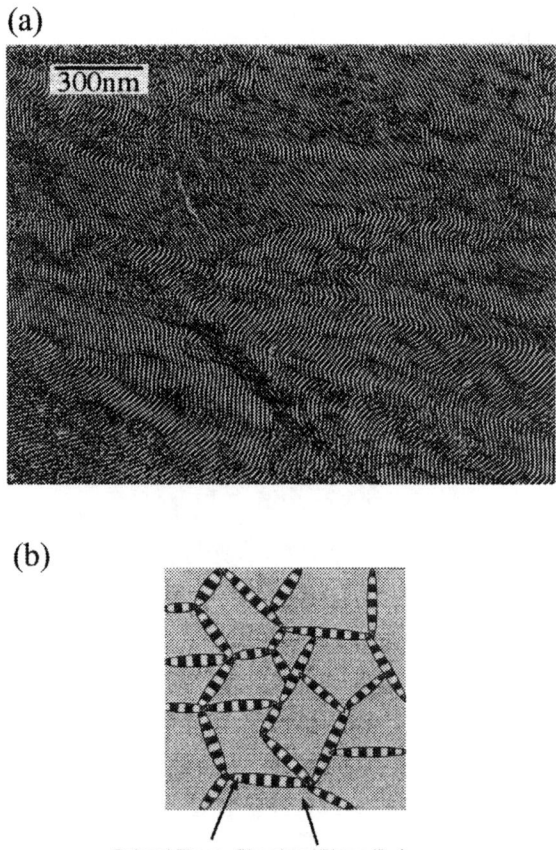

Figure 4.9 (a) TEM micrograph showing structure obtained by prolonged annealing of the sample examined in Figure 4.8. Here, one can see a remaining memory of the anisotropic shape of embryos, although the volume filling of the lamellar microdomains is attained (Reproduced from N. Sakamoto and T. Hashimoto (1998) Macromolecules **31**: 3815–3823, Copyright (1998) with permission from the American Chemical Society.). (b) Schematic illustration to explain this memory effect. As the growth of the randomly oriented embryos proceeds, impingement occurs. Then the volume filling of lamellae takes place in the remaining disordered space so that the final structure retains the memory of the embryos. (Reproduced from T. Hashimoto (2001) Macromol. Symp. **174**: 69–83, Copyright (2001) with permission from Wiley–VCH.)

filling of the lamellar microdomains is attained. This memory effect is explained as follows. As the growth of the randomly oriented embryos proceeds, impingement occurs as illustrated in Figure 4.9(b) [15]. Then the volume filling of lamellae takes place in the remaining disordered space so that the final structure retains the memory of the lamellar embryos.

4.3.5 DETECTION OF LATENT HEAT AT AN ODT OR OOT

Kim *et al.* [83] successfully detected the latent heat at the ODT and hex-to-bcc OOT with DSC for a SIS triblock copolymer, as shown in Figure 4.10. Note that many researchers have reported detection of the ODT using DSC [120–125] but it may be the first report for an OOT. Figure 4.10(a) indicates the results of rheological measurements. The huge drop in G' suggests the occurrence of the OOT while the gradual drop in the loss shear modulus (G'') around 275 °C indicates the ODT. At such transitions, the DSC thermogram exhibits peaks. The exothermic peak around 192 °C and the endothermic peak around 275 °C are due to the OOT and the ODT, respectively. Figure 4.10(c) explains why the hex-to-bcc OOT is characterized by an exothermic latent heat. The key here is undulation of cylinders. Due to an increase in the packing entropy (which was specified as the entropy due to spatial regularity in the packing of microdomains [83], in other words, the position of interfaces) upon the undulation of the interface, the enthalpy gradually increases with the elapsed time during the DSC measurement (i.e., apparently temperature). On the other hand, right after the transition into bcc spheres the ordering improves and hence the entropy immediately decreases. Thus the enthalpy is released abruptly and an exothermic heat flow results at the transition from undulated cylinders to bcc spheres. Without the existence of the undulation mode of cylinders, the change in the enthalpy level may be negligible for the direct transition from cylinders (without undulation) to spheres.

4.4 INFLUENCE OF MELT MICRODOMAINS ON CRYSTALLIZATION IN SEMICRYSTALLINE BLOCK COPOLYMERS QUENCHED FROM THE MELT

4.4.1 OVERVIEW

Spatial confinement provided by microphase separation should play a crucial role in the crystallization of crystalline/amorphous diblock copolymers. Register and coworkers [126–130] extensively studied crystallization kinetics and clarified how crystallization is influenced by the segregation strength, the domain structure, and the mobility of the amorphous chains. Readers who are interested in those general issues could refer to Chapter 6 or ref. [11]. This

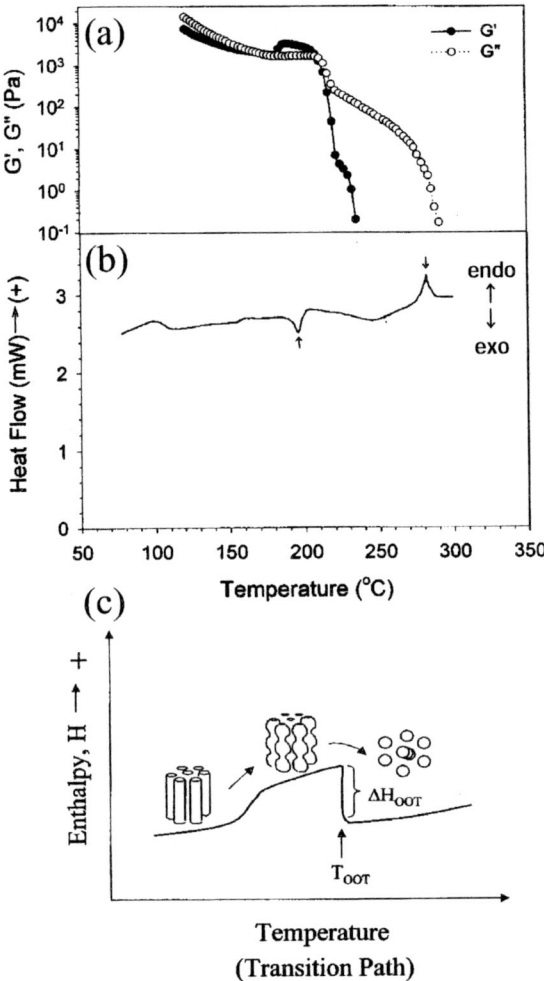

Figure 4.10 (a) Results of rheological measurements for a SIS triblock copolymer bulk sample undergoing a hex-to-bcc transition. The huge drop in G' suggests the occurrence of the OOT while the gradual drop in G'' around 275 °C indicates the ODT. (b) DSC thermogram for this sample showing an exothermic peak around 192 °C and an endothermic peak around 275 °C, which are due to the OOT and the ODT, respectively. (c) Schematic representation for change in enthalpy along with hex-to-bcc transition via undulation of cylinder. (Reproduced from J. K. Kim et al. (1998) Macromolecules **31**: 4045–4048, Copyright (1998) with permission from the American Chemical Society.)

section focuses in particular on preferential orientation of the crystallites that are formed when crystalline/amorphous diblock copolymers are crystallized while maintaining the microdomain structure throughout the crystallization.

Such spatially confined crystallization is only observed in two cases, (1) the glass transition temperature (T_g) of the amorphous block is higher than that of the melting temperature (T_m) of the crystalline block, and (2) the degree of crystallinity is relatively low and the amorphous and crystalline blocks are strongly segregated in the molten state.

The nature of crystallization within microdomains has been a controversial issue in the last decade. In the early 1960s, Skoulios et al. [131] studied poly(ethylene oxide)-b-polystyrene-b-poly(ethylene oxide) (PEO-b-PS-b-PEO) triblock copolymers in a selective solvent for PS. They concluded that the PEO chain is perpendicular to the microdomain interface, investigating a particular case that the block copolymers were microphase separated to form a lamellar structure and where the T_g of the PS block was reduced so as to be lower than T_m of the PEO block by the solvent.

Theoretical studies [132,133] have also stimulated interest in this issue. The free energy for a model of amorphous/crystalline block copolymers has been calculated, using self-consistent mean-field theory. It was assumed that the amorphous blocks are flexible chains, the crystalline blocks being modeled as fully folded chains (i.e., the crystallinity is 100%), and the alternating lamellar microdomain structure is maintained throughout the crystallization. One of the theoretical predictions is that the chain stems align perpendicularly to the microdomain interface. Since then, many studies [134–138] have been undertaken to examine the theoretical prediction. However, it was experimentally difficult to attain the conditions applied in the theories (i.e., 100% crystallinity and lamellar microdomain structure maintained throughout the crystallization). This is because the crystallization usually destroys or heavily deforms the preceding lamellar microdomain structures. To maintain the preceding microdomain structure, the T_g of the amorphous block should be higher than the T_m of the crystalline block. In this case, the crystallization occurs in the molten domain sandwiched between the glassy domains; furthermore, the amorphous block loses micro-Brownian motion and the junction points between the amorphous and crystalline blocks are anchored to the interface of the two domains. Such a system is not appropriate to examine the theories because the theories assume a flexible chain as the amorphous block. Thus scientific interest seems to be shifting from examination of the theory to investigation of the confined crystallization itself.

4.4.2 PREFERENTIAL CRYSTALLITE ORIENTATION AND CONFINEMENT IN LAMELLAR MICRODOMAINS

Zhu et al. [139] studied the crystallization temperature (T_c) dependence of the crystallite orientation for a lamellar-forming poly(ethylene oxide)-b-polystyrene (PEO-b-PS) diblock with M_w (PEO) $= 8.7 \times 10^3$ and $M_{w(PS)} = 9.2 \times 10^3$, T_g of PS $= 62\,°C$ and T_m (PEO) $= 51\,°C$. PEO usually exhibits a monoclinic crystalline form [140] and the fiber pattern (with the *c*-axis or the stem direction parallel to the meridian) has a characteristic feature, showing a strong (120) reflection on the equator. When the lamellae were oriented in the molten state and quickly quenched, the crystallite was randomly oriented in the lamellar microdomain, thus the two-dimensional WAXS pattern was isotropic. However, when the sample was isothermally crystallized between $T_c = -50$ and 35 °C, the WAXS pattern became anisotropic. Figure 4.11 compares azimuthal profiles of the (120) reflection among samples isothermally crystallized at different T_c from -50 to 35 °C, where the azimuthal angle (Φ) is defined so that $\Phi = 0$ along the meridian. They also obtained two-dimensional SAXS data

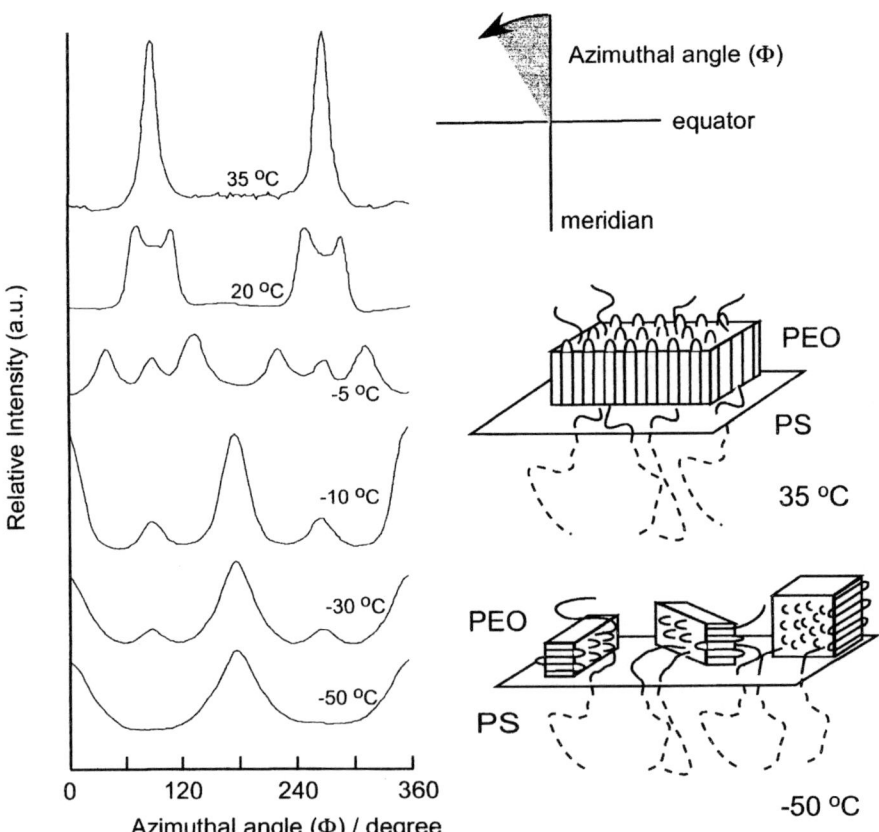

Figure 4.11 Temperature dependence of azimuthal profiles of the (120) reflection and the proposed models for the crystallite orientation at 35 and -50 °C. The azimuthal angle is defined so that the meridian direction has $\Phi = 0$ and the lamellar interface is perpendicular to the meridian. (Reproduced from L. Zhu *et al.* (2000) J. Am. Chem. Soc. **122**: 5957–5967, Copyright (2000) with permission from the American Chemical Society.)

and found that lamellae are formed at all T_c and that the microdomain interface aligns parallel to the equator. At $T_c = 35\,°C$, the (120) reflection has maxima at $\Phi = 90$ and $180°$ (the equatorial direction). This indicates that the c-axis is preferentially oriented perpendicular to the microdomain interface, as illustrated in the figure. With decreasing T_c, the c-axis tends to incline with a particular angle (depending on T_c) with respect to the interface. Finally, the c-axis is oriented parallel to the interface in the range of T_c, $= -10$ to $-50\,°C$. This is a very interesting finding and it remains to be explained why the c-axis orientation is determined by T_c.

PEO is known to achieve a relatively high crystallinity, usually more than 70–80%, thus the microdomain morphology after crystallization should be changed considerably, depending on how the crystallization takes place. This is why the PS glassy microdomain is necessary to maintain the preceding lamellar microdomain structure. In fact, if the amorphous domain is in the rubbery state, it is difficult to maintain the preceding lamellar microdomain structure. Nojima et al. [141] have studied a symmetric poly(ε-caprolactone)-b-polybutadiene (PCL-b-PB) diblock (where the T_g of PB and the T_m of PCL are $-100\,°C$ and $50\,°C$, respectively) and concluded that the crystallization overwhelms the preceding lamellar microdomain structure and alters the microdomain morphology.

Sakurai et al. [142–145] have studied crystallization of polyethylene-b-atactic poly(propylene) (PE-b-PP), where the polyethylene and polypropylene blocks were prepared through hydrogenation of anionically polymerized polybutadiene and 2-methyl-1,3-pentadiene, respectively. Due to this synthetic procedure, the resultant polyethylene has a few mol% of ethyl branches, which reduces the crystallinity to about 40%. Furthermore, the polypropylene is completely atactic, thus amorphous. The total M_w of the PE-b-PP was 1.13×10^5, and M_w of the PE block and that of the PP block were 5.42×10^4 and 5.88×10^4, respectively. This PE-b-PP sample segregates strongly to form an alternating lamellar microdomain structure in the melt [143]. The T_m of the PE block is around $100\,°C$ and the T_g of the PP block is $0\,°C$, therefore, the PE blocks crystallize surrounded by the rubbery PP. Sakurai et al. [143] showed that when planar shear is applied to the PE-b-PP sample at $150\,°C$ (above T_c), the lamellar microdomain can be oriented parallel to the shear direction. When the oriented sample was quenched to $-40\,°C$, the lamellar microdomain can be "frozen" and the orientation of the lamellae is maintained. These facts were proved by small-angle neutron scattering. PE-b-PP has no contrast for X-rays above T_c, because both blocks have a similar electron density. However, once cooled below T_c, the PE block crystallizes, which provides enough contrast for X-rays (also for electrons in TEM). The SAXS profile labeled as "quenched to $-40\,°C$" in Figure 4.12 were obtained for a quenched sample. The first- and third-order diffraction peaks due to lamellar microdomains are observed. The absence of the second-order peak is due to the equal thickness of each lamella. It should be noted that when the Lorentz-corrected plot ($q^2 I$ vs q) is constructed (see the

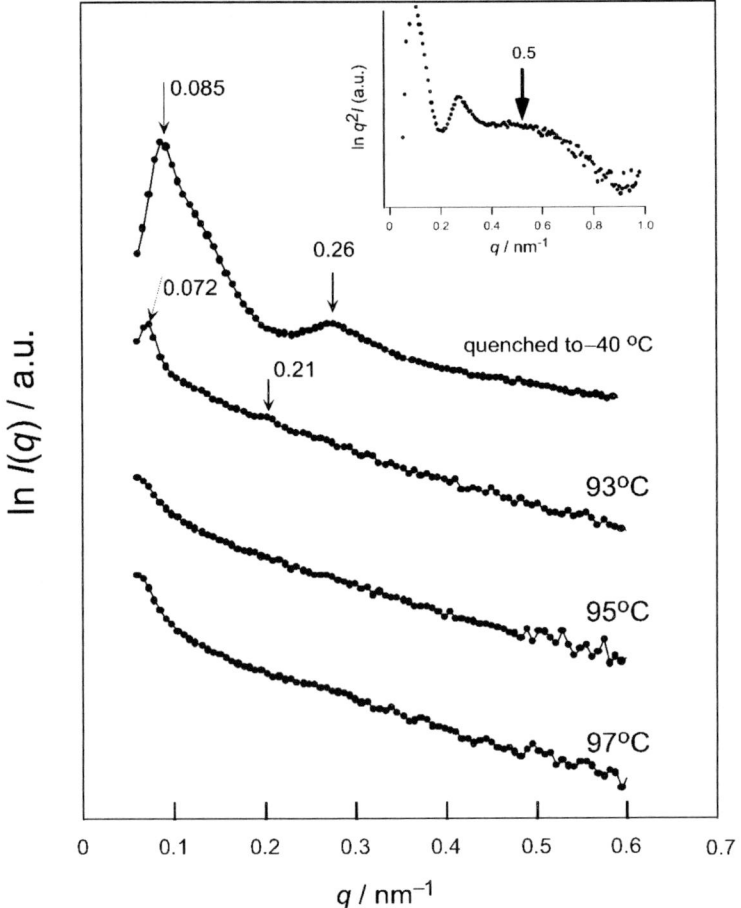

Figure 4.12 Comparison of the SAXS profiles for the quenched and isothermally crystallized samples (PE-*b*-PP diblock copolymer). The arrows indicate the diffraction peak positions (1st and 3rd peaks) due to the lamellar microdomains. The inset shows the $q^2 I$ vs q plot to evaluate the long spacing of the crystalline lamellae [144].

inset of Figure 4.12), the peak indicates the existence of the long spacing due to the PE crystalline lamellae. Thus, it was confirmed that when the PE-*b*-PP sample was quenched, the preceding lamellar microdomain structure was preserved and the PE block could crystallize within the microdomain. This confined crystallization was also proved with TEM [142–144].

When the two-dimensional WAXS pattern was measured for the quenched sample by sending the X-ray beam parallel to the domain interface, the resultant pattern was anisotropic. Figure 4.13 shows the WAXS results. In the equatorial direction (denoted by A), there are strong 110 and weak 020 and 200 reflections. In the meridional direction, there is no 020 reflection and the

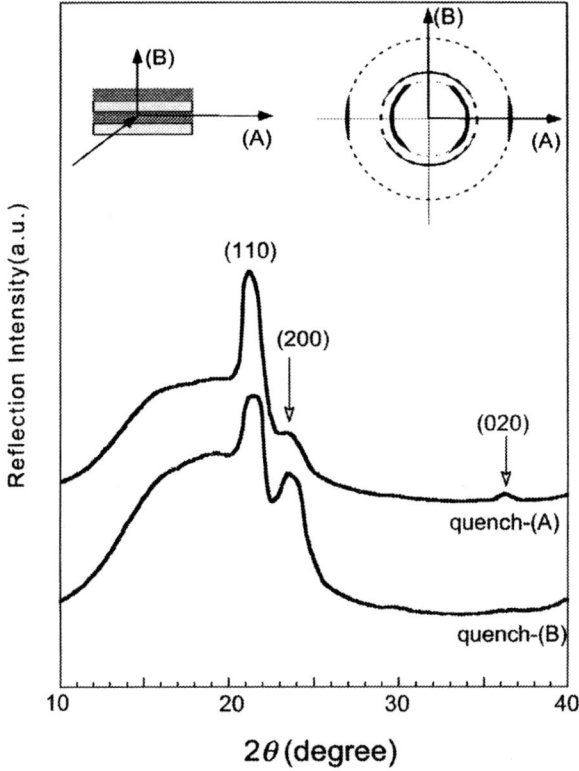

Figure 4.13 Comparison of the WAXS profiles along the equatorial and meridional directions for the quenched samples (PE-*b*-PP diblock copolymer). Also included are schematic representations of the two-dimensional WAXS pattern (upper right) and the X-ray beam coordinates (upper left) [143,144].

intensity of the 200 peak is higher than that along the equator. This anisotropic pattern is quite similar to the fibre pattern observed for uniaxially oriented polyethylene [144]. This fact indicates that the PE crystallite takes an oriented row structure in the lamellar microdomain; namely, the *b*-axis of the PE crystallite is oriented parallel to the lamellar microdomain interface, while the *a*-axis and *c*-axis (chain stem direction) rotate around the *b*-axis, as represented in Figure 4.14. The thickness of the PE lamellar microdomain ($D/2$) and the long spacing of the PE crystallite were evaluated from the SAXS profile in Figure 4.12 to be, respectively, 37 and 13 nm. Therefore, about three crystallite layers can be accommodated within the PE microdomain. When the quenched sample was heated to 150 °C and then isothermally crystallized at 93 °C, the preceding lamellar microdomain was considerably deformed, however, it still remained (which is shown by the fact that the SAXS profile for $T_c = 93$ °C exhibits the first-order and third-order diffraction peaks, as shown in Figure

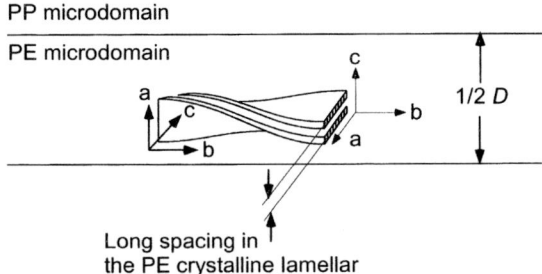

Figure 4.14 Illustration of the oriented row structure observed in the quenched sample (PE-*b*-PP diblock copolymer). The *b*-axis orients in parallel with the domain interface and the other two axes rotate around the *b*-axis [144].

4.12) and the PE crystallite retains orientation, although to a lesser degree. When the sample was isothermally crystallized above 95 °C, the lamellar microdomain structure appeared to be destroyed completely and PE spherulites were observed by optical microscopy and light scattering [142,143]. The spherulite radius was about 70 μm [142], which is much larger than that of the PE microdomain. This fact suggests that the PE crystallite can grow much larger than the PE microdomains. This T_c dependence of the crystallite orientation and morphology can be explained as follows. When the sample is quenched to −40 °C, crystallization takes place quickly. If the crystallization is completed more rapidly than the deformation of the PP microdomain, it cannot be deformed. On the other hand, when the sample is isothermally crystallized at high temperatures, the crystallization takes place gradually enough so as to induce the deformation of the microdomain.

4.4.3 SIMULTANEOUS SAXS/WAXS/*H*v-SALS TECHNIQUE

"Simultaneous" measurements of different techniques have been desirable objectives for some time, and development of these techniques leads to more accurate or novel results than simple comparison of results from separate measurements. In this section, we will briefly review some simultaneous measurement techniques and recent achievements using SAXS/WAXS and H_v-SALS.

Simultaneous measurements together with small-angle X-ray or neutron scattering were performed for more than ten years, as exemplified by the studies of Okamoto *et al.* [146,147] and those of Bates *et al.* [148]. The latter examined the influence of mechanical deformation on the melt-phase behaviour of a block copolymer; poly(ethylene-propylene)-*block*-poly(ethylethylene) (PEP-PEE), which contains 77 vol. % PEP with an overall molecular weight of 1.0×10^5 and a molecular-weight distribution with $M_w/M_n = 1.07$. The order–disorder transition (ODT) temperature was increased by shear deformation in the weak segregation regime, i.e. the microphase separation was induced by the shear. Okamoto *et al.* studied changes in orientation and lattice deformation of spherical microdomains during large-amplitude shear applied to a strongly-segregated asymmetric diblock copolymer melt. For this purpose, they developed [149] and utilized apparatus using image plates for time-resolved 2D SAXS and WAXS measurements. Figure 4.15 shows a typical result [146]. In response to the applied shear deformation, the bcc lattice underwent dynamic lattice deformation under the constraint that the (110) planes orient parallel to the shear plane. They also measured simultaneously the stress during this orientational change. The further development of area detectors, such as CCD cameras, and the utilization of synchrotron radiation as an X-ray source have enabled time-resolved measurements with short intervals (from ms to s) for such studies on the evolution of microdomain structures and rapid structural changes under external fields.

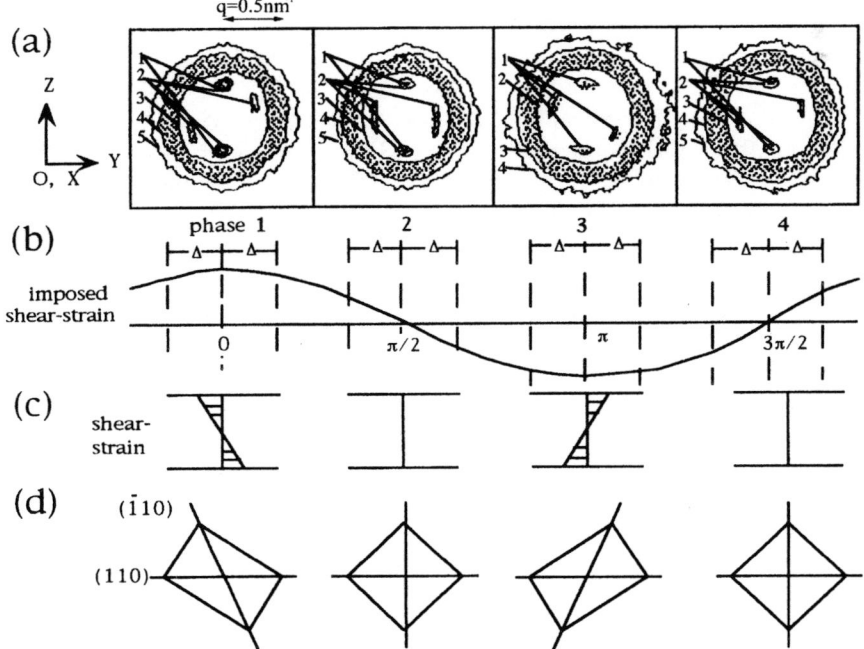

Figure 4.15 2D-SAXS pattern (a) obtained at four representative strain phases as shown in (b) and (c), and the model to explain the four diffraction spots arising from (110) and ($\bar{1}$10) lattice planes (d). The pattern at each phase was obtained for the strain phase of $[-\Delta + \phi_i, \Delta + \phi_i]$ where $\phi_i = 0, \pi/2, \pi$ and $3\pi/2$ for phase 1–4, respectively and $\Delta = 0.194\pi$. The pattern at each phase was obtained by accumulating the SAXS intensity over strain cycles (N) with $80 < N < 150$. The shaded zones in pattern (a) offer visual guides for the scattering maximum and shoulder. The contour lines numbered 1–5 have respectively, logarithm of scattering intensity of 3.60, 3.40, 2.80, 2.40 and 2.20 for the patterns in phases 1 and 2, 3.45, 2.77, 2.32 and 2.10 for those in phase 3, and 3.65, 3.42, 2.74, 2.28 and 2.05 for those in phase 4 [149].

Simultaneous measurement techniques for different kinds of scattering, such as SAXS, WAXS, SALS (Hv-, Vv-) and spectroscopy have been developed. Here Hv- and Vv- mean cross-polarized and parallel-polarized, respectively. Zachmann et al. [150] were the first to perform SAXS/WAXS simultaneous measurements on homopolymer crystallization in 1982. In the early 1990s, they also conducted simultaneous SAXS/WAXS measurements further combined with SALS and/or DSC (differential scanning calorimetry) to measure crystallization and melting behaviour of various homopolymers [69,70,151]. Recently, Okamoto et al. [71,72] applied simultaneous measurements of SAXS/WAXS/Hv-SALS or DSC to a semicrystalline block copolymer. The sample used was a polyethylene-*block*-poly(ethylene-*alt*-propylene) diblock copolymer forming lamellar microdomains. After pre-annealing at 180 °C (much higher than the melting temperature of the polyethylene block, $T_m = 108$ °C), the sample was subjected to a temperature drop (−20 K/min) to a temperature slightly below T_m. Hv-SALS, SAXS and WAXS were recorded as a function of time. The relative invariant (Q) reduced by the value at long annealing time (Q_∞) from Hv-SALS and SAXS are plotted in Figure 4.16, where Q is the intensity integrated over the measurable q region ($Q = \int_0^\infty I(q)q^2 dq$). The crystallinity ($X_{PE}$) was also evaluated by the integrated intensity of the 110 peak in the WAXS profile, which is linearly related to the degree of crystallinity, and this quantity is also shown in Figure 4.16. This figure shows that the Hv-SALS

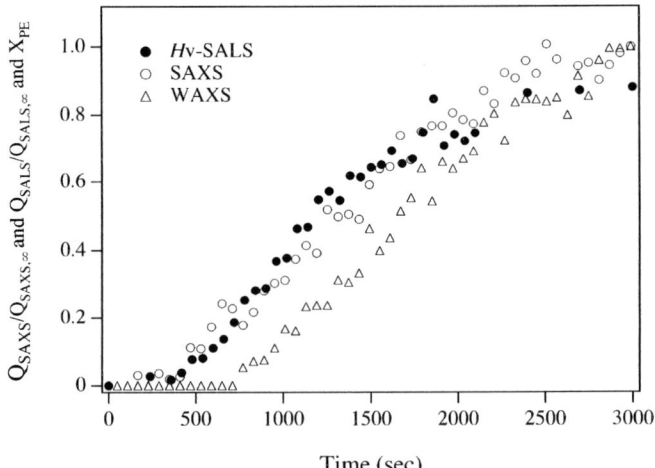

Figure 4.16 SAXS (open circles) and Hv-SALS (filled circles) indicate the relative invariant (Q) reduced by the value measured after sufficiently long annealing time (Q_∞), where Q is the intensity integrated over the measurable q region ($Q = \int_0^\infty I(q)q^2 dq$). WAXS (open triangles) indicates the crystallinity (X_{PE}) evaluated by the integrated intensity of the 110 peak in the WAXS profile, which is linearly related to the degree of crystallinity [71,72].

signal starts to increase almost simultaneously with the SAXS signal, and that WAXS follows, suggesting that long-range density fluctuations (probed by SAXS) and orientational fluctuation (probed by SALS) precede the crystallite nucleation (by WAXS).

It should be rigorously examined whether the order of signal appearance is SALS ≈ SAXS, then WAXS. In the case of homopolymer crystallization, several researchers also reported that the SALS signal increases before the SAXS or WAXS signal start to increase, i.e., density fluctuations with long-range order are generated before crystallization. However, simultaneous measurements have not yet offered a conclusive result because the sensitivity of the detectors for these different techniques is still an issue. Dolbnya et al. [152] developed a 1-dimensional gas microstrip detector for WAXS and SAXS (maximum global count rate of 450 MHz, 4.5×10^5 counts/s/channel). In this situation, they still observed a SAXS signal before WAXS [155], which supports the previous reports on SAXS/WAXS over the last five years [154–157]. Contrary to such results, one can still argue about the detection limit due to statistics [158]. In this case, it is proposed that the experimentally observable induction period is not the time for crystallization to occur, but is merely the time for which crystallinity exceeds some finite value.

Returning to the report by Okamoto et al., it is also interesting that Hv-SALS showed a four-leaf-clover pattern from the beginning in the early stage. This result was similar to that observed by Lee et al. [159]. Okamoto et al. more interestingly reported that long-range order due to density fluctuations was detected and spherulites were formed in the early stage even if the sample had a lamellar microdomain structure with an interlamellar distance of tens of nanometres. The radius was on the order of a micrometre from the beginning and did not change during the crystallization. Surprisingly, the lamellar microdomains were not broken or deformed by the spherulitic formation, as evidenced by SAXS profiles that exhibited multiple scattering maxima up to 4th-order until the end of crystallization. The puzzle here is how the long-range density and orientational fluctuations accommodate within the nanoscaled microdomain structures, this being referred to as crystallization in the confined space of lamellar microdomains. Chen et al. [160] observed crystallite growth in the melt of lamellar microdomains on a macroscopic scale through polarizing optical microscopy. They further reported that the macroscopic growth of crystallites was not observed in cylindrical or spherical microdomains. Thus, it was argued that the crystallites could macroscopically grow exclusively in the lamellar microdomains due to their lateral extent. Although termination of the growth might be expected at any grain boundary, it is still possible to have a continuous through-path at the grain boundary because of the commensuration of the lamellar microdomains. The long-range density fluctuations observed by Okamoto et al. [71,72] may also be accounted for by such an idea.

4.5 FUTURE PERSPECTIVE

Block copolymer systems are rich in microdomain patterns. Nevertheless, the pattern dynamics has not yet been well studied when the equilibrium system is brought to a nonequilibrium state. More studies of pattern dynamics including the growth of grains into single crystals are required in relation to the field of nonlinear science [105,106].

Although pressure effects on the melt-phase behaviour are important to thoroughly understand block copolymers thermodynamics, only the influence on the ODT [161–166] has been examined, and not for OOTs or the morphological phase diagram [167]. Studies of pressure effects on the phase behaviour of block copolymers deserve future investigation, since rich phenomena in the ordered phase diagram have been reported for liquid crystal systems [168].

The existence of the fcc phase in the quiescent bulk of a block copolymer melt has been suggested [25], but critical examination is required. As an external field, magnetic fields have not yet been utilized for orientation of microdomains or for generation of single-crystal grain of the microdomains, although research on this is now underway [169].

It has been clarified that the crystallization temperature of crystalline/amorphous diblock copolymers strongly affects orientation of the crystallite. Such preferential orientation of the crystallite is also observed when crystalline/amorphous diblock copolymers are crystallized in a thin film [142,143]. These studies suggest that the morphology of crystalline/amorphous block copolymers can be controlled at the nanometer scale by combining spatial confinement and the appropriate crystallization temperature.

Simultaneous measurements via SAXS, WAXS and Hv-SALS have revealed concurrent large-scale fluctuations in orientation (further implying spherulite formation) of the optical director upon crystallization in the confined space of the microdomain. Resolving this puzzle may require further studies by simultaneous measurements.

REFERENCES

[1] Molau, G. E. (1970) In *Block Polymers*: Aggarwal SL, Ed., Plenum Press, New York
[2] Hashimoto, T., Fujimura, M., Saijo, K., Kawai, H., Diamant, J. and Shen, M. (1979) In *Multiphase Polymers*: Cooper, S. L. and Estes, G. M., Eds., ACS Advances in Chemistry Series, *American Chemical Society*, Washington, DC, 257–275.
[3] Bates, F. S. and Fredrickson, G. H. (1990) *Annu. Rev. Phys. Chem.* **41**: 525–557.
[4] Bates, F. S. (1991) *Science* **251**: 898–905.
[5] Sakurai, S. (1995) *Trends Polym. Sci.* **3**: 90–98.
[6] Holden, G., Legge, N. R., Quirk, R. and Schroeder, H. E., Eds. (1996) *Thermoplastic Elastomers*, 2nd edition: Hanser, Munich
[7] Hasegawa, H. and Hashimoto, T. (1996) In *Comprehensive Polymer Science*, Aggarwal, S. L. and Russo, S., Eds., Pergamon, Oxford, 497–539.
[8] Sakurai, S. (1996) *Current Trends in Polym. Sci.* **1**: 119–134.

[9] Matsen, M. W. and Schick, M. (1996) *Curr. Opin. Colloid. Interface. Sci.* **1**: 329–336.
[10] Matsen, M. W. and Bates, F. S. (1996) *Macromolecules* **29**: 1091–1098.
[11] Hamley, I. W. (1998) *The Physics of Block Copolymers*, Oxford University Press, Oxford.
[12] Araki, T., Tran-Cong, Q. and Shibayama, M., Eds., (1998) *Structure and Properties of Multi-Phase Polymeric Materials*, Marcel Dekker, New York
[13] Abetz, V. and Goldacker, T. (2000) *Macromol. Rapid. Commun.* **21**: 16–34.
[14] Hamley, I. W. (2001) *J. Phys.: Condens. Matter* **13**: R643–R671.
[15] Hashimoto, T. (2001) *Macromol. Symp.* **174**: 69–83.
[16] Kane, L., Norman, D. A., White, S. A., Matsen, M. W., Satkowski, M. M., Smith, S. D. and Spontak, R. J. (2001) *Macromol. Rapid. Commun.* **22**: 281–296.
[17] Matsen, M. W. (2002) *J. Phys.: Condens. Matter.* **14**: R21–R47.
[18] Leibler, L. (1980) *Macromolecules* **13**: 1602–1617.
[19] Pople, J. A., Hamley, I. W., Fairclough, J. P. A., Ryan, A. J., Komanschek, B. U., Gleeson, A. J., Yu, G.-E. and Booth, C. (1997) *Macromolecules* **30**: 5721–5728.
[20] Hamley, I. W., Daniel, C., Mingvanish, W., Mai, S.-M., Booth, C., Messe, L. and Ryan, A. J. (2000) *Langmuir* **16**: 2508–2514.
[21] Castelletto, V., Hamley, I. W., Holmqvist, P., Rekatas, C., Booth, C. and Grossmann, J. G. (2001) *Colloid. Polym. Sci.* **279**: 621–628.
[22] Thomas, E. L., Kinning, D. J., Alward, D. B. and Henkee, C. S. (1987) *Macromolecules* **20**: 2934–2939.
[23] Bang, J., Brinker, K. L., Burghardt, W. R., Lodge, T. P. and Wang, X. (2002) *Phys. Rev. Lett.* **89**: 215505.
[24] Huang, Y.-Y., Chen, H.-L. and Hashimoto, T. (2003) *Macromolecules* **36**: 764–770.
[25] Imaizumi, K., Ono, T., Kota, T., Okamoto, S. and Sakurai, S. (2003) *J. Appl. Crystallogr.* **36**: 976–998.
[26] Hashimoto, T., Shibayama, M., Kawai, H., Watanabe, H. and Kotaka, T. (1983) *Macromolecules* **16**: 361–371.
[27] Harkless, C. R., Singh, M. A. and Nagler, S. E. (1990) *Phys. Rev. Lett.* **64**: 2285–2288.
[28] Singh, M. A., Harkless, C. R., Nagler, S. E., Shannon, Jr. R. F. and Ghosh, S. S. (1993) *Phys. Rev. B.* **47**: 8425–8435.
[29] Sakamoto, N., Hashimoto, T., Han, C. D., Kim, D. and Vaidya, N. Y. (1997) *Macromolecules* **30**: 1621–1632.
[30] Sakamoto, N., Hashimoto, T., Han, C. D., Kim, D. and Vaidya, N. Y. (1997) *Macromolecules* **30**: 5321–5330.
[31] Kim, J. K., Lee, H. H., Sakurai, S., Aida, S., Masamoto, J., Nomura, S., Kitagawa, Y. and Suda, Y. (1999) *Macromolecules* **32**: 6707–6717.
[32] Han, C. D., Vaidya, N. Y., Kim, D., Shin, G., Yamaguchi, D. and Hashimoto, T. (2000) *Macromolecules* **33**: 3767–3780.
[33] Choi, S., Lee, K. M., Han, C. D., Sota, N. and Hashimoto, T. (2003) *Macromolecules* **36**: 793–803.
[34] Dormidontova, E. E. and Lodge, T. P. (2001) *Macromolecules* **34**: 9143–9155.
[35] Wang, X., Dormidontova, E. E. and Lodge, T. P. (2002) *Macromolecules* **35**: 9687–9697.
[36] Matsen, M. W. and Thompson, R. B. (1999) *J. Chem. Phys.* **111**: 7139–7146.
[37] Matsen, M. W. (2000) *J. Chem. Phys.* **113**: 5539–5544.
[38] Matsen, M. W. and Gardiner, J. M. (2000) *J. Chem. Phys.* **113**: 1673–1676.
[39] Mai, S.-M., Mingvanish, W., Turner, S. C., Chaibundit, C., Fairclough, J. P. A., Heatley, F., Matsen, M. W., Ryan, A. J. and Booth, C. (2000) *Macromolecules* **33**: 5124–5130.

[40] Yang, L., Hong, S., Gido, S. P., Velis, G. and Hadjichristidis, N. (2001) *Macromolecules* **34**: 9069–9073.
[41] Beyer, F. L., Gido, S. P., Büschl, C., Iatrou, H., Uhrig, D., Mays, J. W., Chang, M. Y., Garetz, B. A., Balsara, N. P., Tan, N. B. and Hadjichristidis, N. (2000) *Macromolecules* **33**: 2039–2048.
[42] Matsushita, Y. (1998) In *Structure and Properties of Multi-Phase Polymeric Materials*, Araki, T., Tran-Cong, Q. and Shibayama, M., Eds., Marcel Dekker, New York, 121–154.
[43] Kodama, H. and Doi, M. (1996) *Macromolecules* **29**: 2652–2658.
[44] Ren, S. R., Hamley, I. W., Teixeira, P. I. C. and Olmsted, P. D. (2001) *Phys. Rev. E.* **63**: 041503.
[45] Ren, S. R., Hamley, I. W., Sevink, G. J. A., Zvelindovsky, A. V. and Fraaije, J. G. E. M. (2002) *Macromol. Theory. Simul.* **11**: 123–127.
[46] Watanabe, H. (1998) In *Structure and Properties of Multi-Phase Polymeric Materials*, Araki, T., Tran-Cong, Q. and Shibayama, M., Eds., Marcel Dekker, New York, 317–360.
[47] Vigild, M. E., Chu, C., Sugiyama, M., Chaffin, K. A. and Bates, F. S. (2001) *Macromolecules* **34**: 951–964.
[48] Hermel, T. J., Wu, L., Hahn, S. F., Lodge, T. P. and Bates, F. S. (2002) *Macromolecules* **35**: 4685–4689.
[49] Morrison, F. A., Mays, J. W., Muthukumar, M., Nakatani, A. I. and Han, C. C. (1993) *Macromolecules* **26**: 5271–5273.
[50] Honeker, C. C., Thomas, E. L., Albalak, R. J., Hajduk, D. A., Gruner, S. M. and Capel, M. C. (2000) *Macromolecules* **33**: 9395–9406.
[51] Honeker, C. C. and Thomas, E. L. (2000) *Macromolecules* **33**: 9407–9417.
[52] Park, C., Simmons, S., Fetters, L. J., Hsiao, B., Yeh, F. and Thomas, E. L. (2000) *Polymer* **41**: 2971–2977.
[53] Kotaka, T., Okamoto, M., Kojima, A., Kwon, Y. K. and Nojima, S. (2001) *Polymer* **42**: 1207–1217.
[54] Kotaka, T., Okamoto, M., Kojima, A., Kwon, Y. K. and Nojima, S. (2001) *Polymer* **42**: 3223–3231.
[55] Sakurai, S., Kota, T., Isobe, D., Okamoto, S., Sakurai, K., Ono, T., Imaizumi, K. and Nomura, S. (2003) *J. Macromol. Sci.* in press
[56] Dair, B. J., Avgeropoulos, A., Hadjichristidis, N., Capel, M. and Thomas, E. L. (2000) *Polymer* **41**: 6231–6236.
[57] Sakurai, S., Aida, S., Okamoto, S., Ono, T., Imaizumi, K. and Nomura, S. (2001) *Macromolecules* **34**: 3672–3678.
[58] Sakurai, S., Aida, S., Okamoto, S., Sakurai, K. and Nomura, S. (2003) *Macromolecules* **36**: 1930–1939.
[59] Panyukov, S. and Rubinstein, M. (1996) *Macromolecules* **29**: 8220–8230.
[60] Aida, S., Sakurai, S. and Nomura, S. (2002) *Polymer*, **43**: 2881–2887.
[61] Amundson, K., Helfand, E., Davis, D. D., Quan, X. and Patel, S. S. (1991) *Macromolecules* **24**: 6546–6548.
[62] Serpico, J. M., Wnek, G. E., Krause, S., Smith, T. W., Luca, D. J. and Laeken, A. V. (1992) *Macromolecules* **25**: 6373–6374.
[63] Amundson, K., Helfand, E., Davis, D. D., Quan, X., Smith, S. D. (1993) *Macromolecules* **26**: 2698–2703.
[64] Amundson, K., Helfand, E., Quan, X., Hudson, S. D., Smith, S. D. (1994) *Macromolecules* **27**: 6559–6570.
[65] Morkved, T. L., Lu, M., Urbas, A. M., Ehrichs, E. E., Jaeger, H. M., Mansky, P. and Russell, T. P. (1996) *Science* **273**: 931–933.

[66] Böker, A., Elbs, H., Hänsel, H., Knoll, A., Ludwigs, S., Zettl, H., Urban, V., Abetz, V., Müller, A. H. E. and Krausch, G. (2002) *Phys. Rev. Lett.* **89**: 135502.
[67] Böker, A., Knoll, A., Elbs, H., Abetz, V., Müller, A. H. E. and Krausch, G. (2002) *Macromolecules* **35**: 1319–1325.
[68] Chu, B. and Hsiao, B. S. (2001) *Chem. Rev.* **101**: 1727–1761.
[69] Zachmann, H. G. and Wutz, C. (1993) In *Crystallization of Polymers*, Dosiere, M. Ed., Kluwer Academic, New York, 403–414.
[70] Wutz, C., Bark, M., Cronauer, J., Döhrmann, R. and Zachmann, H. G. (1995) *Rev. Sci. Instrum.* **66**: 1303–1307.
[71] Okamoto, S., Yamamoto, K., Hara, S., Akiba, I., Sakurai, K., Takeuchi, M., Ueno, S., Abe, S., Takahashi, H., Koyama, A., Nomura, M. and Sakurai, S. (2003) to be submitted.
[72] Okamoto, S., Yamamoto, K., Nomura, K., Hara, S., Akiba, I., Sakurai, K., Koyama, A., Nomura, M. and Sakurai, S. (2003) *J. Macromol. Sci.* in press.
[73] Dair, B. J., Honeker, C. C., Alward, D. B., Avgeropoulos, A., Hadjichristidis, N., Fetters, L. J., Capel, M. and Thomas, E. L. (1999) *Macromolecules* **32**: 8145–8152.
[74] Dair, B. J., Avgeropoulos, A., Hadjichristidis, N. and Thomas, E. L. (2000) *J. Mater Sci.* **35**: 5207–5213.
[75] Sakurai, S., Isobe, D., Okamoto, S., Yao, T. and Nomura, S. (2001) *Phys. Rev. E.* **63**: 061803.
[76] Fujimura, M., Hashimoto, T. and Kawai, H. (1978) *Rub. Chem. Tech.* **51**: 215–224.
[77] Cohen, Y., Albalak, R. J., Dair, B. J., Capel, M. S. and Thomas, E. L. (2000) *Macromolecules* **33**: 6502–6516.
[78] Cohen, Y., Brinkmann, M. and Thomas, E. L. (2001) *J. Chem. Phys.* **114**: 984–992.
[79] Hamley, I. W. and Bates, F. S. (1994) *J. Chem. Phys.* **100**: 6813–6817.
[80] Laradji, M., Shi, A.-C., Desai, R. C. and Noolandi, J. (1997) *Phys. Rev. Lett.* **78**: 2577–2580.
[81] Laradji, M., Shi, A.-C., Noolandi, J. and Desai, R. C. (1997) *Macromolecules* **30**: 3242–3255.
[82] Sakurai, S., Momii, T., Taie, K., Shibayama, M., Nomura, S. and Hashimoto, T. (1993) *Macromolecules* **26**: 485–491.
[83] Kim, J. K., Lee, H. H., Gu, Q.-J., Chang, T. and Jeong, Y. H. (1998) *Macromolecules* **31**: 4045–4048.
[84] Kim, J. K., Lee, H. H., Ree, M., Lee, K.-B., Park, Y. (1998) *Macromol. Chem. Phys.* **199**: 641–653.
[85] Qi, S. and Wang, Z.-G. (1998) *Polymer* **39**: 4639–4648.
[86] Ryu, C. Y., Vigild, M. E. and Lodge, T. P. (1998) *Phys. Rev. Lett.* **81**: 5354–5357.
[87] Ryu, C. Y. and Lodge, T. P. (1999) *Macromolecules* **32**: 7190–7201.
[88] Kimishima, K., Koga, T. and Hashimoto, T. (2000) *Macromolecules* **33**: 968–977.
[89] Matsen, M. W. *J. Chem. Phys.* 2001, **114**, 8165–8173.
[90] Jeong, U., Lee, H. H., Yang, H., Kim, J. K., Okamoto, S., Aida, S. and Sakurai, S. (2003) *Macromolecules* **36**: 1685–1693.
[91] Qi, S. and Wang, Z.-G. (1996) *Phys. Rev. Lett.* **76**: 1679–1682.
[92] Qi, S. and Wang, Z.-G. (1997) *Phys. Rev. E.* **55**: 1682–1697.
[93] Krishnamoorti, R., Silva, A. S., Modi, M. A. and Hammouda, B. (2000) *Macromolecules* **33**: 3803–3809.
[94] Krishnamoorti, R., Modi, M. A., Tse, M. F. and Wang, H.-C. (2000) *Macromolecules* **33**: 3810–3817.
[95] Wang, C.-Y. and Lodge, T. P. (2002) *Macromol. Rapid. Commun.* **23**: 49–54.
[96] Wang, C.-Y. and Lodge, T. P. (2002) *Macromolecules* **35**: 6997–7006.
[97] Shi, A.-C., Noolandi, J. and Desai, R. C. (1996). *Macromolecules* **29**: 6487–6504.

[98] Schultz, M. F., Bates, F. S., Almdal, K. and Mortensen, K. (1994) *Phys. Rev. Lett.* **73**: 86–89.
[99] Matsen, M. W. (1998) *Phys. Rev. Lett.* **80**: 4470–4473.
[100] Avrami, M. J. (1939) *J. Chem. Phys.* **7**: 1103–1112.
[101] Schultz, J. M. (2001) *Polymer Crystallization-The Development of Crystalline Order in Thermoplastic Polymers*, Oxford Univ. Press.
[102] Sakurai, S., Umeda, H., Taie, K. and Nomura, S. (1996) *J. Chem. Phys.* **105**: 8902–8908.
[103] Floudas, G., Ulrich, R. and Wiesner, U. (1999) *J. Chem. Phys.* **110**: 652–663.
[104] Qi, S. and Wang, Z.-G. (1999) *J. Chem. Phys.* **111**: 10681–10688.
[105] Kai, S. and Zimmermann, W. (1989) *Prog. Theor. Phys. supp.* **99**: 458–492.
[106] Cross, M. C. and Hohenberg, P. C. (1993) *Rev. Mod. Phys.* **65**: 851–1112.
[107] Wilkes, G. L. and Stein, R. S. (1997) In *Structure and Properties of Oriented Polymers*, Second Edition, Ward IM Ed., Chapman & Hall, London, 44–141.
[108] Balsara, N. P., Perahia, D., Safinya, C. R., Tirrell, M. and Lodge, T. P. (1992) *Macromolecules* **25**: 3896–3901.
[109] Balsara, N. P., Garetz, B. A. and Dai, H. J. (1992) *Macromolecules* **25**: 6072–6074.
[110] Garetz, B. A., Newstein, M. C., Dai, H. J., Jonnalagadda, S. V. and Balsara, N. P. (1993) *Macromolecules* **26**: 3151–3155.
[111] Wang, H., Newstein, M. C., Chang, M. Y., Balsara, N. P. and Garetz, B. A. (2000) *Macromolecules* **33**: 3719–3730.
[112] Lee, H. H., Jeong, W.-Y., Kim, J. K., Ihn, K. J., Kornfield, J. A., Wang, Z.-G. and Qi, S. (2002) *Macromolecules* **35**: 785–794.
[113] Hashimoto, T. and Sakamoto, N. (1995) *Macromolecules* **28**: 4779–4781.
[114] Sakamoto, N. and Hashimoto, T. (1995) *Macromolecules* **28**: 6825–6834.
[115] Hashimoto, T., Sakamoto, N. and Koga, T. (1996) *Phys. Rev. E.* **54**: 5832–5835.
[116] Sakamoto, N. and Hashimoto, T. (1998) *Macromolecules* **31**: 3815–3823.
[117] Koga, T., Koga, T. and Hashimoto, T. (1999) *Phys. Rev. E.* **60**: R1154–R1157.
[118] Koga, T., Koga, T. and Hashimoto, T. (1999) *J. Chem. Phys.* **110**: 11076–11086.
[119] Koga, T., Koga, T., Kimishima, K. and Hashimoto, T. (1999) *Phys. Rev. E.* **60**: R3501–R3504.
[120] Masuda, T., Takigawa, T., Kojima, T. and Ohta, Y. (1989) *J. Rheol.* **33**: 469–480.
[121] Stühn, B. (1992) *J. Polym. Sci. Part B Polym. Phys. Ed* **30**: 1013–1019.
[122] Kasten, H. and Stühn, B. (1995) *Macromolecules* **28**: 4777–4778.
[123] Floudas, G., Hadjichristidis, N., Stamm, M., Likthman, A. E. and Semenov, A. N. (1997) *J. Chem. Phys.* **106**: 3318–3328.
[124] Soenen, H., Liskova, A., Reynders, K., Berghmans, H., Winter, H. H. and Overbergh, N. (1997) *Polymer.* **38**: 5661–5665.
[125] Voronov, V. P., Buleiko, V. M., Podneks, V. E., Hamley, I. W., Fairclough, J. P. A., Ryan, A. J., Mai, S.-M., Kiao, B. X. and Booth, C. (1997) *Macromolecules* **30**: 6674–6676.
[126] Rangarajan, P., Register, R., Adamson, D. H., Fetters, L. J., Bras, W., Naylor, S. and Ryan, A. J. (1995) *Macromolecules* **28**: 1422–1428.
[127] Rangarajan, P., Register, R., Adamson, D. H., Bras, W., Naylor, S. and Ryan, A. J. (1995) *Macromolecules* **28**: 4932–4938.
[128] Quiram, D. J., Register, R., Marchand, G. R. and Ryan, A. J. (1997) *Macromolecules* **30**: 8338–8343.
[129] Loo, Y. L., Register, R. A., Ryan, A. J. and Dee, G. T. (2001) *Macromolecules* **34**: 8968–8977.
[130] Loo, Y. L., Register, R. A. and Ryan, A. J. (2002) *Macromolecules* **35**: 2365–2374.
[131] Skoulios, A. E., Tsouladzem, G. and Franta, E. (1963) *J. Polym. Sci., Part C.* **4**: 507.

[132] Whitmore, M. D. and Noolandi, J. (1998) *Macromolecules* **21**: 1482–1496.
[133] DiMarzio, E. A., Guttman, C. M. and Hoffman, J. D. (1980) *Macromolecules* **13**: 1194–1198.
[134] Douzinas, K. C., Cohen, R. E. and Halasa, A. F. (1991) *Macromolecules* **24**: 4457–4459.
[135] Nojima, S., Yamamoto, S. and Ashida, T. (1995) *Polymer. J.* **27**: 673–682.
[136] Rangarajan, P., Register, R. A. and Fetters, L. J. (1993) *Macromolecules* **26**: 4640–4645.
[137] Hamley, I. W., Fairclough, J. P. A., Terrill, N., Ryan, A. J., Lipic, P. M., Bates, F. S. and Towns-Andrews, E. (1996) *Macromolecules* **29**: 8835–8843.
[138] Ishikawa, S., Ishizu, K. and Fukutomi, T. (1991) *Polymer Communications* **32**: 374–375.
[139] Zhu, L., Cheng, S. Z. D., Caalhoun, B. H., Roderic, Q. G., Quirk, P., Thomas, E. L., Hsiao, B. S., Yeh, F. and Lotz, B. (2000) *J. Am. Chem. Soc.* **122**: 5957–5967.
[140] Tadokoro, H. (1979) *Structure of Crystalline Polymers*, John Wiley & Sons, New York.
[141] Nojima, S., Kato, K., Yamamoto, S. and Ashida, T. (1992) *Macromolecules* **25**: 2237–2242.
[142] Sakurai, K., MacKnight, W. J., Lohse, D. J., Schulz, D. N. and Sissano, J. A. (1994) *Macromolecules* **27**: 4941–4950.
[143] Sakurai, K., MacKnight, W. J., Lohse, D. J., Schulz, D. N., Sissano, J. A., Lin, J. S. and Agamalyan, M. (1996) *Polymer* **37**: 4443–4453.
[144] Sakurai, K., Shinkai, S., Ueda, M., Sakurai, S., Nomura, S., MacKnight, W. J. and Lohse, D. J. (2000) *Macromol. Rapid. Commun.* **212**: 1140–1143.
[145] Ueda, M., Sakurai, K., Okamoto, S., Lohse, D. J., MacKnight, W. J., Shinkai, S., Sakurai, S. and Nomura, S. (2003) *Polymer*, **44**: 6995–7005.
[146] Okamoto, S., Saijo, K. and Hashimoto, T. (1994) *Macromolecules* **27**: 3753–3758.
[147] Okamoto, S., Saijo, K. and Hashimoto, T. (1994) *Macromolecules* **27**: 5547–5555.
[148] Bates, F. S., Koppi, K. A., Tirrell, M., Almdal, K. and Mortensen, K. (1994) *Macromolecules* **27**: 5934–5936.
[149] Hashimoto, T., Okamoto, S., Saijo, K., Kimishima, K. and Kume, T. (1995) *Acta. Polymer.* **46**: 463–470.
[150] Zachmann, H. G. and Gehrke, R. (1986) In *Morphology of Polymers*, edited by Sedlácek B., Walter de Gruyter, New York, 1986, 119.
[151] Stein, R. S., Cronauer, J. and Zachmann, H. G. (1996) *J. Molecular Structure* **383**: 19–22.
[152] Dolbnya, I. P., Alberda, H., Hartjes, F. G., Udo, F., Bakker, R. E., Konijnenburg, M. Homan, M. E., Cerjak, I., Goedtkindt, P. and Bras, W. (2002) *Rev. Sci. Instrum.* **73**: 3754–3758.
[153] Heeley, E., Morgovan, A. C., Bras, W., Dolbyna, I. P., Panine, P., Terrill, N. J., Gleeson, A. J., Fairclough, J. P. A. and Ryan, A. J. (2002) *J. Macromol. Sci. Phys*, in press.
[154] Terrill, N. J., Fairclough, J. P. A., Towns-Andrews, E., Komanschek, B. U., Devine, A., Young, R. J. and Ryan, A. J. (1998) *Polymer* **39**: 2381–2385.
[155] Ryan, A. J., Fairclough, J. P. A., Terrill, N. J., Olmsted, P. D. and Poon, W. C. K. (1999) *Faraday Discuss* **112**: 13–29.
[156] Heeley, E. L., Poh, C. K., Li, W., Maidens, A., Bras, W., Dolbnya, I. P., Gleeson, A. J., Terrill, N. J., Fairclough, J. P. A., Olmsted, P. D., Ristic, R. I., Hounslow, M. J. and Ryan, A. J. (2003) *Faraday Discuss.* **122**: 343–361.
[157] Heeley, E. L., Bras, W., Dolbnya, I. P., Maidens, A., Olmsted, P. D., Fairclough, J. P. A., Ryan, A. J. (2002) *Fibre Diffract. Rev.*, **10**: 63–71.

[158] Wang, Z. -G., Hsiao, B. S., Sirota, E. B., Agarwal, P. and Srinivas, S. (2000) *Macromolecules* **33**: 978–989.
[159] Lee, C. H., Saito, H. and Inoue, T. (1993) *Macromolecules* **26**: 6566–6569.
[160] Chen, H. L., Hsiao, S. C., Lin, T. L., Yamauchi, K., Hasegawa, H. and Hashimoto, T. (2001) *Macromolecules* **34**: 671–674.
[161] Hajduk, D. A., Urayama, P., Gruner, S. M., Erramilli, S., Register, R. A., Brister, K. and Fetters, L. J. (1995) *Macromolecules* **28**: 7148–7156.
[162] Schwahn, D., Frielinghaus, H., Mortensen, K. and Almdal, K. (1996) *Phys. Rev. Lett.* **77**: 3153–3156.
[163] Hajduk, D. A., Gruner, S. M., Erramilli, S., Register, R. A. and Fetters, L. J. (1996) *Macromolecules* **29**: 1473–1481.
[164] Frielinghaus, H., Schwahn, D., Mortensen, K., Almdal, K. and Springer, T. (1996) *Macromolecules* **29**: 3263–3271.
[165] Hasegawa, H., Sakamoto, N., Takeno, H., Jinnai, H., Hashimoto, T., Schwahn, D., Frielinghaus, H., Janssen, S., Imai, M. and Mortensen, K. (1999) *J. Phys. Chem. Solids.* **60**: 1307–1312.
[166] Schwahn, D., Frielinghaus, H., Mortensen, K. and Almdal, K. (2001) *Macromolecules* **34**: 1694–1706.
[167] Sakurai, S., Aida, S., Tamura, T., Kota, T., Okamoto, S., Sakurai, K., Tanaka, N., Kunugi, S. and Nomura, S. (2002) *Polymer* **43**: 1959–1962.
[168] Maeda, Y., Morita, K. and Kutsumizu, S. (2003) *Liq. Cryst.* **30**: 157–164.
[169] Sakurai, S., Funai, E., Munakata, S., Okamoto, S., Yamato, M. and Kimura, T. *In preparation.*

5 Phase Behaviour of Block Copolymer Blends

RICHARD J. SPONTAK[1,2] and NIKUNJ P. PATEL[1]
Departments of [1]*Chemical Engineering and* [2]*Materials Science & Engineering*
North Carolina State University
Raleigh, NC 27695, USA

5.1 INTRODUCTION

5.1.1 PHASE BEHAVIOUR OF NEAT BLOCK COPOLYMERS

As the technological need for lightweight multifunctional materials continues to increase, interest in the design and development of tailored block copolymers likewise increases. Under the right set of conditions, collectively expressed in terms of the thermodynamic incompatibility [1] χN (where χ denotes the temperature-sensitive Flory–Huggins interaction parameter and N is the number of repeat units along the copolymer backbone), these materials self-organize and, depending on factors such as chemical identity and molecular composition [1–6], order to form periodic nanoscale morphologies. In the case of block copolymers composed of A and B repeat units, the *classical* equilibrium morphologies observed to date include A(B) spheres positioned on a body-centered cubic (bcc) lattice in a B(A) matrix ($Q_{Im\bar{3}m}$), A(B) cylinders arranged on a hexagonal lattice in a B(A) matrix (H) and coalternating lamellae (bilayers, L). Examples of nonclassical, or *complex*, morphologies reported for neat bicomponent (AB or ABA) block copolymers include the gyroid [7–11] ($Q_{Ia\bar{3}d}$) and perforated lamellar [12–16] (or lamellar catenoid) morphologies. Unlike their classical analogues, the complex morphologies, commonly described as bicontinuous channel networks, typically occupy relatively small regions of phase space. Due to their structural intricacy, they are also more prone to form nonequilibrium elements during processing than their classical counterparts. While perforated lamellae, for instance, have been observed in a variety of copolymers, the theoretical analyses of Matsen and coworkers [17,18], coupled with the experimental findings of Hajduk *et al.* [19], reveal that this is actually a long-lived metastable, not equilibrium, morphology.

The experimental phase diagrams first reported [3] for the case of poly(styrene-*b*-isoprene) (S-I) diblock copolymers and later [6] for more complex poly(styrene-*b*-isoprene-*b*-ethylene oxide) (S-I-EO) triblock copolymers demand immediate appreciation for the substantial versatility afforded by block copolymers in terms of morphological development. An important and distinguishing characteristic of block copolymer phase diagrams is that each copolymer composition sampled requires careful synthesis of an entirely new macromolecule, in marked contrast to conventional phase diagrams of polymer blends in which blend composition is easily regulated by the physical addition of one constituent species to another. Although the systematic production of such phase diagrams is vital to the fundamental understanding of block copolymer molecular self-organization (which is more fully described in detail elsewhere [20–23]), the prospect of synthesizing new block copolymer molecules with predetermined compositions and molecular weights for specific applications is not always practically appealing. For this reason, numerous efforts over the past decade or so have sought to tailor block copolymer morphologies through the use of multicomponent systems, which are discussed in this chapter. From the extensive data presently available for such systems, it is possible to develop a set of design paradigms that not only serve to enhance the general versatility of self-organized block copolymers, but also permit fundamental inquiry into the increasingly important issue of macromolecular mixing within molecularly confined environments.

5.1.2 PHASE SPACE IN MULTICOMPONENT SYSTEMS

One can envisage two basic strategies by which to probe the possible spatial arrangements of block copolymer molecules in multicomponent systems. In the first methodology, a block copolymer consisting of A and B moieties is physically added to either (i) a single A(B)-selective constituent or a second copolymer to form a binary system or (ii) both A- and B-selective species to generate a ternary system. This design motif derives directly from precursor studies of small-molecule surfactant systems [24,25] and is capable of yielding a wealth of phase behaviour, which is evidenced in the phase diagram reported by Alexandridis *et al.* [26] for a nearly composition-symmetric poly(ethylene oxide-*b*-propylene oxide-*b*-ethylene oxide) (EO_{19}-PO_{44}-EO_{19}) triblock copolymer in the presence of water and *p*-xylene. As Figure 5.1 attests, this phase diagram exhibits all the morphologies previously mentioned with the use of just one copolymer, which confirms that morphological design can be realized through judicious blending of a single copolymer with A- and/or B-selective additives. In the case of Figure 5.1 and related studies [27,28], these additives constitute low molar mass solvents that relate to the inherent hydrophilicity and hydrophobicity of the copolymer molecules. The underlying principles responsible for such morphological diversity readily extend to nonaqueous systems

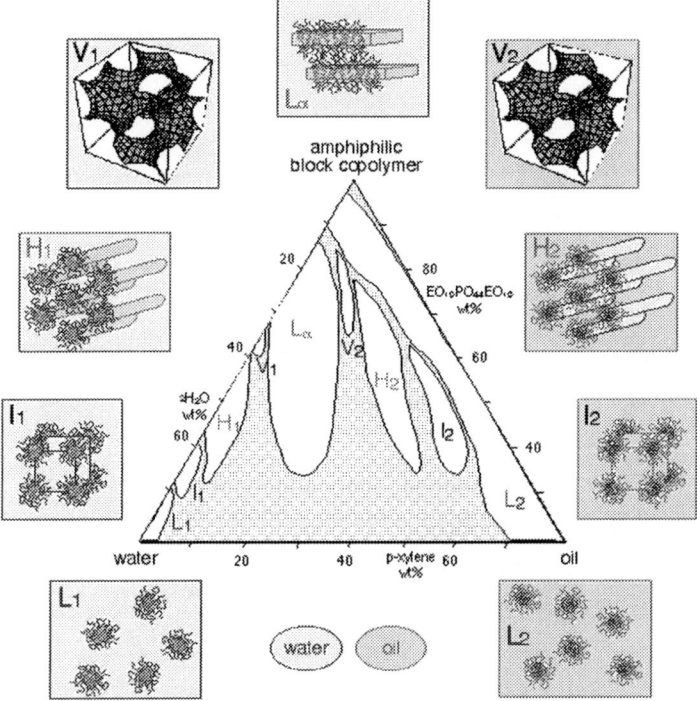

Figure 5.1 Ternary phase diagram of the EO_{19}-PO_{44}-EO_{19}/p-xylene/water system illustrating the rich polymorphism afforded by a single block copolymer in the presence of two (low molar mass) block-selective additives. (Reprinted with permission from Alexandridis, P., Olsson, U. and Lindman, B. Langmuir **14**, 2627, 1998. Copyright (1998) American Chemical Society.)

wherein an organic solvent is block selective (i.e., it is more compatible with one block than with the other) [29–31]. While many block copolymers of technological relevance are intended for use in solventless applications, the general materials-design strategy of physically blending a single copolymer with other additives – e.g., one or two block-selective homopolymers or a second copolymer – opens new avenues to nanostructured materials with specific morphologies or dimensions that can be controllably varied through the rational use of physical parameters.

The second methodology to be considered here imparts greater functionality to the copolymer molecule. Thus far, only block copolymers possessing A and B repeat units arranged in different sequence schemes (or architectures) have been explicitly considered. Although tailored synthesis is nonetheless required, the functionality and morphological diversity of linear AB molecules can be greatly increased through the chemical addition of a third block to form linear ABC triblock copolymers. This approach pioneered by Stadler *et al.* [32] vastly enlarges the scope of morphologies that can be explored within phase space.

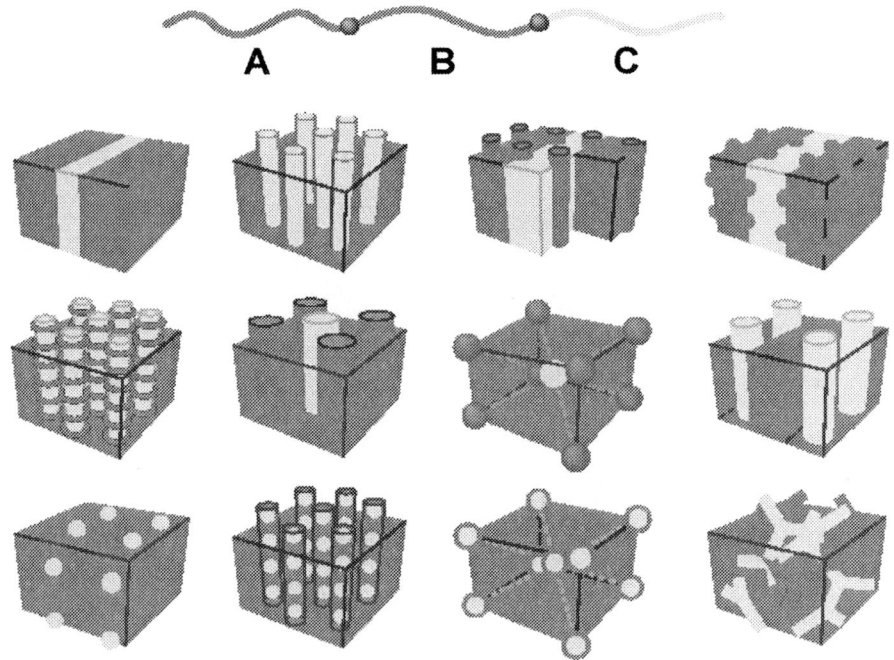

Figure 5.2 A sampling of the diverse morphologies afforded by neat ABC triblock copolymers through systematic variation of molecular parameters such as interblock incompatibility, molecular weight and composition. (Reprinted with permission from Bates, F. S. and Fredrickson, G. H. Phys. Today **52**, 32, 1999. Copyright (1999) American Institute of Physics.)

Representative examples of some of the morphologies accessible through this design route have been described by Zheng and Wang [33] and Abetz [34], and include those displayed [21] for illustrative purposes in Figure 5.2. It should be recognized, however, that the variability available in terms of composition, sequencing, species and molecular weight translates into an immensely enlarged parameter space wherein other, as yet unreported, morphologies may arise. As in the case of the simple AB-type block copolymers, synthesis of designer ABC copolymers is nontrivial, in which case the blending strategy alluded to earlier may likewise be employed to generate morphologies such as those depicted in Figure 5.2 [35]. The possibilities made available through this approach provide a tremendous opportunity for both theoretical and experimental investigation, and require a basic understanding of the factors governing miscibility in these systems.

5.1.3 FACTORS REGULATING MISCIBILITY OF BLENDS

Early studies of commercial block copolymers in the presence of a single block-selective homopolymer for toughening applications often show evidence of macrophase separation [36], in which the copolymer molecules form a dispersed ordered phase within a homopolymer matrix. In this case, the homopolymer molecules, which tended to be much larger in molecular weight than the corresponding block(s) of the copolymer, could not penetrate the dense brush created by the self-organized copolymer block(s), resulting in the formation of a so-called *dry* brush [37,38]. This conformational entropic penalty ultimately drives the system toward macrophase separation between the copolymer and homopolymer constituents of the blend. The benchmark swelling and transitional studies performed by Thomas and coworkers [39–41] and Hashimoto and coworkers [38,42,43] on binary copolymer/homopolymer blends have established the compositions and molecular-weight ratios required to ensure production of miscible blends wherein the homopolymer molecules penetrate (*wet*) and swell the relevant copolymer brush, ultimately becoming incorporated within the copolymer nanostructure. From such experimental and complementary theoretical [44–47] studies, two general rules regarding the design of miscible copolymer/homopolymer blends are identified: (i) the molecular-weight ratio of the homopolymer (M_{hA}) to that of the corresponding block in the copolymer (M_A) should be less than unity, and (ii) the fraction (ϕ) of homopolymer that can be added to a blend increases as M_{hA}/M_A (hereafter referred to as α) decreases. It immediately follows that block copolymers should not tend to macrophase separate from selective solvents. While this limit is experimentally observed [26–31] and is important in its own right, we restrict the focus of the present work to multicomponent block copolymer systems composed of a copolymer with at least one other macromolecule.

In the case of a block copolymer distributed between both parent homopolymers, many recent studies have investigated the emulsifying attributes [48–58], as well as interfacial elasticity [59–62] and adhesion [63,64], of the copolymer as a macromolecular surfactant. Here, added copolymer molecules, examined in terms of concentration, copolymer composition/architecture and thermal behaviour, are envisaged to locate along the interface formed by the macrophase-separated homopolymers, which generally tend to be thermodynamically immiscible due to a combination of endothermic mixing ($\Delta h_{mix} > 0$) and a near-negligible entropy of mixing ($\Delta S_{mix} \sim 0$). Similar results can be achieved through reactive compatibilization [65–68], but this topic is not considered further. Moreover, Velankar et al. [69] have likewise demonstrated that copolymer-promoted compatibilization under shear is strongly influenced by external factors such as flow-induced interfacial tension gradients. Recent efforts by Macosko and coworkers [70] have successfully extended the strategy developed for binary copolymer/homopolymer blends to probe the material-related factors governing morphology development in ternary blends of poly(cyclohexyl methacrylate) (hCH), poly(methyl methacrylate) (hM) and poly(styrene-*b*-methyl methacrylate) (S-M) diblock copolymers differing in M_{hM}/M_M and ϕ_{SM}. Their results for blends in which $M_{hM}/M_M = 0.14$ and $\phi_{SM} = 20\,\text{wt}\%$ (i) provide direct visual evidence for copolymer localization along the hCH/hM interface at relatively low $\phi_{SM}(< 10\,\text{wt}\%)$ and (ii) confirm a reduction in dispersion size by about an order of magnitude with increasing ϕ_{SM} up to 30 wt%. The series of transmission electron microscopy (TEM) images displayed in Figure 5.3 clearly shows a steady progression from compatibilized macroscale dispersions to micelles as the concentration of block copolymer is increased. At low copolymer fractions, the stained copolymer molecules are also found to form discrete micelles within the hM-rich dispersions, indicating that not all of the copolymer molecules locate at the homopolymer/homopoly-

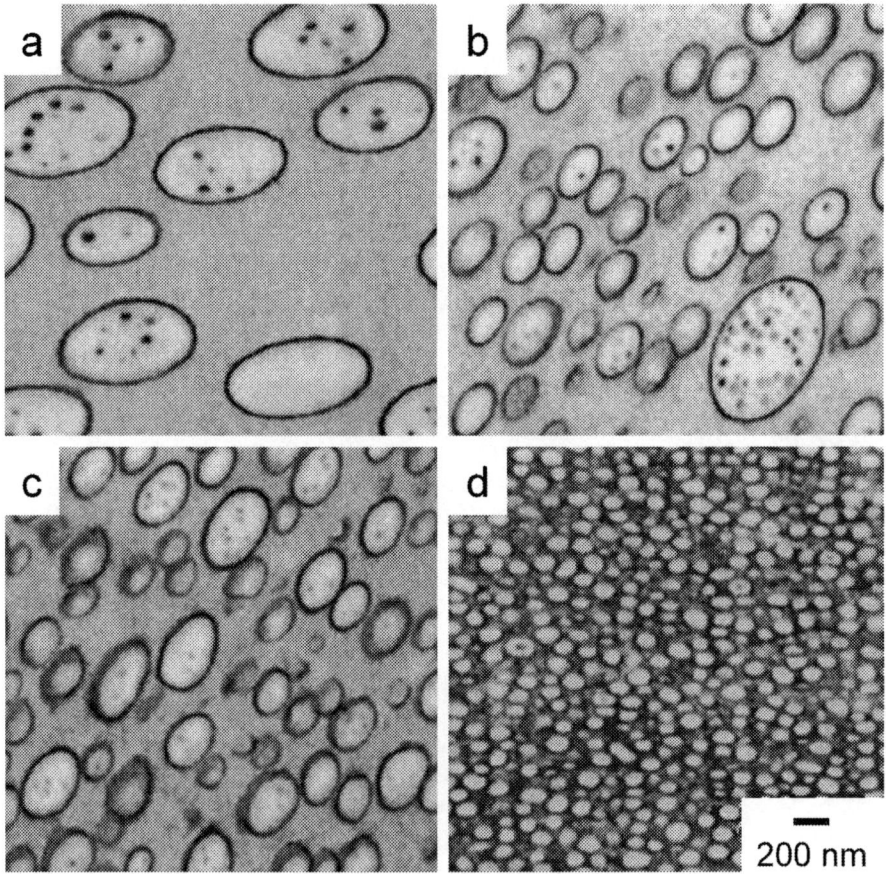

Figure 5.3 TEM images demonstrating the effect of copolymer concentration (in wt %)—2 (a), 5 (b), 10 (c) and 30 (d)—on morphology in ternary blends composed of poly(cyclohexyl methacrylate), 20 wt % poly(methyl methacrylate) and a symmetric poly(styrene-b-methyl methacrylate) (S-M) diblock copolymer for which $\alpha = 0.14$ (α is defined as p. 165). The copolymer is selectively stained and appears dark in these images. (Adapted from Adedeji, A., Lyu, S. and Macosko, C. W. Macromolecules **34**, 8663, 2001, and reprinted with permission. Copyright (2001) American Chemical Society.)

mer interface and promote a reduction in interfacial tension between the two homopolymers. This observation is important in the development of paradigms capable of rendering well-defined and uniform morphologies wherein the copolymer molecules are used efficiently. Mezzenga *et al.* [71] also report that novel high internal phase emulsions (HIPEs) can be produced in ternary copolymer/homopolymer/homopolymer blends through systematic variation of conventional parameters such as solvent quality (neutral versus selective), polymer molecular weight and blend composition. Jamieson and coworkers [72–74] have likewise found that the attraction of block copolymer molecules to interfaces can be controllably enhanced through the use of thermodynamically attractive moieties that exhibit exothermic mixing, but this approach requires the use of particular chemical species and, for this reason, is not considered further here. Similarly, related discussion of adhesion [63,64] typically measured by welding a block copolymer thin film between two homopolymer films and monitoring crack propagation is not included in this chapter. Topics to be addressed in the following sections include contemporary advances in AB/ABA/ABC block copolymer/homopolymer binary and ternary blends, as well as block copolymer/(block or random) copolymer binary blends, with an emphasis on the material factors that tend to yield miscible blends.

5.2 BLOCK COPOLYMER/HOMOPOLYMER BLENDS

5.2.1 DIBLOCK COPOLYMER/HOMOPOLYMER BINARY BLENDS

Incorporation of parent homopolymer A (hA) to an AB diblock copolymer can result in the formation of either dry or wet copolymer brushes, depending on the magnitude of M_{hA}/M_A, ($= \alpha$). At small values of α, hA molecules are solubilized throughout their host copolymer microphase, indicating that the segmental density distribution of hA across the microphase is broad. In this case, the increase in translational entropy of the hA molecules exceeds the slight reduction in conformational entropy of the A blocks, which must stretch to permit interpenetration of the hA molecules. As the value of α is increased at constant ϕ_{hA}, however, the hA molecules tend to remain unmixed from the corresponding copolymer blocks due to a high conformational entropic penalty and, thus, localize far from the interface that separates adjacent microphases. This spatial arrangement yields hA segmental density distributions that exhibit a relatively sharp maximum positioned near the center of the host microphase [44–47,75,76]. Eventually, a fraction of hA molecules may separate altogether to form a separate phase, as discussed in the previous section. If the value of ϕ_{hA} is increased at constant α where α is relatively small, then the hA molecules may ultimately change the packing arrangement of chains along the interface and consequently induce a change in interfacial curvature. By doing so, it is then possible to effect a transition from one morphology to another by varying the homopolymer fraction in the copolymer/homopolymer blend. Careful accounting of the parameters required to promote a desired morphological transition yields the following design variables: the molecular-weight disparity (α), the overall blend composition (ϕ, expressed in terms of hA or AB), the composition of the copolymer (f, given in terms of A or B), and the intrinsic thermodynamic incompatibility of the copolymer (χN). If a nonparent block-selective homopolymer (hC) is added [72,77–81], then another parameter – the enthalpic interaction between hC molecules and the compatible block of the copolymer – must be considered. Omission of the subscripts on ϕ and f implies that they correspond to the same moiety.

Since experimental studies must necessarily sample a relatively small region of parameter space to remain feasible, we first turn our attention to theoretical efforts capable of systematically varying the parameters of interest, namely, α, ϕ, f and χN. While several predictive treatments [44,45,47,82–85] and simulation protocols [86] have been proposed for the specific case of block copolymer/homopolymer blends, detailed phase diagrams generated by the self-consistent field (SCF) formalism of Matsen [44,45] are provided in Figure 5.4 to illustrate the importance of all the parameters listed above. In Figures 5.4a and b, α is set equal to unity, whereas χN is varied in the weak-segregation limit from 10.0 to 11.0. When the phase diagram at $\chi N = 10.0$ is viewed along the ordinate (f), it resembles the experimental phase diagram of neat diblock copolymers, with the starkest difference being a large biphasic region at low f. Other heterogeneous regions are also present in this and the remaining phase diagrams shown in Figure 5.4, but they are not labeled due to their small size. As χN is increased in Figure 5.4b, the disordered phase nearly disappears, and the ordered morphologies become predominantly stable, at low ϕ. Moreover, the stability region of the lamellar microphase, in particular, enlarges under these conditions, thereby causing the ordered envelope to extend over a markedly larger range in f. Increasing α from 1.0 (Figure 5.4b) to 1.5 (Figure 5.4d) at constant χN, however, is accompanied by dramatic expansion of the biphasic region, which reflects the dry-brush scenario previously discussed. Alternatively, a reduction in α (Figure 5.4c) promotes greater stability of the ordered microphases over a wide range of both ϕ and f. Theoretical phase diagrams such as these are particularly valuable in that they not only capture the underlying physics involved in adding a parent homopolymer to a diblock copolymer, but also explicitly demonstrate the relative importance of the parameters that must be considered to achieve a specific morphology.

The experimental phase diagrams reported by Winey et al. [40,41] for several series of diblock copolymer/homopolymer blends constitute the first systematic account of tailoring block copolymer morphology via physical blending. Since that time, numerous independent studies have used this approach to investigate the phase behaviour of diblock copolymer/homopolymer blends. Of particular interest are the stability and dimensional characteristics of complex bicontinuous

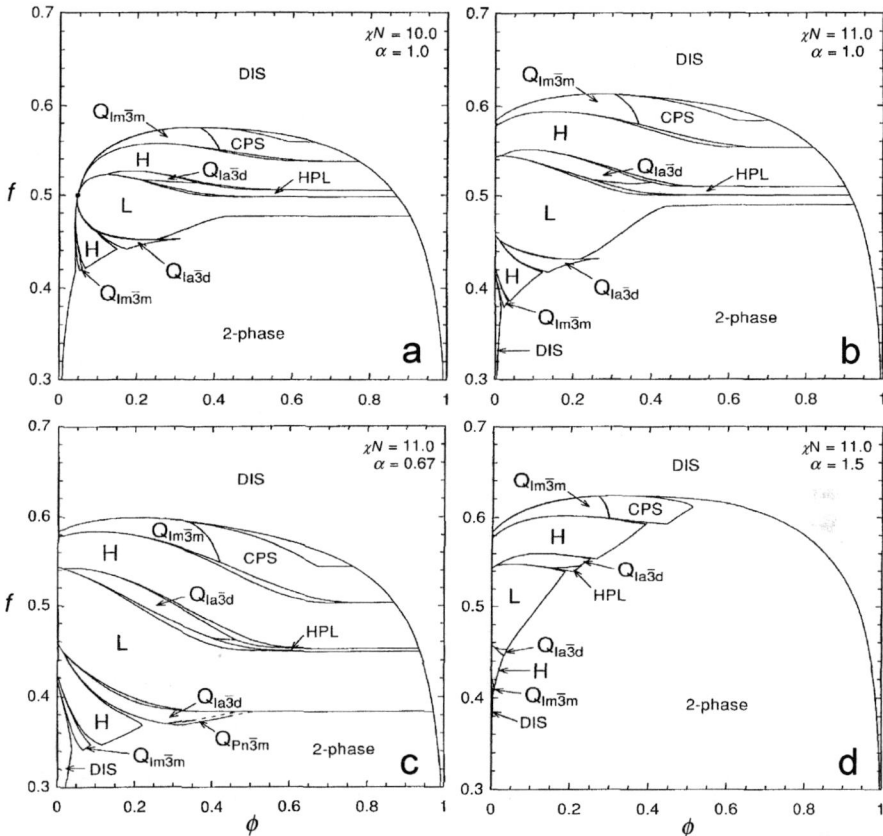

Figure 5.4 SCF phase diagrams of diblock copolymer/homopolymer blends in which the fraction of A repeat units in the copolymer (f) is presented as a function of homopolymer volume fraction (ϕ) at different combinations of copolymer incompatibility (χN) and molecular weight ratio (α): (a) $\chi N = 10.0$, $\alpha = 1.0$; (b) $\chi N = 11.0$, $\alpha = 1.0$; (c) $\chi N = 11.0, \alpha = 0.67$; and (d) $\chi N = 11.0, \alpha = 1.5$. (Compiled from Matsen, M. W. Macromolecules **28**, 5765 (1995) and reprinted with permission. Copyright 1995 American Chemical Society.)

morphologies [87,88], such as the gyroid and perforated lamellar morphologies, which occur naturally over a relatively limited composition range in neat copolymer systems. The experimental phase diagram generated by Bodycomb *et al.* [89] for miscible S-I/hS blends in which $\alpha = 0.55$ reveals that the blend morphology expectedly changes in the following order as the copolymer concentration is reduced: lamellae → gyroid → cylinders → disordered spheres. An interesting feature of their phase diagram is the existence of an order–order transition (OOT) between the gyroid and cylindrical morphologies in this blend series. Upon slow cooling across the order–disorder transition (ODT), the gyroid morphology is produced at all temperatures over the composition range indicated. Rapid cooling, however, results in the formation of a stable cylindrical morphology at temperatures below the OOT. Heating the blend from the cylindrical morphology promotes a transformation to the gyroid morphology, but cooling the gyroid morphology does not generate cylinders, which suggests the same type of kinetic limitation observed by Hajduk et al. [90] in the transition from gyroid to lamellae in unary block copolymer systems. In a related vein, Ahn and Zin [91] report that the molecular weight of a homopolymer added to a lamellar diblock copolymer strongly influences the formation of the metastable perforated lamellar morphology at compositions intermediate between those yielding the lamellar and gyroid morphologies, thereby confirming that the extent of brush wetting (dictated by α) affects the degree of interfacial chain-packing frustration and the complex morphology ultimately stabilized [17,18].

Most experimental studies of diblock copolymer/homopolymer blends employ copolymers synthesized via living anionic polymerization (to ensure good molecular weight control and low polydispersity) from a combination of either methyl methacrylate (M), styrene (S) or (hydrogenated) diene (e.g., butadiene, B, or isoprene, I) monomers. For this reason, attempts designed to elucidate the factors that govern morphological development in diblock copolymer/homopolymer blends continue to rely heavily on a relatively small family of materials. Even within this subset, the phase behaviour of diblock copolymer/homopolymer blends can be unexpectedly composition- and species-dependent. The findings of Vaidya and Han [92], who examined a wide array of blends composed of either a lamellar/spherical S-I or lamellar S-B copolymer with hS, hB or hI, indicate that the resultant phase diagrams are sensitive to both α (especially for the blends containing the spherical S-I copolymer) and the homopolymer used to form the blend (hB or hI versus hS). Extension of this approach to other, chemically dissimilar systems is thus needed to confirm its general applicability. Lammertink et al. [93], for instance, have demonstrated that they could controllably produce both miscible and biphasic morphologies in their binary blends of poly(styrene-b-ferrocenyldimethylsilane) (S-F) diblock copolymers in the presence of either hS or poly(ferrocenyldimethylsilane) (hF). At concentrations of 66 and 68 vol% F in two different copolymer series, their blends exhibit the gyroid morphology, as evidenced by the TEM image in Figure 5.5a and the corresponding small-angle x-ray scattering (SAXS) patterns in Figure 5.5b. While it is interesting that these blends do not, however, exhibit a reverse gyroid morphology in which the matrix is S-rich, it is not very surprising since the formation of bicontinuous morphologies requires subtle interplay between enthalpic repulsion and chain-packing frustration along the interface [17,18]. David et al. [94], on the other hand, have examined the morphological characteristics of poly(styrene-b-1,3-cyclohexadiene) (S-CHD) diblock copolymers in the presence of hS and find that their blends exhibit a core-shell cylindrical morphology wherein the CHD blocks form cylindrical annuli. This morphology differs substantially from the classical and complex morphologies generally established for microphase-ordered block copolymers

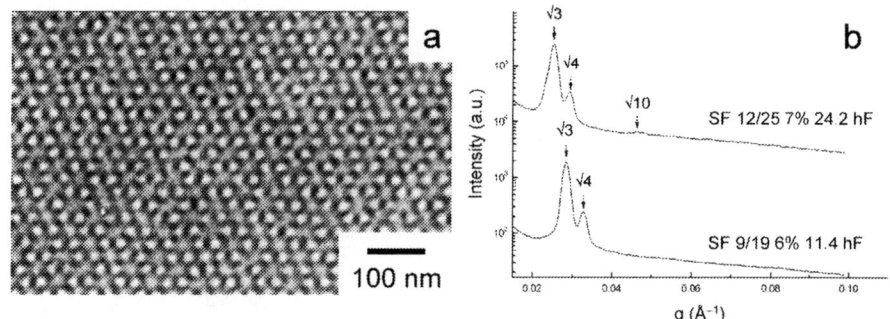

Figure 5.5 TEM image (a) and corresponding SAXS patterns (b) of the gyroid morphology generated in a binary blend of an asymmetric poly(styrene-b-ferrocenyldimethylsilane) (S-F) diblock copolymer with 6 wt % poly(ferrocenyldimethylsilane) (hF). The image in (a) shows the [111] projection. Included for comparison is a SAXS pattern obtained from a complementary blend composed of a slightly higher molecular weight S-F copolymer and homopolymer with 7 wt % hF. (Adapted from Lammertink, R. G. H., Hempenius, M. A., Thomas, E. L. and Vancso, G. J. J. Polym. Sci. B: Polym. Phys. **37**, 1009 (1999) and reprinted with permission. Copyright 1999 Wiley-Interscience.)

and reflects the inherent morphology of the neat S-CHD copolymer. Even in relatively conventional diblock copolymer/homopolymer blends, insightful findings continue to be reported. By exploring the high-temperature behaviour of homopolymer-rich blends composed of hB and a symmetric B-EO copolymer, Huang et al. [95] have discovered the existence of a face-centered cubic (fcc) spherical morphology between the bcc morphology and the ODT.

In a separate study of semicrystalline B-EO/hB blends, thermal calorimetry has been used to detect morphological transitions on the basis of their thermal signatures. According to the data of Chen et al. [96] displayed in Figure 5.6, the neat copolymer exhibits a lamellar morphology and a normal freezing (crystallization) temperature (T_f) of its EO block that is about 5 °C below that of a hEO homopolymer of equal molecular weight. The difference in thermal signatures between the homopolymer and matched copolymer is attributed to block confinement within the microphase-ordered morphology. Addition of hB ($\alpha = 0.52$) to the copolymer initially results in a slight reduction in T_f. As ϕ_{hB} is increased further, however, T_f drops precipitously (by ~55 °C) as the blend morphology changes from cocontinuous lamellae to dispersed EO cylinders, which serves to more severely constrain the ability of the EO blocks to crystallize. A second discontinuity in T_f is induced at still higher values of ϕ_{hB} where the EO blocks order into spheres on a bcc lattice. The overall confinement-induced reduction in T_f over the range of blend compositions explored is substantial, ~70 °C. Similar findings have been reported by Xu et al. [97] for binary blends consisting of a series of poly(ethylene oxide-b-butylene oxide) (EO-BO) copolymers and hBO, whereas Liu et al. [98] have found that addition of crystallizable polytetrahydrofuran (hTHF) to a (THF-M) diblock copolymer

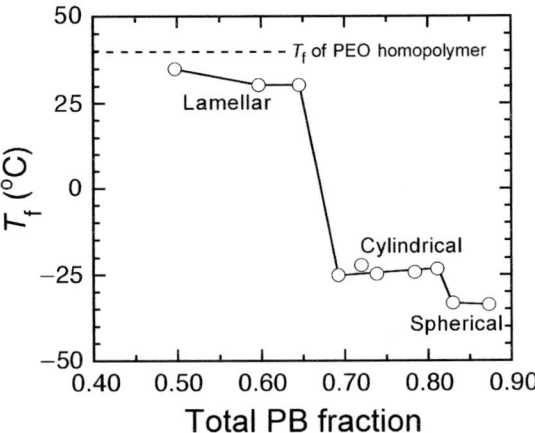

Figure 5.6 Morphology-induced dependence of the crystallization temperature (T_f) of the EO block on total polybutadiene (PB) concentration in binary blends of a polybutadiene homopolymer and a B-EO diblock copolymer. Data obtained from a poly(ethylene oxide) homopolymer are included (dashed line) for reference. (Reprinted with permission from Chen, H.-L., Hsiao, S.-C., Lin, T.-L., Yamauchi, K., Hasegawa, H. and Hashimoto, T. Macromolecules **34** 671, (2001). Copyright 2001 American Chemical Society.)

improves the crystallizability of the THF block. These results clearly demonstrate that blend composition and, hence, homopolymer distribution play a prominent role in morphology and property development by regulating the ability of the semicrystalline copolymer blocks to self-organize and subsequently crystallize. In their related investigation, Chen et al. [99] provide direct visualization of how crystallization of the EO block can distort the morphology of miscible B-EO/hB blends. Huang et al. [100] and Zhu et al. [101] have likewise shown that the morphology of microphase-ordered S-EO/hS blends can be used to direct the growth kinetics, thermodynamic stability and orientation of EO crystals. Another example where nanoscale confinement effects constitute a non-negligible consideration in the design of copolymer/homopolymer blends is in the fabrication of molecularly thin films. The dry- and wet-brush conditions alluded to earlier are schematically depicted in Figure 5.7 and illustrate the impact of homopolymer distribution on lamellar swelling [102]. Orso and Green [103] have verified that the interlamellar spacing (D) of copolymer/homopolymer thin films is given by the same expression proposed by Hamdoun et al. [104] for inorganic nanoparticles in a block copolymer matrix, namely,

$$D_{local} = \frac{D_0}{1-\phi} \qquad (5.1)$$

if α is large and the homopolymer molecules localize along the midplane of their host lamellae (Figure 5.7a). Here, D_0 is the spacing of the neat copolymer and ϕ

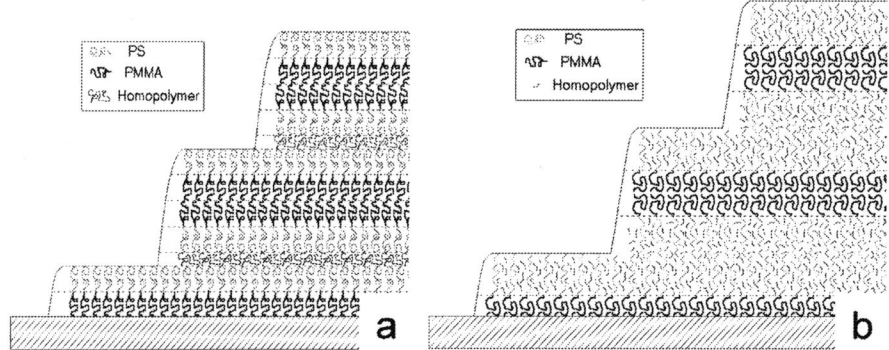

Figure 5.7 Schematic illustrations of diblock copolymer/homopolymer blends in molecularly thin films. In (a), the homopolymer molecules are long relative to their host blocks and localize along the lamellar midplane, whereas those in (b) are short and distribute more uniformly. (Adapted from Smith, M. D., Green, P. F., Saunders, R. Macromolecules **32**, 8392 (1999), and reprinted with permission. Copyright 1999 American Chemical Society.)

denotes the volume fraction of added homopolymer. If α is relatively small so that the homopolymer molecules are more evenly *distr*ibuted (Figure 5.7b), then D can be written as

$$D_{\text{distr}} = \frac{D_0 g^{1/3}}{1 - \phi} \quad (5.2)$$

where

$$g = \frac{f + (1-f)\phi^2}{f(1-\phi)^2} \quad (5.3)$$

The conformational attributes and swelling behavior of diblock copolymer/homopolymer blends in thin-film geometries are addressed in more detail by Green and Limary [105] and Retsos *et al.* [106].

All the systems described up to this point consist of an ordered block copolymer to which homopolymer is added. Interest also exists in AB/hA systems wherein a compositionally symmetric copolymer is structurally disordered to ascertain the effects of M_{hA} and ϕ_{hA} on intermolecular interactions (collectively expressed through an effective χ) and miscibility. Tanaka and Hashimoto [107] have used SAXS to study such blends systematically varying both M_{hA} and ϕ_{hA} to demonstrate that the temperature dependence of χ in these blends is sensitive to both parameters. For block copolymers that undergo microphase ordering by an enthalpically driven mechanism, χ can be conveniently written as $A + B/T$, where $B > 0$. Increasing M_{hA} at constant ϕ_{hA} or, conversely, increasing ϕ_{hA} at constant M_{hA} is observed to promote a strikingly similar change in $\chi(T)$. In the case of conventional diblock copolymers with pure blocks, χN must be sufficiently small to achieve disorder, thereby requiring either (i) low molecular weight (small N), (ii) high temperature (low χ), or (iii) chemically similar moieties (low χ). Alternatively, Laurer et al. [108] have explored the morphologies of disordered diblock copolymers consisting of random blocks differing in composition in the presence of a parent homopolymer. By using random-copolymer blocks containing styrene and diene repeat units in different monomer ratios, it is possible to tailor the effective χ between the blocks and thus produce high molecular weight block copolymers based on S and I that are disordered. Their results reveal a progression of blend morphologies that systematically vary from channel-like hA dispersions at low M_{hA} and ϕ_{hA} to simultaneously microphase/macrophase-separated dispersions at high M_{hA} and ϕ_{hA}. Han et al. [109] have explored the morphological characteristics and phase behaviour of binary blends composed of a disordered low molecular weight S-I diblock copolymer and hS. Their results, as discerned by a combination of dynamic rheology, SAXS and TEM, indicate that the apparent existence of a bicontinuous morphology reflects frozen-in composition fluctuations of the copolymer near its ODT.

5.2.2 DIBLOCK COPOLYMER/HOMOPOLYMER TERNARY BLENDS

As mentioned earlier, the addition of small quantities of a diblock copolymer to two (often parent) homopolymers has been routinely implemented as an effective means by which to achieve compatibilization in a biphasic blend (see Figure 5.3). In this case, the copolymer behaves as a low molar mass surfactant and forms a monolayer between the two immiscible homopolymers to effectively reduce interfacial tension, thereby reducing the size of macroscopic dispersions by increasing interfacial area. Sung et al. [110] have shown that, at relatively low copolymer concentrations, the spinodal temperature, which identifies the stable two-phase region of an immiscible polymer blend decreases in linear fashion with increasing copolymer concentration. They also report that the extent of this reduction is dependent on component molecular weight, a blend characteristic explicitly examined in the SCF analyses of Komura et al. [111] and Janert and Schick [112], as well as the Monte Carlo simulations of Kim and Jo [113]. On the basis of predictions regarding interfacial interaction and elasticity, Thompson and Matsen [114] propose that the relative molecular size ratio of homopolymer to copolymer in a compatibilized blend is optimized at $\sim 80\%$. The role of copolymer architecture/sequencing in both blend compatibilization and microstructural development has likewise been the subject of several independent studies [115–117]. Balsara and coworkers [118,119] have focused their efforts specifically on the phase behaviour of ternary polyolefin blends, which serve as model systems for comparison with theory due to their basic level of molecular interactions. Their results have established the homopolymer and copolymer conditions that identify incipient ordering in ternary blends [120], as well as the presence of a polymeric microemulsion [121,122]. The discovery

[123] and design strategy [124] of a bicontinuous microemulsion (BμE) phase in ternary copolymer/homopolymer blends at copolymer concentration levels typically on the order of 10–20 vol% is certainly one of the most intriguing developments in ternary block copolymer systems and helps to promote direct comparison with the elegant phase behaviour of low molar mass surfactant systems [125–127]. A TEM image of a ternary poly(ethylene)/poly(ethylene-*alt*-propylene)/poly[ethylene-*b*-(ethylene-*alt*-propylene)] (hE/hEP/E-EP) system containing 10 vol% E-EP copolymer is provided [125] in Figure 5.8 and confirms the existence of a layered morphology with no long-range order. This result is consistent with complementary small-angle scattering profiles that reveal only a single pronounced peak for the BμE morphology.

In this complex morphology, the copolymer molecules form monolayers that divide space nearly equally, in which case they must be reasonably flexible. This constraint may, depending on the value of χ between the A and B moieties, require the use of relatively low molecular weight copolymers [128]. Since the translational entropy of the homopolymer molecules is responsible for preventing monolayer attraction (and the onset of macrophase separation) and is predicted to vary inversely with homopolymer molecular weight, it again follows that α must be much less than unity for a BμE to be stabilized. With these

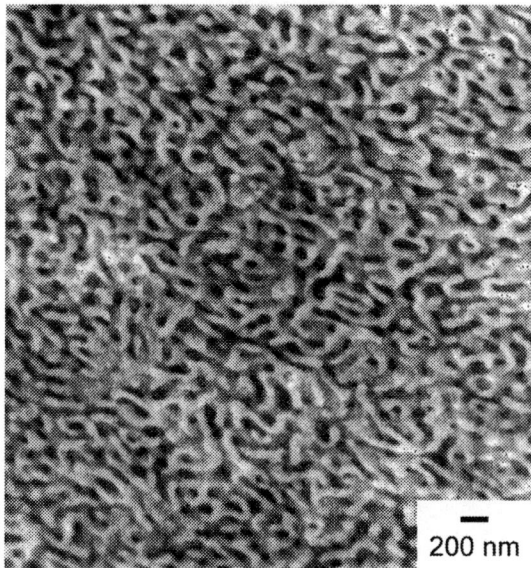

Figure 5.8 TEM image of a bicontinuous microemulsion (BμE) phase produced in a ternary poly(ethylene)/poly(ethylene-*alt*-propylene)/poly[ethylene-*b*-(ethylene-*alt*-propylene)] (hE/hEP/E-EP) blend. Note the presence of clearly defined channels without long-range order. (Reprinted with permission from Hillmyer, M. A., Maurer, W. W., Lodge, T. P., Bates, F. S. and Almdal, K. J. Phys. Chem. B **103**, 4814, 1999. Copyright (1999) American Chemical Society.)

considerations in mind, polymeric microemulsions have been observed between the lamellar (bilayered) morphology and the macrophase-separated region of the phase diagram in close proximity to the ODT. This position in phase space reflects the layered nature of the BμE morphology and its proximity to the biphasic region. While unable to organize into an ordered lamellar morphology, the copolymer monolayers are sufficiently attracted to each other so as to expel homopolymer without permitting macrophase separation. Several relevant experimental phase diagrams reported by Hillmyer et al. [125] on the basis of dynamic rheology, small-angle scattering and cloud point measurements are shown in Figure 5.9. In all three cases, the stability region for the BμE morphology is seen to exist as a relatively narrow channel located between the lamellar and macrophase-separated regimes at temperatures below the ODT of the blend. Similar findings have been reported by Corvazier et al. [129] for microemulsions produced in hS/hI/S-I ternary blends with 79–93 vol% homopolymer. The location of the BμE phase in Figure 5.9 closely coincides with the conditions corresponding to a critical point referred to as an isotropic Lifshitz point, which is classified as the intersection of the loci of ODT points and the loci of phase-separation critical points [124,130]. At this condition, the thermodynamic driving forces favouring microphase and macrophase separation in the blend are balanced. If we consider the particular case of a symmetric AB diblock copolymer of chain length N_{AB} in a ternary A/B/AB blend containing homopolymers of equal chain length $N_{hA} = N_{hB} = \beta N_{AB}$, where $\beta = \alpha/2$ in the present example, then the total homopolymer volume fraction, Φ_L, and thermodynamic incompatibility, $(\chi N_{AB})_L$, at which the Lifshitz point occurs are given by [131]

$$\Phi_L = \frac{1}{(1 + 2\beta^2)} \tag{5.4}$$

and

$$(\chi N_{AB})_L = \frac{2(1 + 2\beta^2)}{\beta} \tag{5.5}$$

In their effort to facilitate the design of macromolecular BμE phases, Fredrickson and Bates [124] provide the corresponding, more general, set of conditions

---

Figure 5.9 Experimental phase diagrams showing the location of the BμE phase (dashed lines) in three diblock copolymer/homopolymer ternary blends: (a) poly(ethylene)/poly(ethylene-*alt*-propylene)/poly[ethylene-*b*-(ethylene-*alt*-propylene)], (b) poly(ethylethylene)/poly(dimethylsiloxane)/poly(ethylethylene-*b*-(dimethylsiloxane) and (c) poly(ethylene)/poly(ethylene oxide)/poly(ethylene-*b*-ethylene oxide). The lamellar (L), hexagonal cylindrical (H) and phase-separated (PS) regimes are labeled, and the data displayed as a function of homopolymer content (θ_H) have been collected by a combination of rheology (x), small-angle neutron scattering (circles) and cloud-point measurements (squares). (Reprinted with permission from Hillmyer, M. A., Maurer, W. W., Lodge, T. P., Bates, F. S. and Almdal, K. J. Phys. Chem. B **103**, 4814, 1999. Copyright (1999) American Chemical Society.)

Phase Behaviour of Block Copolymer Blends

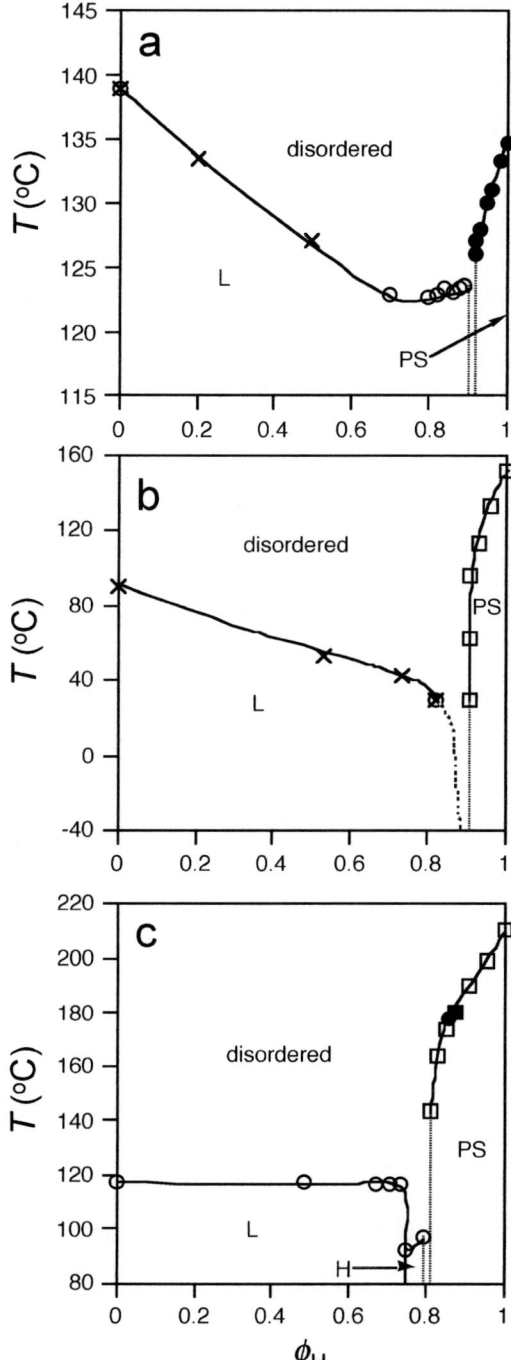

yielding Lifshitz points in asymmetric hA/hB/AB ternary blends. A detailed scattering analysis of isotropic Lifshitz critical behaviour has been recently undertaken by Schwahn *et al.* [132–134] and is complemented by the theoretical efforts of Kudlay and Stepanow [135]. The SCF predictions of Thompson and Matsen [136] indicate that high molecular weight (more incompatible) macromolecules could likewise be used to produce the BμE morphology if the copolymer used to stabilize the morphology is polydisperse, which would serve to broaden the copolymer/homopolymer interface, enhance the flexibility of the monolayers and subsequently reduce (i) the loss of homopolymer configurational entropy and (ii) the attraction between copolymer monolayers. In addition to theoretical efforts addressing the stability of the BμE phase [137], a recent study [138] also suggests the existence of an aperiodic lamellar phase in ternary diblock copolymer/homopolymer blends.

5.2.3 MULTIBLOCK COPOLYMER/HOMOPOLYMER BLENDS

Thus far, only copolymer/homopolymer blends composed of simple diblock copolymer molecules have been considered, although their ABA copolymer counterparts exhibit very similar phase behaviour [3,4,9,10,139–142]. A transmission electron microtomography (TEMT) image [143,144] of the gyroid morphology formed in an ordered S-I-S copolymer is presented in Figure 5.10a and is seen to compare favourably with its theoretical analogue displayed in Figure 5.10b. Many commercially relevant copolymer systems (such as

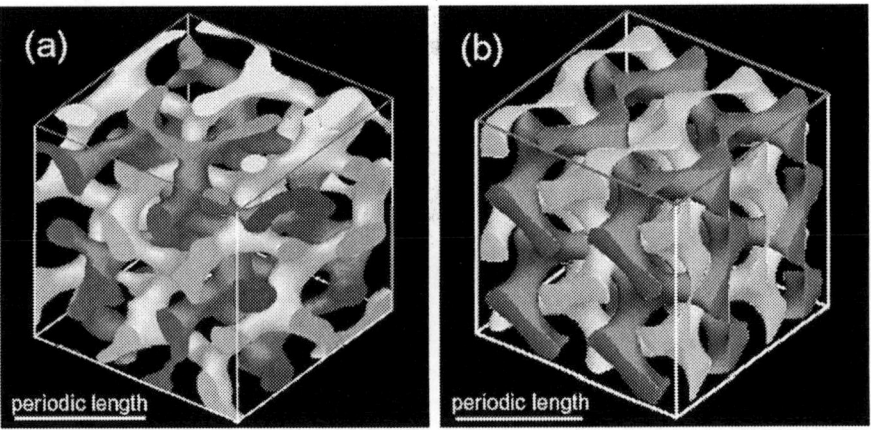

Figure 5.10 TEMT image (a) and computed Schoen surface (b) of the gyroid morphology in a microphase-ordered ABA triblock copolymer containing 32 vol % A [143,144]. The minority, nonintersecting A channels are displayed as light and dark gray, whereas the B matrix is transparent to facilitate viewing. The periodic length (bar) discerned from Fourier analysis is about 74 nm. The top cross section of the cube identifies the (001) plane.

thermoplastic elastomers [36,145]) consist of ABA triblock copolymers in which the A endblocks are either glassy or semicrystalline and the B midblock is a rubbery material with a low (subambient) glass transition temperature (T_g). Upon microphase separation, the A blocks form solid microdomains that, at equilibrium, serve to anchor the B blocks at both termini. If the two termini reside in the same microdomain, their impact on property development derives principally from chain entanglement and, from this standpoint, is physically similar to that of diblock copolymers. If, however, a B block connects two neighbouring microdomains, it becomes a bridge that can form the basis of a molecular network. The presence of bridged midblocks in a block copolymer system can vastly improve the mechanical properties of the system. For this reason, several experimental [146–149] and theoretical [150–152] studies have sought to quantitate the average bridging fraction (v_b) in neat microphase-ordered triblock copolymers. On the basis of the dielectric relaxation measurements of Watanabe and coworkers [146–149] and the SCF predictions of Matsen and Schick [151], v_b is expected to lie between 0.40 and 0.45 for moderately incompatible copolymers exhibiting the lamellar morphology. In the limiting case of an ABC copolymer wherein the A and C blocks are incompatible, $v_b = 1.00$. Higher-order multiblock copolymers with more than one midblock capable of forming bridges offer a substantially greater challenge in this vein [153–158], particularly if the blocks are randomly coupled so that the block lengths are polydisperse [159]. Even with perfectly alternating multi-block copolymers of the form $(AB)_n$ wherein the block lengths are relatively uniform, attempts to form miscible blends with either parent homopolymer have met with minimal success due to the propensity of the copolymer molecules to self-organize and exclude homopolymers of relatively low molecular weight [160,161].

5.2.3.1 ABA Triblock Copolymer Systems

Incorporation of homopolymer A (hA) or an A-compatible homopolymer into a microphase-ordered ABA copolymer to form a miscible blend tends to obey the same design paradigms established for AB/hA blends, since the A endblocks of the triblock copolymer form dense brushes in the same fashion as their diblock analogues. Several recent studies [162–164] have investigated the (dis)-ordering and (de)micellization behaviour of compositionally asymmetric ABA triblock copolymers in the presence of low molecular weight hA and report that the (de)micellization temperature decreases with increasing hA concentration. Addition of S-compatible poly(xylenyl ether) to a poly[styrene-*b*-(ethylene-*co*-butylene)-*b*-styrene] (S-EB-S) copolymer is found [165] to result in improved thermo-mechanical properties and an increase in the ODT, which qualitatively agrees, in principle, with the predicted phase diagrams provided in Figure 5.4. Incorporation of hB into a lamellar ABA copolymer is expected to result in a more complicated segmental distribution [166]. A representative distribution of hB in a lamellar ABA copolymer, predicted from SCF analysis [167], is provided for illustrative purposes in Figure 5.11a, and the accompanying effect of hB addition on v_b is displayed as a function of blend composition for several different α ratios in Figure 5.11b. Since a finite fraction of the B midblocks of the copolymer remain bridged across each B-rich microdomain, the hB molecules must distribute more uniformly within their host microdomain than they would otherwise without bridges (as in AB/hB blends). This constraint, coupled with the physical reduction in the size of the B microdomains (since the B midblocks must either form bridges, in which case they span the entire width of a microdomain, or loops, in which case they effectively behave as single-tethered chains of half molecular weight), therefore requires the value of α to be smaller in ABA/hB blends than in comparable AB/hB blends to retain miscibility. As in comparable AB/hB blends, addition of hB to ordered ABA copolymers can, under the right combination of α and ϕ_{hB}, either swell the B-rich microdomains or induce transitions to other morphologies or macrophase separation [168]. In either case, substantial changes in bulk properties, such as the mechanical and thermal characteristics, are manifested [167]. If the midblock is crystallizable (as with S-EO-S copolymers), the same confined crystallization effects previously discussed with regard to diblock copolymer/homopolymer blends are observed. In this case, addition of hEO can, depending on its molecular weight, reduce or enhance the crystallinity of the EO copolymer block, as evidenced in Figure 5.12 by the thermal and gas-transport signatures of S-EO-S/hEO blends containing a lamellar S-EO-S triblock copolymer and either an amorphous or semicrystalline hEO [169].

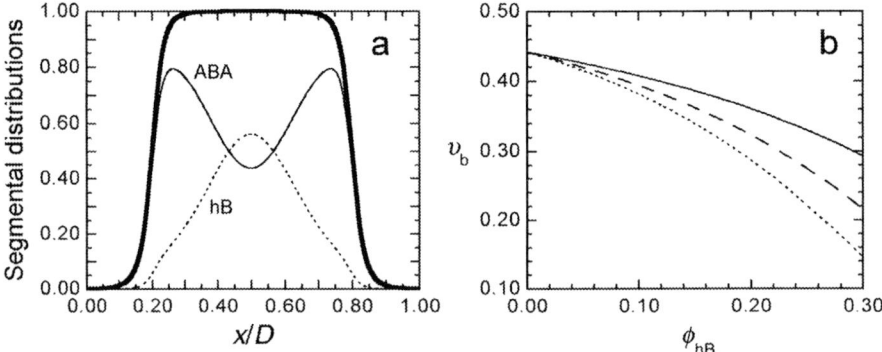

Figure 5.11 Addition of a midblock-selective homopolymer (hB) to a lamellar ABA triblock copolymer: (a) segmental density distributions and (b) effect on bridging fraction (v_b) [167]. In (a), the distributions of B units deposited from the copolymer (thin solid line) and homopolymer (dotted line) are displayed and labeled. In (b), the dependence of v_b on hB concentration is shown for three different α values (calculated on the basis of half the molecular weight of the B midblock): 0.29 (solid line), 0.59 (dashed line) and 1.18 (dotted line).

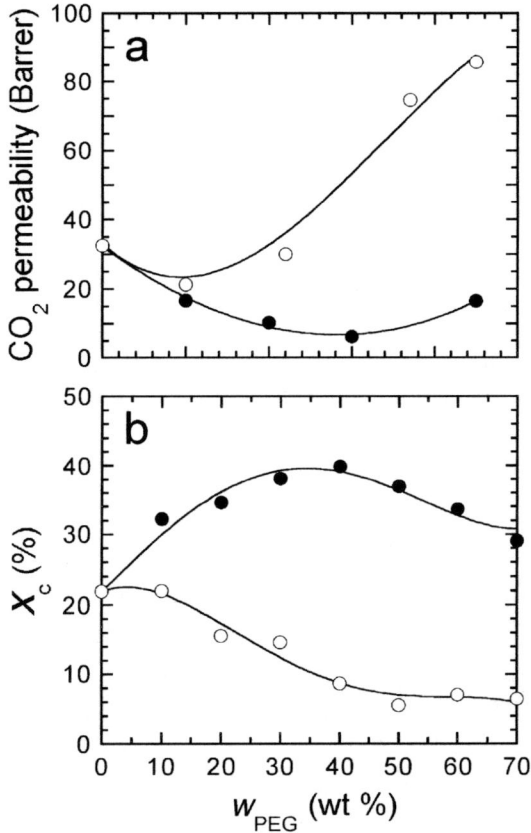

Figure 5.12 Dependence of CO_2 permeability (a) and % crystallinity, X_c, (b) on blend composition in binary blends of a lamellar poly(styrene-b-ethylene oxide-b-styrene) (S-EO-S) triblock copolymer with two added poly(ethylene glycol)s (PEGs) of different molecular weight (in g/mol): 400 (amorphous, open circles) and 4600 (semicrystalline, filled circles) [169].

5.2.3.2 ABC Triblock Copolymer Systems

As alluded to earlier, the parameter variability afforded by neat ABC triblock copolymers is extensive and has greatly expedited the development [21,32] and prediction [33,170] of a large number of exciting new copolymer morphologies. This parameter space can be further enlarged through the addition of a single homopolymer, which results in three new parameters: the choice of homopolymer (hA, hB or hC), α and ϕ. One may reasonably expect that the design of ABC/hA(hC) blends would obey, to some extent, the paradigms previously established for AB/hA(hB) blends [34,35]. While this is generally true, a characteristic of microphase-ordered ABC copolymers exhibiting three distinct microphases is that each molecule must traverse two different interfaces, in contrast to a diblock copolymer with only one interface/molecule. Thus, addition of a single homopolymer at the A- or C-rich microphases of an ABC copolymer may affect the adjacent interface differently. To illustrate this point, Lescanec et al. [171] have added 10 vol% of hS to a compositionally symmetric poly(2-vinyl pyridine-b-isoprene-b-styrene) (P-I-S) triblock copolymer. At a low value of α ($=0.27$), the resultant hexagonal morphology appears to be faceted along its axial projection (Figure 5.13a), which is schematically depicted in terms of a Wigner–Seitz cell. In this case, the P endblocks comprising the core are surrounded by an I-rich inner layer and an S-rich outer layer. Increasing α to 3.4 yields a similar, but rounded, I-rich layer (Figure 5.13b) due to incorporation of longer hS chains in the outer layer. These large hS molecules can only be accommodated in a hexagonally-packed microdomain arrangement if the vertices of the I microdomains in Figure 5.13a become rounded. This observation is contrary to what is expected in miscible diblock copolymer/homopolymer blends wherein low molecular weight homopolymers have more impact on interfacial curvature in the copolymer morphology than high molecular weight homopolymer molecules due to brush wetting. The principal difference between the two blends is that the I midblocks are anchored at both the P-I and I-S junctions in the present system, in which case they are effectively immobilized and cannot change their interfacial curvature very much along the I-S junction without profoundly affecting chain packing along the P-I interface. This constraint implies that, while the A and C blocks in an ABC copolymer can freely stretch upon addition of hA or hC, respectively, the ability of the B block to

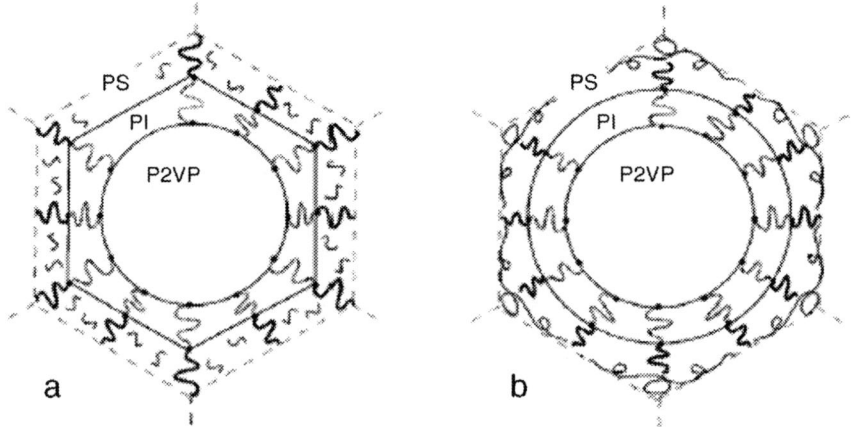

Figure 5.13 Schematic illustration showing the location of added homopolymer C to (a) low molecular weight and (b) high molecular weight, in a microphase-ordered ABC triblock copolymer. Note the subtle change in morphology from hexagonal to rounded microdomains promoted by the increase in homopolymer molecular weight. (Reprinted with permission from Lescanec, R. L., Fetters, L. J. and Thomas, E. L. Macromolecules **31**, 1680, 1998. Copyright (1998) American Chemical Society.)

stretch upon addition of hB is limited, which is corroborated by the experimental findings of Suzuki *et al.* [172].

Thus, due to the added interfacial constraints accompanying the ABC molecular architecture, incorporation of hA(hC) into microphase-ordered ABC copolymers does not precisely follow the same guidelines identified for AB/hA(hB) blends. Sugiyama *et al.* [173] have, for instance, generated core-shell variations of the cylindrical and gyroid morphologies in their blends of a nearly symmetric poly(styrene-*b*-isoprene-*b*-dimethylsiloxane) (S-I-D) triblock copolymer with either hS or hD. The Monte Carlo simulations of Dotera [174], on the other hand, suggest the existence of numerous bicontinuous morphologies, including the double-diamond, in ABC/hA/hC blends. Even with such variations, the paradigm of adding a single homopolymer to a microphase-ordered ABC block copolymer remains a viable and expedient route to relatively complex tricomponent morphologies. If the middle block is a random, not tapered, sequence of the repeat units comprising the two end blocks, the resultant A(A/B)B copolymer can likewise be envisaged as an ABC triblock copolymer with relatively low interblock incompatibilities. Binary copolymer/homopolymer blends containing such materials have been found [175,176] to exhibit complex morphologies, such as those shown in Figure 5.14a, in which a S-(S-*r*-I)-I copolymer with 40 wt% random midblock is blended with hS so that the total styrene content of each blend is 90 wt%. A TEMT image of the sponge-like morphology coexisting with swollen lamellar bilayers in Figure 5.14a is provided in Figure 5.14b. The corresponding mean (H) and Gaussian (K) curvature distributions in Figures 5.14c and d, respectively, reveal that this morphology possesses a zero area-averaged mean curvature and a negative area-averaged Gaussian curvature (indicating a hyperbolic topology), which are both consistent with the requirements for a minimal surface. According to a global topological analysis [177] of the full TEMT image, the coordination of this morphology is, for the most part (> 90%), 3 (channels/vertex) and its genus is ~ 2. At sufficiently high homopolymer concentrations, this bicontinuous morphology degenerates into micelles that coexist with isolated bilayer sheets [177].

5.2.4 NONLINEAR BLOCK COPOLYMER/HOMOPOLYMER BLENDS

While most fundamental studies of model block copolymers have focused on diblock and triblock architectures due primarily to synthesis considerations, several nonlinear designs, such as the star, miktoarm and single-graft motifs, have likewise been investigated. Self-organization of such copolymers has yielded morphologies that, in some cases, differ markedly from those observed in comparable linear copolymer systems [178,179]. Recall that a morphological transformation in conventional copolymer/homopolymer blends reflects a

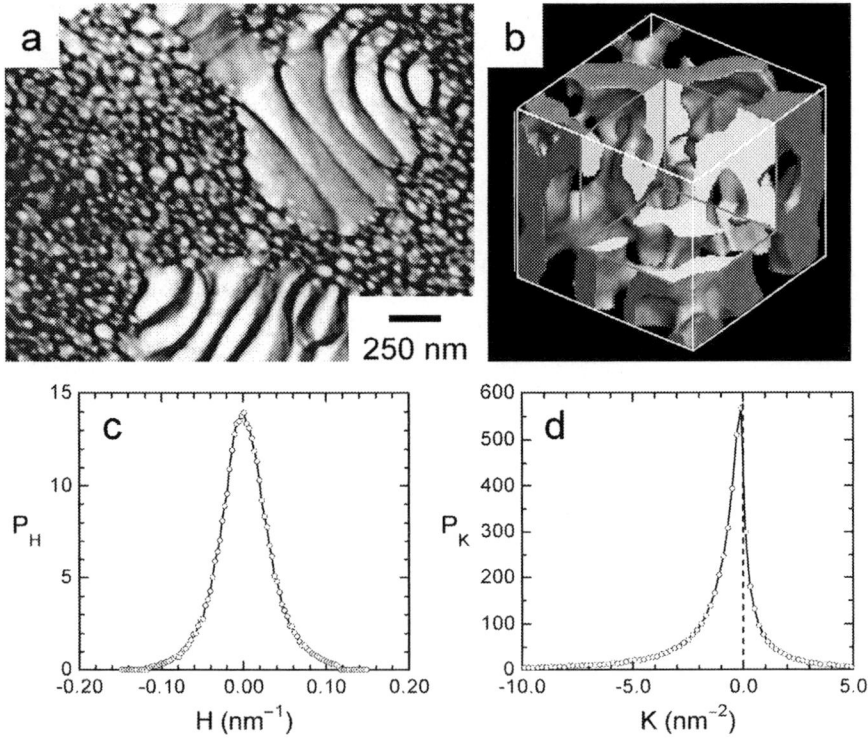

Figure 5.14 Characteristics of a sponge-like morphology produced in a binary blend of a lamellar A(A/B)B triblock copolymer (40 wt % A/B midblock) with homopolymer A [175,176]. This morphology is observed in (a) to coexist with swollen lamellae at an overall blend composition of 90 wt % A, and a TEMT image of the morphology [177] is provided in (b). The mean (H) and Gaussian (K) curvature distributions computed directly from the TEMT image are included in (c) and (d), respectively.

change in interfacial curvature due to a change in chain packing along one side of the interface due to the presence of homopolymer molecules that wet the host copolymer brush. In the case of ordered, miscible AB/hB blends, for instance, the interface present in the neat copolymer curves toward the microphase with a less densely packed interface (A in this example), and the extent to which it curves depends on the wettability of the homopolymer (related to α) and the population of homopolymer available (related to ϕ_{hB}). In ordered A_2B single-graft copolymers, however, the morphology consists of two A blocks/molecule residing in the A microphase and one B block/molecule in the B microphase, with all three blocks covalently linked together in the interfacial region. An increase in interfacial chain packing in the B microphase due to addition of hB is not expected to promote the same change in interfacial curvature encountered in linear block copolymer/homopolymer blends due to the inherently higher density of A segments residing within the interface in blends containing A_2B

copolymer molecules. For this reason, the I_2-S/hS blends investigated by Yang et al. [180] exhibit morphological boundaries that are shifted in composition from their S-I/hS analogues and form sheet-like morphologies, such as perforated lamellae. Similar behaviour has been observed by Avgeropoulos et al. [181] in comparable blends containing lamellar miktoarm star block copolymers. These nonlinear materials not only illustrate the role of interfacial chain packing on interfacial curvature in binary block copolymer/homopolymer blends, but also produce relatively uncommon morphological features, such as T-junction grain boundaries [182].

5.3 BLOCK COPOLYMER/COPOLYMER BLENDS

Although block copolymer/homopolymer blends provide tremendous versatility in terms of tailored polymer nanostructures, they are nonetheless subject to the limitations regarding α and ϕ discussed in the previous section and explicitly illustrated in Figure 5.4. A facile means by which to overcome such limitations is through the use of a second copolymer as a cosurfactant, as initially demonstrated by Hadziioannou and Skoulios [183]. Whereas an imbibed homopolymer will tend to localize to an α-dependent extent within its host microphase, an added copolymer will tend to be more spatially confined, especially if its blocks are sufficiently incompatible to induce microphase separation. As alluded to earlier in the case of ABC copolymers, the use of two block copolymers to control the morphology of copolymer/copolymer blends greatly enlarges the parameter space that can be feasibly explored. The chemical identity (e.g., A, B or C blocks), composition, molecular weight and architecture (e.g., diblock or triblock) of each copolymer in a given copolymer/copolymer blend, as well as the blend composition, can all be systematically varied to yield stunning results that provide not only fundamental insight into molecular self-organization, but also novel morphologies that might not be easily, if at all, accessible through the use of designer copolymers or copolymer/homopolymer blends. The enthalpic and entropic considerations required to accurately describe the thermodynamics and phase behaviour of AB, ABA and ABC block copolymers play important roles in determining whether such a blend will form a mixed (single-phase) morphology or an immiscible morphology composed of copolymer-rich macrophases. Another possibility to consider in the design of binary block copolymer/copolymer blends is that the copolymer used as an additive may be a random, rather than block, copolymer. In this section, we examine a variety of blends composed of two block copolymers: two diblock copolymers possessing a common chemical species – $(AB)_\alpha/(AB)_\beta$ or AB/AC – or an ordered diblock copolymer mixed with an ordered triblock copolymer – AB/ABA or AB/ABC. Blends consisting of two ordered triblock copolymers or an ordered block copolymer and a random copolymer are also briefly discussed.

5.3.1 DIBLOCK COPOLYMER/DIBLOCK COPOLYMER BLENDS

5.3.1.1 $(AB)_\alpha/(AB)_\beta$ Blends

In the case of $(AB)_\alpha/(AB)_\beta$ diblock copolymer blends composed of two different copolymers (denoted by α and β), two blend strategies become evident. The first approach requires that the composition of each copolymer is identical so that the molecular weight can be systematically varied to traverse the block copolymer phase diagram along the χN (incompatibility) axis under isoplethic conditions. Numerous independent studies have shown that the scaling relationship between the interlamellar spacing (D) of neat AB diblock copolymers and the molecular weight (M) can be used as a convenient means by which to assign a neat copolymer to a segregation regime (weak, intermediate or strong) [184,185]. In the series of compositionally symmetric S-I diblock copolymers investigated by Kane et al. [186], D is found to scale as $M^{0.71}$, which puts these materials in the intermediate- or strong-segregation regimes. Addition of each of these copolymers to a copolymer of higher molecular weight yields completely miscible blends as ε, defined as N_β/N_α, ranges from 0.20 to 0.57. The ratio of the interlamellar spacing of the blend to that of the high molecular weight copolymer ($D_{\alpha\beta}/D_\alpha$) obtained from both TEM and SAXS analyses is provided as a function of blend composition for each blend series in Figure 5.15 and demonstrates that the spacing of a miscible $(AB)_\alpha/(AB)_\beta$ diblock copolymer blend is not accurately represented as the equivalent spacing of a pure diblock copolymer of average molecular weight (indicated by the dashed lines). On the basis of the strong-segregation theory developed for neat triblock copolymers by Zhulina and Halperin [150], the free-energy (F) of a bidisperse copolymer blend exhibiting the lamellar morphology can be written [186,187] to include the unequal nonuniform stretching of the constituent (compositionally identical) copolymer molecules. Minimization of F with respect to $D_{\alpha\beta}$, followed by division of the analogous expression for D_α, yields

$$\frac{D_{\alpha\beta}}{D_\alpha} = \frac{\varepsilon + x(1-\varepsilon)}{[\varepsilon + x^3(1-\varepsilon)]^{1/3}} \quad (5.6)$$

where x denotes the mole fraction of the high molecular weight copolymer in the blend. Predictions derived from Equation (5.6) are included in Figure 5.15 and, with no adjustable parameters, show favourable agreement with the data. Recent comparative efforts by Court and Hashimoto [188] indicate that this result is equivalent to that derived from the strong-segregation theory proposed by Zhulina and Birshtein [189] for a mixture of bidisperse brushes on a planar surface. Matsen [190] has provided a more refined SCF approach to describe such molecular mixing by properly accounting for chain interdigitation along the lamellar midplane. His SCF framework is also capable of predicting the onset of macrophase separation, which has been observed [186,191,192] in

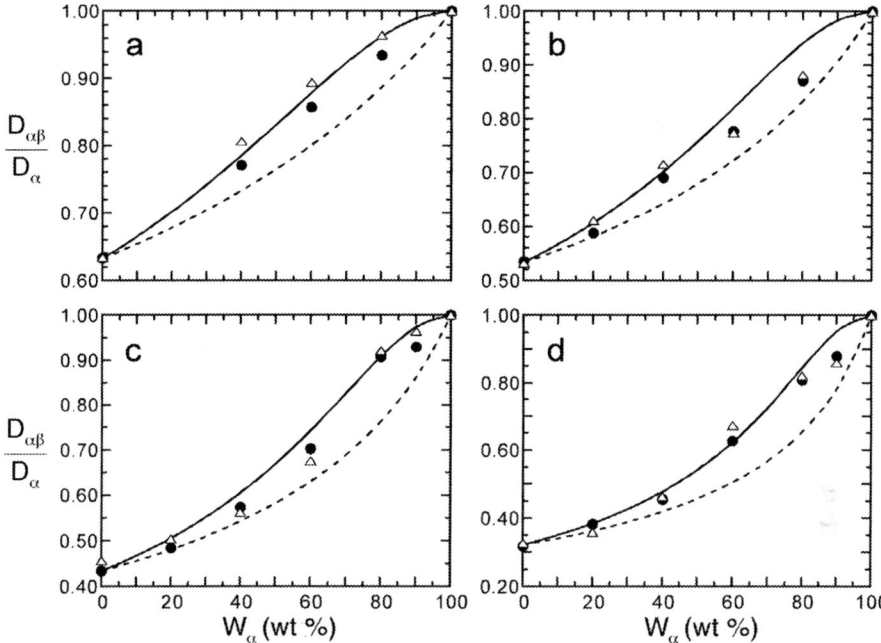

Figure 5.15 The normalized interlamellar period ($D_{\alpha\beta}/D_{\alpha}$) presented as a function of blend composition in four symmetric $(AB)_{\alpha}/(AB)_{\beta}$ copolymer blends varying in ε: (a) 0.57, (b) 0.45, (c) 0.33 and (d) 0.20 [186]. Data have been collected by TEM (open triangles) and SAXS (filled circles). The dashed lines correspond to predictions based on average molecular weight, whereas the solid lines are obtained from Eq. (5.6).

$(AB)_{\alpha}/(AB)_{\beta}$ diblock copolymer blends when ε becomes sufficiently small (ca. 0.1 or less). In this case, the blends tend to exhibit partial miscibility in which single-phase blends form at low and high x, whereas two-phase blends develop at intermediate compositions. Within the two-phase composition window, a fraction of the low molecular weight component resides in, and dilutes, the phase formed by the high molecular weight copolymer (see the SAXS data provided in Figure 5.16), thereby promoting a net reduction in $D_{\alpha\beta}$ relative to D_{α}. Yamaguchi *et al.* [193–195] have recently provided an extraordinarily detailed series of studies addressing the phase behaviour of, and chain location in, blends composed of nearly symmetric (lamellar) diblock copolymers in terms of molecular weight, composition and temperature considerations. Complementary dynamic density-functional simulations provided by Morita *et al.* [196] help to elucidate the dynamics of, as well as the competition between, microphase and macrophase separation in such blends.

The second strategy to be considered with regard to $(AB)_{\alpha}/(AB)_{\beta}$ diblock copolymer blends involves copolymers that have comparable molecular weights but differ in composition. In this case, the primary objective is to control

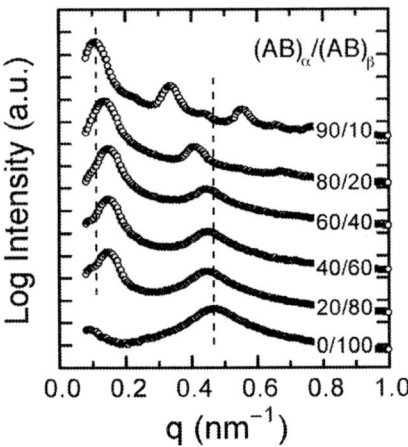

Figure 5.16 SAXS patterns acquired from a symmetric $(AB)_\alpha/(AB)_\beta$ copolymer blend in which $\varepsilon = 0.10$ and macrophase separation occurs, as evidenced by the relative invariance of the scattering peaks and accompanying TEM analysis [186].

interfacial curvature and, hence, morphological development through physical blending. The $(AB)_\alpha/(AB)_\beta$ blends previously described undergo nearly identical changes in the interfacial packing of A and B blocks as x is varied. By choosing copolymers with different compositions (morphologies), it is possible to control, in systematic fashion, the extent to which the A and B blocks pack along the interface and, consequently, the corresponding interfacial curvature. Zhao et al. [197] and Spontak et al. [198] have demonstrated that this strategy can be used to generate all the intermediate morphologies, including the gyroid, lying between those of the constituent copolymers. An interesting feature of these blend morphologies is that they form at blend compositions that are comparable, if not identical, to those of the neat copolymers, which suggests that the phase diagram of a block copolymer blend composed of copolymers differing in molecular composition should resemble that of the neat parent copolymers. This *one-component approximation* proposed by Matsen and Bates [199] is evident in the predicted phase diagram provided in Figure 5.17a, which displays the thermodynamic incompatibility (χN) as a function of blend composition (ϕ) for two diblock copolymers having identical segment lengths (b) and numbers (N), and the compositions listed in the figure caption. The remarkable similarity between this SCF diagram and that of a neat diblock copolymer is immediately evident. An interesting feature of this phase diagram is the existence of biphasic regions, which are more clearly seen in Figure 5.17b. In this figure, the copolymer compositions are varied at single values of ϕ and χN to provide guidance for generating intermediate "classical" (noncomplex) morphologies (spheres, cylinders and lamellae). The complementary theoretical

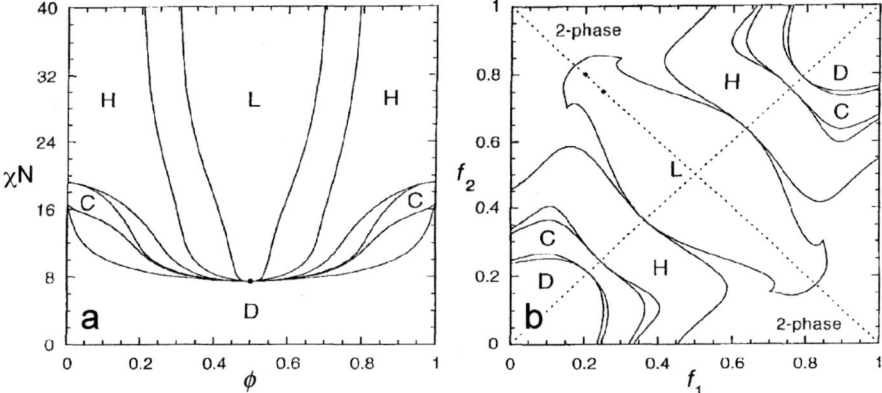

Figure 5.17 SCF phase diagrams of binary $(AB)_\alpha/(AB)_\beta$ copolymer blends based on the one-component approximation of Matsen and Bates [199]. In (a), the thermodynamic incompatibility (χN) is provided as a function of the volume fraction of the β copolymer (ϕ) for copolymers with $f_1 = 1 - f_2 = 0.25$ (the subscripts 1 and 2 refer to copolymers α and β, respectively). In (b), the effect of copolymer compositions (f_1 and f_2) on phase stability at $\chi N = 20$ and $\phi = 0.5$ is shown. (Compiled from Matsen, M. W. and Bates, F. S. Macromolecules **28**, 7298, 1995, and reprinted with permission. Copyright (1995) American Chemical Society.)

approach of Shi and Noolandi [200] yields comparable results. Development of the gyroid morphology, in particular, has attracted attention in such blends, since it occurs over a very narrow composition range in neat diblock copolymers. Sakurai *et al.* [201] have performed a rigorous study aimed at elucidating the effects of blend composition and temperature on the stability of the gyroid morphology in $(AB)_\alpha/(AB)_\beta$ diblock copolymer blends, whereas Hashimoto and coworkers [202,203] have explored the phase behaviour of diblock copolymer blends in which the constituent copolymers differ substantially in composition. In particular, the phase diagram prepared by Court and Hashimoto [202] (see Figure 5.18) is derived from blends composed of one copolymer with a spherical morphology and three lamellar copolymers differing in molecular weight, and conclusively demonstrates that the composition window over which a bicontinuous morphology develops depends on ε. While all the studies alluded to thus far address diblock copolymer blends in bulk systems, Koneripalli *et al.* [204] have also explored the phase behavior of blends consisting of symmetric diblock copolymers in thin-film geometries. Moreover, most studies of $(AB)_\alpha/(AB)_\beta$ blends justifiably focus on near-equilibrium morphologies and phase behaviour to deduce the underlying physicochemical principles governing molecular self-organization. Through a careful comparison of experimental data and theoretical predictions, Lipic *et al.* [205], however, provide valuable insight into the manifestation of nonequilibrium effects in such systems.

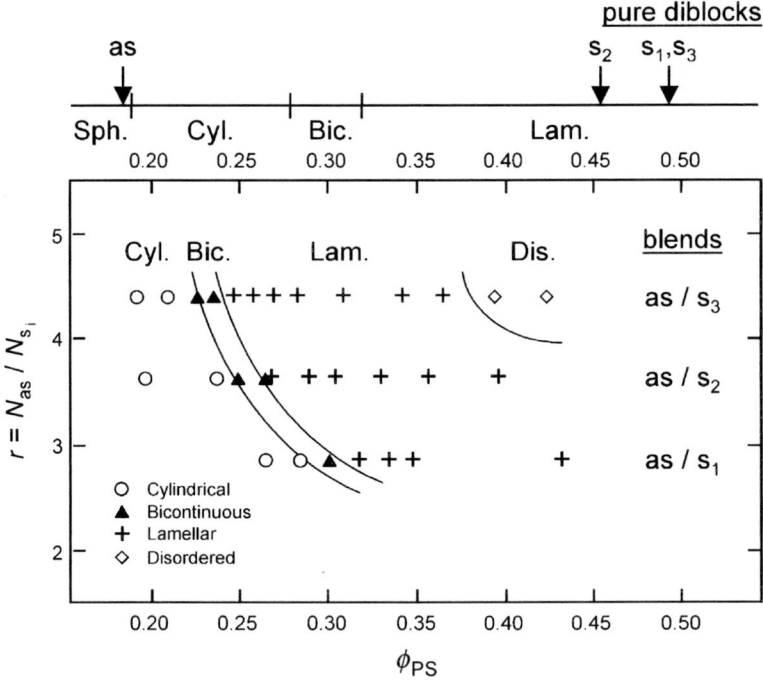

Figure 5.18 Experimental phase diagram of three $(AB)_\alpha/(AB)_\beta$ copolymer blends differing in composition and molecular weight. Molecular characteristics of the neat copolymers (as, s_1, s_2 and s_3) are provided at the top, and regions of phase stability are denoted by the labeled symbols. (Reprinted with permission from Court, F. and Hashimoto, T. Macromolecules **34**, 2536, 2001. Copyright (2001) American Chemical Society.)

5.3.1.2 AB/AC Blends

Using $(AB)_\alpha/(AB)_\beta$ diblock copolymer blends to generate intermediate morphologies provides tremendous impetus for exploring AB/AC blends as a facile route to the elegant morphologies afforded by ABC triblock copolymers (see Figure 5.2). As independent experimental [35,206–209] studies indicate, however, such blends may remain miscible, but are prone to undergo macrophase separation even though the copolymer molecules possess a common block. In fact, Olmsted and Hamley [210] have reported that binary AB/AC copolymer blends can exhibit multiple Lifshitz points. Kimishima *et al.* [211] have attempted to probe the conditions governing macrophase separation in AB/AC diblock copolymer blends by hydrogenating a lamellar S-I diblock copolymer to different degrees, thereby producing a series of S-(I-*r*-EP) copolymers. In the event that hydrogenation is nearly complete and the I block is converted to an EP block, macrophase separation between the S-I and S-EP copolymers is observed to occur, resulting in well-defined copolymer grains with boundaries that are oriented along the lamellar parallel (Figure 5.19a) and lamellar normal (Figure 5.19b). This observation is in agreement with the results of Jeon *et al.* [212]. At lower hydrogenation levels, however, Kimishima *et al.* [211] observe miscible blend morphologies, the characteristics of which are sensitive to the incompatibility between the I-*r*-EP and EP blocks of the copolymer pairs. At 40% hydrogenation, for example, curved EP microdomains reside within I-*r*-EP lamellae (Figure 5.19c), whereas cocontinuous EP and I-*r*-EP lamellae form at 60% hydrogenation (Figure 5.19d). Thus, by systematically tuning χ between the chemically dissimilar B and C blocks in AB/AC diblock copolymer blends, it is possible to achieve single, ordered microphases consisting of both copolymer species. Frielinghaus *et al.* [207]

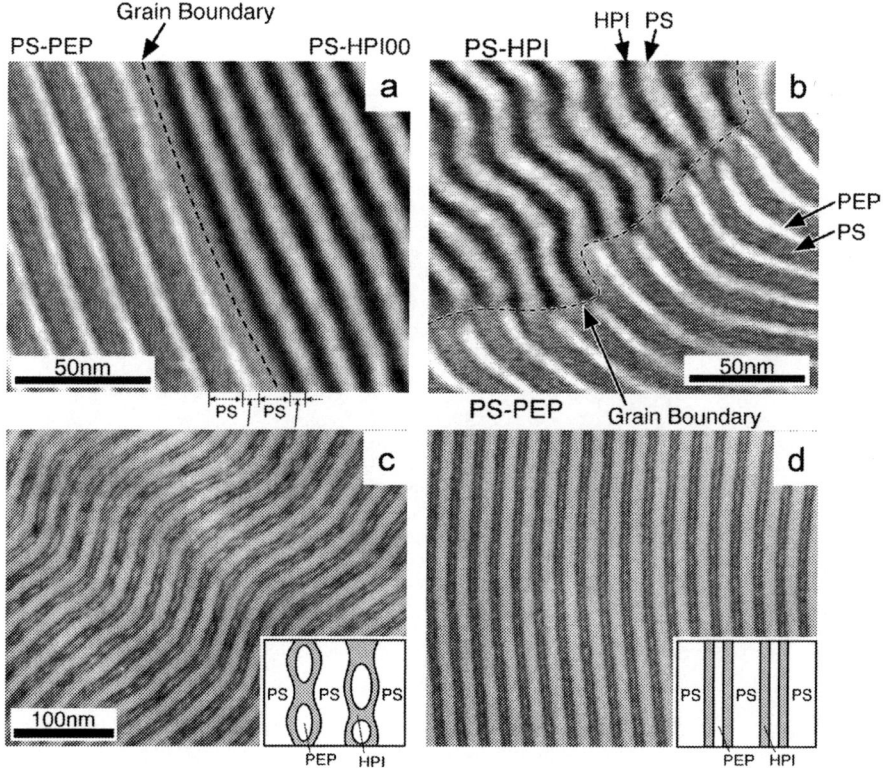

Figure 5.19 TEM images collected from binary blends of matched S-I diblock copolymers in which one of the copolymers is fully hydrogenated to form the corresponding S-EP copolymer and the other is either untreated (a,b) or partially hydrogenated (40 % in c and 60 % in d) to form an intermediate S-(I-*r*-EP) copolymer. Note the grain boundaries in the immiscible blends displayed in (a) and (b), and the single-phase morphologies generated in (c) and (d). The microphases are labeled, and schematic diagrams of the morphologies in (c) and (d) are included. (Compiled from Kimishima, I., Jinnai, H. and Hashimoto, T. Macromolecules **32**, 2585, 1999, and reprinted with permission. Copyright (1999) American Chemical Society.)

have systematically examined the phase behaviour of binary blends composed of S-I and I-EO diblock copolymers differing in I content. In this benchmark study, blends with compositionally symmetric copolymers ($f_I = 0.5$) exhibit two ordered phases at low temperatures, one ordered/one disordered phase at slightly higher temperatures, two disordered phases at intermediate temperatures and one disordered phase at high temperatures, as illustrated in the experimental phase diagram presented in Figure 5.20a. The complementary theoretical phase diagram provided in Figure 5.20b shows favourable agreement with the data. Their investigation likewise finds that at sufficiently high I fractions ($f_I = 0.7$), macrophase separation can be suppressed (see Figures 5.20c and 5.20d for the corresponding experimental and predicted phase diagrams) and yield a single ordered phase composed of both copolymers. They

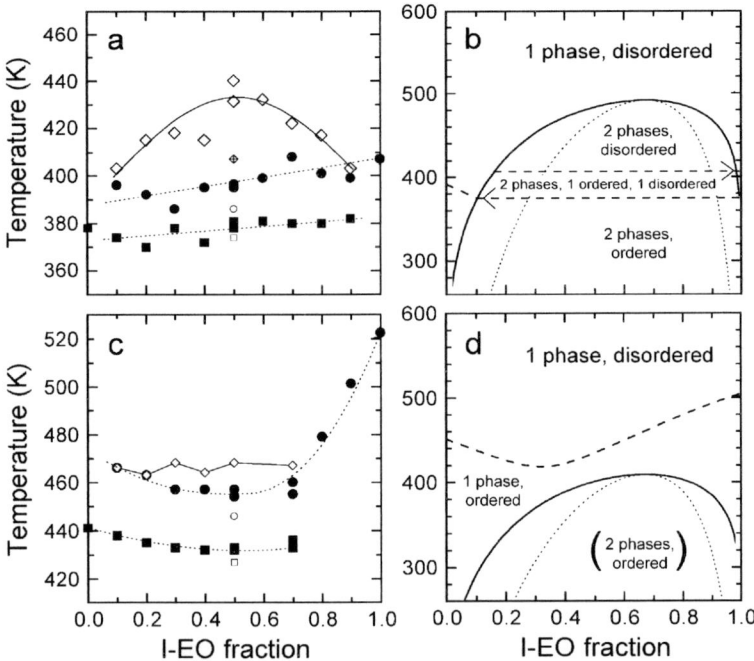

Figure 5.20 Experimental phase diagrams (a,c) of binary $(AB)_\alpha/(AB)_\beta$ blends composed of S-I and I-EO diblock copolymers and corresponding theoretical predictions (b,d). In (a,b), the I fraction is about 0.5 in each copolymer, whereas this fraction is about 0.7 in (c,d). The experimental phase diagrams identify the conditions corresponding to macrophase separation (diamonds), microphase separation of the I-EO-rich phase (circles) and microphase separation of the S-I-rich phase (squares). The predicted phase diagrams show the binodal (solid lines) and spinodal (dotted lines) conditions, as well as microphase separation events (dashed lines). (Compiled from Frielinghaus, H., Hermsdorf, N., Sigel, R., Almdal, K., Mortensen, K., Hamley, I. W., Messé, L., Corvazier, L., Ryan, A. J., van Dusschoten, D., Wilhelm, M., Floudas, G. and Fytas, G. Macromolecules **34**, 4907, 2001, and reprinted with permission. Copyright (2001) American Chemical Society.)

attribute this change in phase behaviour to the presence of a Lifshitz point, which is predicted in this system. The observation that microphase separation can dominate over macrophase separation in composition-controlled AB/AC diblock copolymer blends nicely complements the findings of Kimishima *et al.* [211]. Since most studies addressing the phase behaviour of AB/AC blends confirm the propensity of such blends to undergo macrophase separation into coexisting ordered AB and AC microphases, these blends cannot be used, in a general sense, to emulate ABC triblock copolymers in the same fashion as $(AB)_\alpha/(AB)_\beta$ blends can generate materials with intermediate characteristic dimensions (constant f, variable N) or morphologies (constant N, variable f).

5.3.2 DIBLOCK COPOLYMER/TRIBLOCK COPOLYMER BLENDS

The blends discussed in the previous section involve two diblock copolymers in which both blocks are pinned at only one end. We now consider binary blends of AB diblock copolymers with higher-order triblock copolymers, including those that are chemically identical – ABA copolymers – and those with a third chemically dissimilar species – ABC copolymers.

5.3.2.1 AB/ABA Blends

Recall that the fraction of bridged B blocks in microphase-ordered ABA triblock copolymers is less than unity and that these blocks can, with the proper choice of A and B moieties, provide the copolymers with shape memory. Addition of an ordered AB copolymer could be used to alter the fraction of bridged midblocks and, hence, the intrinsic mechanical properties, as well as the morphological characteristics, of ABA copolymers [167,168]. Segmental distributions generated from SCF theory and shown in Figure 5.21 illustrate the importance of the size of the B block in the AB copolymer relative to that of the B block in the ABA copolymer in two miscible blends in which the A blocks have identical N. If the AB copolymer possesses a relatively short B block, it will reside near the interface, thereby forcing the B segments of the ABA copolymer to fill space near the center of the B-rich microphase (Figure 5.21a). Conversely, long B blocks deposited by the AB copolymer will occupy the center of the microphase and force B segments of the triblock copolymer to lie near the interface (Figure 5.21b). Without necessarily altering interfacial curvature, this bidisperse block arrangement will influence the fraction of bridged B midblocks (v_b), which is displayed as a function of AB copolymer volume fraction in Figure 5.21c. By driving the midblocks of the ABA copolymer to the microphase center, the diblock copolymer with the short B block effectively promotes an increase in v_b, whereas the opposite is predicted when the B blocks of the AB copolymer force the midblocks away from the center

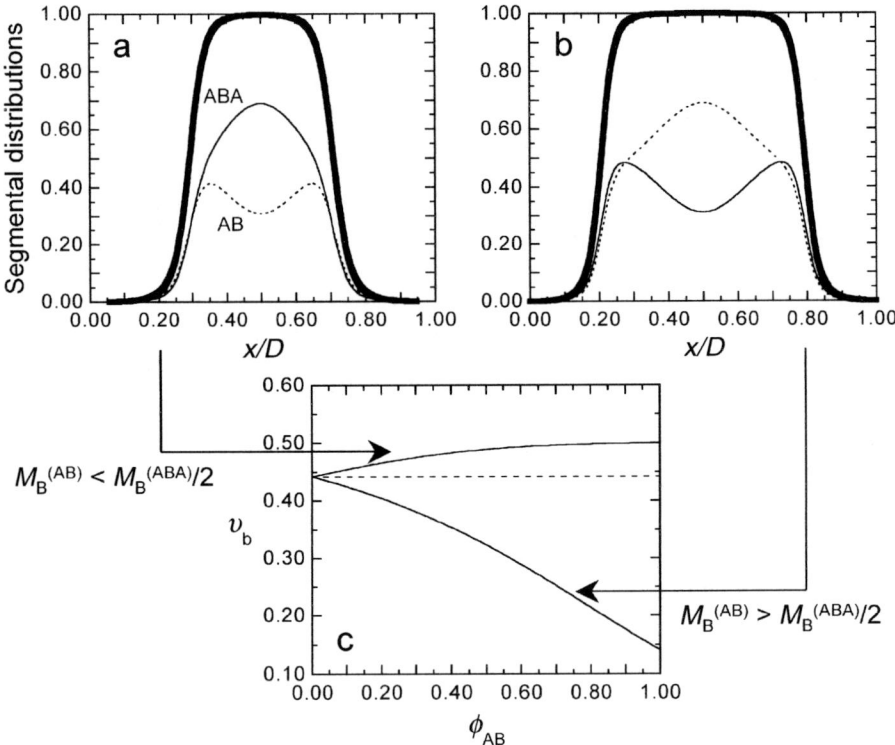

Figure 5.21 Segmental density distributions of AB/ABA copolymer blends in which the B block of the AB copolymer is (a) shorter and (b) longer than the B midblock of the ABA copolymer. Segments deposited from the AB (dotted line) and ABA (thin solid line) copolymers are labeled. The accompanying effect of copolymer blending on the fraction of bridged midblocks (v_b) is included in (c) [167].

towards the interface. These molecular-level results help to explain the variation in mechanical properties realized in AB/ABA copolymer blends [167]. Related studies [168] have demonstrated that, in the same spirit as $(AB)_\alpha/(AB)_\beta$ blends, miscible AB/ABA copolymer blends of comparable molecular weight can be used to generate intermediate morphologies. As with $(AB)_\alpha/(AB)_\beta$ blends, substantial molecular weight disparity will induce macrophase separation in AB/ABA blends.

5.3.2.2 AB/ABC Blends

As with AB/AC diblock copolymer blends, the principal reasons for exploring the phase behaviour of AB/ABC block copolymer blends are the possibilities of (i) generating the wide variety of elegant morphologies already afforded by

ABC copolymers and (ii) establishing new morphologies (without synthesizing new copolymers) and the corresponding chain-packing arrangements responsible for such motifs. Since the parameter space available in the design of such blends is intrinsically large, we first consider the simple case of forming a miscible lamellar blend by mixing a lamellar diblock copolymer with a lamellar ABC copolymer. Abetz [34] has recently detailed the copolymer morphologies and blend compositions used to explore this limiting case, and we consider here only a small subset for illustrative purposes. When a lamellar ABC triblock copolymer microphase orders, the lamellae are inherently arranged in alternating fashion according to the following sequence ... ABCCBAABCCBA ..., which is referred to as *centrosymmetric*. Addition of a compositionally symmetric poly(styrene-*b*-butadiene-*b*-*tert*-methyl methacrylate) (S-B-T) triblock copolymer to a near compositionally symmetric S-T diblock copolymer, however, results in a noncentrosymmetric lamellar morphology with the sequence ... SBT TS SBT TS ... at a blend composition of 60/40 S-B-T/S-T. Such macroscopically polarizable materials are attractive for their electrical and optical properties [213]. Alternatively, mixing a comparable poly(styrene-*b*-butadiene-*b*-methyl methacrylate) (S-B-M) triblock copolymer with the same S-T diblock copolymer yields a double-lamellar centrosymmetric morphology of the form ... MBS ST TS SBM ... at the same blend composition. These two examples clearly show that tuning the compositions, molecular weights and up to two interaction parameters/molecule (through judicious choice of chemical species or temperature) permits substantially greater flexibility in the design of AB/ABC blends relative to AB/AC blends, which often tend to macrophase separate, and neat ABC triblock copolymers, which may be incapable of forming some of the morphologies attainable in their blends.

Birshtein *et al.* [214–216] have considered the thermodynamics of lamellar AB/ABC block copolymer systems in their endeavour to develop theoretical guidelines to assist in the design of such blends. While they have provided analytical expressions for the free energy per chain (F) of several mixed superlattice morphologies, only the one derived for a mixed centrosymmetric lamellar morphology is provided below for descriptive purposes:

$$\frac{F}{kT} = (\gamma_{AB} + \gamma_{BC})\sigma + \left(\frac{\pi^2 a^4}{8p\sigma^2}\right) K(x) + x \ln x + (1-x) \ln(1-x) \quad (5.7)$$

where

$$K(x) = N_A\left[\varepsilon_A + (1-\varepsilon_A)x^3\right] + N_B\left[\varepsilon_B(1-x\tau)^3 + \frac{12x^2}{\pi^2}(\varepsilon_B + x(1-\varepsilon_B))\right] + x^3 N_C \quad (5.8)$$

Here, k is the Boltzmann constant, T denotes absolute temperature, γ_{ij} is the interfacial tension between the i and j moieties, σ is the interfacial area per chain, a

is the length of a repeat unit (assumed to be the same for all moieties), p is the ratio of persistence length to a, ap is the Kuhn segment length (also assumed to be constant), N_i is the number of statistical segments of block i ($i =$ A, B or C) in the ABC copolymer, and x is the mole fraction of the ABC copolymer in the blend. The term τ identifies the width of the region within the B microphase wherein only bridged midblocks (no free ends) reside. It is related to x according to $x = (\pi/2)(1 - x\tau)\tan(\pi\tau/2)$. In nomenclature similar to that used to describe $(AB)_\alpha/(AB)_\beta$ diblock copolymer blends, the block size ratios ε_A and ε_B are written as $N_A^{(AB)}/N_A$ and $N_B^{(AB)}/N_B$, where the superscripted (AB) refers to the AB copolymer. Examination of Equation (5.7) reveals that F is sensitive to γ_{AB} and γ_{BC} (which relate to the Flory–Huggins parameters χ_{AB} and χ_{BC}, respectively), the sizes of the blocks in both copolymers and the blend composition. While these adjustable parameters are physically comparable to those identified in the block copolymer blends previously portrayed here, their increased number provides for much greater flexibility in terms of materials design.

While the case of mixing two lamellar AB and ABC copolymers to produce either a centrosymmetric or noncentrosymmetric lamellar blend is a natural starting point, the far-reaching versatility afforded by AB/ABC block copolymer blends lies in the controllable formation of curved interfaces due to nonuniform chain packing [35], which is schematically depicted in Figure 5.22. Consider a blend composed of the S-B-M triblock copolymer used in the previous example above and a nearly compositionally symmetric B-M diblock copolymer of lower molecular weight. An equimass blend of these two copolymers corresponds to the scheme shown in Figure 5.22 with S, B and M color-

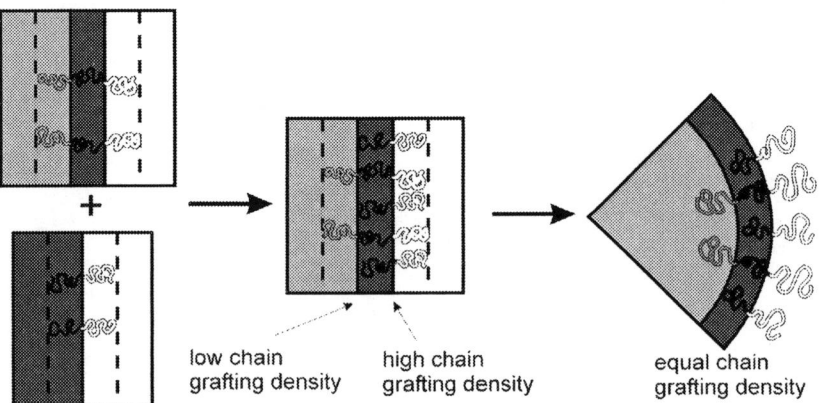

Figure 5.22 Schematic diagram of the strategy behind blending two lamellar ABC and BC copolymers (A, B and C blocks are color coded as light gray, dark gray and white, respectively) to produce a curved interface due to the unequal chain density along the A/B interface and the driving force to achieve uniform volume filling. (Reprinted with permission from Abetz, V. and Goldacker, T. Macromol. Rapid Commun. **21**, 16, 2000. Copyright (2000) Wiley-Interscience.)

coded as light gray, dark gray and white, respectively. According to TEM analysis [217], this blend exhibits a core-shell cylindrical morphology (Figure 5.23a) in which the cylindrical core consists of S and the cylindrical shell is composed of B (which appears dark in Figure 5.23a due to selective staining) in a matrix of M. At a lower concentration ($\sim$ 20 wt%) of the B-M diblock, the change in interfacial curvature is not as pronounced, in which case a core-shell gyroid morphology forms (see Figure 5.23b for the corresponding TEM image along the [110] projection and Figure 5.23c for a corresponding image simulation) [217]. This strategy of designing complex ternary morphologies through the use of AB/ABC copolymer blends has been successfully utilized to generate core-shell analogues of all the morphologies observed in neat diblock copolymers: spheres, cylinders, gyroid and lamellae. A highly detailed compilation of the variety of morphologies that have been achieved through the controlled use of AB/ABC blends is provided by Abetz and Goldacker [35].

5.3.3 DIBLOCK COPOLYMER/RANDOM COPOLYMER BLENDS

While the compatibilizing efficacy of ABR random copolymers in ternary ABR/hA/hB blends has been the subject of investigation, we consider only binary blends consisting of an AB diblock copolymer and an ABR random copolymer in this section. Unlike miscible AB/hA blends in which the hA chain can be solubilized within the A microdomains of the AB copolymer or miscible $(AB)_\alpha/(AB)_\beta$ blends in which the A and B blocks of each copolymer reside in their respective microphases, the bicomponent ABR chain cannot be readily

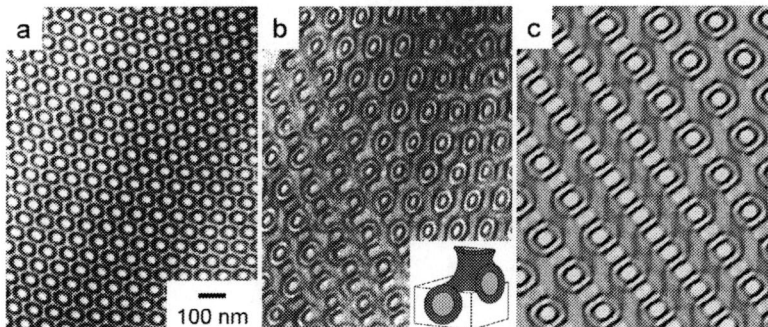

Figure 5.23 Examples of two nonlamellar morphologies – (a) core-shell cylinders and (b) core-shell gyroid – generated by blending a nearly symmetric B-M diblock copolymer with a symmetric S-B-M triblock copolymer (see Figure 5.22). The blend compositions are 52/48 B-M/S-B-M in (a) and 21/79 B-M/S-B-M in (b). A simulated projection of the gyroid morphology is included for comparison in (c). (Adapted from Goldacker, T. and Abetz, V. Macromolecules **32**, 5165, 1999, and reprinted with permission. Copyright (1999) American Chemical Society.)

incorporated into either microphase of the diblock copolymer without incurring both enthalpic and entropic penalties. The theoretical studies of Lee and Zin [218] predict that such blends can exhibit three different phases: (i) one in which the ABR copolymer is solubilized within the mesophase of the AB copolymer, (ii) one in which the AB copolymer mesophase coexists with an isotropic liquid phase and (iii) a single liquid phase. The first region occurs at relatively low concentrations of the ABR copolymer due to a non-negligible enthalpy of mixing. The biphasic mesophase+liquid regime can extend to surprisingly high ABR copolymer concentrations (in excess of 75 vol% at the computational conditions reported), which can be increased further (>90 vol%) with increasing molecular weight of the ABR copolymer. A more modest increase in ABR concentration can be achieved by changing the composition. In all cases, addition of the ABR random copolymer to the AB diblock copolymer serves to reduce the order–disorder transition (ODT) temperature at which the block copolymer mesophase dissolves into a structureless liquid. Complementary experimental studies [219] indicate that the solubility limit of a compositionally symmetric poly(styrene-r-butadiene) (SBR) copolymer in a lamellar S-B diblock copolymer is about 15 wt%, and suggest that the SBR copolymer may localize within the interfacial region afforded by the microphase-ordered block copolymer.

5.3.4 TRIBLOCK COPOLYMER/TRIBLOCK COPOLYMER BLENDS

Sakurai *et al.* [220,221] have extended the materials design paradigm developed for $(AB)_\alpha/(AB)_\beta$ diblock copolymer blends to $(ABA)_\alpha/(ABA)_\beta$ triblock copolymer blends as an efficient means by which to control the mechanical properties of thermoplastic elastomers. The strategy of using AB/ABC copolymer blends to controllably alter microdomain swelling and interfacial curvature through nonuniform chain packing to achieve existing or novel ternary morphologies is also readily extended to more complex systems, such as those composed of two ABC triblock copolymers. An excellent example of how $(ABC)_\alpha/(ABC)_\beta$ blends can be used to generate complex morphologies is displayed in Figure 5.24 [222]. In this figure, the same centrosymmetric lamellar S-B-M triblock copolymer discussed with regard to AB/ABC blends in Section 5.3.2.2. is blended with a lower molecular weight S-B-M copolymer with a shortened midblock, which forms cylinders along the S/M interface (the "cylinder at lamellar interface," or lc, morphology). An 82/18 w/w blend of these two copolymers is found to produce the "knitting pattern" (kp) morphology, wherein the S and M microphases appear as undulating lamellae and the B segments self-organize into channels that orient both parallel and perpendicular to the undulating lamellae, as seen in Figure 5.24. Formation of this complex morphology, which has been observed in neat ABC triblock copolymers [223], through the blending of two triblock copolymers clearly attests to the potential of this physical approach to designer nanostructured polymeric materials.

Phase Behaviour of Block Copolymer Blends

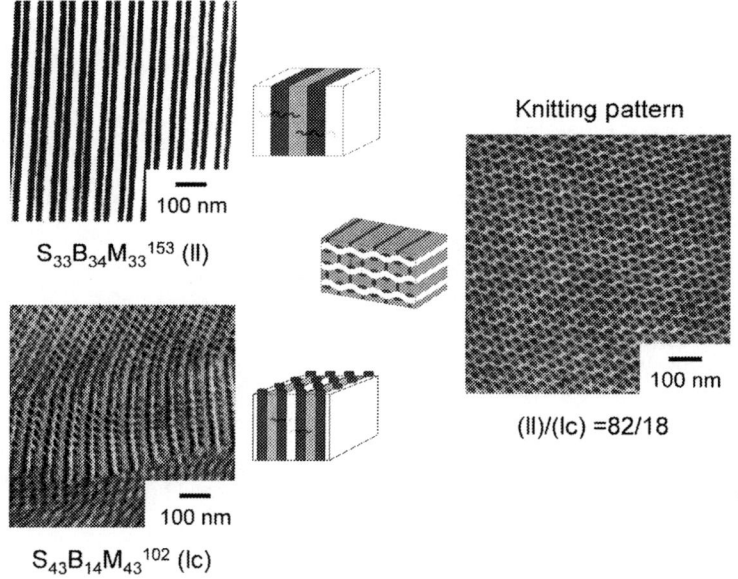

Figure 5.24 Blending two S-B-M triblock copolymers – one possessing an alternating lamellar (ll) morphology and the other having the "cylinder at lamellar interface" (lc) morphology – as an alternative (physical) means by which to generate the intricate "knitting pattern" (kp) morphology. Details regarding the copolymers and the composition of the blend are all displayed in the figure. (Reprinted with permission from Goldacker, T. and Abetz, V. Macromol. Rapid Commun. **20**, 415, 1999. Copyright (1999) Wiley-Interscience.)

5.4 SOLVATED BLOCK COPOLYMER SYSTEMS

Numerous studies have investigated the phase behaviour and properties of linear di/triblock copolymers in the presence of a low molar mass solvent. Extensive efforts by Alexandridis and coworkers [26–28] have repeatedly demonstrated the wealth of morphologies available in aqueous systems containing amphiphilic triblock copolymers, such as those possessing at least one hydrophilic EO block. Förster et al. [224] have likewise shown that complex nanostructural elements, such as the systematic progression of micelles to randomly perforated membranes of the L_3 phase to lamellar bilayers, can be controllably generated in hydrated B-EO diblock copolymers (see Figure 5.25). Lodge and coworkers [29,31] have used rheological, scattering and birefringence measurements to confirm that the addition of a neutral or selective organic solvent to a microphase-ordered diblock copolymer systematically yields the entire spectrum of morphologies observed in the melt. Laurer et al. [30] have likewise generated all the classical block copolymer morphologies in blends of S-I-S and its hydrogenated S-EP-S analogue in a block-selective solvent. Two interesting variations on this theme reflect the addition of an ordered block copolymer to

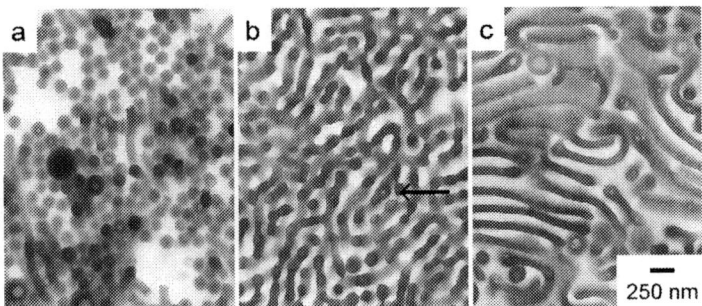

Figure 5.25 TEM images of the morphologies formed by a B-EO diblock copolymer in aqueous environments containing different copolymer concentrations (in wt %): (a) 0.30, (b) 0.50 and (c) 0.70. As the copolymer concentration increases, the morphologies transform from a micellar (L_1) phase in (a) to the sponge (L_3) phase in (b) and ultimately to the lamellar (L_α) phase in (c). Holes exist in the bilayers in the sponge phase (see arrow). (Adapted from Förster, S., Berton, B., Hentze, H. P., Krämer, E., Antonietti, M. and Lindner, P., Macromolecules **34**, 4610, 2001, and reprinted with permission. Copyright (2001) American Chemical Society.)

either a thermoset or asphalt. The seminal work of Bates and coworkers [225,226] has demonstrated that block copolymers can microphase order into various morphologies in the presence of a thermosetting polymer during cure. Additional phase studies of diblock and triblock copolymer in the presence of a thermosetting polymer have been performed by Mijovic *et al.* [227], Dean *et al.* [228] and Guo *et al.* [229,230] Ritzenthaler *et al.* [231,232] have recently extended this expanding field of nanostructured thermosets by employing an ordered S-B-M triblock copolymer, whereas Girard-Reydet *et al.* [233] have investigated the morphological characteristics of multicomponent hS/hM/ epoxy blends with and without the corresponding S-M diblock copolymer. Similarly, incorporation of a self-organizing block copolymer to asphalt can promote interesting phase behaviour. Since the early studies of copolymer-reinforced asphalt by Kraus [234], ongoing efforts have sought to elucidate the phase behaviour of [235,236], as well as structure–property relationships in [237], such systems.

In addition to self-organizing into nanoscale morphologies, solvated block copolymers have also been observed to form relatively large (micrometer-scale) and surprisingly tough vesicles, collectively referred to as *polymersomes* [238–240]. While the diversity of studies addressing solvent-regulated ordering of block copolymers is simply too vast to be considered with the proper level of attention it deserves here, we instead consider the phase and mechanical behaviour of a blend of two chemically identical block copolymers in the presence of a common selective solvent. We also explore contemporary topics in which a low molar mass solvent is used as a carrier medium by which to diffuse another component into an existing microphase-ordered block copolymer.

5.4.1 BLOCK COPOLYMER MIXED GELS

Addition of a midblock-selective solvent to a microphase-ordered ABA triblock copolymer has long been established as a means by which to generate physical gels stabilized by incompatible A-rich microdomains. Physical gels are defined [241] as liquid-rich systems possessing a dynamic elastic shear modulus (G') that is not only greater in magnitude than the corresponding dynamic viscous modulus (G'') but also frequency invariant. If the solvent is relatively nonvolatile under application conditions, these thermoreversible systems are relevant in a wide variety of technologies requiring, for example, vibration dampening, shape memory or adhesion [242]. Over the concentration range wherein the copolymer molecules are sufficiently swollen to form spherical A micelles on a lattice, the magnitude of G' reflects a combination of entangled (looped) midblocks in the "flowered-micelle" regime and bridged midblocks that connect neighbouring micelles [243]. In this regime, G' scales with copolymer concentration (C) as C^n, where $n > 1$. As the solvent concentration is increased and long-range micellar order gives way to liquid-like order due to an accompanying increase in the intermicellar distance, the contribution of entangled midblocks to G' diminishes and ultimately disappears as only bridged midblocks remain to form a continuous molecular network. In this semidilute limit, $G' \sim C$. While copious studies of midblock-swollen triblock copolymers have addressed the effects of block copolymer composition and molecular weight, temperature and deformation (abbreviated reviews are available elsewhere [244,245]), relatively few have systematically investigated the effect of adding a second copolymer. Addition of a matched AB diblock copolymer to a solvated ABA triblock copolymer can result in interesting results, such as comicellization. Yang *et al.* [246] have used dynamic light scattering to ascertain that their copolymers formed mixed AB/ABA micelles in an aqueous medium. This result, coupled with the dynamic rheological measurements on a different system [247,248] provided in Figure 5.26, indicate that the incorporation of B tails from the AB copolymer in the coronae of AB/ABA micelles not only increases the extent to which the micelles interact (as evidenced by a modest increase in G' at low AB concentrations) but also induces intermicellar interactions (as evidenced by a measurable G') at copolymer concentrations below the critical gel concentration of the ABA copolymer. This observation is consistent with previous solution studies [249] of hydrophobically modified copolymers in the presence of surfactants, as well as theoretical frameworks of grafted chains capable of looping on an impenetrable surface [250]. Using a combination of rheological and scattering methods, Vega *et al.* [251] have generated a phase diagram for solvated AB/ABA block copolymer blends (see Figure 5.27) in which they identify a biphasic envelope composed of non-networked copolymer flocs at low total (AB + ABA) and AB copolymer concentrations. An increase in the total copolymer concentration at constant AB concentration ultimately favours the formation of a molecular network and, thus, a physical gel. Conversely, an increase in AB concentration at constant total

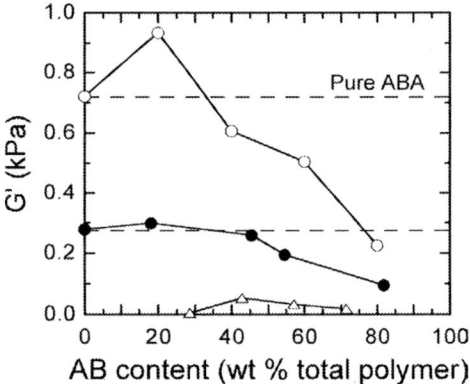

Figure 5.26 Dependence of the dynamic elastic modulus (G') on AB diblock copolymer content in bidisperse AB/ABA blends dissolved in a B-selective solvent [247,248]. The total (AB + ABA) polymer concentrations (in wt %) are 15 (open circles), 11 (filled circles) and 7 (open triangles). Note that the systems with 11 and 15 wt % total polymer concentrations yield maxima in G' as the AB content is increased, whereas the system with 7 wt % total polymer concentration yields a measurable G' indicative of a physical gel only at intermediate AB loading levels.

copolymer concentration can, depending on the value of the total copolymer concentration, result in a physical gel or, at sufficiently high AB concentrations, a liquid. This result is in favourable agreement with the rheological measurements provided in Figure 5.26. As demonstrated by Quintana *et al.* [252] and Kŏnák and Helmstedt [253], comicellization can also be induced in solutions containing chemically dissimilar diblock and triblock copolymers differing markedly in molecular weight.

5.4.2 BLOCK COPOLYMER/HOMOPOLYMER MESOBLENDS

As alluded to earlier, an increase in the solvent concentration of a midblock-swollen ABA copolymer promotes morphological changes that are consistent with increasing interfacial curvature, as well as a systematic reduction in modulus. According to the SCF results presented in Figure 5.11 for ABA/hB blends, the bridging fraction (v_b) is predicted to decrease with increasing hB content. Since a midblock-selective solvent establishes the low molecular weight limit for a hB homopolymer, it immediately follows that v_b decreases with increasing solvent concentration in ABA copolymer solutions. Zhulina and Halperin [150] have suggested an alternate design strategy by which to prepare block copolymer gels without sustaining such a reduction in v_b. In their so-called *mesogels*, the B-selective solvent is diffusively imbibed into a pre-existing microphase-ordered block copolymer with rigid A microdomains to yield nonequilibrium morphologies. Their approach has successfully yielded triblock copolymer gels that exhibit markedly higher moduli than their counterparts prepared by

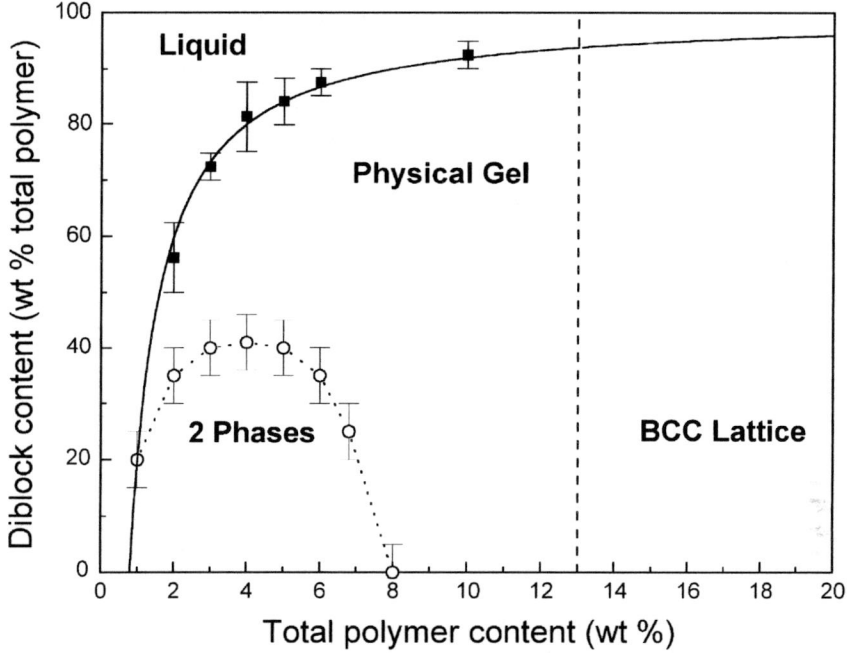

Figure 5.27 Experimental phase diagram of AB/ABA block copolymer blends dissolved in a B-selective solvent according to sol-gel and cloud-point measurements. A phase-separated regime is found to exist at low AB and total (AB + ABA) polymer concentrations. At higher AB or total polymer concentrations, the system forms a physical gel network. A further increase in AB content ultimately disrupts the ABA network and induces liquid-like behaviour. (Reprinted with permission from Vega, D. A., Sebastian, J. M., Loo, Y.-L. and Register, R. A. J. Polym. Sci. B: Polym. Phys. **39**, 2183, 2001. Copyright (2001) Wiley-Interscience.)

conventional equilibrium protocols [254]. On the basis of this principle, a similar strategy has been introduced [255,256] to produce analogous ABA/hB *mesoblends*, the procedure for which is depicted in Figure 5.28a. In this case, immersion of a microphase-ordered ABA copolymer in a B-selective solvent containing hB molecules results in nonequilibrium ABA/hB blends, such as the one displayed in Figure 5.28b. The solubility of hB within a lamellar ABA matrix at ambient temperature is sensitive to factors such as M_{hB}, M_B and the concentration of hB in solution, in addition to the (in)compatibility of the solvent with the A and B blocks of the copolymer.

5.4.3 BLOCK COPOLYMER TEMPLATING MEDIA

Using a solvent carrier to diffuse a new molecular species into a microphase-ordered block copolymer provides a plethora of new possibilities in terms of nanostructured materials development. One such possibility that continues to

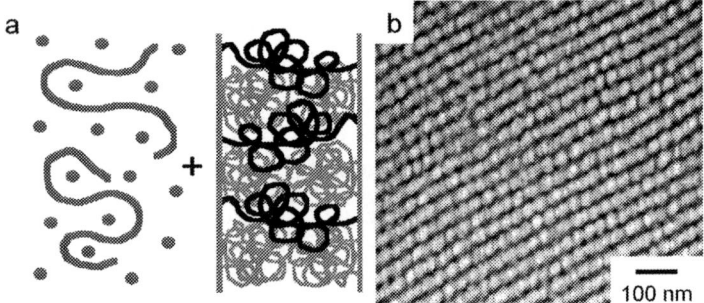

Figure 5.28 Illustration of a block copolymer/homopolymer mesoblend (a) and a TEM image obtained from such a material (b). In (a), a microphase-ordered ABA copolymer is immersed in a B-selective solvent containing a B-compatible homopolymer, which subsequently diffuses into the swollen ABA matrix [255,256]. In (b), the S-I-S/hI mesoblend displayed has sorbed ca. 4.5 wt % hI, for which α (calculated on the basis of half the molecular weight of the B midblock) is 0.59.

attract tremendous attention employs amphiphilic block copolymers in aqueous media to template ceramic materials that are stable at high temperatures and resistant to solvent attack. In this approach, a water-borne ceramic precursor is imbibed into a microphase-ordered block copolymer and converted to the corresponding ceramic through heat treatment. The copolymer is subsequently removed by either dissolution or pyrolysis. If, for example, a parent copolymer possesses hydrophobic cylinders on a hexagonal lattice in a hydrophilic matrix, this approach yields a mesoporous silica exhibiting cylindrical pores [257,258]. Such materials are of interest in developing (i) membranes for separations processes and (ii) a fundamental understanding of physical phenomena (e.g., wetting) in well-defined nanoscale environments [259]. Wiesner and coworkers [260,261] have shown that, depending on the loading level of the precursor and, hence, the degree of microphase swelling, a variety of morphologies can be controllably formed in I-EO diblock copolymers prior to formation of the templated ceramic. A series of particulate (spheres, cylinders and plates), as well as mesoporous (bicontinuous and cylindrical), ceramic materials have been successfully fashioned in this manner, as illustrated by the mechanistic pathways shown in Figure 5.29. Yang et al. [262] have further demonstrated that this methodology can be extended to generate ceramic materials that are patterned over several length scales. For a complete description of this blossoming field, the interested reader is referred to the review by Simon et al. [261].

5.5 CONCLUDING REMARKS

The ability of block copolymers to spontaneously self-organize into defined nanostructures has attracted tremendous attention in a remarkably diverse

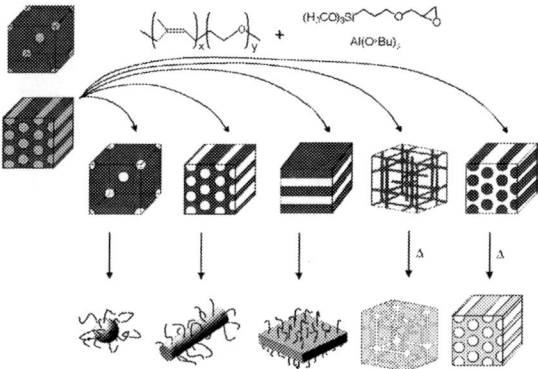

Figure 5.29 Strategy for developing inorganic nanoscale objects and mesoporous media from microphase-ordered block copolymer templates. In this case, an I-EO diblock copolymer is imbibed with a ceramic precursor such as 3-(glycidyloxypropyl)trimethoxysilane (GLYMO) and aluminum *sec*-butoxide so that the corresponding silicate can be subsequently formed within the confined environment of the copolymer matrix. Dissolution of the copolymer results in discrete, polymer-covered ("hairy") objects, whereas calcination at elevated temperatures yields mesoporous ceramic materials. (Reprinted with permission from Simon, P. F. W., Ulrich, R., Spiess, H. W. and Wiesner, U. Chem. Mater. **13**, 3464, 2001. Copyright (2001) American Chemical Society.)

range of fields. While numerous efforts have previously sought to establish theoretical and experimental guidelines to explain the phase behaviour and properties of relatively simple diblock copolymers, ongoing studies tend to focus on multicomponent mixtures of di/tri/multiblock copolymers with either low molar mass solvents or other macromolecules. For this reason, it is important to understand the material and environmental factors governing the phase behaviour of such systems, as well as the design paradigms required to achieve a particular morphology. In this compilation, we have elected not to explicitly consider the behaviour and properties of neat block copolymers in most cases, since these topics are covered in detail elsewhere [20–23]. Rather, we have endeavoured to describe and compare the wide variety of block copolymer blends that have been recently reported in the literature. These blends include diblock and triblock copolymers in the presence of one or two homopolymers, as well as blends of two block copolymers differing in composition, molecular weight or architecture. As much of the metallurgical side of materials science has benefited from the development and use of metal alloys, the field established around block copolymers will likewise invariably profit from the tremendous versatility afforded by the blends considered here. A fundamental understanding of *how* chain packing and interfacial tension can be used to controllably alter the interfacial curvature in, for instance, a binary AB/ABC block copolymer blend to yield a nonparent, perhaps complex, morphology is a prerequisite to the enlightened design of multicomponent copolymer-containing materials with possibly new nanoscale morphologies and properties that have yet to be discovered. If one

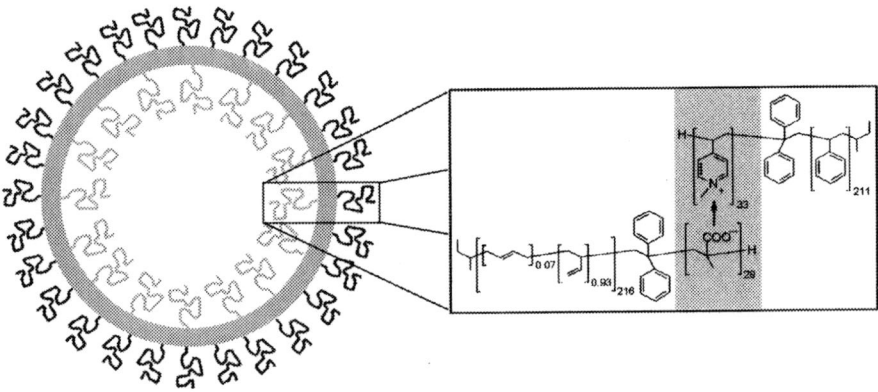

Figure 5.30 Schematic illustration of a polyion complex (PIC) vesicle generated by blending a poly(1,2-butadiene-*b*-cesium methacrylate) block ionomer with an oppositely charged poly(styrene-*b*-1-methyl-4-vinylpyridinium iodide) block ionomer. Note that the internal and external polymer brushes are chemically dissimilar, demonstrating the fine tunability and enhanced versatility of this class of materials. (Adapted from Schrage, S., Sigel, R. and Schlaad, H. Macromolecules **36**, 1417, 2003, and reprinted with permission. Copyright (2003) American Chemical Society.)

adds to this scenario the use of nonlinear (e.g., single-graft or miktoarm) block copolymers or nonequilibrium process strategies, it becomes evident that there remains much to explore in the vast parameter space afforded by block copolymer blends. Another challenge and opportunity on the forefront of block copolymer blend research is the use of block ionomers [263] or hydrogen-bonding copolymers [264] to promote supramolecular ordering. This growing field offers enormous flexibility to the modular design of unique superstructures, such as polyion complex vesicles produced by coupling oppositely-charged ionomers composed of poly(1,2-butadiene-*b*-cesium methacrylate) and poly(styrene-*b*-1-methyl-4-vinylpyridinium iodide) (see Figure 5.30).

ACKNOWLEDGMENTS

This work was supported by the U.S. Department of Energy under Contract No. DE-FG02-99ER14991.

REFERENCES

1. Leibler, L. *Macromolecules* **13**, 1602 (1980).
2. Bates, F. S. and Fredrickson, G. H. *Annu. Rev. Phys. Chem.* **41**, 525 (1990).
3. Khandpur, A. K., Förster, S., Bates, F. S., Hamley, I. W., Ryan, A. J., Bras, W., Almdal, K. and Mortensen, K. *Macromolecules* **28**, 8796 (1995).

4. Zhao, J., Majumdar, B., Schulz, M. F., Bates, F. S., Almdal, K., Mortensen, K., Hajduk, D. A. and Gruner, S. M. *Macromolecules* **29**, 1204 (1996).
5. Floudas, G., Vazaiou, B., Schipper, F., Ulrich, R., Wiesner, U., Iatrou, H. and Hadjichristidis, N. *Macromolecules* **34**, 2947 (2001).
6. Bailey, T. S., Pham, H. D. and Bates, F. S. *Macromolecules* **34**, 6994 (2001).
7. Schulz, M. F., Bates, F. S., Almdal, K. and Mortensen, K. *Phys. Rev. Lett.* **73**, 86 (1994).
8. Hadjuk, D. A., Harper, P. E., Gruner, S. M., Honeker, C. C., Kim, G., Thomas, E. L. and Fetters, L. J. *Macromolecules* **27**, 4063 (1994).
9. Laurer, J. H., Hajduk, D. A., Fung, J. C., Sedat, J. W., Smith, S. D., Gruner, S. M., Agard, D. A. and Spontak, R. J. *Macromolecules* **30**, 3938 (1997).
10. Avgeropoulos, A., Dair, B. J., Hadjichristidis, N., and Thomas, E. L. *Macromolecules* **30**, 5634 (1997).
11. Schick, M. *Physica A* **251**, 1 (1998).
12. Disko, M. M., Liang, K. S., Behal, S. K., Roe, R. J. and Jeon, K. J. *Macromolecules* **26**, 2983 (1993).
13. Spontak, R. J., Smith, S. D., and Ashraf, A. *Polymer* **34**, 2233 (1993).
14. Förster, S., Khandpur, A. K., Zhao, J., Bates, F. S., Hamley, I. W., Ryan, A. J. and Bras, W. *Macromolecules* **27**, 6922 (1994).
15. Burger, C., Micha, M. A., Oestreich, S., Förster, S. and Antonietti, M. *Europhys. Lett.* **42**, 425 (1998).
16. Zhu, L., Huang, P., Cheng, S. Z. D., Ge, Q., Quirk, R. P., Thomas, E. L., Lotz, B., Wittmann, J.-C., Hsiao, B. S., Yeh, F. J. and Liu, L. Z. *Phys. Rev. Lett.* **86**, 6030 (2001).
17. Matsen, M. W. and Bates, F. S. *Macromolecules* **29**, 7641 (1996).
18. Matsen, M. W. and Bates, F. S. *J. Chem. Phys.* **106**, 2436 (1997).
19. Hajduk, D. A., Takenouchi, H., Hillmyer, M. A., Bates, F. S., Vigild, M. E. and Almdal, K. *Macromolecules* **30**, 3788 (1997).
20. Hamley, I. W. *The Physics of Block Copolymers*, Oxford University Press, NY, (1998).
21. Bates, F. S. and Fredrickson, G. H. *Phys. Today* **52**, 32 (1999).
22. Hudson, S. D. and Jamieson, A. M. in *Polymer Blends* Vol. 1: Formulation (D. R. Paul and C. B. Bucknall, eds.) Wiley-Interscience, New York, (2000), Chap. 15.
23. Hadjichristidis, N., Pispas, S. and Floudas, G. *Block Copolymers: Synthetic Strategies, Physical Properties, and Applications*, Wiley-Interscience, New York, (2003).
24. Piirma, I. Polymeric Surfactants Marcel Dekker, Inc. New York, (1992).
25. Hasegawa, H. *Curr. Opin. Colloid Interface Sci.* **3**, 264 (1998).
26. Alexandridis, P., Olsson, U. and Lindman, B. *Langmuir* **14**, 2627 (1998).
27. Alexandridis, P., Olsson, U., Linse, P. and Lindman, B. in *Amphiphilic Block Copolymers: Self-Assembly and Applications* (P. Alexandridis and B. Lindman, eds.) Elsevier Science B. V., Amsterdam, (2000), Chap. 8.
28. Alexandridis, P. and Spontak, R. J. *Curr. Opin. Colloid Interface Sci.* **4**, 130 (1999).
29. Hanley, K. J. and Lodge, T. P. *J. Polym. Sci. B: Polym. Phys.* **36**, 3101 (1998).
30. Laurer, J. H., Khan, S. A., Spontak, R. J., Satkowski, M. M., Grothaus, J. T., Smith, S. D. and Lin, J. S. *Langmuir* **15**, 7947 (1999).
31. Lodge, T. P., Pudil, B. and Hanley, K. J. *Macromolecules* **35**, 4707 (2002).
32. Stadler, R., Auschra, C., Beckmann, J., Krappe, U., Voigt-Martin, I. and Leibler, L. *Macromolecules* **28**, 3080 (1995).
33. Zheng, W. and Wang, Z.-G. *Macromolecules* **28**, 7215 (1995).
34. Abetz, V. in *Encyclopedia of Polymer Science and Technology*, 3rd ed. (J. I. Kroschwitz, ed.) John Wiley & Sons, Inc., Hoboken, New Jersey, (2003), Vol. 1, pp. 482–523.

35. Abetz, V. and Goldacker, T. *Macromol. Rapid Commun.* **21**, 16 (2000).
36. Holden, G., Legge, N. R., Quirk, R. and Schroeder, H. E. (eds.) *Thermoplastic Elastomers* 2nd ed., Hanser, Munich, (1996).
37. Leibler, L. *Makromol. Chem. Macromol. Symp.* **16**, 1 (1988).
38. Hashimoto, T., Tanaka, H. and Hasegawa, H. *Macromolecules* **23**, 4378 (1990).
39. Winey, K. I., Thomas, E. L. and Fetters, L. J. *Macromolecules* **24**, 6182 (1991).
40. Winey, K. I., Thomas, E. L. and Fetters, L. J. *J. Chem. Phys.* **95**, 9367 (1991).
41. Winey, K. I., Thomas, E. L. and Fetters, L. J. *Macromolecules* **25**, 422, 2645 (1992).
42. Tanaka, H., Hasegawa, H. and Hashimoto, T. *Macromolecules* **24**, 240 (1991).
43. Han, C. D., Baek, D. M., Kim, J., Kimishima, K. and Hashimoto, T. *Macromolecules* **25**, 3052 (1992).
44. Matsen, M. W. *Phys. Rev. Lett.* **74**, 4225 (1995).
45. Matsen, M. W. *Macromolecules* **28**, 5765 (1995).
46. Shull, K. R. and Winey, K. I. *Macromolecules* **25**, 2637 (1992).
47. Banaszak, M. and Whitmore, M. D. *Macromolecules* **25**, 2757 (1992).
48. Lyatskaya, Y., Gersappe, D. and Balazs, A. C. *Macromolecules* **28**, 6278 (1995).
49. Cigana, P. and Favis, B. D. *Polymer* **39**, 3373 (1998).
50. Cavanaugh, T. J., Buttle, K., Turner, J. N. and Nauman, E. B. *Polymer* **39**, 4191 (1998).
51. Xu, S., Zhao, H., Tang, T., Dong, L. and Huang, B. *Polymer* **40**, 1537 (1999).
52. Chun, S. B. and Han, C. D. *Macromolecules* **32**, 4030 (1999).
53. Hlavatá, D., Horák, Z., Hromádková, Lednicky, F. and Pleska, A. *J. Polym. Sci. B: Polym. Phys.* **37**, 1647 (1999).
54. Van Puyvelde, P., Velankar, S. and Moldenaers, P. *Curr. Opin. Colloid Interface Sci.* **6**, 457 (2001).
55. Wang, Y. and Hillmyer, M. A. *J. Polym. Sci. A: Polym. Chem.* **39**, 2755 (2001).
56. Retsos, H., Margiolaki, I., Messaritaki, A. and Anastasiadis, S. H. *Macromolecules* **34**, 5295 (2001).
57. Marić, M. and Macosko, C. W. *J. Polym. Sci. B: Polym. Phys.* **40**, 346 (2002).
58. Shi, T., Wen, G., Jiang, W., An, L. and Li, B. *Eur. Polym. J.* **39**, 551 (2003).
59. Matsen, M. W. and Schick, M. *Macromolecules* **26**, 3878 (1993).
60. Laradji, M. and Desai, R. C. *J. Chem. Phys.* **108**, 4662 (1998).
61. Ohta, T. and Nonomura, M. *Eur. Phys. J. B* **2**, 57 (1998).
62. Müller, M. and Gompper, G. *Phys. Rev. E* **66**, 041805 (2002).
63. Edgecombe, B. D., Stein, J. A., Frechét, J. M. J., Xu, Z. and Kramer, E. J. *Macromolecules* **31**, 1292 (1998).
64. Bernard, B., Brown, H. R., Hawker, C. J., Kellock, A. J. and Russell, T. P. *Macromolecules* **32**, 6254 (1999).
65. Majumdar, B. and Paul, D. R. in *Polymer Blends Vol. 1: Formulation* (D. R. Paul and C. B. Bucknall, eds.) Wiley-Interscience, New York, (2000), Chap. 17.
66. Charoensirisomboon, P., Inoue, T. and Weber, M. *Polymer* **41**, 4483 (2000).
67. Hayashi, M., Grüll, H., Esker, A. R., Weber, M., Sung, L., Satija, S. K., Han, C. C. and Hashimoto, T. *Macromolecules* **33**, 6485 (2000).
68. Baker, W., Scott, C. and Hu, G.-H. (eds.) Reactive Polymer Blending, Hanser, München, (2001).
69. Velankar, S., Van Puyvelde, P., Mewis, J. and Moldenaers, P. *J. Rheol.* **45**, 1007 (2001).
70. Adedeji, A., Lyu, S. and Macosko, C. W. *Macromolecules* **34**, 8663 (2001).
71. Mezzenga, R., Fredrickson, G. H. and Kramer, E. J. *Macromolecules* **36**, 4457 (2003).
72. Adedeji, A., Jamieson, A. M. and Hudson, S. D. *Macromolecules* **28**, 5255 (1995).
73. Adedeji, A., Hudson, S. D. and Jamieson, A. M. *Polymer* **38**, 737 (1997).

74. Prahsarn, C. and Jamieson, A. M. *Polymer* **38**, 1273 (1997).
75. Chen, H.-L., Lin, S.-Y., Huang, Y.-Y., Chiu, F.-C., Liou, W. and Lin, J. S. *Macromolecules* **35**, 9434 (2002).
76. Rharbi, Y., Zhang, J., Spiro, J. G., Chen, L., Winnik, M. A., Vavasour, J. D., Whitmore, M. D., Zhang, J.-X. and Jérôme, R. *Macromolecules* **36**, 1241 (2003).
77. Xie, R., Yang, B. and Jiang, B. *J. Polym. Sci. B: Polym. Phys.* **34**, 1489 (1996).
78. Zhao, J. Q., Pearce, E. M. and Kwei, T. K. *Macromolecules* **30**, 7119 (1997).
79. Iizuka, N., Bodycomb, J., Hasegawa, H. and Hashimoto, T. *Macromolecules* **31**, 7256 (1998).
80. Iizuka, N., Bodycomb, J., Hasegawa, H. and Hashimoto, T. *Macromolecules* **31**, 7256 (1998).
81. Lee, J. H., Balsara, N. P., Chakraborty, A. K., Krishnamoorti, R. and Hammouda, B. *Macromolecules* **35**, 7748 (2002).
82. Whitmore, M. D. and Vavasour, J. D. *Acta Polym.* **46**, 341 (1995).
83. Likhtman, A. E. and Semenov, A. N. *Macromolecules* **30**, 7273 (1997).
84. Janert, P. K. and Schick, M. *Macromolecules* **31**, 1109 (1998).
85. Vavasour, J. D. and Whitmore, M. D. *Macromolecules* **34**, 3471 (2001).
86. Binder, K. and Müller, M. *Curr. Opin. Colloid Interface Sci.* **5**, 315 (2000).
87. Spontak, R. J., Smith, S. D. and Ashraf, A. *Macromolecules* **26**, 956 (1993).
88. Hasegawa, H. *Sen'i Gakkaishi* **55**, 82 (1999).
89. Bodycomb, J., Yamaguchi, D. and Hashimoto, T. *Macromolecules* **33**, 5187 (2000).
90. Hajduk, D. A., Ho, R. M., Hillmyer, M. A., Bates, F. S. and Almdal, K. *J. Phys. Chem. B* **102**, 1356 (1998).
91. Ahn, J.-H. and Zin, W.-C. *Macromolecules* **35**, 10238 (2002).
92. Vaidya, N. Y. and Han, C. D. *Polymer* **43**, 3047 (2002).
93. Lammertink, R. G. H., Hempenius, M. A., Thomas, E. L. and Vancso, G. J. *J. Polym. Sci. B: Polym. Phys.* **37**, 1009 (1999).
94. David, J. L., Gido, S. P., Hong, K., Kunlun, H., Zhou, J., Mays, J. W. and Tan, N. B. *Macromolecules* **32**, 3216 (1999).
95. Huang, Y.-Y., Chen, H.-L. and Hashimoto, T. *Macromolecules* **36**, 764 (2003).
96. Chen, H.-L., Hsiao, S.-C., Lin, T.-L., Yamauchi, K., Hasegawa, H. and Hashimoto, T. *Macromolecules* **34**, 671 (2001).
97. Xu, J.-T., Turner, S. C., Fairclough, J. P. A., Mai, S.-M., Ryan, A. J., Chaibundit, C. and Booth, C. *Macromolecules* **35**, 3614 (2002).
98. Liu, L.-Z., Xu, W., Li, H., Su, F. and Zhou, E. *Macromolecules* **30**, 1363 (1997).
99. Chen, H.-L., Li, H.-C., Huang, Y.-Y. and Chiu, F.-C. *Macromolecules* **35**, 2417 (2002).
100. Huang, P., Zhu, L., Cheng, S. Z. D., Ge, Q., Quirk, R. P., Thomas, E. L., Lotz, B., Hsiao, B. S., Liu, L. and Yeh, F. *Macromolecules* **34**, 6649 (2001).
101. Zhu, L., Mimnaugh, B. R., Ge, Q., Quirk, R. P., Cheng, S. Z. D., Thomas, E. L., Lotz, B., Hsiao, B. S., Yeh, F. and Liu, L. *Polymer* **42**, 9121 (2001).
102. Smith, M. D., Green, P. F., Saunders, R. *Macromolecules* **32**, 8392 (1999).
103. Orso, K. A. and Green, P. F. *Macromolecules* **32**, 1087 (1999).
104. Hamdoun, B., Ausserre, D., Cabuil, V. and Joly, S. *J. Phys. II* **6**, 503 (1996).
105. Green, P. F. and Limary R. *Adv. Colloid Interface Sci.* **94**, 53 (2001).
106. Retsos, H., Terzis, A. F., Anastasiadis, S. H., Anastassopoulos, D. L., Toprakcioglu, C., Theodorou, D. N., Smith, G. S., Menelle, A., Gill, R. E., Hadziioannou, G. and Gallot, Y. *Macromolecules* **35**, 1116 (2002).
107. Tanaka, H. and Hashimoto, T. *Macromolecules* **24**, 5398 (1991).
108. Laurer, J. H., Ashraf, A., Smith, S. D. and Spontak, R. J. *Langmuir* **13**, 2250 (1997).

109. Han, C. D., Vaidya, N. Y., Yamaguchi, D. and Hashimoto, T. *Polymer* **41**, 3779 (2000).
110. Sung, L., Nakatani, A. I., Han, C. C., Karim, A., Douglas, J. F. and Satija, S. K. *Physica B* **241–243**, 1013 (1998).
111. Komura, S., Kodama, H. and Tamura, K. *J. Chem. Phys.* **117**, 9903 (2002).
112. Janert, P. K. and Schick, M. *Macromolecules* **30**, 137 (1997).
113. Kim, S. H. and Jo, W. H. *J. Chem. Phys.* **110**, 12193 (1999).
114. Thompson, R. B. and Matsen, M. W. *J. Chem. Phys.* **112**, 6863 (2000).
115. Fredrickson, G. H. and Bates, F. S. *Eur. Phys. J. B* **1**, 71 (1998).
116. Liang, H. *Macromolecules* **32**, 8204 (1999).
117. Ko, M. J., Kim, S. H. and Jo, W. H. *Polymer* **41**, 6387 (2000).
118. Balsara, N. P. *Curr. Opin. Solid State Mater. Sci.* **3**, 589 (1998).
119. Jeon, H. S., Lee, J. H. and Balsara, N. P. *Macromolecules* **31**, 3328 (1998).
120. Jeon, H. S., Lee, J. H. and Balsara, N. P. *Phys. Rev. Lett.* **79**, 3274 (1997).
121. Jeon, H. S., Lee, J. H., Balsara, N. P. and Newstein, M. C. *Macromolecules* **31**, 3340 (1998).
122. Lee, J. H., Jeon, H. S., Balsara, N. P. and Newstein, M. C. *J. Chem. Phys.* **108**, 5173 (1998).
123. Bates, F. S., Maurer, W. W., Lipic, P. M., Hillmyer, M. A., Almdal, K., Mortensen, K. Fredrickson, G. H. and Lodge, T. P. *Phys. Rev. Lett.* **79**, 849 (1997).
124. Fredrickson, G. H. and Bates, F. S. *J. Polym. Sci. B: Polym. Phys.* **35**, 2775 (1997).
125. Hillmyer, M. A., Maurer, W. W., Lodge, T. P., Bates, F. S. and Almdal, K. *J. Phys. Chem. B* **103**, 4814 (1999).
126. Washburn, N. R., Lodge, T. P. and Bates, F. S. *J. Phys. Chem. B* **104**, 6987 (2000).
127. Morkved, T. L., Stepanek, P., Krishnan, K., Bates, F. S. and Lodge, T. P. *J. Chem. Phys.* **114**, 7247 (2001).
128. Matsen, M. W. *J. Chem. Phys.* **110**, 4658 (1999).
129. Corvazier, L., Messé, L., Salou, C. L. O., Young, R. N., Fairclough, J. P. A. and Ryan, A. J. *J. Mater. Chem.* **11**, 2864 (2001).
130. Bates, F. S., Maurer, W., Lodge, T. P., Schulz, M. F., Matsen, M. W., Almdal, K. and Mortensen, K. *Phys. Rev. Lett.* **75**, 4429 (1995).
131. Broseta, D. and Fredrickson, G. H. *J. Chem. Phys.* **93**, 2927 (1990).
132. Schwahn, D., Mortensen, K., Frielinghaus, H. and Almdal, K. *Phys. Rev. Lett.* **82**, 5056 (1999).
133. Schwahn, D., Mortensen, K., Frielinghaus, H. and Almdal, K. *Physica B* **276–278**, 353 (2000).
134. Schwahn, D., Mortensen, K., Frielinghaus, H., Almdal, K. and Kielhorn, L. *J. Chem. Phys.* **112**, 5454 (2000).
135. Kudlay, A. and Stepanow, S. *Macromol. Theory Simul.* **11**, 16 (2002).
136. Thompson, R. B. and Matsen, M. W. *Phys. Rev. Lett.* **85**, 670 (2000).
137. Kodama, H., Komura, S. and Tamura, K. *Europhys. Lett.* **53**, 46 (2001).
138. Naughton, J. R. and Matsen, M. W. *Macromolecules* **35**, 8926 (2002).
139. Mayes, A. M. and Olvera de la Cruz, M. *J. Chem. Phys.* **91**, 7228 (1989).
140. Matsen, M. W. and Thompson, R. B. *J. Chem. Phys.* **111**, 7139 (1999).
141. Matsen, M. W. *J. Chem. Phys.* **113**, 5539 (2000).
142. Mai, S.-M., Mingvanish, W., Turner, S. C., Chaibundit, C., Fairclough, J. P. A., Heatley, F., Matsen, M. W., Ryan, A. J. and Booth, C. *Macromolecules* **33**, 5124 (2000).
143. Jinnai, H., Nishikawa, Y., Spontak, R. J., Smith, S. D., Agard, D. A. and Hashimoto, T. *Phys. Rev. Lett.* **84**, 518 (2000).
144. Jinnai, H., Kajihara, T., Watashiba, H., Nishikawa, Y. and Spontak, R. J. *Phys. Rev. E*, **64**, 10803(R) (2001). [Erratum: *Phys. Rev. E*, **64**, 69903 (2001).]

145. Spontak, R. J. and Patel, N. P. *Curr. Opin. Colloid Interface Sci.* **5**, 334 (2000).
146. Watanabe, H. *Macromolecules* **28**, 5006 (1995).
147. Watanabe, H., Sato, T., Osaki, K., Yao, M.-L. and Yamagishi, A. *Macromolecules* **30**, 5877 (1997).
148. Karatasos, K., Anastasiadis, S. H., Pakula, T. and Watanabe, H. *Macromolecules* **33**, 523 (2000).
149. Watanabe, H., Sato, T. and Osaki, K. *Macromolecules* **33**, 2545 (2000).
150. Zhulina, E. B. and Halperin, A. *Macromolecules* **25**, 5730 (1992).
151. Matsen, M. W. and Schick, M. *Macromolecules* **27**, 187 (1994).
152. Li, B. Q. and Ruckenstein, E. *Macromol. Theory Simul.* **7**, 333 (1998).
153. Smith, S. D., Spontak, R. J., Satkowski, M. M., Ashraf, A. and Lin, J. S. *Phys. Rev. B* **47**, 14555 (1993).
154. Matsushita, Y., Mogi, Y., Mukai, H., Watanabe, J. and Noda, I. *Polymer* **35**, 246 (1994).
155. Smith, S. D., Spontak, R. J., Satkowski, M. M., Ashraf, A., Heape, A. K. and Lin, J. S. *Polymer* **35**, 4527 (1994).
156. Matsen, M. W. *J. Chem. Phys.* **102**, 3884 (1995).
157. Spontak, R. J. and Smith, S. D. *J. Polym. Sci. B: Polym. Phys.* **39**, 947 (2001).
158. Rasmussen, K. Ø., Kober, E. M., Lookman, T. and Saxena, A. *J. Polym. Sci. B: Polym. Phys.* **41**, 104 (2003).
159. Slot, J. J. M., Angerman, H. J. and ten Brinke G. *J. Chem. Phys.* **109**, 8677 (1998).
160. Spontak, R. J., Smith, S. D. and Ashraf, A. *Macromolecules*, **26**, 5118 (1993).
161. Laurer, J. H., Hajduk, D. A., Dreckötter, S., Smith, S. D. and Spontak, R. J. *Macromolecules*, **31**, 7546 (1998).
162. Lee, S.-H. and Char, K. *ACS Symp. Ser.* **739**, 496 (2000).
163. Vaidya, N. Y., Han, C. D., Kim, D., Sakamoto, N. and Hashimoto, T. *Macromolecules* **34**, 222 (2001).
164. Choi, S., Lee, K. M., Han, C. D., Sota, N. and Hashimoto, T. *Macromolecules* **36**, 793 (2003).
165. Baetzold, J. P. and Koberstein, J. T. *Macromolecules* **34**, 8986 (2001).
166. Lee, S.-H., Koberstein, J. T., Quan, X., Gancarz, I., Wignall, G. D. and Wilson, F. C. *Macromolecules* **27**, 3199 (1994).
167. Kane, L., Norman, D. A., White, S. A., Matsen, M. W., Satkowski, M. M., Smith, S. D. and Spontak, R. J. *Macromol. Rapid Commun.* **22**, 281 (2001).
168. Norman, D. A., Kane, L., White, S. A., Smith, S. D. and Spontak, R. J. *J. Mater. Sci. Lett.* **17**, 545 (1998).
169. Patel, N. P. Ph. D. Dissertation, North Carolina State University (2003).
170. Drolet, F. and Fredrickson, G. H. *Phys. Rev. Lett.* **83**, 4317 (1999).
171. Lescanec, R. L., Fetters, L. J. and Thomas, E. L. *Macromolecules* **31**, 1680 (1998).
172. Suzuki, J., Furuya, M., Iinuma, M., Takano, A. and Matsushita, Y. *J. Polym. Sci. B: Polym. Phys.* **40**, 1135 (2002).
173. Sugiyama, M., Shefelbine, T. A., Vigild, M. E. and Bates, F. S. *J. Phys. Chem. B* **105**, 12448 (2001).
174. Dotera, T. *Phys. Rev. Lett.* **89**, 205502 (2002).
175. Laurer, J. H., Fung, J. C., Sedat, J. W., Smith, S. D., Samseth, J., Mortensen, K., Agard, D. A. and Spontak, R. J. *Langmuir* **13**, 2177 (1997).
176. Laurer, J. H., Smith, S. D., Samseth, J., Mortensen, K. and Spontak, R. J. *Macromolecules* **31**, 4975 (1998).
177. Jinnai, H., Nishikawa, Y., Ito, M., Smith, S. D., Agard, D. A. and Spontak, R. J. *Adv. Mater.* **14**, 1615 (2002).
178. Pochan, DJ., Gido, S. P., Pispas, S., Mays, J. W., Ryan, A. J., Fairclough, J. P. A., Hamley, I. W. and Terrill, N. J. *Macromolecules* **29**, 5091 (1996).

179. Pochan, D. J., Gido, S. P., Pispas, S. and Mays, J. W. *Macromolecules* **29**, 5099 (1996).
180. Yang, L., Gido, S. P., Mays, J. W., Pispas, S. and Hadjichristidis, N. *Macromolecules* **34**, 4235 (2001).
181. Avgeropoulos, A., Dair, B. J., Thomas, E. L. and Hadjichristidis, N. *Polymer* **43**, 3257 (2002).
182. Burgaz, E. and Gido, S. P. *Macromolecules* **33**, 8739 (2000).
183. Hadziioannou, G. and Skoulios, A. *Macromolecules* **15**, 258, 267 (1982).
184. Melenkevitz, J. and Muthukumar, M. *Macromolecules* **24**, 4199 (1991).
185. Matsen, M. W. and Bates, F. S. *Macromolecules* **29**, 1091 (1996).
186. Kane, L., Satkowski, M. M., Smith, S. D. and Spontak, R. J. *Macromolecules* **29**, 8862 (1996).
187. Spontak, R. J. *Macromolecules* **27**, 6363 (1994).
188. Court, F. and Hashimoto, T. *Macromolecules* **35**, 2566 (2002).
189. Zhulina, E. B. and Birshtein, T. M. *Polymer* **32**, 1299 (1991).
190. Matsen, M. W. *J. Chem. Phys.* **103**, 3268 (1995).
191. Papadakis, C. M., Mortensen, K. and Posselt, D. *Macromol. Symp.* **149**, 99 (2000).
192. Yamaguchi, D., Shiratake, S. and Hashimoto, T. *Macromolecules* **33**, 8258 (2000).
193. Yamaguchi, D. and Hashimoto, T. *Macromolecules* **34**, 6495 (2001).
194. Yamaguchi, D., Hasegawa, H. and Hashimoto, T. *Macromolecules* **34**, 6506 (2001).
195. Yamaguchi, D., Bodycomb, J., Koizumi, S. and Hashimoto, T. *Macromolecules* **32**, 5884 (1999).
196. Morita, H., Kawakatsu, T., Doi, M. Yamaguchi, D., Takenaka, M. and Hashimoto, T. *Macromolecules* **35**, 7473 (2002).
197. Zhao, J., Majumdar, B., Schulz, M. F., Bates, F. S., Almdal, K., Mortensen, K., Hajduk, D. A. and Gruner, S. M. *Macromolecules* **29**, 1204 (1996).
198. Spontak, R. J., Fung, J. C., Braunfeld, M. B., Sedat, J. W., Agard, D. A., Kane, L., Smith, S. D., Satkowski, M. M., Ashraf, A., Hajduk, D. A. and Gruner, S. M. *Macromolecules* **29**, 4494 (1996).
199. Matsen, M. W. and Bates, F. S. *Macromolecules* **28**, 7298 (1995).
200. Shi, A. C. and Noolandi, J. *Macromolecules* **28**, 3103 (1995).
201. Sakurai, S., Irie, H., Umeda, H., Nomura, S., Lee, H. H. and Kim, J. K. *Macromolecules* **31**, 336 (1998).
202. Court, F. and Hashimoto, T. *Macromolecules* **34**, 2536 (2001).
203. Yamaguchi, D., Takenaka, M., Hasegawa, H. and Hashimoto, T. *Macromolecules* **34**, 1707 (2001).
204. Koneripalli, N., Levicky, R., Bates, F. S., Matsen, M. W., Satija, S. K., Ankner, J. and Kaiser, H. *Macromolecules* **31**, 3498 (1998).
205. Lipic, P. M., Bates, F. S. and Matsen, M. W. *J. Polym. Sci. B: Polym. Phys.* **37**, 2229 (1999).
206. Vaidya, N. Y. and Han, C. D. *Macromolecules* **33**, 3009 (2000).
207. Frielinghaus, H., Hermsdorf, N., Sigel, R., Almdal, K., Mortensen, K., Hamley, I. W., Meesé, L., Corvazier, L., Ryan, A. J., van Dusschoten, D., Wilhelm, M., Floudas, G. and Fytas, G. *Macromolecules* **34**, 4907 (2001).
208. Kim, S. H., Lee, H. S., Lee, M. S. and Jo, W. H. *Macromol. Chem. Phys.* **203**, 2188 (2002).
209. Papadakis, C. M., Busch, P., Weidisch, R., Eckerlebe, H. and Posselt, D. *Macromolecules* **35**, 9236 (2002).
210. Olmsted, P. D. and Hamley, I. W. *Europhys. Lett.* **45**, 83 (1999).
211. Kimishima, I., Jinnai, H. and Hashimoto, T. *Macromolecules* **32**, 2585 (1999).
212. Jeon, H. G., Hudson, S. D., Ishida, H. and Smith, S. D. *Macromolecules* **32**, 1803 (1999).

213. Goldacker, T., Abetz, V., Stadler, R., Erukhimovich, I. and Leibler, L. *Nature* **398**, 137 (1999).
214. Birshtein, T. M. and Polotsky, A. A. *Macromol. Theory Simul.* **9**, 115 (2000).
215. Birshtein, T. M., Zhulina, E. B., Polotsky, A. A., Abetz, V. and Stadler, R. *Macromol. Theory Simul.* **8**, 151 (1999).
216. Birshtein, T. M., Polotsky, A. A. and Abetz, V. *Macromol. Theory Simul.* **10**, 700 (2001).
217. Goldacker, T. and Abetz, V. *Macromolecules* **32**, 5165 (1999).
218. Lee, H.-K. and Zin, W.-C. *Macromolecules* **33**, 2894 (2000).
219. Kim, D.-C., Lee, H.-K., Sohn, B.-H. and Zin, W.-C. *Macromolecules* **34**, 7767 (2001).
220. Sakurai, Isobe, D., Okamoto, S. and Nomura, S. *J. Mater. Sci. Res. Int.* **7**, 225 (2001).
221. Sakurai, Isobe, D., Okamoto, S. and Nomura, S. *J. Macromol. Sci.-Phys.* **B41**, 387 (2002).
222. Goldacker, T. and Abetz, V. *Macromol. Rapid Commun.* **20**, 415 (1999).
223. Breiner, U., Krappe, U., Thomas, E. L. and Stadler, R. *Macromolecules* **31**, 135 (1998).
224. Förster, S., Berton, B., Hentze, H. P., Krämer, E., Antonietti, M. and Lindner, P., *Macromolecules* **34**, 4610 (2001).
225. Hillmyer, M. A., Lipic, P. M., Hajduk, D. A., Almdal, K. and Bates, F. S. *J. Am. Chem. Soc.* **119**, 2749 (1997).
226. Lipic, P. M., Bates, F. S. and Hillmyer, M. A. *J. Am. Chem. Soc.* **120**, 8963 (1998).
227. Mijovic, J., Shen, M., Sy, J. W. and Mondragon, I. *Macromolecules* **33**, 5235 (2000).
228. Dean, J. M., Lipic, P. M., Grubbs, R. B., Cook, R. F. and Bates, F. S. *J. Polym. Sci. B: Polym. Phys.* **39**, 2996 (2001).
229. Guo, Q., Thomann, R., Gronski, W. and Thurn-Albrecht, T. *Macromolecules* **35**, 3133 (2002).
230. Guo, Q., Thomann, R., Gronski, W., Staneva, R., Ivanova, R. and Stühn, B. *Macromolecules* **36**, 3635 (2003).
231. Ritzenthaler, S., Court, F., David, L., Girard-Reydet, E., Leibler, L. and Pascault, J.-P. *Macromolecules* **35**, 6245 (2002).
232. Ritzenthaler, S., Court, F., Girard-Reydet, E., Leibler, L. and Pascault, J.-P. *Macromolecules* **36**, 118 (2003).
233. Girard-Reydet, E., Sévignon, A., Pascault, J.-P., Hoppe, C. E., Galante, M. J., Oyanguren, P. A. and Williams, R. J. J. *Macromol. Chem. Phys.* **203**, 947 (2002).
234. Kraus, G. *Rubber Chem. Tech.* **55**, 1389 (1982).
235. Ho, R.-M., Adedeji, A., Giles, D. W., Hajduk, D. A., Macosko, C. W. and Bates, F. S. *J. Polym. Sci. B: Polym. Phys.* **35**, 2857 (1997).
236. Varma, R., Takeichi, H., Hall, J. E., Ozawa, Y. F. and Kyu, T. *Polymer* **43**, 4667 (2002).
237. Adedeji, A., Grünfelder, T., Bates, F. S., Macosko, C. W., Stroup-Gardiner, M. and Newcomb, D. E. *Polym. Eng. Sci.* **36**, 1707 (1996).
238. Discher, B. M., Won, Y. Y., Ege, D. S., Lee, J. C. M., Bates, F. S., Discher, D. E. and Hammer, D. A. *Science* **284**, 1143 (1999).
239. Haluska, C. K., Gozdz, W. T., Dobereiner, H. G., Förster, S. and Gompper, G. *Phys. Rev. Lett.* **89**, 238302 (2002).
240. Kukula, H., Schlaad, H., Antonietti, M. and Förster, S. *J. Am. Chem. Soc.* **124**, 1658 (2002).
241. Kavanagh, G. M. and Ross-Murphy, S. B. *Prog. Polym. Sci.* **23**, 533 (1998).
242. Flanigan, C. M., Crosby, A. J. and Shull, K. R. *Macromolecules* **32**, 7251 (1999).

243. Raspaud, E., Lairez, D., Adam, M. and Carton, J.-P. *Macromolecules* **29**, 1269 (1996).
244. Mortensen, K. *Curr. Opin. Colloid Interface Sci.* **3**, 12 (1998).
245. Hamley, I. W. *Curr. Opin. Colloid Interface Sci.* **5**, 342 (2000).
246. Yang, Z., Yang, Y.-W., Zhou, Z.-K., Attwood, D. and Booth, C. *J. Chem. Soc., Faraday Trans.* **92**, 257 (1996).
247. Spontak, R. J., Wilder, E. A. and Smith, S. D., *Langmuir* **17**, 2294 (2001).
248. Wilder, E. A., White, S. A., Smith, S. D. and Spontak, R. J., *ACS Symp. Ser.* **833**, 248 (2003).
249. Annable, T., Buscall, R. and Ettelaie, R. in *Amphiphilic Block Copolymers: Self-Assembly and Applications* (P. Alexandridis and B. Lindman, eds.) Elsevier Science B. V., Amsterdam, (2000), Chap. 12.
250. Skvortsov, A. M., Pavlushkov, I. V., Gorbunov, A. A. and Zhulina, E. B. *J. Chem. Phys.* **105**, 2119 (1996).
251. Vega, D. A., Sebastian, J. M., Loo, Y.-L. and Register, R. A. *J. Polym. Sci. B: Polym. Phys.* **39**, 2183 (2001).
252. Quintana, J. R., Hernáez, E. and Katime, I. *Polymer* **43**, 3217 (2002).
253. Koňák, C. and Helmstedt, M. *Macromolecules* **36**, 4603 (2003).
254. King, M. R., White, S. A., Smith, S. D. and Spontak, R. J. *Langmuir* **15**, 7886 (1999).
255. Roberge, R. L., Patel, N. P., White, S. A., Thongruang, W., Smith, S. D. and Spontak, R. J. *Macromolecules* **35**, 2268 (2002).
256. Stevens, J. E., Thongruang, W., Patel, N. P., Smith, S. D. and Spontak, R. J. *Macromolecules* **36**, 3206 (2003).
257. Zhao, D. Y., Feng, J. L., Huo, Q. S., Melosh, N., Fredrickson, G. H., Chmelka, B. F. and Stucky, G. D. *Science* **279**, 548 (1998).
258. Yang, P. D., Zhao, D. Y., Margolese, D. I., Chmelka, B. F. and Stucky, G. D. *Nature* **396**, 152 (1998).
259. Gelb, L. D., Gubbins, K. E., Radhakrishnan, R. and Sliwinska-Bartkowiak, M. *Rep. Prog. Phys.* **62**, 1573 (1999).
260. Templin, M., Franck, A., DuChesne, A., Leist, H., Zhang, Y. M., Ulrich, R., Schadler, V. and Wiesner, U. *Science* **278**, 1795 (1997).
261. Simon, P. F. W., Ulrich, R., Spiess, H. W. and Wiesner, U. *Chem. Mater.* **13**, 3464 (2001).
262. Yang, P., Deng, T., Zhao, D., Feng, P., Pine, D., Chmelka, B. F., Whitesides, G. M. and Stucky, G. *Science* **282**, 2244 (1998).
263. Schrage, S., Sigel, R. and Schlaad, H. *Macromolecules* **36**, 1417 (2003).
264. Jiang, S., Göpfert, A. and Abetz, V. *Macromolecules* (in press).

6 Crystallization Within Block Copolymer Mesophases

YUEH-LIN LOO[1] and RICHARD A. REGISTER[2]

[1] Chemical Engineering, Center for Nano- & Molecular Science and Technology, and Texas Materials Institute, University of Texas at Austin, Austin TX
[2] Chemical Engineering and Princeton Materials Institute, Princeton University, Princeton, NJ

6.1 INTRODUCTION

The phase behavior of model noncrystallizing block copolymers is fairly well understood today [Hamley 1998a]: microphase separation takes place when the interblock segregation is sufficiently high, and results in the formation of periodic nanoscale structures – such as spheres, cylinders, gyroid and lamellae – depending on the relative block lengths. When crystallizable segments are incorporated within a block copolymer, however, the structure that the material adopts is more difficult to predict *a priori*. These materials – semicrystalline block copolymers – possess two mechanisms that can drive phase separation, i.e., microphase separation and crystallization, and the interplay between these two results in both morphological richness and kinetic complexity.

The development of solid-state structure in semicrystalline block copolymers has been studied extensively over the past decade. Most of the earlier studies are covered in a recent review by Hamley [1999]; the present chapter complements this earlier review by highlighting recent advances in this field, with a particular focus on how the processes of microphase separation and crystallization interact. We begin by providing a brief overview of synthetic routes to near-monodisperse semicrystalline block copolymers. We then enumerate the key experimental techniques used to examine the crystallization behavior and morphology of semicrystalline block copolymers. The remainder of the chapter focuses on the solid-state structures that these materials exhibit, the pathways by which these structures develop, and the impact of melt microphase separation on polymer crystallization kinetics.

6.2 SYNTHESIS

Our focus in this chapter is on "model" block copolymers: those with well-defined molecular architectures, i.e., where the block sequences (AB vs. ABA vs. ABC . . .) are practically identical across the ensemble of chains, and where the individual blocks possess narrow chain length distributions. Throughout this chapter, block copolymer chemistries will be denoted as "A/B", and particular diblock copolymers as "A/B n/m", where "A" is the abbreviation for the monomer comprising the crystallizable block (e.g., "CL" for ε-caprolactone), "B" is the monomer comprising the amorphous block (e.g., "S" for styrene), and "n" and "m" are the crystallizable and amorphous block molecular weights, in kg/mol (rounded to the nearest kg/mol). This notation immediately connotes the approximate volume fraction of A block, and hence suggests the likely melt morphology.

Polymers with the desired well-defined architectures are synthesized by "controlled" or "living" polymerizations, where spontaneous termination and transfer events are largely or completely suppressed, and where initiation is relatively rapid. The best known of these synthetic routes is anionic polymerization [Morton 1983; Young et al. 1984], and indeed, this has been the route employed to prepare the majority of the model crystallizable block copolymers studied to date. However, other synthetic methodologies – such as cationic and ring-opening metathesis polymerization – have also been successfully employed to prepare well-defined crystallizable block copolymers, and use of these and other synthetic routes will doubtless grow in the coming decades as our control over the initiation, propagation, and termination steps advances.

6.2.1 POLYETHYLENE BLOCKS

Because ethylene polymerizes only very slowly with carbanionic initiators [Hay 1978], the incorporation of "polyethylene" segments into block copolymers is typically achieved through the anionic polymerization of butadiene, followed by hydrogenation. Hydrogenation of polybutadiene comprised wholly of 1,4 units (added to the chain through a conjugated addition; any cis/trans ratio) would produce a material structurally identical to linear (high-density) polyethylene. Unfortunately, anionic polymerization is incapable of producing a wholly-1,4 microstructure. Typical polymerization conditions (alkyllithium initiator, aliphatic hydrocarbon solvent, 60 °C) yield roughly 8 % 1,2 units (formed through nonconjugated addition) distributed statistically [Krigas 1985] along the chain. Upon hydrogenation, the 1,4 units become pairs of ethylene mers, while the 1,2 units become butene mers in the chain, yielding a material that resembles a statistical ethylene-butene copolymer with 8 wt% butene, or roughly 20 ethyl branches per 1000 backbone carbons.

The ethyl branches contributed by these butene units are typically excluded from the polymer crystals [Hosoda *et al.* 1990], so hydrogenated polybutadienes exhibit melting temperatures and degrees of crystallinity substantially lower than those for linear polyethylene homopolymers of similar molecular weights. For example, hydrogenated polybutadienes with approximately 20 ethyl branches per 1000 backbone carbons typically exhibit a degree of crystallinity (weight fraction) of 0.35 and a peak melting temperature near 100 °C [Howard and Crist 1989; Rangarajan *et al.* 1993]. Another feature that distinguishes hydrogenated polybutadienes from polyethylene is that hydrogenated polybutadienes do not undergo substantial crystal thickening on annealing, nor does the crystal thickness depend substantially on the crystallization history (e.g., temperature of isothermal crystallization). Since hydrogenated polybutadiene is a statistical copolymer, the thicknesses of the crystals that form are limited by the statistical distribution of crystallizable ethylene sequences between butene mers; further annealing does not extend the crystals beyond the thickness determined by the length of the ethylene segments. The average E crystal thickness in a polymer with 20 ethyl branches per 1000 backbone carbons is approximately 5 nm [Rangarajan *et al.* 1993].

Though these differences between hydrogenated polybutadiene and truly linear polyethylene are important, we will (for convenience) refer to hydrogenated polybutadiene as "polyethylene" (E) throughout this chapter, since the only systematic studies to date of the solid-state structures in near-monodisperse polymers containing polyethylene-like blocks have employed hydrogenated polybutadiene. However, we note that ring-opening methathesis polymerization of cycloolefins (notably cyclobutene [Wu and Grubbs 1994] and cyclopentene [Trzaska *et al.* 2000]) has recently been used to prepare well-defined unbranched chains that, upon hydrogenation, yield truly linear polyethylene. These highly crystalline polyethylene-containing block copolymers, presently under investigation at Princeton, will provide an interesting contrast to otherwise similar block copolymers based on hydrogenated polybutadiene.

6.2.2 POLY(ETHYLENE OXIDE) BLOCKS

Block copolymers containing poly(ethylene oxide), EO, can also be synthesized by anionic polymerization, typically in ether solvents with sodium- or potassium-based initiators [Lotz and Kovacs 1966]. However, if the amorphous block is based on an all-hydrocarbon monomer such as styrene or butadiene, the EO block must be synthesized second (last). Since the chain structure of the EO block is perfectly regular, crystallinity within the EO block can be high (>80%), and extended-chain or integrally folded crystals become possible. The melting temperature (T_m) of high molecular weight EO homopolymer is in the vicinity of 75 °C, but T_m is greatly suppressed at the molecular weights commonly used in block copolymers ($T_m < 60$ °C).

6.2.3 POLY(ε-CAPROLACTONE) BLOCKS

Poly(ε-caprolactone), CL, can be synthesized through anionic polymerization with alkyllithium initiators [Nojima *et al.* 1992], though as with EO, if the first block is a hydrocarbon, the CL block must be synthesized second. CL homopolymers generally melt in the vicinity of 60 °C, though T_m is lower at the melting points typically employed for block copolymers. CL is not only biocompatible (as is EO), but also biodegradable. Thus, determining and controlling the structures of EO- and CL- containing block copolymers can potentially impact the growing field of biomaterials.

6.3 TOOLS FOR EXAMINING BLOCK COPOLYMER CRYSTALLIZATION

6.3.1 CALORIMETRY

Differential scanning calorimetry (DSC) is the most common technique for quantifying the thermal transitions associated with semicrystalline block copolymers, including melting and freezing of the crystallizable block and the glass transition of the amorphous block. Aside from locating the thermal transitions, calorimetry can also be used to quantify their strengths. The degree of crystallinity (X_c) can be determined from DSC measurements as:

$$X_c = \frac{\Delta H_f}{\Delta H_{f, 100\%}} \cdot W \qquad (6.1)$$

where W represents the weight fraction of the crystallizable block; ΔH_f is the heat of fusion obtained from calorimetry; and $\Delta H_{f, 100\%}$ is the heat of fusion of the homopolymer corresponding to the crystallizable block if it were 100% crystalline.

6.3.2 X-RAY SCATTERING

Both small- and wide-angle X-ray scattering (SAXS and WAXS) are useful for examining the structures exhibited by semicrystalline block copolymers. SAXS is used for probing structures on the 1–100 nm length scale, making it well suited to examine microdomain and crystallite structure, periodicity, and orientation. WAXS is used for examining finer structures, generally spanning 0.1–1 nm, and so is useful for studying crystalline unit cell structure and orientation. Tandem SAXS and WAXS on macroscopically oriented specimens is a powerful combination for elucidating how polymer chains crystallize with respect to the microdomain interface, as reviewed in detail later in this chapter. Researchers have also combined synchrotron-based time-resolved SAXS and/or WAXS with in-situ calorimetry to monitor structural changes and to track crystallization kinetics in real time. These experiments have been instrumental in providing mechanistic details of the crystallization process in semicrystalline block copolymers, as reviewed in the latter portion of this chapter.

6.3.3 MICROSCOPY

To examine semicrystalline block copolymer morphology by electron microscopy, the specimens need to be sectioned and stained to provide sufficient imaging contrast. Though challenging, the development of new sample-preparation protocols has recently enabled the examination of individual crystals within block copolymer microdomains [Loo et al. 2000b]. Optical microscopy is a long-established technique for investigating crystalline superstructure, especially the presence and structure of spherulites. Recently, atomic force microscopy (AFM) has been applied to study the structures of thin supported films of crystallizable block copolymers. Optical and AFM measurements are nondestructive and can be conducted in or near real time so the process of crystallization can be tracked.

6.4 STRUCTURES FORMED BY CRYSTALLIZATION FROM MICROPHASE-SEPARATED MELTS

The incorporation of crystallizable moieties within a block copolymer architecture can greatly enhance the morphological richness of the system. For example, a "structure-within-structure" hierarchical morphology can be obtained in semicrystalline block copolymers if crystal formation occurs within the nanoscale domains formed by microphase separation in the melt. Ultimately, the competition between microphase separation and crystallization sets the final structure. Since the energy of crystallization (of order 100 J/g) is significantly larger than that associated with microphase separation (of order 1 J/g), it might seem that crystallization would always dominate the structure of semicrystalline block copolymers: that is, for any melt morphology, drastic structural rearrangement might be expected to occur so as to permit crystals to grow to their maximum extent, eradicating the microphase-separated structure as crystallization proceeds. Such freely growing crystals could even form a spherulitic superstructure.

Indeed, most early experiments [Ryan *et al.* 1995; Rangarajan *et al.* 1995; Floudas and Tsitsilianis 1997, Floudas *et al.* 1999; Hillmyer *et al.* 1996; Nojima *et al.* 1992, 1993] were consistent with this argument: a crystalline alternating lamellar morphology, sometimes organized into spherulites on a larger length scale, always resulted – independent of the melt structure prescribed by microphase separation. However, a glassy amorphous component should effectively preserve the microphase-separated structure into the solid state: provided the glass transition temperature of the amorphous block is higher than the freezing temperature of the crystallizable block, the melt structure can be vitrified on cooling, forcing crystallization to occur within the microphase-separated structure formed by self-assembly in the melt. The following section highlights recent results on semicrystalline-glassy block copolymer systems. Subsequently, this chapter will cover experimental results on semicrystalline-rubbery systems where crystallization occurs prior to the vitrification of the amorphous block, as it has recently been demonstrated that strong interblock segregation is indeed sufficient to preserve the melt microdomain structure even in the absence of a vitreous block. Concurrently, we also discuss aspects of chain folding and preferential crystal orientation that result from crystallization within or around anisotropic microdomains.

6.4.1 SEMICRYSTALLINE-GLASSY BLOCK COPOLYMERS

Semicrystalline-glassy systems which have been examined extensively include [hydrogenated poly(1,4-butadiene)]-*b*-polystyrene, E/S [Cohen *et al.* 1990]; [hydrogenated poly(1,4-butadiene)]-*b*-poly(vinyl cyclohexane), E/VCH [Loo *et al.* 2000b, 2000c, 2001; Weimann *et al.* 1999; Hamley *et al.* 1996; Bates *et al.* 2001]; and poly(ethylene oxide)-*b*-polystyrene, EO/S [Lotz and Kovacs 1969; Zhu *et al.* 2000, 2001, 2002; Huang *et al.* 2001]. In these cases, as the block copolymer is cooled from its microphase separated melt, it first encounters the glass transition temperature of the amorphous component; cooling the block copolymer further induces freezing of the crystallizable component within the now-vitrified microdomain structure. The formation of spherulites is effectively suppressed, resulting in optically clear materials – in stark contrast to semicrystalline homopolymers, which are generally translucent or opaque due to spherulite formation.

While it is no surprise that a vitreous matrix can preserve the self-assembled structure established in the melt, how the system accommodates the density change that accompanies crystallization was unclear until recently. Three possibilities immediately suggest themselves. First, individual crystallized microdomains could simply exist under significant hydrostatic tension. If this were the case, however, a large melting temperature suppression compared to a homopolymer of the crystallizable block would be observed, which is readily ruled out through calorimetry data, at least in the E/VCH system [Loo *et al.*

2001]. Second, cavitation could occur within the crystallizing domains in order to relieve this hydrostatic tension. The third possibility is that the system undergoes macroscopically affine contraction to accommodate the volume change. For a series of E/VCH semicrystalline-glassy block copolymers forming E spheres, cylinders, and lamellae, dilatometry experiments [Loo *et al.* 2001] demonstrate that the third possibility is correct: the specimen's macroscopic volume changes during crystallization and melting agree well with those expected from the block copolymers' compositions and the degrees of crystallinity (X_c) measured by DSC through Equation (6.1). Though the VCH matrix is vitreous throughout crystallization and melting, it can deform as necessary so that the E domains neither cavitate nor exist under great overall tension.

Crystallization within spheres and cylinders

Recently, Loo *et al.* [2000b] modified an established RuO_4-staining procedure for saturated polymers to enable the resolution of individual crystals within E/VCH block copolymers by TEM. E crystallization within spherical microdomains (E/VCH 5/22) led to the formation of 5-nm thick crystals – the same thickness as achieved in hydrogenated polybutadiene of the same composition, since this thickness is set by exclusion of the ethyl branches from the crystal. Crystallization within cylinders (E/VCH 10/26) led to highly anisotropic, ribbon-like crystals running down the cylindrical microdomains, as shown in Figure 6.1. This crystal anisotropy was also reflected in the orientation of the E unit cells, which can be readily probed by WAXS on specimens macroscopically oriented through the application of extensional flow [Quiram *et al.* 1998]. For E/VCH 10/26, the E crystals had their *c* axes (chain axis) generally aligned with the cylinder radius, and their *b* axes generally aligned with the cylinder axes (both with some preferred tilt). The *b* axis is the fast growth axis for polyethylene, providing a simple explanation for the observed orientation: crystals that have their *b* axes aligned with the cylinder axis are the crystals that grow fastest, thus dominating the ensemble-average orientation measured by WAXS. The tilt was attributed to the need to accommodate the noncrystalline material at the crystal surface [Quiram *et al.* 1998].

Huang *et al.* [2001] conducted an extensive study of EO crystal orientation as a function of crystallization temperature within the cylinders of an EO/S diblock (EO/S 9/9 blended with S homopolymer to produce EO cylinders). When crystallized at $-50\,°C$, where the nucleation density is quite high and the resulting crystals quite small, no preferred orientation was present. At $T_c = 10\,°C$ and above, the crystals were oriented so that the fast-growth direction for EO (the [120] direction) was aligned with the cylinder axis, as shown in Figure 6.2. When crystallized between $-30\,°C$ and $0\,°C$, a preferred tilt of the fast growth direction to the cylinder axis was observed, as Quiram *et al.* [1998]

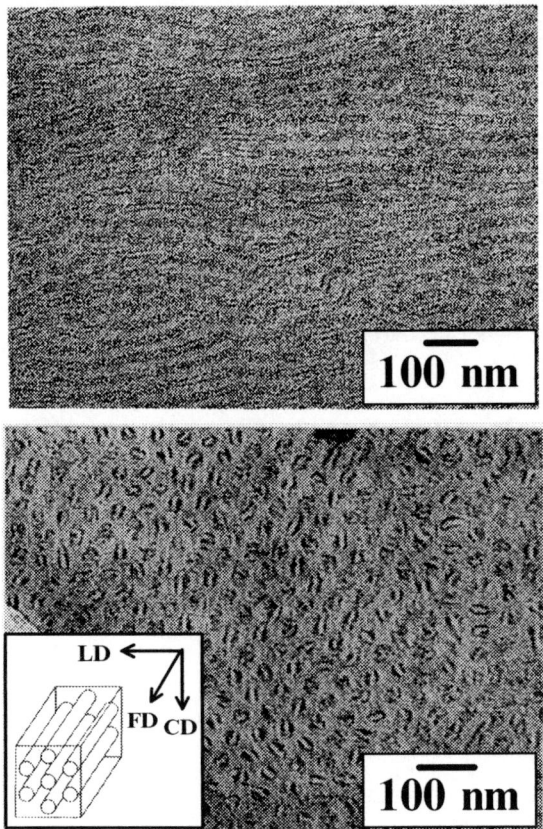

Figure 6.1 TEM images of a crystallized, oriented specimen of E/VCH 10/26. Cylinder orientation is schematized in the inset. RuO_4 staining preferentially stains the amorphous E region (dark), leaving the E crystals and VCH matrix essentially unstained. Top: section cut in the FD/LD plane, showing crystals running horizontally within cylinders; darker "stripes" of amorphous E just above and below each E crystal can be seen. Bottom: section cut in the CD/LD plane, showing the cylinder cross section; the single crystal within each cylinder appears as a short bright stripe surrounded by a darker oval of stained amorphous E. [Reprinted with permission from Loo et al. 2000b.]

observed in E/VCH; Huang et al. [2001] attribute the observed tilt to changes in the energy barriers to EO crystal-stem deposition at the growing crystal surface – changes induced by the confined geometry. The striking consistency between the results for the E/VCH and EO/S systems suggests that the observed orientations are general for crystallization within block copolymer cylinders in the presence of a glassy matrix, and that for polymers yet to be developed, we can anticipate a general alignment of the fast-growth axis of the crystalline component with the cylinder axis.

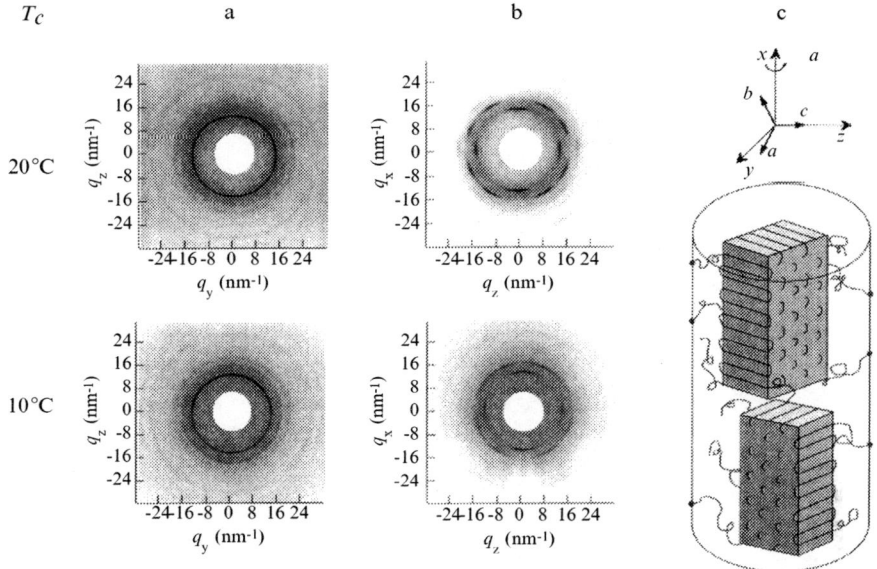

Figure 6.2 Right: schematic of crystal orientation in S/EO diblocks, forming EO cylinders, when crystallized at relatively shallow undercoolings (low crystal nucleation densities). The fast-growth axis is aligned with the cylinder axis, and the crystal stems are aligned with the cylinder radius (perpendicular to the microdomain interface). The remainder of the figure shows the WAXS data from which the crystal orientation was deduced, following isothermal crystallization at the indicated temperature, T_c. Left: X-ray beam parallel to cylinder axis, no orientation observed. Center: X-ray beam perpendicular to the cylinder axis, showing preferred scattering of the (120) planes (inner ring) on the meridian and the equator. [Reprinted with permission from Huang *et al.*, *Macromolecules* (2001), **34**, 6649–6657. Copyright (2001) American Chemical Society.]

Crystallization within lamellae

For lamellar E/VCH diblocks, TEM on RuO_4-stained specimens directly reveals that the crystallites are oriented with their thin dimension (crystal stem) *parallel* to the lamellar interface [Loo *et al.* 2001]. This orientation had been discerned earlier for E/VCH diblocks by Hamley *et al.* [1996] through parallel SAXS/WAXS measurements on specimens macroscopically oriented via large-amplitude oscillatory shear, and is schematized in Figure 6.3. Indeed, such an orientation in E-based block copolymers had been determined even earlier by Douzinas and Cohen [1992], using a combination of SAXS and WAXS pole figure analyses on a series of polyethylene-poly(ethylethylene) diblocks, E/EE, where EE is rubbery at the E freezing point. This same orientation was also found subsequently by Cohen *et al.* [1994] in an E/S diblock having a polydisperse but truly linear (high-density) E block, such that the S block

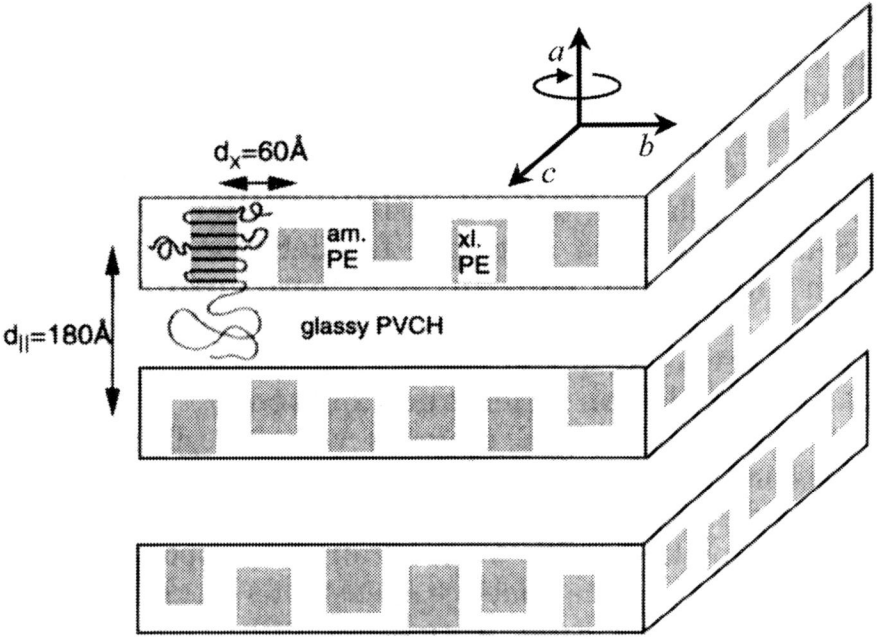

Figure 6.3 Schematic of crystal orientation in lamellar E/VCH diblocks, where the featureless layers are VCH and the E layers consist of E crystallites (gray) and amorphous E regions, alternating horizontally. The inset at top shows the orientation of the orthorhombic E unit cell. [Reprinted with permission from Hamley *et al.*, *Macromolecules* (1996), **29**, 8835–8843. Copyright (1996) American Chemical Society.]

is above its glass transition temperature at the freezing point of the E block. (Semicrystalline-rubbery diblocks are discussed in the following section.)

The inset in Figure 6.3 shows the preferred directions for the basis vectors of the orthorhombic E unit cell, as deduced from the WAXS data (by Hamley *et al.* [1996], and earlier for E/EE and E/S diblocks by Cohen and coworkers [Douzinas and Cohen 1992; Cohen *et al.* 1994]). The observed crystal orientation in these lamellar E/VCH diblocks also aligns the fast-growth (*b*) axis in the direction where crystals can grow for an extended distance (i.e., *b* axis parallel to the lamellar surface), as discussed above for crystallization within cylinders. Within cylinders, alignment of the *b* axis with the cylinder axis automatically specifies the orientation of the *a* and *c* axes to within the rotational symmetry expected for a cylinder: the *a* and *c* axes must both lie parallel to the cylinder radius. But for lamellae, placing the *b* axis parallel to the lamellar surface does not uniquely describe the orientation of the unit cell, since either the *a* or *c* axes (or neither) can be preferentially aligned with the lamellar normal. The observed orientation, with the *a* axis parallel to the lamellar normal, is one that accommodates the general incommensurability between the E crystal spacing and the E microdomain thickness; aligning the *c* axis direction with the lamellar microdomain normal would require that precisely one, or two, or three, etc. E crystals be stacked within the predefined E microdomain.

For E-based polymers crystallizing from *homogeneous* melts (polyethylene-poly(ethylene-*alt*-propylene), E/EP), Rangarajan *et al.* [1993] had previously inferred a different orientation, where the *c* axis is aligned with the lamellar normal (exchange *a* and *c* in Figure 6.3). Indeed, this structure had been deduced even earlier for EO/S single crystals grown from dilute solution [Lotz and Kovacs 1966], and was the basis for the first theoretical treatment of the domain spacing scaling in crystalline-amorphous diblocks [DiMarzio *et al.* 1980]. Of course, when crystallization proceeds from homogeneous melts, there is no predefined lamellar space within which the E crystals must fit; the thickness of the E-rich domain is consequently defined by the number of crystals that span it. This orientation was subsequently confirmed [Rangarajan *et al.* 1995] by measurement of the sign of the birefringence in the quadrants of the spherulites formed by diblocks crystallizing from homogeneous or weakly segregated melts; the quadrant birefringences matched those for spherulites formed by E homopolymers, where the chain axes run tangentially in the spherulite. While these different orientations initially seemed to defy a unified explanation, they are now easily understood by considering only two points: 1) the desire to orient the fast-growth axis in a direction where crystals can grow unobstructed, and 2) for lamellae, the desire to avoid any commensurability constraint, which can be achieved in homogeneous or weakly segregated diblocks by setting or resetting the microdomain layer thickness during crystallization, or in semicrystalline-glassy diblocks by orienting the *c* (crystal stem) axis parallel to the microdomain interface, so that the crystallites can stack without constraint.

Zhu *et al.* [2000, 2001] systematically investigated the crystal orientation within the lamellar EO/S 9/9 diblock, crystallized at various temperatures. Again, at the lowest crystallization temperatures (highest nucleation densities), no preferential orientation developed. When crystallized between -50 and $-10\,°C$, an orientation analogous to that in Figure 6.3 was observed: both the fast-growth direction and the *c* axis lie in the plane of the lamellae. At the highest crystallization temperatures, $35\,°C$ and above, the low molecular weight S matrix begins to devitrify; a change in the lamellar spacing upon crystallization was seen by SAXS, and the EO crystals oriented their *c* axes parallel to the lamellar normal, as previously inferred by Rangarajan *et al.* [1993] for E/EP diblocks crystallizing from homogeneous melts (no constraints; the case of a "soft", or devitrified, matrix will be discussed in the following section). At crystallization temperatures between -10 and $35\,°C$, the *c* axes are aligned with a preferential tilt (between zero and $90°$, dependent on crystallization temperature) with respect to the lamellar normal. Again, the agreement between the E/VCH and EO/S systems when crystallized far below the matrix glass transition temperature is gratifying: the crystals align such that both the chain axis and fast-growth direction lie in the plane of the lamellae.

Crystallization within the matrix

SAXS/WAXS and TEM studies have also been conducted on semicrystalline-glassy block copolymer systems where the crystallizable component constitutes the continuous matrix. In these materials, crystallization occurs around preformed glassy microdomains, and the crystals are expected to be interconnected, as in homopolymer spherulites. While the presence of glassy microdomains does not preclude the formation of spherulites, it can influence the crystal growth habit, leading to intriguing preferential orientation of these matrix crystals. Simultaneous SAXS/WAXS on an E-rich E/VCH block copolymer where the VCH cylinders are glassy during crystallization of the matrix revealed that not only do these amorphous cylinders induce orientation, but that the E crystals actually exhibit a higher mode of preferred orientation than when E crystallizes *inside* cylinders [Loo *et al.* 2000c]. First, as when the crystals are confined inside cylinders, the fast-growth direction (b axis for E) is aligned with the cylinder axis, allowing the unobstructed growth of long crystals parallel to the cylinders. Second, the same crystals are simultaneously oriented in the plane of the cylinder *radii*, as shown in Figure 6.4; their crystal stems (c axis direction) lie parallel to the (10) planes of the hexagonal lattice formed by the cylindrical microdomains, rather than showing simple rotational isotropy of the c axis as observed when crystallization occurs within cylinders [Quiram *et al.* 1998]. Equivalently, since the macrolattice [10] direction is three-fold degenerate, the crystal stems can be viewed as lying parallel to the planes connecting the axes of adjacent cylinders (these planes are in/out of the page in Figure 6.4). Such preferential orientation in the plane of the cylinder radii results from the similarity in length scale between the intercrystal spacing and the intercylinder spacing. The observed orientation is the one that imposes the least constraint on the crystal thickness and spacing, as it orients the c axis (crystal stem direction) parallel to the cylinder center-to-center direction, though there is a corresponding limitation on the lateral extent of the crystals (in the a axis direction). As is the case for E/VCH crystallization within lamellae (Figure 6.3), the crystals orient themselves so as to relieve (as far as possible) the commensurability constraint imposed by the lattice of microdomains. As the cylinders become more widely spaced, and this constraint becomes less severe, the orientation in the plane of the cylinder radii should disappear and an orientation identical to that for crystallization within cylinders is anticipated: pure rotational symmetry about the b axis, which should remain aligned with the cylinder axis. Such an orientation has been observed by Park *et al.* [2000] for crystallization within the matrix of an E/EP/S triblock (S cylinders), where the E block is only 15% of the polymer's mass and where the E and EP blocks are not microphase separated in the melt.

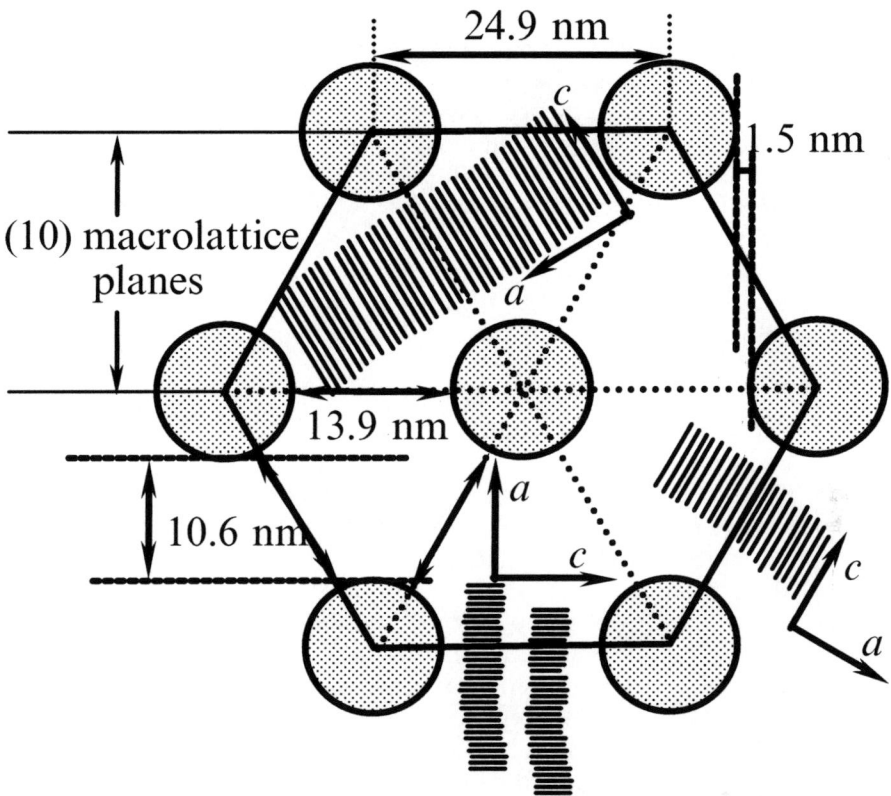

Figure 6.4 Schematic of the two dimensions of crystal orientation when the E matrix crystallizes around glassy cylinders. View is down the cylinder axis; cylinder cross sections appear as circles. E crystals have their *b* axes aligned with the cylinder axes (in/out of page); the *a* and *c* axis directions are shown for the experimentally observed second dimension of orientation (*a* and *c* axes orientations are three-fold symmetric, due to the hexagonal symmetry of the block copolymer macrolattice). The numerical dimensions shown correpond to a particular polyethylene-polystyrene diblock (E/S 3/13, where the S block is not vitreous at the E freezing point), but the mode of orientation holds for E/VCH as well. [Reprinted with permission from Loo *et al.*, *Macromolecules* (2000), **33**, 8361–8366. Copyright (2000) American Chemical Society.]

Crystallization in other morphologies

Crystallization of the E blocks within the channels of the gyroid structure has been demonstrated in an E/VCH diblock [Loo *et al.* 2001]. However, the cubic symmetry of the gyroid mesophase – like that for the body-centered-cubic spherical phase – necessarily yields a global isotopy of the crystals, making this system unsuited for imparting crystal orientation through confinement. Recently, Zhu *et al.* [2002] have examined crystallization of EO within the minority component of a perforated lamellar phase in an EO/S diblock (EO/S

11/17). As in the case of growth within unperforated lamellae [Zhu et al. 2000, 2001], specimens crystallized at moderate undercoolings show a general alignment of the fast-growth direction for EO (the [120] direction) in the plane of the lamellae. However, the hexagonal perforations present obstacles to crystal growth, and at shallow undercoolings, there is a preferred alignment of the EO fast-growth direction with the (10) planes of the macrolattice formed by these hexagonal perforations. (With reference to the schematic in Figure 6.4, if the circles there are taken to represent the cross sections of the S perforations through the EO layer, then the c axes of the EO crystals are rotated by 30° from those schematized in Figure 6.4.) This orientation permits the crystals to grow for an extended distance laterally, by avoiding the S perforations.

6.4.2 SEMICRYSTALLINE-RUBBERY BLOCK COPOLYMERS

As noted above, many early experiments indicated that without a glassy amorphous component, microphase-separated block copolymers undergo enormous structural rearrangement during crystallization, resulting in the formation of lamellar crystallites and even a spherulitic superstructure. The systems studied covered a broad range of chemistries and physical properties of the constituent blocks, including poly(ε-caprolactone)-*b*-polybutadiene, CL/B [Nojima et al. 1992, 1993]; poly(ethylene oxide)-*b*-polyisoprene, EO/I [Floudas et al. 1999]; polyethylene-*b*-poly(ethylethylene), E/EE [Ryan et al. 1995]; polyethylene-*b*-atactic polypropylene, E/aP [Sakurai et al. 1994]; polyethylene-*b*-head-to-head polypropylene, E/hhP [Rangarajan et al. 1995]; and poly(ethylene oxide)-*b*-poly(butylene oxide), EO/BO [Ryan et al. 1997]. With so many examples on record, it seemed plausible that a crystallization-driven solid-state morphology was inevitable in the absence of a vitreous amorphous block.

However, the studies noted above did not cover strongly segregated polymers; indeed, many of the block copolymers examined had thermally accessible order–disorder transitions (ODTs) in the melt. For a given chemistry and volume fraction of the crystallizable block (e.g., spheres of E), segregation strength can be increased either by employing an amorphous block that has a stronger melt incompatibility with the crystallizable block, or by increasing polymer molecular weight, or both. The first direct demonstration [Loo et al. 2000a] of crystal confinement solely through strong segregation employed a sphere-forming block copolymer of polyethylene-*b*-poly(styrene-*r*-ethylene-*r*-butene), E/SEB, where a 70 wt% styrene content in the SEB block provided a strong incompatibility with E in the melt, allowing segregation strengths more than triple that at the ODT to be accessed at reasonable molecular weights. The glass transition temperature of a random SEB terpolymer of this composition is 25 °C, still well below the freezing temperature of the E block so that the matrix remains rubbery during crystallization. SAXS and TEM experiments conducted on a high molecular weight sphere-forming diblock (E/SEB 9/55) revealed that the melt structure can be faithfully preserved into the solid state, even for extremely slow crystallization conditions. A TEM image [Loo et al. 2002] of E/SEB 9/55 is shown in Figure 6.5, where the SEB matrix is stained dark with RuO_4. This specimen was crystallized very slowly: isothermally at 70 °C, where the crystallization half-time is approximately one hour.

Since the sphere diameter is only 25 nm and the crystallization process extends over hours, it seems unlikely that the morphology developed under these conditions reflects any kinetic limitation imposed by hindered diffusion of block copolymer chains; more likely, crystallization confined to spheres is the equilibrium morphology of E/SEB 9/55. Confining polymer crystallization within individual 25-nm spherical microdomains also has profound implications for the nucleation mechanism and growth kinetics, as will be discussed later in this chapter. Diminishing the segregation strength in these E/SEB diblocks by reducing the molecular weight recovers the expected "crystal breakout" behavior [Loo et al. 2002], though the extent of structural rearrangement shows an interesting dependence on segregation strength and crystallization conditions (more extensive breakout at lower segregation strengths and/or slower crystallization rates).

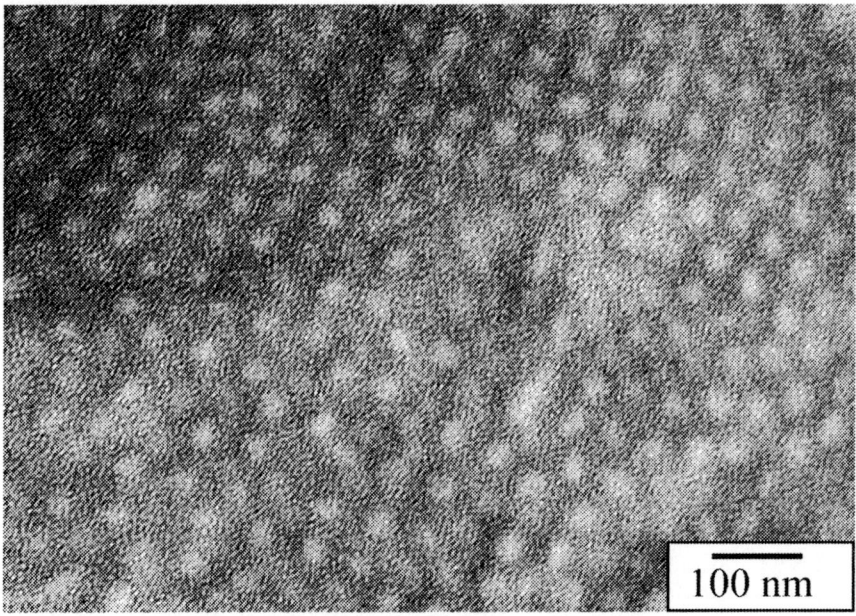

Figure 6.5 TEM image of a microtomed section of E/SEB 9/55, where the SEB matrix is stained dark with RuO_4. Specimen was crystallized isothermally at 70 °C for 3.5 h; the crystallization half-time at 70 °C is approximately one hour. [Reprinted with permission from Loo et al., *Macromolecules* (2002), **35**, 2365–2374. Copyright (2002) American Chemical Society.]

Similarly, a strongly segregated E/SEB diblock containing cylinders of E (E/SEB 17/52) shows confinement of the E crystals within the pre-existing cylinders [Loo et al. 2002], as well as the same orientation of the crystals (E b-axis parallel to cylinder axis) found in the glassy-matrix E/VCH diblocks. Similar results for crystal orientation, implying confinement within cylinders, were found previously by Quiram et al. [1997a, 1998] in high molecular weight polyethylene-*b*-poly(3-methyl-1-butene), E/MB, diblocks forming cylinders of E; the interaction energy density for the E/MB pair is approximately 1/3 that for E/SEB, making E/MB 17/45 much less strongly segregated than E/SEB 17/52. As the segregation strength was further reduced by lowering the E/MB molecular weight from E/MB 17/45, "crystal breakout" was again observed by SAXS [Quiram 1997a] – and as with the sphere-forming E/SEB polymers studied subsequently by Loo et al. [2002], the extent of breakout depended on both segregation strength and crystallization conditions.

Confined crystallization has since been found in sphere- and cylinder-forming semicrystalline-rubbery diblocks of other chemistries as well. Chen et al. [2002] applied SAXS and TEM to confirm the retention of EO spheres formed in the melt by blending an EO/B diblock (EO/B 6/20) with B homopolymer. Xu et al. [2002a, 2002b] recently reported a comprehensive DSC and time-resolved SAXS study of EO/BO diblocks blended with BO homopolymer to vary the composition and thus morphology. They also found that while EO crystallization could readily be confined within spheres, confinement within cylinders was achieved only for the most strongly segregated systems examined [Xu et al. 2002a], and confinement within lamellae was never achieved; in the lamellar case, a substantial step change increase in the lamellar spacing between melt and solid was always observed, indicating extensive rearrangement of the layered EO-BO structure formed in the melt.

Crystallization within lamellae

The spacing increase upon crystallization noted by Xu et al. [2002a] for lamellar EO/BO diblocks is in fact a common feature when crystallization is not effectively confined [Ryan et al. 1995; Rangarajan et al. 1995, 1997; Chen et al. 2001b]. It results from the compromise structure adopted by lamellar semicrystalline block copolymers: an equilibrium degree of chain folding in the crystalline block, coupled with a stretching of the random coil conformation of the amorphous block so as to preserve equal areas per block (crystalline and amorphous) on both sides of the lamellar interface [DiMarzio et al. 1980]. The ability of nonvitreous layers to alter their thickness through the diffusion of polymer chains relieves the commensurability constraint discussed above for semicrystalline/glassy lamellar diblocks; consequently the preferred structure is not that of Figure 6.3, but rather one where both the chain axis and fast-growth direction lie in the plane of the lamellae, as shown schematically in Figure 6.6.

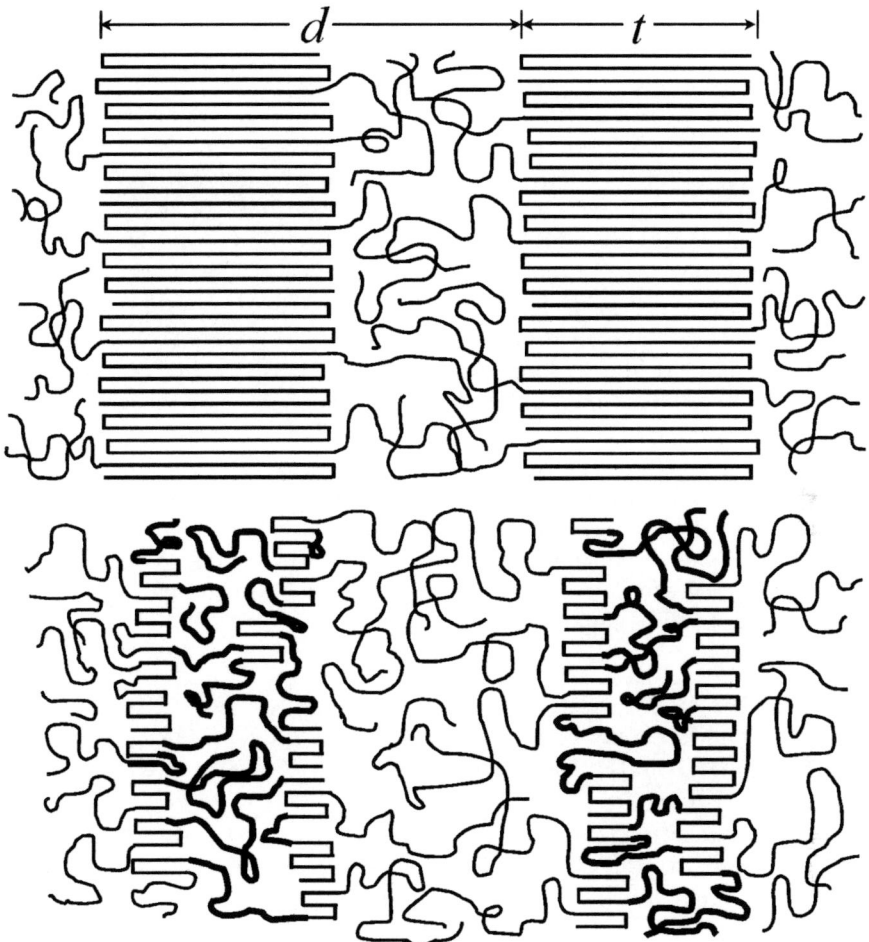

Figure 6.6 Schematic lamellar structure of crystalline-amorphous diblocks crystallizing either from homogeneous melts, or under conditions where alteration of the lamellar period d is facile (rubbery matrix, slow crystallization conditions). Horizontal line segments represent crystal stems formed by the crystalline block, while meandering curves represent the amorphous blocks. Top panel represents the situation for high-crystallinity blocks (e.g., EO), where fully crystalline lamellae of thickness t formed by the crystallizable block alternate with layers of the amorphous block. Bottom panel represents the situation for lower-crystallinity blocks (e.g., hydrogenated polybutadiene, "E"), where crystal thickness is uncorrelated with the thickness of the domain formed by the crystallizable block – a large fraction of which is amorphous, as shown by the heavier meandering chains (amorphous sections of crystallizable block) lying between the thin crystals.

The structure of Figure 6.6 has been definitively confirmed for an EO/B 6/5 diblock by Hong *et al.* [2001a, 2001b, 2001c]. For bulk specimens, TEM coupled with selected-area electron diffraction clearly showed that the material presents a well-ordered lamellar morphology in the solid state, and that the EO crystal stems lie perpendicular to the lamellar EO-B interface. By studying the same polymer as a thin film (1–3 lamellar periods thick) supported on a Si wafer, the increase in lamellar period upon EO crystallization could be directly observed through the change in optical interference color. Opitz *et al.* [2002] investigated a similar system, a diblock copolymer of EO and atactic polybutene (EO/aB 4/4) supported on a Si wafer, where aB is also rubbery at the EO freezing point. By applying X-ray reflectivity, they were able to confirm and precisely quantify the increase in lamellar spacing that occurs upon crystallization.

Even more interestingly, Hong *et al.* [2001a, 2001b] found that when a crystallization front moved through their trilayer EO/B films, it did so in all three layers and with the crystals in orientational registry – but with a time lag between the crystallization of successive layers, suggesting that EO crystals nucleated in one lamella can eventually crystallize the material in adjacent lamellae. Hong *et al.* suggest that such "spreading" of crystals from one layer to another could be facilitated by the presence of pre-existing edge or screw dislocations in the molten film, which would permit a growing EO crystal to branch into different EO lamellae. This idea of "spreading growth", and its facilitation by defects, strongly influences crystallization kinetics as well, as discussed in the following section.

6.5 CRYSTALLIZATION KINETICS AND NUCLEATION MECHANISMS IN THE PRESENCE OF MELT MESOPHASES

Polymer crystallization proceeds by nucleation and growth; for a recent review of homopolymer crystallization, including kinetics and mechanism, see Schultz [2001]. If the polymer is free of contaminants, homogeneous nucleation initiates crystallization. Homogeneous nucleation requires the formation of a critical nucleus; the process is thermally activated and has a large energy barrier. A large undercooling is therefore expected for homogeneous nucleation to occur at a substantial rate. In semicrystalline homopolymers, crystallization is more commonly initiated heterogeneously, by dust particles or other impurities present in the melt. These impurities eliminate the need to form a critical nucleus, so the undercooling associated with crystallization proceeding from heterogenous nucleation is typically much smaller than from homogeneous nucleation.

Crystallization in bulk polymers typically follows a sigmoidal time evolution, described by the Avrami equation:

$$y = 1 - \exp(-kt^n), \tag{6.2}$$

where y is the fraction of the ultimate crystallinity that is achieved at time t; k is the rate constant, and n is the Avrami exponent. In principle, the Avrami exponent provides information about the nucleation mode and crystal growth habit. For example, instantaneous nucleation followed by isothermal three-dimensional growth (as for spherulites) yields $n = 3$, and indeed, most crystallization data for bulk polymers are adequately described by the Avrami equation with $n = 2-4$. While spherulites are a specific case, sigmoidal crystallization kinetics ($n > 1$) are generally expected for *any* process where the growing crystals "spread" with time: where the amount of material deposited, in successive time intervals, onto the structure formed from a single nucleus increases steadily with time, until the crystallizable material is depleted and the crystallization rate falls. (Again considering the particular case of spherulites, successively deposited spherical shells of the same thickness have progressively greater volume, producing the initial autoacceleration in $y(t)$ reflected in Equation (6.2)).

Restricting crystallization on a nanometer length scale necessarily impacts how crystallization is initiated and how it proceeds. Consider first the case of spheres, where the overall number density of impurities (typically of order $10^9 \, \text{cm}^{-3}$) is many orders of magnitude below the number density of microdomains (typically $10^{17} \, \text{cm}^{-3}$, depending on molecular weight). If crystals are indeed confined to individual spheres, then the overwhelming majority of these spheres must nucleate homogeneously. Moreover, since the diameter of individual microdomains is only a few tens of nanometers, crystal growth over the full spatial extent of the microdomain should be essentially instantaneous once nucleated. In this case, the rate of crystallization will simply be proportional to the fraction of microdomains that have not yet nucleated, yielding an Avrami exponent $n = 1$ and nonsigmoidal kinetics (rate decreases continuously with time). This idea was first advanced by Lotz and Kovacs [1969], though the techniques available at that time precluded the precise kinetic study coupled with structural characterization needed to confirm this picture. Recently, isothermal crystallizations tracked by synchrotron-based, time-resolved simultaneous SAXS/WAXS, have provided valuable insight into the nucleation modes and growth habits during the crystallization of semicrystalline block copolymers, as discussed in the following section.

6.5.1 ISOTHERMAL CRYSTALLIZATION WITHIN SPHERES

Loo *et al.* applied time-resolved SAXS/WAXS to study crystallization in a range of E-containing semicrystalline-glassy [2001] and semicrystalline-rubbery [2002] block copolymers, where the E block formed spheres or cylinders; many of these were the same polymers whose solid-state structures (confined crystallization vs. breakout) were reviewed earlier in this chapter. Considering first the sphere-forming diblocks, complete confinement could be achieved either through a glassy matrix (E/VCH 5/22) or through strong interblock segregation (E/SEB 9/55). The progress of crystallization could be tracked easily through the increase in the WAXS peak intensity for the (110) reflection of orthorhombic E, or through the growth of the SAXS feature near $q = 0.5\,\text{nm}^{-1}$, which arises from the formation of crystallites, or through the change in the principal SAXS peak intensity as the electron density difference between spheres and matrix changes when the spheres densify through crystallization. These features can all be seen in the SAXS and WAXS patterns shown as insets in Figure 6.7, which presents the crystallization kinetics for the strongly segregated E/SEB 9/55. The SAXS and WAXS patterns evolve in parallel, and both fit well to first-order kinetics ($n = 1$), with a half-time that is identical to within measurement error. The observation of first-order kinetics in E/SEB 9/55 simultaneously confirms complete confinement of the growing crystals by the isolated microdomains, and a homogeneous nucleation mechanism.

The well-ordered microdomain morphologies that block copolymers present allow an unambiguous and precise calculation of the number density of spherical microdomains, so that the overall crystallization rate can be directly trans-

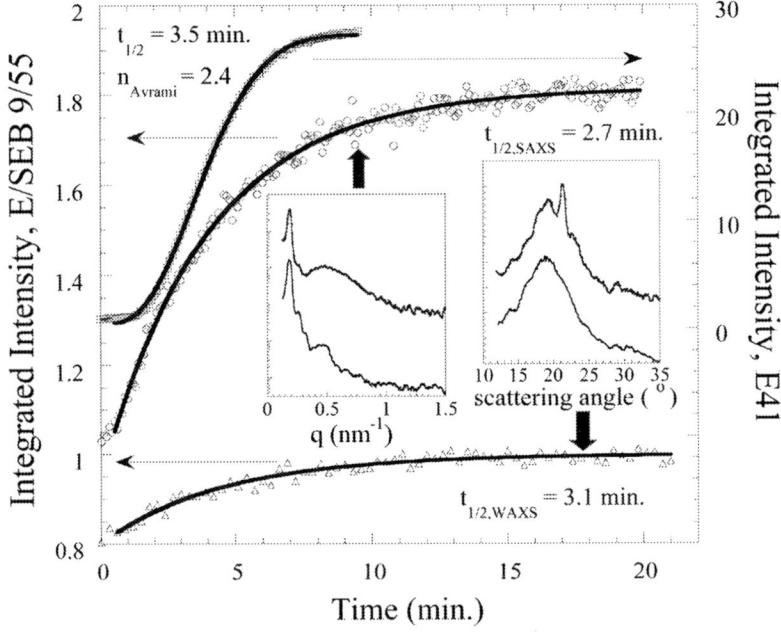

Figure 6.7 Isothermal crystallization (67 °C) of strongly segregated sphere-forming E/SEB 9/55 monitored by time-resolved SAXS/WAXS. Insets show the SAXS (left) and WAXS (right) patterns for E/SEB 9/55 both at the start (top trace in each panel) and end (bottom trace) of crystallization. E (110) peak intensity is shown as the lower data set (△), integrated SAXS intensity near $q = 0.5\,\text{nm}^{-1}$ is shown as the middle data set (○), both read on the left axis. Both are fit to first-order kinetics (solid curves) with half-times of 3 min. For comparison, the crystallization behavior of a hydrogenated polybutadiene homopolymer (E41) is shown as the top data set (□), read on the right axis. The kinetics are qualitatively different (sigmoidal vs. first-order), and to achieve a similar half-time, a much higher crystallization temperature was required for E41 (95 vs. 67 °C). [Reprinted with permission from Loo *et al.* 2000a].

lated into a quantitative value of the homogeneous nucleation rate of the polymer forming the spheres [Loo et al. 2002]. Varying the isothermal crystallization temperature thus provides the temperature dependence of the homogeneous nucleation rate. Figure 6.8 shows the results of such a calculation, including data from the glassy-matrix E/VCH 5/22, the strongly segregated E/SEB 9/55, and classic data of Koutsky et al. [1967] for the homogeneous nucleation rate in linear polyethylene, measured by dispersing the polymer into micrometer-size droplets in a suspending liquid. The slopes of all three data sets are similar, indicating that the nucleation rate increases by approximately a factor of 3 for each additional 1 °C of undercooling. The three data

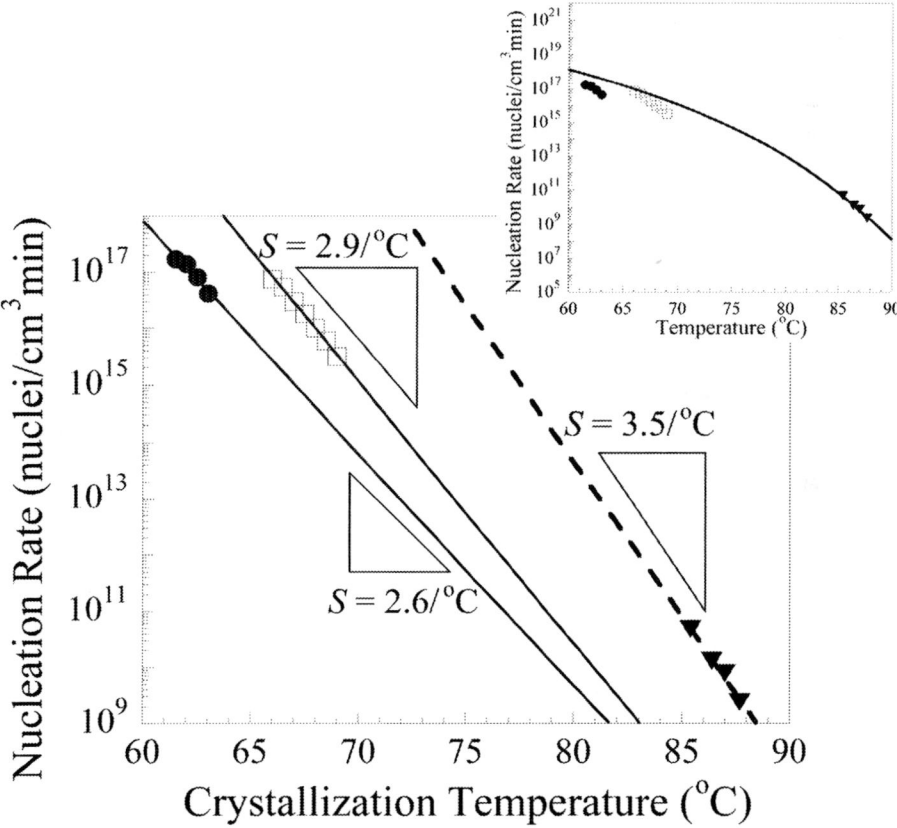

Figure 6.8 Homogeneous nucleation rates for polyethylene extracted from measurements on bulk E/VCH 5/22 (●), bulk E/SEB 9/55 (□), and a suspension of linear polyethylene droplets (▼, data of Koutsky et al. [1967]). Slopes S of log(rate) vs. T_c are indicated for each of the three data sets. Inset shows the same data plotted on an expanded rate scale, with the prediction of classical nucleation theory shown as the solid curve (prediction uses the material parameters given by Koutsky et al. [1967]).

sets do not quite collapse onto a single master curve within experimental error, as shown in the inset. This may reflect modest differences in nucleation rate due to chemical microstructure (linear polyethylene vs. hydrogenated polybutadiene), as well as to the effect that "tethering" the block copolymer chains to the microdomain interfaces might have on the energetics of critical nucleus formation. That said, the agreement between the three sets is remarkable, with the block copolymer data extending the classical rate measurements by over 20 °C in undercooling and nearly seven decades in rate, due to the much smaller volumes of block copolymer microdomains as compared with suspended polymer droplets. This approach to measuring homogeneous nucleation rates should be straightforward to apply to any polymer that can be incorporated into a block copolymer.

Finally, Reiter et al. [2001] recently presented exciting atomic force microscopy (AFM) experiments on thin films (one microdomain thick) of an EO/aB diblock (EO/aB 4/21, where aB is atactic polybutene) supported on a Si wafer substrate. The elasticity differences between amorphous aB, amorphous EO, and crystalline EO permitted the clear resolution and identification of amorphous and crystallized EO spheres within the aB matrix, as shown in Figure 6.9. By imaging the array of spheres after various times of isothermal crystallization and simply counting the crystallized and uncrystallized spheres to determine the fraction remaining uncrystallized, the authors showed – in real space – that each spherical microdomain crystallizes independently, and generally follows first-order kinetics. Some deviations from simple first-order kinetics were observed towards the end of the crystallization process (last 10% of the domains), where the rate slowed substantially; similar deviations from precise first-order kinetics are seen in the isothermal crystallization kinetics of bulk EO/aB 4/21 measured by DSC [Röttele et al. 2003]. These deviations indicate that there is some variation in the nucleation rate across the population of EO spheres, with some spheres having a distinctly lower rate of nucleation than the average, though the origin of this variation remains unclear.

6.5.2 ISOTHERMAL CRYSTALLIZATION WITHIN CYLINDERS

Confined crystallization in cylindrical microdomains is also expected to result in first-order crystallization kinetics, provided the cylinders are not "connected" at their ends through grain boundaries between regions of cylinders with different orientation. Though the length of the cylinders – typically of order 1 micrometer – is much greater than their diameter, the number density of microdomains still vastly exceeds the typical density of homogeneous nuclei, and even a 1-μm growth distance can be covered rapidly at the deep undercooling at which homogeneous nucleation is effective. Both glassy-matrix E/VCH [Loo et al. 2001] and strongly segregated E/SEB diblocks [Loo et al. 2002] forming E cylinders indeed exhibited first-order crystallization kinetics, with

Crystallization Within Block Copolymer Mesophases 235

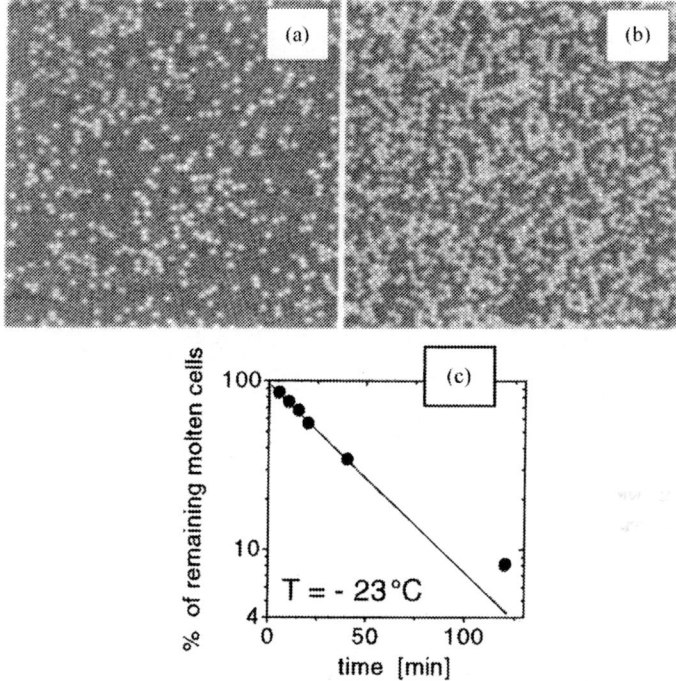

Figure 6.9 AFM phase images of a single-microdomain array of sphere-forming EO/aB 4/21. White circles represent crystallized EO spheres; dark circles are amorphous EO spheres. Panel A shows the array after a 5-min isothermal hold at −23 °C, while panel B shows the array after 15 min at −23 °C. By counting the uncrystallized spheres that remain after various crystallization intervals, the kinetic plot shown in panel C was created; the straight line represents the fit to first-order kinetics (Avrami $n = 1$). [Reprinted with permission from Reiter et al. 2001].

rates approximately a factor of 30 faster than for their sphere-forming counterparts, reflecting the correspondingly larger volume per microdomain (cylinder vs. sphere).

Curiously, however, the same high molecular weight E/MB diblocks (E/MB 17/45 and E/MB 23/63) that appeared to retain the melt morphology (cylinders of E) upon crystallization [Quiram et al. 1997a] showed sigmoidal crystallization kinetics ($n = 1.7–3.4$ [Quiram et al. 1997b]), rather than the expected first-order kinetics. Similar findings are evident in the data of Shiomi et al. [2002] for cylinder-forming EO/B 5/10, where SAXS demonstrated that the hexagonal macrolattice is clearly retained even for relatively slow crystallizations, yet Avrami exponents $n = 2.2–3.1$ were measured. These puzzling results were explained by Loo et al. [2002], after obtaining TEM images of E/MB 17/45 that showed occasional crystals traversing from one cylinder to another. Thus, a single nucleus can crystallize the material initially in several cylinders, and this "spreading" habit – created through infrequent "poke through" events – produces the observed $n > 1$. Loo et al. [2002] termed this regime "templated" crystallization: the overall morphology of hexagonally packed cylinders is retained; the individual cylinders are effective in guiding the growing crystals (fast growth axis aligned with cylinder axis); but the crystallization kinetics are sigmoidal, and the overall crystallization rate is faster – often by orders of magnitude – than for analogous polymers (E/VCH and E/SEB) where E crystallization is wholly confined, and each cylinder must be separately nucleated.

Indeed, this strong dependence of rate on the connectivity of the crystallizable domains also manifests itself in the freezing point (T_f) measured by DSC at a constant rate of cooling. Typically, it is found that T_f decreases in the order lamellae > cylinders > spheres for polymers in the same chemical family, including EO/B [Chen et al. 2001a], E/VCH [Loo et al. 2001], and EO/BO [Xu et al. 2002a]. Frequently, a substantial depression of T_f from its value for the homopolymer, or a lamellar block copolymer, is taken as evidence of confined crystallization. However, this interpretation is likely oversimplified, especially for systems with rubbery matrices [Müller et al. 2002]. For example, lower molecular weight E/SEB sphere-forming diblocks show a depression of T_f by some 25 °C from the value for E homopolymers and E/VCH lamellar diblocks, but also exhibit sigmoidal crystallization kinetics and extensive "breakout" during isothermal crystallization [Loo et al. 2002]. Indeed, the T_f values for lower molecular weight E/SEB sphere-forming diblocks are actually slightly *lower* than for their higher molecular weight analogs, where confined crystallization is observed for all crystallization conditions. In the lower molecular weight diblocks, crystallization is still nucleated homogeneously, so during dynamic cooling, more time is required to nucleate each of the smaller spheres formed at lower molecular weights. But at the slower rates characteristic of isothermal crystallization (vs. dynamic cooling), each crystal can grow to span the material originally contained within several spheres, producing extensive breakout and sigmoidal kinetics. Two points should be drawn from these observations: first, that freezing points measured during dynamic cooling are only loosely related to the state of confinement during isothermal crystallization, and second, that direct structural measurements (e.g., by in-situ SAXS or post-facto TEM) are essential to properly confirm the state of confinement imposed on the crystals.

Based on the results from these E/MB and E/SEB block copolymers forming spheres or cylinders of E, Loo et al. [2002] were able to compile a "classification map" where the normalized interblock segregation strength (the ratio of interblock segregation at the crystallization temperature to that at its order–disorder transition temperature) is plotted against the volume fraction of the crystallizable component, as shown in Figure 6.10. In the case of sphere-formers, crystallization is effectively confined within the microdomains when the normalized interblock segregation strength is high. Below the threshold segregation strength of 3 (normalized), dramatic structural rearrangement is observed on

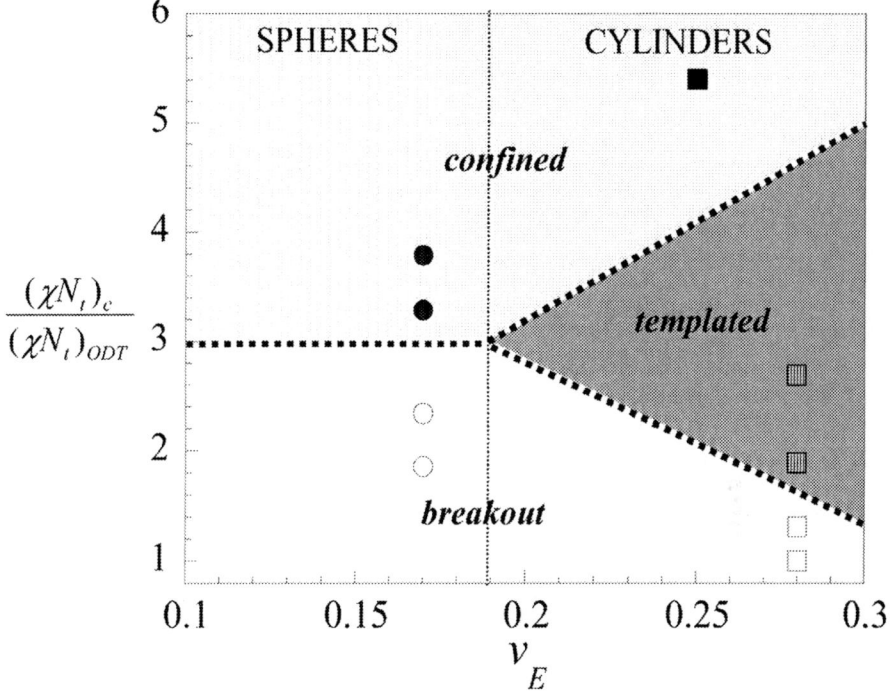

Figure 6.10 Classification map of crystallization modes in E-based semicrystalline diblocks with rubbery matrices. Segregation strength at the crystallization temperature, normalized to that at the ODT, is indicated on the y axis. Volume fraction of ethylene block (v_E) in each diblock is shown on the x axis; polymers with $v_E < 0.19$ form spheres of E (circles), those with $v_E > 0.19$ form cylinders of E (squares). Open symbols denote complete destruction of the melt mesophase upon crystallization ("breakout"); filled symbols denote complete confinement, as evidenced through first-order kinetics (Avrami $n = 1$); symbols with a vertical hatch denote templated crystallization, where SAXS indicates a general retention of the cylindrical melt morphology but sigmoidal crystallization kinetics ($n > 1$) indicate a "spreading" growth habit. [Reprinted with permission from Loo et al., Macromolecules (2002), **35**, 2365–2374. Copyright (2002) American Chemical Society.]

crystallization. For cylinder-formers, structural rearrangement is again observed at weak interblock segregation (< 1.5, normalized) and confined crystallization is again observed at strong interblock segregation (> 4). However, "templated" crystallization occurs between these two limits; here, crystallization generally occurs within the microdomains determined by microphase separation, but crystallization produces local distortions to the regular microdomains established in the melt, extending even to occasional interconnections between cylinders that permit a single nucleus to crystallize the material originally in many cylinders. Xu et al. [2002b] also found that the EO/BO system hewed closely to this same classification map, including the positions of the dividing lines (especially the critical value of 3 for the normalized segregation strength needed to confine crystallization within spheres). Given the substantial chemical and physical differences between the EO/BO, E/SEB, and E/MB systems, the classification map in Figure 6.10 should be a useful guide for determining the conditions needed to confine crystallization in any semicrystalline-rubbery block copolymer.

6.5.3 ISOTHERMAL CRYSTALLIZATION IN INTERCONNECTED MICRODOMAINS

Both the "templated" crystallization in cylinders described by Loo et al. [2002], and the orientationally registered crystals in thin films of the lamellar EO/B 6/5 described by Hong et al. [2001a, 2001b] point out the strong impact that interconnections between microdomains can have on the crystallization process. When growing crystals can percolate through all the E-rich microdomains present in the melt, then the crystallization kinetics are not expected to differ qualitatively from those of homopolymers – even when the microphase-separated morphology established in the melt is preserved into the solid. For example, Loo et al. [2001] investigated crystallization within the gyroid channels in a semicrystalline-glassy diblock (E/VCH 8/13), and found unremarkable sigmoidal kinetics ($n = 1.7$) despite faithful preservation of the gyroid structure by the glassy matrix.

As noted by Hong et al. [2001a, 2001b], lamellae present a particularly interesting case. Like cylinders, the lamellae in an idealized block copolymer grain are totally unconnected. However, the practical difficulty in isolating lamellae from each other is even greater than for cylinders, because of the larger volume per lamella (vs. cylinder); an extremely low defect density would be required to observe isolated crystallization within lamellae. Consequently, lamellar block copolymers are generally reported to exhibit sigmoidal crystallization kinetics, even when the other block is vitreous; for example, Hamley et al. [1998b] reported an Avrami $n = 3$ for E/VCH 8/7. Still, this result might be expected to depend strongly on the defect density in the mesophase structure, which is difficult to characterize independently. By contrast, Loo et al. [2001] observed a two-step crystallization process in a lamellar semicrystalline-glassy diblock (E/VCH 12/8) similar to the E/VCH 8/7 studied by Hamley et al. [1998b]; the (minority) higher-temperature crystallization process followed sigmoidal crystallization kinetics, while the (majority) lower-temperature process followed first-order kinetics. This two-step process reveals the presence of two distinct populations of E lamellae: a minority interconnected population (perhaps through grain boundaries or screw dislocations), and a majority population of isolated E lamellae, each of which must be independently and homogeneously nucleated.

These results demonstrate the exquisite sensitivity of crystallization kinetics to microdomain topology, particularly interconnection of the domains formed by the crystallizable component. Compared with conventional methods such as mechanical testing or gas-transport measurements, the crystallization kinetics – both Avrami exponent and undercooling – can reveal even low levels of microdomain-connecting defects, at the level of the percolation threshold (two connecting defects per microdomain).

6.6 CONCLUDING REMARKS

The presence of a melt mesophase can profoundly influence both the solid-state structure and the crystallization kinetics in semicrystalline block copolymers. The impact is especially great when the crystallizable component is dispersed into discrete domains (spheres or cylinders), and when the material is strongly segregated, or when the amorphous matrix is vitreous at the crystallizable component's freezing point. Under these conditions, homogeneous nucleation initiates crystallization, and measuring the overall crystallization rate of the block copolymer permits the extraction of the crystallizable component's homogeneous nucleation rate. By contrast, block copolymers crystallizing from homogeneous or weakly segregated melts, where neither component is vitreous, show structures and crystallization kinetics qualitatively similar to those of semicrystalline homopolymers. The degree of confinement can be described through a classification map, where each polymer is categorized by its melt morphology (spheres, cylinders, lamellae) and by its segregation strength at the point of crystallization, normalized to that at its order–disorder transition.

When the crystallizable component crystallizes within or around anisotropic microdomains (cylinders or lamellae), a strong orientation of the crystal stems typically results. A range of orientations can be achieved, depending on the nature of the microdomains (cylinders vs. lamellae), and for lamellae, the ability (or lack thereof) for the lamellar microdomain to adjust its thickness to accommodate the crystals. Tuning the volume fraction of crystallizable material and the segregation strength permits one to reliably direct crystal orientation in polymers having a range of block chemistries.

The ability to confine crystallization to spheres or cylinders through strong interblock segregation opens the possibility of preparing new thermoplastic elastomers from crystalline/amorphous/crystalline triblocks, which would possess excellent solvent resistance imparted by the crystalline endblocks coupled with optical clarity, since spherulite formation is suppressed. Also, while most work to date has focused on materials having one crystalline and one amorphous block, even richer morphological possibilities would be presented by materials having two crystallizable blocks, since the solid-state structure could potentially be governed by the melt mesophase (through confinement or templating), or via "breakout" crystallization of one or the other block. Thus, while our understanding of these complex materials has expanded greatly over the past decade, we look forward to future advances with equal anticipation.

ACKNOWLEDGEMENTS

The authors wish to thank the National Science Foundation, Polymers Program, for generous support of their work on crystallizable block copolymers through Grants DMR97-11436 and DMR02-20236. Y. L. L. would also like to thank the Camille and Henry Dreyfus New Faculty Award Program for funding. The authors have also benefited from an extensive and long-standing collaboration with Professor Anthony J. Ryan, University of Sheffield, especially in investigating the crystallization kinetics in these complex systems.

REFERENCES

Bates F. S., Fredrickson G. H., Hucul D., and Hahn S. F. (2001) PCHE-based pentablock copolymers: Evolution of a new plastic. *AIChE J.* **47**:762–765.

Chen H.-L., Hsiao S.-C., Lin T.-L., Yamauchi K., Hasegawa H., and Hashimoto T. (2001a) Microdomain-tailored crystallization kinetics of block copolymers. *Macromolecules* **34**:671–674.

Chen H.-L., Wu J.-C., Lin T.-L., and Lin J. S. (2001b) Crystallization kinetics in microphase-separated poly(ethylene oxide)-block-poly(1,4-butadiene). *Macromolecules* **34**:6936–6944.

Chen H.-L., Li H.-C., Huang Y.-Y., and Lin J. S. (2002) Crystallization-induced deformation of spherical microdomains in block copolymer blends consisting of a soft amorphous phase. *Macromolecules* **35**:2417–2422.

Cohen R. E., Cheng P.-L., Douzinas K., Kofinas P., and Berney C. V. (1990) Path-dependent morphologies of a diblock copolymer of polystyrene/hydrogenated polybutadiene. *Macromolecules* **23**:324–327.

Cohen R. E., Bellare A., and Drzewinski M. A. (1994) Spatial organization of polymer chains in a crystallizable diblock copolymer of polyethylene and polystyrene. *Macromolecules* **27**:2321–2323.

DiMarzio E. A., Guttman C. M., and Hoffman J. D. (1980) Calculation of lamellar thickness in a diblock copolymer, one of whose components is crystalline. *Macromolecules* **13**:1194–1198.

Douzinas K. C. and Cohen R. E. (1992) Chain folding in EBEE semicrystalline diblock copolymers. *Macromolecules* **25**:5030–5035.

Floudas G. and Tsitsilianis C. (1997) Crystallization kinetics of poly(ethylene oxide) in poly(ethylene oxide)-polystyrene-poly(ethylene oxide) triblock copolymers. *Macromolecules* **30**:4381–4390.

Floudas G., Ulrich R., and Wiesner U. (1999) Microphase separation in poly(isoprene-b-ethylene oxide) diblock copolymer melts. I. Phase state and kinetics of the order-to-order transitions. *J. Chem. Phys.* **110**:652–663.

Hamley I. W., Fairclough J. P. A., Terrill N. J., Ryan A. J., Lipic P. M., Bates F. S., and Towns-Andrews E. (1996) Crystallization in oriented semicrystalline diblock copolymers. *Macromolecules* **29**:8835–8843.

Hamley I. W. (1998a) The Physics of Block Copolymers. (Oxford: Oxford University Press).

Hamley I. W., Fairclough J. P. A., Bates F. S., and Ryan A. J. (1998b) Crystallization thermodynamics and kinetics in semicrystalline diblock copolymers. *Polymer* **39**:1429–1437.

Hamley I. W. (1999) Crystallization in block copolymers. *Adv. Polym. Sci.* **148**:114–137.

Hay J. N., and McCabe J. F. (1978) Block copolymers of ethylene. I. Butadiene. *J. Polym. Sci.: Polym. Chem. Ed.* **16**:2983–2900.
Hillmyer M. A., Bates F. S., Almdal K., Mortensen K., Ryan A. J., and Fairclough J. P. A. (1996) Complex phase behavior in solvent-free nonionic surfactants. *Science* **271**:976–978.
Hong S., MacKnight W. J., Russell T. P., and Gido S. P. (2001a) Orientationally registered crystals in thin film crystalline/amorphous block copolymers. *Macromolecules* **34**:2398–2399.
Hong S., MacKnight W. J., Russell T. P., and Gido S. P. (2001b) Structural evolution of multilayered, crystalline-amorphous diblock copolymer thin films. *Macromolecules* **34**:2876–2883.
Hong S., Yang L., MacKnight W. J., and Gido S. P. (2001c) Morphology of a crystalline/amorphous diblock copolymer: Poly((ethylene oxide)-*b*-butadiene). *Macromolecules* **34**:7009–7016.
Hosoda S., Nomura H., Gotoh Y., and Kihara H. (1990) Degree of branch inclusion into the lamellar crystal for various ethylene/α-olefin copolymers. *Polymer* **31**:1999–2005.
Howard P. R., and Crist B. (1989) Unit-cell dimensions in model ethylene butene-1 copolymers. *J. Polym. Sci. B: Polym. Phys.* **27**:2269–2282.
Huang P., Zhu L., Cheng S. Z. D., Ge Q., Quirk R. P., Thomas E. L., Lotz B., Hsiao B. S., Liu L., and Yeh F. (2001) Crystal orientation changes in two-dimensionally confined nanocylinders in a poly(ethylene oxide)-*b*-polystyrene/polystyrene blend. *Macromolecules* **34**:6649–6657.
Koutsky J. A., Walton A. G., and Baer E. (1967) Nucleation of polymer droplets. *J. Appl. Phys.* **38**:1832–1839.
Krigas T. M., Carella J. M., Struglinski M. J., Crist B., Graessley W. W., and Schilling F. C. (1985) Model copolymers of ethylene with butene-1 made by hydrogenation of polybutadiene: Chemical composition and selected physical properties. *J. Polym. Sci. B: Polym. Phys.* **25**:509–520.
Loo Y.-L., Register R. A., and Ryan A. J. (2000a) Polymer crystallization in 25-nm spheres. *Phys. Rev. Lett.* **84**:4120–4123.
Loo Y.-L., Register R. A., and Adamson D. H. (2000b) Direct imaging of polyethylene crystallites within block copolymer microdomains. *J. Polym. Sci. B: Polym. Phys.* **38**:2564–2570.
Loo Y.-L., Register R. A., and Adamson D. H. (2000c) Polyethylene crystal orientation induced by block copolymer cylinders. *Macromolecules* **33**:8361–8366.
Loo Y.-L., Register R. A., Ryan A. J., and Dee G. T. (2001) Polymer crystallization confined in one, two, or three dimensions. *Macromolecules* **34**:8968–8977.
Loo Y.-L., Register R. A., and Ryan A. J. (2002) Modes of crystallization in block copolymer microdomains: Breakout, templated, and confined. *Macromolecules* **35**:2365–2374.
Lotz B., and Kovacs A. J. (1966) Propriétés des copolymères biséquencés polyoxyéthylène-polystyrène. I. Préparation, composition, et étude microscopique des monocristaux. *Koll Z.u.Z. Polym.* **209**:97–114.
Lotz B., and Kovacs A. J. (1969) Phase transitions in block copolymers of polystyrene and polyethylene oxide. *ACS Polym. Prepr.* **10(2)**:820–825.
Morton M. (1983) Anionic Polymerization: Principles and Practice (New York: Academic Press).
Müller A. J., Balsamo V., Arnal M. L., Jakob T., Schmalz H., and Abetz V. (2002) Homogeneous nucleation and fractionated crystallization in block copolymers. *Macromolecules* **35**:3048–3058.

Nojima S., Kato K., Yamamoto S., and Ashida T. (1992) Crystallization of block copolymers. 1. Small-angle X-ray scattering study of an ε-caprolactone-butadiene diblock copolymer. *Macromolecules* **25**:2237–2242.
Nojima S., Nakano H., and Ashida T. (1993) Crystallization behaviour of a microphase-separated diblock copolymer. *Polymer* **34**:4168–4170.
Opitz R., Lambreva D. M., and de Jeu W. H. (2002) Confined crystallization of ethylene-oxide butadiene diblock copolymers in lamellar films. *Macromolecules* **35**:6930–6936.
Park C., De Rosa C., Fetters L. J., and Thomas E. L. (2000) Influence of an oriented glassy cylindrical microdomain structure on the morphology of crystallizing lamellae in a semicrystalline block terpolymer. *Macromolecules* **33**:7931–7938.
Quiram D. J., Register R. A., and Marchand G. R. (1997a) Crystallization of asymmetric diblock copolymers from microphase-separated melts. *Macromolecules* **30**:4551–4558.
Quiram D. J., Register R. A., Marchand G. R., and Ryan A. J. (1997b) Dynamics of structure formation and crystallization in asymmetric diblock copolymers. *Macromolecules* **30**:8338–8343.
Quiram D. J., Register R. A., Marchand G. R., and Adamson D. H. (1998) Chain orientation in block copolymers exhibiting cylindrically confined crystallization. *Macromolecules* **31**:4891–4898.
Rangarajan P., Register R. A., and Fetters L. J. (1993). Morphology of semicrystalline block copolymers of ethylene-(ethylene-*alt*-propylene). *Macromolecules* **26**:4640–4645.
Rangarajan P., Register R. A., Fetters L. J., Bras W., Naylor S., and Ryan A. J. (1995) Crystallization of a weakly segregated polyolefin diblock copolymer. *Macromolecules* **28**:4932–4938.
Rangarajan P., Haisch C. F., Register R. A., Adamson D. H., and Fetters L. J. (1997) Influence of semicrystalline homopolymer addition on the morphology of semicrystalline diblock copolymers. *Macromolecules* **30**:494–502.
Reiter G., Castelein G., Sommer J.-U., Röttele A., and Thurn-Albrecht T. (2001) Direct visualization of random crystallization and melting in arrays of nanometer-size polymer crystals. *Phys. Rev. Lett.* **87**:226101.
Röttele A., Thurn-Albrecht T., Sommer J.-U., and Reiter G. (2003) Thermodynamics of formation, reorganization, and melting of confined nanometer-sized polymer crystals. *Macromolecules* **36**:1257–1260.
Ryan A. J., Hamley I. W., Bras W., and Bates F. S. (1995) Structure development in semicrystalline diblock copolymers crystallizing from the ordered melt. *Macromolecules* **28**:3860–3868.
Ryan A. J., Fairclough J. P. A., Hamley I. W., Mai S.-M., and Booth C. (1997) Chain folding in crystallizable block copolymers. *Macromolecules* **30**:1723–1727.
Sakurai K., MacKnight W. J., Lohse D. J., Schulz D. N., and Sissano J. A. (1994) Blends of amorphous-crystalline block copolymers with amorphous homopolymers. 2. Synthesis and characterization of poly(ethylene-propylene) diblock copolymer and crystallization kinetics for the blend with atactic polypropylene. *Macromolecules* **27**:4941–4951.
Schultz J. M. (2001). Polymer crystallization: The Development of Crystalline Order in Thermoplastic Polymers (Oxford: Oxford University Press).
Shiomi T., Takeshita H., Kawaguchi H., Nagai M., Takenaka K., and Miya M. (2002) Crystallization and structure formation of block copolymers containing a rubbery amorphous component. *Macromolecules* **35**:8056–8065.
Trzaska S. T., Lee L.-B. W., and Register R. A. (2000) Synthesis of narrow-distribution "perfect" polyethylene and its block copolymers by polymerization of cyclopentene. *Macromolecules* **33**:9215–9221.

Weimann P. A., Hajduk D. A., Chu C., Chaffin K. A., Brodil J. C., and Bates F. S. (1999) Crystallization of tethered polyethylene in confined geometries. *J. Polym. Sci. B: Polym. Phys.* **37**:2053–2068.

Wu Z., and Grubbs R. H. (1994) Synthesis of narrow dispersed linear polyethylene and block copolymers from polycyclobutene. *Macromolecules* **27**:6700–6703.

Xu J.-T., Turner S. C., Fairclough J. P. A., Mai S.-M., Ryan A. J., Chaibundit C., and Booth C. (2002a) Morphological confinement on crystallization in blends of poly(oxyethylene-*block*-oxybutylene) and poly(oxybutylene). *Macromolecules* **35**:3614–3621.

Xu J.-T., Fairclough J. P. A., Mai S.-M., Ryan A. J., and Chaibundit C. (2002b) Isothermal crystallization kinetics and melting behavior of poly(oxyethylene)-*b*-poly(oxybutylene)/poly(oxybutylene) blends. *Macromolecules* **35**:6937–6945.

Young R. N., Quirk R. P., and Fetters L. J. (1984) Anionic polymerizations of non-polar monomers involving lithium. *Adv. Polym. Sci.* **56**:1–90.

Zhu L., Cheng S. Z. D., Calhoun B., Ge Q., Quirk R. P., Thomas E. L., Hsiao B. S., Yeh F., and Lotz B. (2000) Crystallization temperature-dependent crystal orientations within nanoscale confined lamellae of a self-assembled crystalline-amorphous diblock copolymer. *J. Am. Chem. Soc.* **122**:5957–5967.

Zhu L., Calhoun B. H., Ge Q., Quirk R. P., Cheng S. Z. D., Thomas E. L., Hsiao B. S., Yeh F., Liu L., and Lotz B. (2001) Initial-stage growth controlled crystal orientations in nanoconfined lamellae of a self-assembled crystalline-amorphous diblock copolymer. *Macromolecules* **34**:1244–1251.

Zhu L., Huang P., Chen W. Y., Ge Q., Quirk R. P., Cheng S. Z. D., Thomas E. L., Lotz B., Hsiao B. S., Yeh F., and Liu L. (2002) Nanotailored crystalline morphology in hexagonally perforated layers of a self-assembled PS-*b*-PEO diblock copolymer. *Macromolecules* **35**:3553–3562.

7 Dynamical Microphase Modelling with Mesodyn

HANS FRAAIJE, AGUR SEVINK, ANDREI ZVELINDOVSKY
Soft Condensed Matter Group,
Leiden Institute of Chemistry, Leiden Unversity,
PO Box 9502, 2300 RA Leiden,
The Netherlands

7.1 INTRODUCTION

Mesodyn provides a general framework to calculate the dynamics of mesoscale pattern formation in a variety of block polymer mixtures and solutions, in the framework of extended Flory–Huggins mean-field theory. Mesodyn was developed originally at Akzo Nobel in the early 1990s, in an attempt to solve a stability problem in waterborne coatings. Since then, it has been greatly expanded to cover many different systems and external conditions, following two large industrial European research projects CAESAR and Mesodyn, led by BASF.

The stability problem in waterborne coatings is typical for polymeric systems: when homopolymers and solvents or other polymers are mixed the usual situation is a macroscopic phase separation into a dilute and concentrated polymer solution. Such macrophase separation is almost always a nuisance, and ways are sought to prevent it from happening. On the other hand, when one uses (block) copolymers, instead of homopolymers, the system may phase separate internally – this is *micro*phase separation – and form a mesoscale pattern, with typical length scale 1–1000 nm. Morphologies formed include: a distribution of micelles, lamellae, or cylinders, or more exotic structures such as vesicles and bicontinuous phases.

Almost every day, academic and industrial colloid and polymer scientists find examples of new block copolymer morphologies, and investigate possible new applications. The applications vary just as much as the systems: from traditional high-impact polymer materials, to novel high-performance elastomers, to advanced drug-delivery polymer capsules and biochips, artificial skin and smart gels, contact lenses and electro-optical polymer displays.

Without exception, in all practical applications of block copolymers and block copolymer solutions, the mesoscale pattern determines the performance. And also, in almost all of these systems the dynamical pathway of formation is of crucial importance, since the systems easily become trapped in local free-energy minima. The dynamics of the patterns is very sluggish: the typical time scale for formation, or change in performance upon shift in conditions, is at least of the order of seconds; if not much larger. The large length and time scales are typical for problems in colloid and polymer science, but they are very large compared to typical motions on molecular scales. Molecular simulations with atomic force fields can be applied to those cases where one has a good idea of the morphologies, and is interested in a submesoscale molecular structure, embedded in the mesoscale scaffold; but otherwise the simulation method must be adapted to reach the mesoscale. This is where Mesodyn comes in. It is optimized to do just one thing: the dynamic calculation of the polymer morphology. It has guaranteed proper thermodynamic behavior (in the mean field Flory–Huggins framework), using realistic and readily available parameters such as Flory–Huggins χ parameters, monomer charge, polymer-chain architectures and mixture composition.

The language of Mesodyn is that of traditional colloid and polymer theory, and differs in some important respects from the typical framework of molecular simulations. One now has free-energy models, rather than molecular Hamiltonians; coarse-grained interaction parameters rather than force-field parameters, concentration variables rather than atomic-position variables, and some invented hydrodynamic or diffusion scheme for updating dynamical variables in time. Mesodyn merges two classical theories: the Flory–Huggins theory for polymer phase diagrams, and the Ginzburg–Landau model for dynamic pattern formation.

In the present introduction to Mesodyn we assume the reader has had some exposure to statistical thermodynamics and Flory–Huggins theory, but otherwise we do not suppose familiarity with the typical functional mathematical language of colloid and polymer physics. The introduction chapter to this book contains an extensive list of references to morphologies in block copolymer systems, and mean-field calculations, as in Hamley's book [1]. Here we focus on the work done in our own group [2,3]. General introductory books are those of de Gennes [4] and Doi and Edwards [5]. An excellent recent review paper dealing with dynamical field models is available [6].

In the theory section we summarize the continuum Flory–Huggins theory for inhomogeneous solutions and the dynamical equations. In the application section we present a few examples of morphology calculations in polymer surfactant solutions and block copolymer thin films.

7.2 THEORY

7.2.1 PRINCIPLE

The classic Flory–Huggins theory is derived based on a lattice model for polymer configurations [7–9]. It is a so-called mean-field model in which all thermodynamic properties are calculated from the behavior of a single polymer chain in the average interaction field generated by its neighbors. The mean-field approach is restricted in its application to relatively flexible polymers in concentrated solution and melts, but nevertheless has been applied with tremendous success to a wide variety of phase-separation problems.

It is illustrative to recall when and why a single-chain mean-field model gives a realistic picture of a polymer system. In a dense homogeneous solution of flexible polymer, each chain will sample a number of configurations in a certain volume much larger than the atomic volume of all monomers of the same chain combined. Thus, many chains penetrate the volume element of the coil. When the number of chains in the volume element is very large, the monomer–monomer interactions are uncorrelated from the chain–chain interactions, and the chain generating the coil behaves as if it is embedded in the average field from the other chains.

In an inhomogeneous polymer system, the same argument applies. For example, imagine a triblock copolymer tethered in space by two adjoining micelles in a concentrated solution (Figure 7.1).

In one illustrated case, the polymer surfactant is a "reverse" surfactant, with the hydrophobic blocks at the two ends of the polymer. The two hydrophobic blocks stick in two different micellar cores, and the hydrophilic middle block binds the two micelles together. Such a system would be a good example of a reversible polymer surfactant gel [10]. It will probably be very viscous and on a short time scale even strongly elastic. Such gels find many applications in personal and health care products as slow-release agents, since the micellar cores may act as reservoirs for small organic drug molecules. Notice that the "regular" surfactant with sequence hydrophilic-hydrophobic-hydrophilic

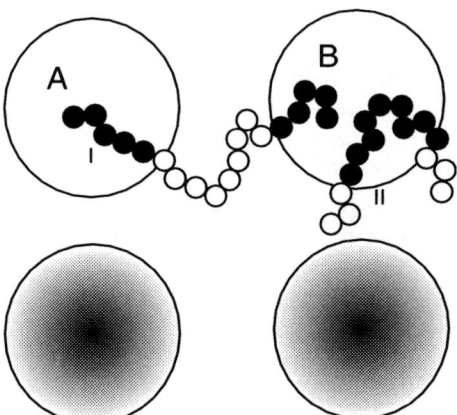

Figure 7.1 Top: Micelles of triblock polymer surfactant, I reverse, II regular. Hydrophobic (●) and hydrophylic (○) beads indicated. Bottom: concentration field of hydrophobic blocks.

(model for Pluronic L64, $E_3P_9E_3$, see application section), cannot physically connect two micelles, and thus such a gel will be much less viscous or elastic – or will not behave like a reversible gel at all.

In the illustrated micellar system, each polymer molecule is obviously constrained in sampling the conformations compared to an ideal homogeneous system, because of the tethering in space. If the constraints are very strong, only a few conformations will remain, and the mean-field picture will break down. But, nevertheless, in almost all of these applications the interactions are weak, the gel is soft and fragile, enough of the coil volume remains open for other chains and solvent, and the mean field will still be a very good approximation.

It is not trivial to decide from a given detailed molecular architecture if and when a single chain mean field is reasonable or not, only rough guidelines can be given: (1) the concentration of the polymer should be above the overlap concentration, (2) the polymer must be flexible, so that the coil volume does not deviate too much from that of a random coil and (3) the coil must be large enough, otherwise too few of the neighboring chains can penetrate the coil volume element. For example, a dilute solution of short and stiff oligo-carbohydrate molecule with 10 monomers falls (very far) outside the range of applicability of the mean-field model. A concentrated solution of low molecular weight surfactants or lipids will also be outside the range of applicability (although less than the previous example), because the chains are too short. A concentrated solution of flexible polymer surfactant, a porous layer of weakly charged polyelectrolyte, or a melt of copolymer will do fine.

7.2.2 FIELD MODEL

The thermodynamic model inherent in Mesodyn is a 3D continuum extension of the Flory–Huggins model. This implies that, if proper limits are taken, the classical mean-field behavior is exactly reproduced. But the details of the model are slightly different. We do not impose the lattice restriction, and the system is not necessarily in equilibrium. Consider again the system in Figure 7.1. We imagine that we follow the micelles during a certain period of time by a suitable microscope. The microscope has high enough spatial resolution, so we can see the individual micelles as a concentration field image (like the grayscale micelle images at the bottom in Figure 7.1.). The collective dynamics of the micelles will be extremely sluggish with a typical collective correlation time of at least a few seconds (depending of course on molecular compositions), much larger than the internal chain relaxation time τ_C. The key assumption in the Mesodyn approach is that on the coarse-grained time scale with time increment $\gg \tau_C$, all chains sample all possible conformations with proper statistical Boltzmann weight; that is the chains are in *local* equilibrium. Effectively the micelles form a slowly changing external potential U such that, given the instantaneous distribution of the micelles in space, the free energy is minimal. From the

Boltzmann weight one can calculate all the interesting parameters: the entropy, correlation functions and so on, provided of course one knows U. But the potential can be calculated by iteration. From the Boltzmann weights one can calculate the average polymer concentration, or monomer concentration profiles. In turn, since the micelles themselves are composed of polymer, the micellar distribution is also determined from the same Boltzmann distribution by proper addition of weights. Thus we have a so-called consistent field method: the spatial distribution of the micelles generates $U(\mathbf{r})$, which generates the Boltzmann weight, which generates the monomer distribution, which generates the micellar distribution, etc. When the set of weights is consistent, that is, when the generated concentration fields exactly match those of the micellar pattern, one can calculate the partition function ϕ and related properties through statistical thermodynamics. For example, the concentration field and external potential are related through the derivative of the partition function

$$\rho_I(\mathbf{r}) = -nkT \frac{\delta \ln \phi}{\delta U_I(\mathbf{r})}, \qquad (7.1)$$

where n is the number of chains, and I is the index for the bead type. In the Boltzmann weight calculation we neglect the interchain correlations, since we assume that the mean field governs the intermolecular energetic interactions. The net free energy is the sum of the ideal free energy for the collection of all types i of single chains, and the mean-field contribution:

$$F = -kT \sum_i \ln \frac{\phi_i^{ni}}{n_i!} - \sum_I \int_V \mathrm{d}\mathbf{r}\, U_I \rho_I + F_{\mathrm{MF}}, \qquad (7.2)$$

where i counts the different types of polymer chains, and I the different bead types, $I \geq i$. e.g. in a mixture of an block copolymer AB and solvent C, $i = AB, C$ and $I = A, B, C$.

The system is not necessarily in *global* equilibrium. The Boltzmann weight calculation only generates the proper potential, conjugate to the concentration field, and therefore only takes into account the entropy loss due to the confinement in the micelles. It may be that energy can be gained, for example by turning the micelles into cylinders, or perhaps the energy is lower when the intermicellar distance is made smaller. Global equilibrium is reached only in a stationary state, such that any infinitesimal change in the concentration field results in exact balance of energy and entropy, gain and loss. In a mathematical sense, there are infinitely many small changes $\delta \rho_I(\mathbf{r})$ in concentration field possible. Some shifts will correspond to a shape change of the micelles; others will lead to a reduction in intermicellar distance. The net free-energy change is:

$$\delta F = \sum_I F[\rho_I + \delta \rho_I] - F[\rho_I] = \sum_I \int_V \mathrm{d}\mathbf{r}\, \frac{\delta F}{\delta \rho_I} \delta \rho_I \qquad (7.3)$$

$$\frac{\delta F}{\delta \rho_I} = -U_I + \frac{\delta F_{MF}}{\delta \rho_I} \tag{7.4}$$

$$\frac{\delta F}{\delta \rho_I} = 0 \quad \forall I: \text{ global equilibrium} \tag{7.5}$$

$$\frac{\delta^2 F}{\delta \rho_I(\mathbf{r}) \delta \rho_J(\mathbf{r}')} > 0 \quad \forall I, J: \text{ stable minimum.} \tag{7.6}$$

The intrinsic chemical potential is denoted by the functional derivative $\delta F/\delta \rho_I(\mathbf{r})$. It is easy to see that any shift $\delta \rho_I(\mathbf{r})$ with the same sign as $\delta F/\delta \rho_I$ (component wise), leads to an increase in free energy, and thus cannot be the result of an internal spontaneous process. On the other hand, when the shift is the result of a flux, in accordance with the laws of linear nonequilibrium thermodynamics, it will reduce the free energy. For example, for a system with one component, when the flux $\mathbf{J} = -L \nabla \delta F/\delta \rho$, with L a number > 0, in a small time step δt, $\delta \rho = \delta t L \nabla^2 \delta F/\delta \rho$ (conservation of mass), then the free-energy change is $\delta F = \delta t L \int_V d\mathbf{r} \delta F/\delta \rho \nabla^2 \delta F/\delta \rho$, and with Gauss theorem neglecting surface integrals $\delta F = -\delta t L \int_V d\mathbf{r} [\nabla \delta F/\delta \rho]^2 < 0$. The same argument applies to the more elaborate diffusion models listed below.

In global equilibrium with stable minimum, the gradients of the intrinsic chemical potential are zero, *and* any second-order perturbation must lead to an increase in free energy, so that the derivative matrix $\delta^2 F/\delta \rho_I(\mathbf{r}) \delta \rho_J(\mathbf{r}')$ must be positive definite. Notice that there are very many such minima possible; almost all of them will be metastable – except for the global minimum, the morphology with the lowest free energy. But this morphology is very difficult to achieve, and the system rather becomes trapped in a metastable state.

7.2.3 DETAILS OF THE FREE-ENERGY MODEL

In the equilibrium limit $\delta F/\delta \rho_I = 0$ corresponds exactly to the set of equations of equilibrium polymer self-consistent field models. The equilibrium mean-field models come in many flavors: the models have been reinvented several times, in different scientific disciplines. One can perform calculations analytically (known as the random phase approximation, or RPA) [4,11], numerically with a lattice chain model (Scheutjens–Fleer), with a molecular detailed single-chain Hamiltonian [13], or numerically with discrete Gaussian chains (Mesodyn) or continuous wormlike chains (see e.g. [14]). The earliest well-documented numerical analyis of the self-consistent field model is due to Helfand, in the early 1970s, originally applied to block copolymer layers and polymer-blend interfaces [15–17]. In Helfand's model too, the chain is represented as a wormlike continuous chain. The difference between all these models is to a large extent irrelevant. For long enough chains they are all identical, for short chains none of them will apply. There are some differences in ease of use. The analytical RPA approach is the "bread and butter" of polymer physics – if one is well versed in field theory the method can be applied with advantage. The language of the Scheutjens–Fleer lattice-chain model is very close to that of the original Flory–Huggins model: one imagines polymers as walks on a regular lattice. The Scheutjens–Fleer model has been applied very successfully to numerous problems in polymer adsorption in colloid and surface science [12].

The discrete Gaussian chain model (Mesodyn) and the continuous chain model only differ in minor details of the Boltzmann weight calculations; both of them have been applied primarily to melts and concentrated solution in bulk. In Mesodyn, a chain is represented as a necklace of beads, each bead representing a large number of monomers. Each bead is thought to be a tiny random coil; the beads are connected via harmonic springs. In the continuous-chain model, the chain is a flexible worm, also with harmonic springs between consecutive chain elements.

The details of the model are as follows. The chain Hamiltonian for the necklace of beads is

$$H = \frac{3kT}{2a^2} \sum_{s=2}^{N} (\mathbf{R}_s - \mathbf{R}_{s-1})^2, \qquad (7.7)$$

with a the bead size. Note that the average distance between consecutive beads is zero: each bead is a tiny coil, and thus can be penetrated by monomers from the adjacent bead. The partition function ϕ of a single chain is

$$\phi[U] = N \int_{V^N} d\mathbf{R}_1 \ldots d\mathbf{R}_N e^{-\beta \left[H + \sum_{s=1}^{N} U_s(\mathbf{R}_s) \right]}, \qquad (7.8)$$

with N a normalization constant such that $\phi(0) = V/\Lambda^3$ (ideal gas limit), and $\beta \equiv 1/kT$. The density functional for bead type I of the chain is the ensemble average of a microscopic density operator

$$\rho_I[U](\mathbf{r}) = \sum_{x=1}^{N} \theta_{Ix} \rho_s(\mathbf{r}) \qquad (7.9)$$

$$\rho_s(\mathbf{r}) = n < \delta(\mathbf{r} - \mathbf{R}_x) > \qquad (7.10)$$

$$<\delta(\mathbf{r} - \mathbf{R}_x)> = \frac{\int_{V^N} d\mathbf{R}_1 \cdots d\mathbf{R}_N \delta(\mathbf{r} - \mathbf{R}_x) e^{-\beta \left[H + \sum_{s=1}^{N} U_s(\mathbf{R}_s) \right]}}{\int_{V^N} d\mathbf{R}_1 \cdots d\mathbf{R}_N e^{-\beta \left[H + \sum_{s=1}^{N} U_s(\mathbf{R}_s) \right]}}, \qquad (7.11)$$

where $\theta_{Ix} = 1$ when bead x is of type I, and 0 otherwise.

In the heart of all ideal chain models is an efficient algorithm for the calculation of the partition function and concentrations fields. The algorithm is either called the matrix scheme (in 1D lattice theories), and/or propagator scheme (for discrete chains in continuum), or Edwards diffusion equation (continuous chains in continuum). The algorithm uses the phantom character of the ideal chain. Any point along the chain, somewhere in space, can be regarded as the product of two weighted random walks, one in the direction from $1 \to N$, and the inverse from $N \to 1$. For example, for the discrete chain we define the recurrence relations for propagator functions G and G^i

$$G_s(\mathbf{r}) = e^{-\beta U_s(\mathbf{r})} \sigma[G_{s-1}](\mathbf{r}) \qquad s: 1 \to N \qquad (7.12)$$

$$G^i_s(\mathbf{r}) = e^{-\beta U_s(\mathbf{r})} \sigma[G^i_{s+1}](\mathbf{r}) \qquad s: N \to 1 \qquad (7.13)$$

$$G_0(\mathbf{r}) = 1 \qquad (7.14)$$

$$G^i_{N+1}(\mathbf{r}) = 1, \qquad (7.15)$$

with the linkage operator as a Gaussian filter

$$\sigma[f](\mathbf{r}) = \left(\frac{3}{2\pi a^2}\right)^{\frac{3}{2}} \int_V d\mathbf{r}' e^{-\frac{3(\mathbf{r}-\mathbf{r}')^2}{2a^2}} f(\mathbf{r}'). \qquad (7.16)$$

It is not difficult to see that

$$\phi \Lambda^3 = \int_V d\mathbf{r} G_N(\mathbf{r}) = v \int_V d\mathbf{r} G^i_1(\mathbf{r}) \qquad (7.17)$$

$$\rho_s(\mathbf{r}) = n \frac{G_s(\mathbf{r}) \sigma[G^i_{s+1}](\mathbf{r})}{\int_V d\mathbf{r} G_N(\mathbf{r})} \quad \text{composition law.} \qquad (7.18)$$

It is apparent from the recursion, that the computational cost of a calculation scales linearly with the length of the chains: double the chain length and the calculation of the partition function is twice as expensive. These recurrence relations are crucial for efficient numerical evaluations, but otherwise do not affect the thermodynamics.

The mean-field free energy is

$$F_{MF} = \frac{1}{2} \int_V \int_V \varepsilon_{IJ}(\mathbf{r}-\mathbf{r}') \rho_I \rho_J d\mathbf{r} d\mathbf{r}' + \frac{\kappa}{2} \int_V d\mathbf{r} \left(\sum_I v \rho_I(\mathbf{r}) - 1\right)^2, \qquad (7.19)$$

with the mean-field excluded volume parameter v, the compressibility parameter κ, and the interaction kernels

$$\varepsilon_{IJ}(\mathbf{r}-\mathbf{r}') \equiv \varepsilon^0_{IJ} \left(\frac{3}{2\pi a^2}\right)^{\frac{3}{2}} e^{-\frac{3(\mathbf{r}-\mathbf{r}')^2}{2a^2}}. \qquad (7.20)$$

The compressibility term allows for small (harmonic) deviations of a few % from average total density – the system is thus slightly compressible, which has some advantages for the numerical calculations. It is also possible to adjust the free energy such that the system is *exactly* incompressible, by including an additional constraint via a Lagrange multiplier.

The bare interactions are related to the dimensionless Flory–Huggins χ-parameters through

$$\chi_{IJ} = \frac{\beta}{2v}\left(\varepsilon_{II}^0 + \varepsilon_{JJ}^0 - \varepsilon_{IJ}^0 - \varepsilon_{JI}^0\right). \tag{7.21}$$

7.2.4 DYNAMICS

The dynamical model is that of convection-diffusion. By assumption, on the coarse-grained time scale the chain distribution function is relaxed constantly, since all internal modes are in equilibrium. The simplification is enormous, since now we do not have to consider memory effects associated with unrelaxed chain conformations. In fact, we simply combine the flux equations of linear nonequilibrium thermodynamics

$$\text{Diffusion:} \quad \mathbf{J} = -L\nabla \frac{\delta F}{\delta \rho}$$
$$\text{Convection:} \quad \mathbf{J} = \mathbf{v}\rho, \tag{7.22}$$

(L is a positive definite Onsager coefficient and $\mathbf{v}$ the velocity field), with the law for the conservation of mass

$$\frac{\partial \rho}{\partial t} + \nabla \cdot \mathbf{J} = 0. \tag{7.23}$$

At this point, we do not yet consider hydrodynamics: in the example below where we shear a morphology, the velocity field is imposed from the outside. For most mesoscale polymer systems considered by the molecular modeling community, relaxation by internally driven hydrodynamics is relatively unimportant. Also, it is not easy to find a general-purpose model. At this point we do not have a good expression for the local stress; if we do, we can extend the approach better to chemical engineering applications such as extrusion.

There are a few catches to the diffusion model – several models are possible for the Onsager coefficient. To a first approximation one can simply set the Onsager coefficient L as a constant, but obviously such models neglect the fundamental property that net flux should be proportional to force *and* concentration. Slightly better is therefore to take a constant friction coefficient, and set $L = M\rho$ (this is the local coupling or 'mixed dynamics' algorithm in Mesodyn). But also this approximation is not entirely consistent, and neglects the extension of the chain. In the inhomogeneous system, each chain samples many conformations, and one should in fact add all thermodynamic forces for each conformation with the proper Boltzmann weight [18,19]. For example, consider again Figure 7.1. Suppose micelle A is in equilibrium, so that the intrinsic chemical potential $\delta F/\delta \rho$ is constant in the neighborhood of A. According to local coupling models the fluxes in A will be zero, since the gradients of the intrinsic chemical potential are zero locally. But now, if micelle B is out of equilibrium and experiences a force in a certain direction, because of the mutual bridging connections micelle A will also experience a force in the same direction, and move accordingly. This effect is captured by the collective Rouse dynamics model, in which the thermodynamic forces are weighted with a long-range kernel derived from the monomer–monomer correlations – this is the collective Rouse dynamics or external potential dynamics algorithm in Mesodyn. A similar, but more involved model can be derived for reptation.

$$L = \begin{cases} M & \text{constant} \\ M\rho & \text{local coupling} \\ \int_V d\mathbf{r}' \Lambda(\mathbf{r}', \mathbf{r})(\cdot) & \text{general (Rouse, reptation).} \end{cases} \quad (7.24)$$

The literature on the various dynamical models in not extensive, also very few experimental results are available to check whether it is worthwhile to derive more elaborate models for the Onsager coefficients.

A second catch is the noise. If one observes the movements of a colloidal particle, the Brownian motion will be evident. There may be a constant drift in the dynamics, but the movement will be irregular. Likewise, if one observes a phase-separating liquid mixture on the mesoscale, the concentration levels would not be steady, but fluctuating. The thermodynamic mean-field model neglects all fluctuations, but they can be restored in the dynamical equations, similar to added noise in particle Brownian dynamics models. The result is a set of stochastic diffusion equations, with an additional random noise source η [20]. In principle, the value and spectrum of the noise is dictated by a fluctuation dissipation theorem, but usually one simply takes a white noise source.

Finally, one realizes of course that in the compound system all fields of all components interact thermodynamically *and* dynamically, which is reflected in the choice for the Onsager coefficients and the interaction model. If we now put everything together, we have the general equation

$$\frac{\partial \rho_I}{\partial t} + \nabla \cdot \mathbf{v}\rho_I = \sum_J \nabla \cdot \int_V d\mathbf{r}' \Lambda_{IJ}(\mathbf{r}', \mathbf{r}) \nabla_{\mathbf{r}'} \frac{\delta F}{\delta \rho_I} + \eta_I. \quad (7.25)$$

7.3 APPLICATIONS

We discuss three illustrative applications from our own work: the phase diagram of a polymer surfactant solution with and without shear, and block copolymer thin-film formation.

7.3.1 POLYMER SURFACTANT SOLUTION

The surfactant in this case, L64, is a member of the Pluronic family (marketed by BASF); these are triblocks composed of poly(ethyleneoxide) (PEO) and poly(propyleneoxide) (PPO) blocks. Some of these surfactants are popular in drug delivery, others are used in washing powders and personal care products such as toothpaste. The surfactants are 'soft' – they are mild to the skin. The amphiphilic power is modest too. The hydrophilic block PEO is only slightly less hydrophobic than the PPO block. In fact, the solubility of PEO is an unresolved mystery in itself, maybe related to the cage structure of the hydrated ethylene oxide monomer. Poly (methylene oxide) is insoluble, PEO is soluble, PPO and poly (butylene oxide) and higher are all insoluble.

The molecular structure of L64 is shown in Figure 7.2. The triblock polymer is $EO_{13}PO_{30}EO_{13}$, first a hydrophilic block, then a long hydrophobic block and then again the hydrophilic block. The mesoscale simulations proceed by calculating the lyotropic phases in a narrow concentration range 50–70 % [21]. From accurate experimental data [22–24], in this interval four phases are known, and the structure factors have been measured: micellar phase, cylindrical or hexagonal phase, a bicontinuous phase and a lamellar phase.

The original paper [21] reproduced all important characteristics of the phases: the lyotropic order, the phase boundaries, the size and the structure factor (Figure 7.4).

In the presented simulations we started from a homogeneous solution, then quenched the solution to the proper Flory–Huggins values for the inhomogeneous system. During the subsequent collective diffusive relaxation, the free energy goes down and order parameters go up. The indicated results (Figure 7.3) are snapshots of the isodensity surfaces of the propylene oxide monomer

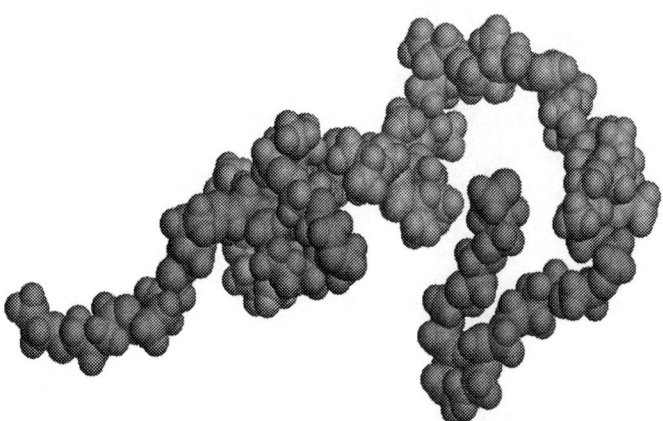

Figure 7.2 Molecular model of Pluronic L64, $(EO)_{13}(PO)_{30}(EO)_{13}$.

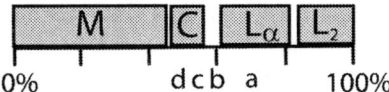

Figure 7.3 Experimental phase diagram of Pluronic L64 in water. Polymer weight fraction indicated. M: Micellar phase, C: Hexagonal (cylindrical) phase, L_α: lamellar phase, L_2: water-lean continuous phase. Region between M and C: mixed M + C, in between C and L_α: mixed C + L_α + L', with L' bicontinuous. Adapted from [23]. Labels a–d refer to simulations, Figure 7.4.

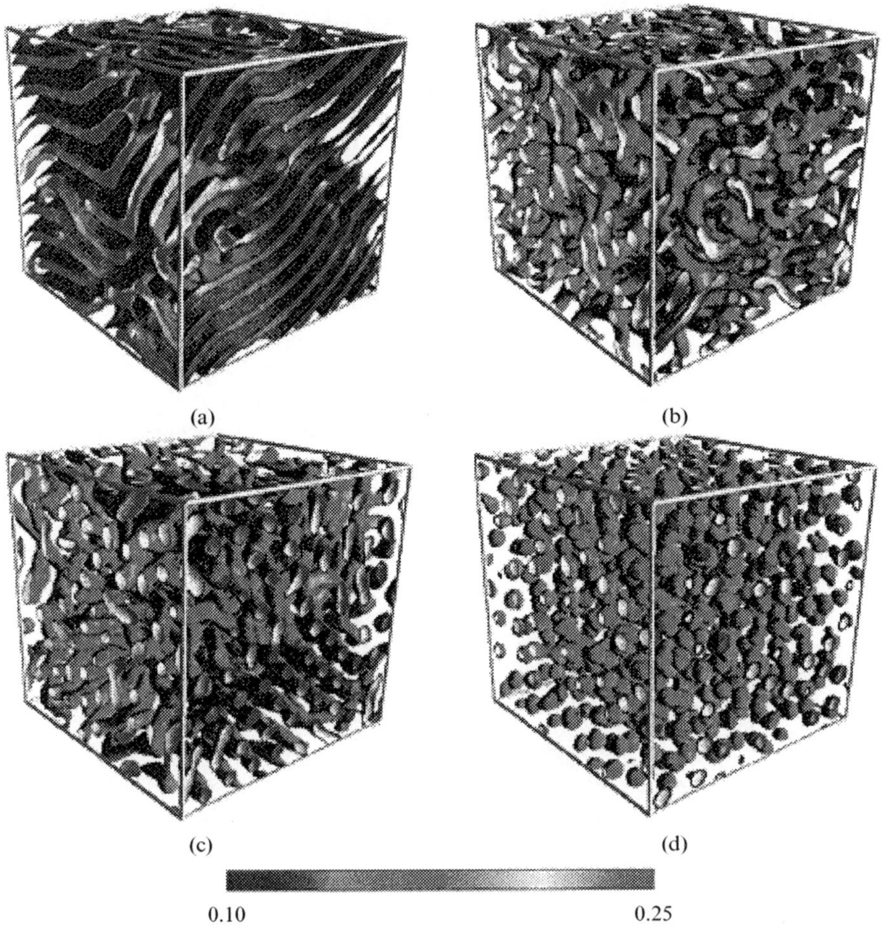

Figure 7.4 Snapshots of mesoscale structures in L64 polymer surfactant solutions at dimensionless time = 4200. PO isosurfaces with EO surface distribution in color. (a) 70 %, isolevel PO = 0.59, (b) 60 %, isolevel PO = 0.5, (c) 55 %, isolevel PO = 0.46, (d) 50 %, isolevel PO = 0.42. EO as indicated in the color legend. Figure from [21].

concentration at a certain value of dimensionless time, defined by $\tau = k_B T M h^{-2} t$, with M the mobility, h the mesh size (related to bead size through $a/h = 1.1543$, for details of numerics see [3]) and t is 'real' time.

In the mesoscale modeling we have addressed the necessary parametrization issues as follows. First, the molecular model is converted into a representative Gaussian chain. From a practical point of view it is desirable to use as few beads as possible, since the computational cost scales linearly with N, but not too few beads, since then we would lose sufficient resolution of the blocks along the chain. For the purpose of the single-chain density calculations a "real" molecular detail polymer chain (such as the model in Figure 7.2) can be replaced by a Gaussian chain, provided the response functions (or correlators) are the same. The calculation of response functions for Gaussian chains is easy (and takes only a fraction of a second on a PC), while the correlation calculation for the molecular model is more cumbersome, but nevertheless can be done. In this way, we have found that the linear response curves are indistinguishable if we replace each 3–4 monomers by one Gaussian bead. Thus the Gaussian chain is determined as $E_3P_9E_3$, where "E" is an ethylene oxide bead and "P" a propylene oxide bead – this is the regular surfactant depicted in Figure 7.1. In the theory section we have remarked that each bead should have a large number of monomers, so that it behaves as a tiny random coil. The small number of monomers per bead we have used here is probably the lower limit of applicability – it would be better to use a larger monomer/bead ratio. By consequence of the physical size of the bead, the solvent is represented by a single particle, with the same excluded volume as the polymer bead, and with unresolved internal structure.

Second, having established the molecular chain, the Flory–Huggins parameters need to be determined. This is not a trivial matter. We remarked already that PEO is a strange polymer, which should be insoluble as a member of an insoluble homologous series, but is not. This is reflected in a strong dependence of the molecular-interaction parameters on factors such as concentration and temperature, and chain length. Recent comparison of molecular-dynamics simulation and a thermodynamic model, points to a strong influence of hydrogen-bond network formation [26], with competition between water–water and ethylene oxide–water bonds. It is at present a challenge to calculate the Flory–Huggins parameters from first principles (such as molecular-modelling or force-field models). But fortunately, since PEO (or its cousin PEG and also PPO) is a well-studied polymer, a large body of experimental data is available that allows us to proceed in a semiempirical fashion. Already in the early 1950s, Flory–Huggins parameters were calculated from vapor pressure data on PEG and PPO homopolymer solutions [25]. In the concentration interval 50–70% the χ-parameters are nearly constant, with (for homopolymers of length similar to the blocks in the surfactant) $\chi_{EW} = 1.4$ and $\chi_{PW} = 1.7$. We used here the standard Flory–Huggins expression for the solvent vapour pressure, $\ln p/p^0 = \ln(1-\theta) + (1-1/N)\theta + \chi\theta^2$, with an important twist: rather than inserting N as the number of monomers, as is commonly done, we intepreted N as the number of beads. Correspondingly, the effective bead-based χ-parameters are surprisingly high: Interpreted in a naive fashion, it would seem to imply that a very long polymer of the ethylene oxide monomer is insoluble, in contrast with the common observation that, for example, PEG is soluble in all compositions. But, as we have remarked before, the numbers are semiempirical and dependent on various factors, including chain length. As a result of the semiempirical fit, we have a reasonable model for the solvent-controlled swelling of the hydrophilic and hydrophobic blocks, and thus the relative domain volumes. It is the ratio of these volumes that determines the particular phase, and is probably the most imprtant factor in getting the right lyotropic phase diagram. We could not find experimental data of similar quality to estimate the E-P bead–bead interaction. Hamley and coworkers [27] determined the Flory–Huggins parameter in melts of poly (ethylene oxide)-poly(propylene oxide) diblocks through small-angle X-ray scattering, and estimated a low value ≈ 0.1, on a *monomer* basis (using our monomer/bead ratio 3–4, this would imply ≈ 0.3–0.4 on a bead basis). However, such an estimate for a melt does not necessarily reflect the effective behavior in solution, as we have remarked before. From group-contribution methods [28], taking into account the 3–4 monomers in each bead, we roughly estimated the effective Flory–Huggins parameter between beads as 3–5. In the simulation we used $\chi_{EP} = 3.0$; this is almost an order of magnitude larger than Hamley's finding.

7.3.2 POLYMER SURFACTANT SOLUTION UNDER SHEAR

The results agree very well with the established experimental phase diagram, both in the size scale of the domains as in the detected lyotropic phases. But the phases are not perfect. Is this also correct? Indeed, also in experimental systems the phases are almost always less than perfect, and ways are sought to apply 'external' agents, such as surfaces or shear fields to force the phases into perfect symmetry. Hence, we sheared the L64 hexagonal phase (55%) by adding a directional convective term to the dynamical equations (steady shear). The results (Figures 7.5–7.7, from [29]) clearly show the adjustment of the microphase to the imposed shear field. From an initial random-oriented cylindrical phase, the system develops into an almost perfect array of hexagons, in the direction of shear. A significant detail is the orientation of the hexagons: they are slightly tilted with respect to the shear gradient. There are several clues in the experimental literature indicating that this is indeed correct.

In the initial stages of shear the reordering proceeds via breakup of the structures, which leads to an increase of the anisotropy. The pieces are tilted in the direction of flow. Then oblong micelle-like structures coalesce to form new cylinders that align in the direction of shear. This reorientation is

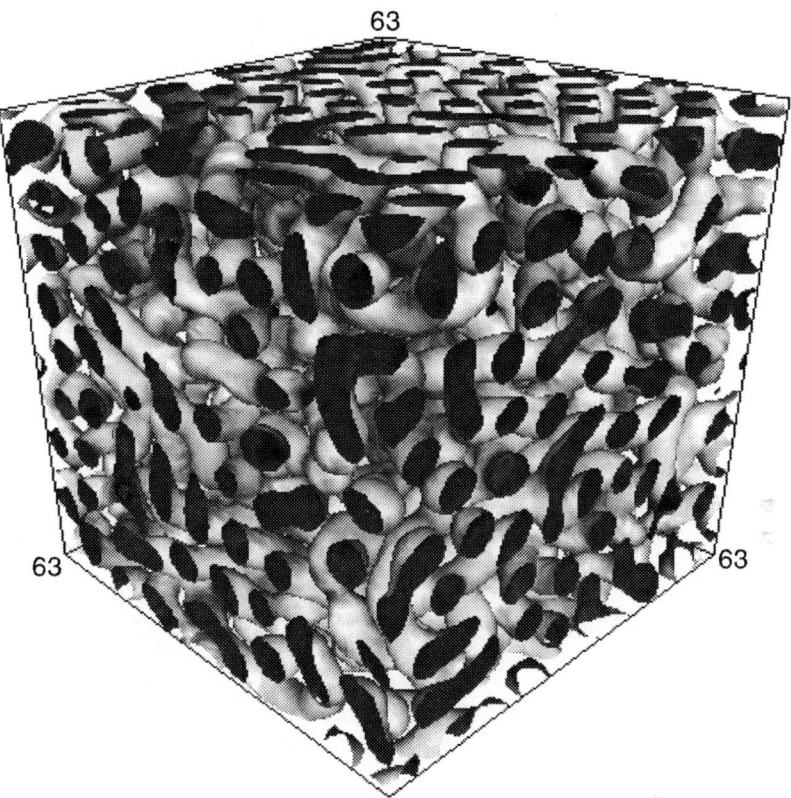

Figure 7.5 Disordered hexagonal phase for a 55 % L64 solution.

reflected in the rapid increase in the anisotropy factor. The alignment process is clearly observed in the projections of the $3D$ structure factor. The yx-projection is squeezed into a line and the yz-projection forms a circle. The position of the primary peak remains the same during shearing. The characteristic features of the last stage of reorientation are defect annihilation and reordering of hexagonal clusters. Initially there are many hexagonal clusters with different orientations. In the final stage the system forms a few big clusters with nearly the same orientation. Experiments on the same Pluronic surfactant solution [30] – with almost the same concentration – 53 % demonstrate exactly the same alignment of cylinders in the direction of flow as in our simulation. The 10 plane of the hexagonal lattice is experimentally found to be parallel to the shear plane. Our simulation gives the orientation of the main cluster that is 10° off. For another triblock copolymer system the perpendicular lattice orientation was found together with the same orientation of cylinders along the flow direction [31].

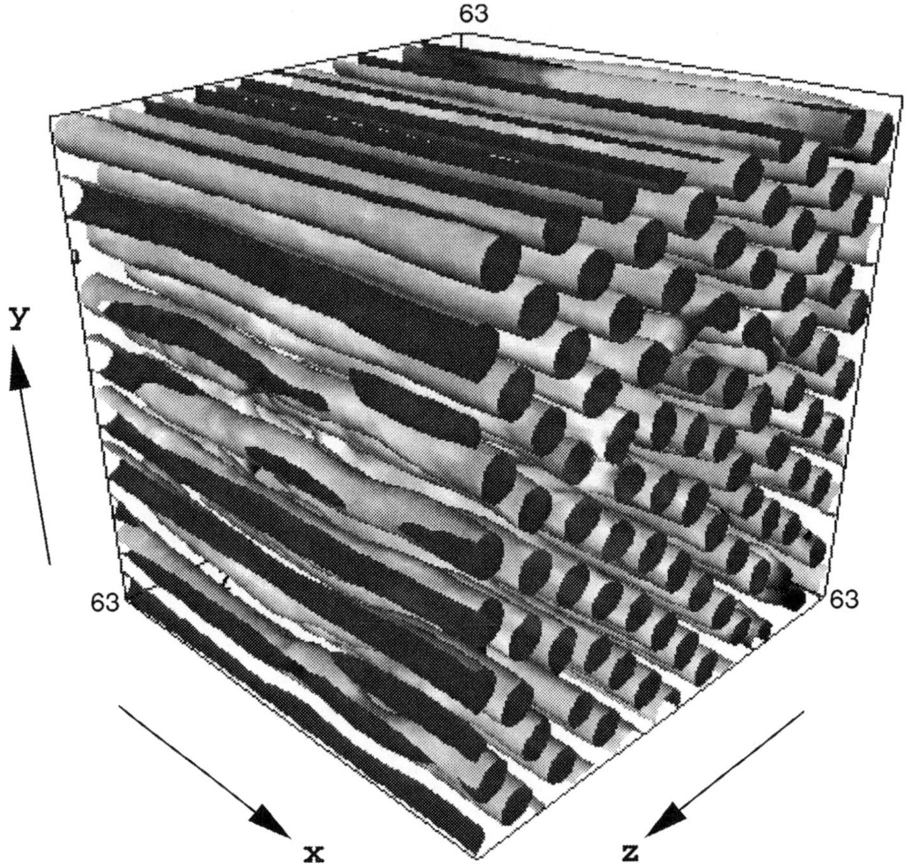

Figure 7.6 Sheared hexagonal phase of L64.

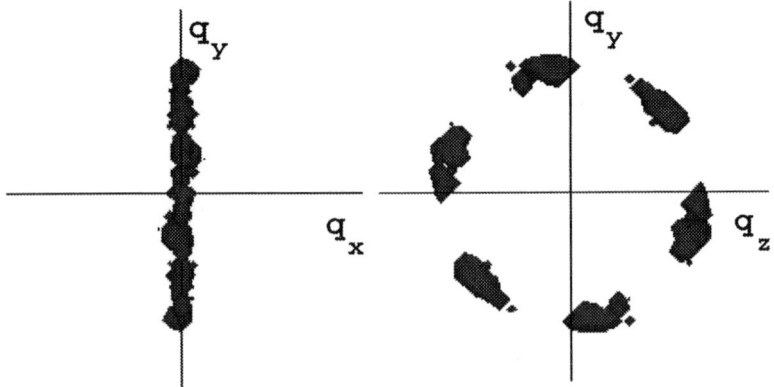

Figure 7.7 Fourier transforms of the sheared hexagonal phase.

7.3.3 THIN-FILM FORMATION

In the soft condensed matter literature, an impressive agreement between self-consistent-field theory and experiment was unravelled some time ago by Matsen and Schick [32]. The comparison concerned the microphase diagram of diblock copolymers, with the classical sequence of microphases, depending on block ratio: lamellar, cylindrical and micellar, with gyroid in between lamellar and cylindrical. The microdomain structure in the bulk is determined mainly by the molecular architecture, in particular the ratio of block lengths and the interaction between the two components (blocks). At interfaces and in thin films an additional driving force for structure formation exists, because one component typically has a lower interfacial energy than the other. This phenomenon belongs to a class of interfaces of modulated phases, and are specific to the particular system and/or route of film preparation. As a result, no general agreement is reached about the underlying fundamentals. In thin films, additional constraints exist. Here, the microdomain structure has to adjust to two boundary surfaces and a certain film thickness, which can be a noninteger multiple of the "natural" bulk domain spacing. Both constraints together cause a complex and interesting phase behavior.

We calculated the microphases of a thin film of neat three-block polystyrene – polybutadiene – polystyrene (SBS) deposited on a solid substrate [33]. The parametrization protocol was somewhat different from the case of L64, now using data on the microphase-separation temperature of a bulk melt to determine the critical Flory–Huggins value, rather than solution data.

Figure 7.8 shows experimental data on morphologies in the thin SBS film, and the comparison with the simulation. Here, the thickness of the simulated block copolymer film was imposed following a gradient in the x-direction, in the experimental system steps on the surface with variable thickness were detected and analysed. The agreement between experiment and simulation is very good: a sequence of 9 morphologies is found, in one simulation, in the right order. The sequence runs from completely disordered very thin films, to films with cylinders parallel to the surface, with exotic intermediate structures such as perforated lamellae and perpendicular cylinders. Well-defined microdomain patterns have formed, which change systematically as a function of the gradually changing film thickness (at steps between terraces). In particular, boundaries between different structures correspond to height contour lines. A major fraction of the surface displays bright stripes, which are indicative of polystyrene cylinders oriented parallel to the surface. In thinner regions of the film two additional patterns are found: One is characterized by hexagonally ordered dark spots, indicative of polybutadiene microdomains in an otherwise continuous polystyrene layer, i.e., a perforated lamella (PL). The slopes between neighboring terraces display a hexagonal pattern of bright dots, indicative of polystyrene cylinders oriented perpendicular to the surface (C). Finally, the thinnest parts of the films display no lateral structure at all, indicative of

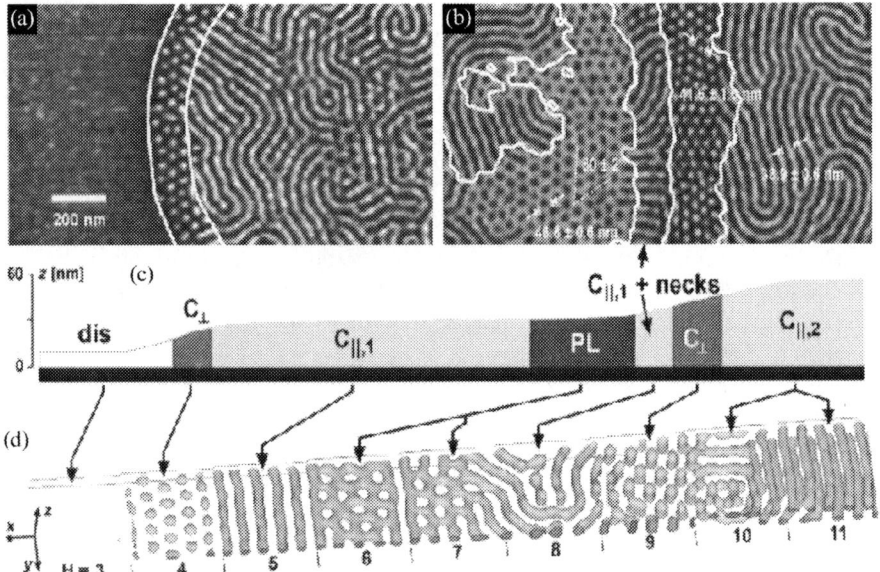

Figure 7.8 A comparison of the experimental results (a) and the MesoDyn simulation (b). Grid points with increasing film thickness $H(x)$, $\varepsilon_{AB} = 6.5$, and $\varepsilon_M = 6.0$. The latter corresponds to a preferential attraction of B beads to the surface. The isodensity surface $\rho_A = 0.5$ is shown. Figure from [33].

either a disordered (dis) phase or a lamellar wetting layer (W). In thicker films, the sloped regions between terraces display stripes as well. We note that these phases were all reported earlier and for various experimental conditions and cylinder-forming block copolymers. In the present experiments and simulations, however, all phases appear in a single system and under identical experimental conditions. This finding indicates that the film thickness is an important control parameter.

7.4 CONCLUSION

The dynamical mean-field approach is a powerful tool to predict and analyse block copolymer morphologies in a variety of systems. Extensions in further directions are being pursued in our group and elsewhere: the development of a proper rheological model, the development of models for stiff-flexible polymer morphologies, charged systems and reactive systems. All these extensions can be treated to some extent at the mean-field level. What remains a challenge for the distant future however, is the prediction of morphologies in the case that specific interactions dominate, such as packing factors in rigid molecules, or hydrogen bonding. Such factors are important when one considers molecular systems of biological origin, or in supramolecular chemistry or the prediction of morphologies controlled by specific interactions.

REFERENCES

1. I. W. Hamley, *The Physics of Block Copolymers*, Oxford University Press, Oxford, 1998
2. J. G. E. M. Fraaije, *J. Chem. Phys.* **99**:9202–9212, 1993
3. J. G. E. M. Fraaije, B. A. C. van Vlimmeren, N. M. Maurits, M. Postma, O. A. Evers, C. Hoffmann, P. Altevogt and G. Goldbeck-Wood, *J. Chem. Phys.* **106**(10):4260–4269, 1997
4. P.-G. de Gennes. *Scaling Concepts in Polymer Physics*. Cornell University, Ithaca, NY, 1979
5. M. Doi and S. F. Edwards, *The Theory of Polymer Dynamics*, Clarendon Press, Oxford, 1986
6. G. H. Fredrickson, V. Ganesan, F. Drolet, *Field-theoretic computer simulation methods for polymers and complex fluids*, Macromolecules, **35**(1): 16–39, 2002
7. T. Hill, *An Introduction to Statistical Thermodyamics*, Addison-Wesley, Reading, 1962
8. P. J. Flory, *Principles of Polymer Chemistry*, Ithaca, N.Y., Cornell, 1953
9. P. J. Flory, *The Statistical Mechanics of Chain Molecules*, John Wiley, New York, 1969
10. *Amphiphilic Block Copolymers, Self-assembly and Applications*, P. Alexandridis, B. Lindman eds., Elsevier, Amsterdam, 2000
11. L. Leibler, *Macromolecules* **13**(6):1602–1617, 1980
12. G. J. Fleer, M. A. Cohen Stuart, J. M. H. M. Scheutjens, T. Cosgrove. and B. Vincent, *Polymers at Interfaces*, Chapman Hall, London, 1993
13. I. Szleifer, A. Benshaul, W. Gelbart, *J. Chem. Phys.* **94**(12):5081–5089, 1990
14. M. Doi, *Introduction to Polymer Physics*, Clarendon Press, Oxford, 1996
15. E. Helfand, *J. Chem. Phys.* **62**(3):999–1005,1975
16. E. Helfand, Z. R. Wasserman, *Macromolecules* **9**:879–888, 1976
17. E. Helfand, Y. Tagami, *Theory of the interface between immiscible polymers*, *J. Polym. Sci. B-Polym. Phys.* **34**(12):1947–1952, 1996 (reprinted from *J. Polym. Sci., Polym. Lett.* **9**:741–746, 1971)
18. M. M. Maurits and J. G. E. M. Fraaije, *J. Chem. Phys.* **107**(15):5879–5889, 1997.
19. K. Kawasaki and K. Sekimoto *Macromolecules* **22**:3063–3075, 1989.
20. C. W. Gardiner, *Handbook of Stochastic Methods*, Springer-Verlag, Berlin, 2nd edition, 1990
21. B. A. C. van Vlimmeren, N. M. Maurits, A. V. Zvelindovsky, G. J. A. Sevink and J. G. E. M. Fraaije, *Macromolecules*, **32**:646–656, 1999
22. P. Alexandridis, T. A. Hatton, Colloid Surface A **96**(1–2):1–46, 1995
23. P. Alexandridis, U. Olsson, B. Lindmann, *Macromolecules* **28**:7700–7710, 1995
24. P. Alexandridis, U. Olsson, B. Lindmann, *J. Phys. Chem.* **100**:280–288, 1996
25. N. M. Malcol, J. S. Rowlinson *Trans. Faraday Soc.* **53**:921–931, 1957
26. E. Dormidontova, *Macromolecules* **35**:987–1001, 2002
27. I. W. Hamley, V. Castelletto, Z. Yang, C. Price, C. Booth, *Macromolecules* **34**:4079–4081, 2001
28. D. W. Krevelen, *Properties of Polymers, Their Correlation with Chemical Structure. Their Numerical Estimation and Prediction from Additive Group Contribution*, Elseier, Amsterdam, 1990

29. A. V. M. Zvelindovsky, B. A. C. van Vlimmeren, G. J. A. Sevink, N. M. Maurits, J. G. E. M. Fraaije, *J. Chem. Phys.* **109**(20): 8751–8754, 1998
30. G. Schmidt, W. Richtering, P. Lindner, P. Alexandridis, P., *Macromolecules* **31**, 2293–2298, 1998
31. G. Hadziioannou, A. Mathis, and A. Skoulios, *Colloid Polym. Sci.* **257**:15, 1979
32. M. W. Matsen and M. Schick, *Phys. Rev. Lett.* **72**(16):2660–2663, 1994
33. A. Knoll, A. Horvat, K. S. Lyakhova, G. Krausch, G. J. A. Sevink, A. V. Zvelindovsky, and R. Magerle, *Phys. Rev. Lett.* **89**, 035501 (2002)

8 Self-Consistent Field Theory of Block Copolymers

AN-CHANG SHI
*Department of Physics and Astronomy, McMaster University
Hamilton, Ontario Canada L8S 4M1*

8.1 INTRODUCTION

Block copolymers, in which two or more chemically different subchains form a single molecule, are a fascinating class of soft materials with unique structural and mechanical properties [1,2]. Interest in block copolymers has grown considerably in recent years because of their ability to self-assemble into a variety of ordered structures with domain sizes in the nanometer range. The self-assembly of block copolymers is governed by a delicate balance of the interaction energy and the chain stretching. The repulsive interaction between the chemically different blocks drives the system to phase separate, whereas the connectivity of the copolymer chains prevents macroscopic phase separation. As a result of these competing trends, block copolymer systems self-organize into many complex structures. For diblock copolymers, these structures range from lamellar (lam), hexagonal-packed cylinder (hex) and body-centered cubic sphere (bcc) phases to complex bicontinuous cubic (gyroid) phases (Figure 8.1). These structures can be controlled by varying the chemical composition of the block copolymer or the segregation between blocks (via temperature or molecular weight) [3].

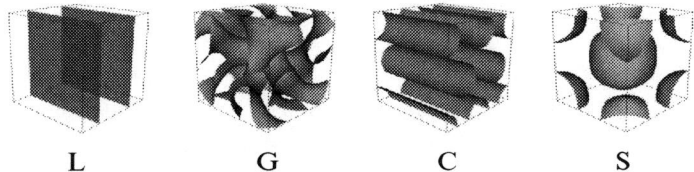

L G C S

Figure 8.1 Schematic illustration of four ordered equilibrium structures formed in diblock copolymer melts. The structures are from left to right: lamellar (L), bicontinuous gyroid (G) cylindrical (C), and spherical (S) phases.

From a theoretical point of view, the biggest challenge is to predict phase behavior and morphologies of block copolymers with a particular molecular architecture. Ideally the theory should take the molecular parameters as input and be able to predict thermodynamically stable phases and the phase-transition boundaries among them. Towards this goal, a variety of theoretical methods have been developed to study the phases and phase behavior of block copolymer systems. One of the most successful theoretical frameworks for block copolymers is the self-consistent field theory (SCFT) that has its origin in work by Edwards in the 1960s [4]. This theoretical framework was explicitly adapted to treat block copolymers by Helfand in 1975 [5], and later important contributions to the theory were made by, among others, Hong and Noolandi [6]. From a mathematical point of view, block copolymer theory presents a complex problem, and exact solutions are difficult to obtain. Different approaches to study the block copolymer theory have been developed. In 1980 Leibler [7] proposed an approximate analytical theory by assuming that the free-energy functional can be expanded around the homogeneous state. Therefore the theory is valid in the weak-segregation regime. A similar theory was introduced by Ohta and Kawasaki [8] in 1986. Another approximate analytical theory, valid in the strong-segregation regime, was introduced by Semenov in 1985 [9]. These approximate theories have been applied to a variety of block copolymer systems, leading to valuable insights into the phases and phase transition in block copolymers. An alternative approach to block copolymer theory is to solve the mean-field equations exactly using numerical techniques. The earliest attempts to obtain numerical solutions were made by Helfand and coworkers [10]. Later Shull [11] and Whitmore and coworkers [12,13] constructed the phase diagrams of block copolymer melts and solutions using approximate numerical techniques. The current state-of-the-art numerical approach to solving the mean-field equations was developed by Matsen and Schick in 1994 [13]. This method utilizes the crystalline symmetry of the ordered phases and provides exact numerical solutions to the mean-field equations. This technique has been applied to a variety of block copolymer systems [3]. In recent years, new developments in the self-consistent field theory have been made. Examples are theory for Gaussian fluctuations in ordered phases [15,16] and numerical techniques for solving the mean-field theory in real space [17–19]. Numerical methods to simulate the block copolymer field theory have also been proposed [20,21]. It can be concluded that the self-consistent field theory based on the Gaussian model of the polymer chains forms a powerful basis for the study of block copolymers. In this chapter a systematic derivation of the self-consistent field theory is presented, emphasizing the theoretical development and techniques.

Block copolymer theory is a very active research area in polymer science and progress is being made constantly. This makes it virtually impossible to include all the current theoretical developments in a short review chapter. Instead, in this chapter we focus on the framework and techniques based on field-theoretical methods. In particular, we will try to present a systematic derivation of the field-theoretical formulation of block copolymer theory. The theory is examined using the mean-field approximation and the method of solving the mean-field equations is given. Gaussian fluctuations around the mean-field solutions are examined. It is hoped that this approach will serve as a reference source for readers who are interested in applying the SCFT to block copolymer systems. There is a large body of literature on block copolymer theories, including a number of valuable reviews [3,21–23]. In the presentation of the material in this chapter, we rely heavily on these review articles and on our own work, and we have made no attempt to be comprehensive. We apologize in advance for our failure to cite all of the relevant and important literature. By focusing on the self-consistent field theory, many other theoretical developments, such as the PRISM integral-equation theory [24] and a variety of simulation studies [25], are not discussed in this chapter.

8.2 THEORETICAL FRAMEWORK: SELF-CONSISTENT FIELD THEORY OF BLOCK COPOLYMERS

For block copolymers, recent theoretical studies of their phase behavior are mostly based on the so-called "standard model", in which the polymer chains are modeled as flexible Gaussian chains, and the interactions between the different monomers are modeled by short-range contact potentials [3]. Furthermore, the hard-core repulsive interactions are approximated by the incompressibility condition. The advantage of this simple model is that the thermodynamic behavior of the system can be formulated in terms of a field theory, the self-consistent field theory (SCFT) [4–6,15,21–23], which allows systematic studies using a variety of techniques. In particular, the mean-field approximation of the theory has led to a powerful theoretical framework, the self-consistent mean-field theory (SCMFT) for the study of block copolymer equilibrium phase behavior. The solutions of the SCMFT have provided a quantitative phase diagram for diblock copolymers, which is in good agreement with experiments. Further development of the SCFT includes the study of anisotropic Gaussian fluctuations around ordered phases [15,16], which provides an understanding on the nature of these fluctuations, as well as a technique to study the stability, kinetic pathways, and scattering functions of ordered structures. In what follows, a review of the theory of block copolymers is given, emphasizing the field-theory technique and its applications to the study of Gaussian fluctuations. There are a number of valuable reviews on self-consistent field theory [21–23]. The following theoretical development is based on the work of Shi *et al.* [15]. The presentation is similar to those used by Schmid [22] and by Fredrickson *et al.* [21].

We formulate the theory using diblock copolymer melts as an example. Extension to other block copolymers as well as block copolymer blends and solutions is straightforward. We use a canonical ensemble approach and consider n_c copolymer chains in a volume V. Each copolymer chain is built from

N monomers of species $\alpha = A, B$. The degrees of polymerization of the block are $N_A = f_A N$ and $N_B = f_B N$ with $f_A + f_B = 1$. Each block has an associated Kuhn length $b_\alpha = \sigma_\alpha b$, where b is a reference Kuhn length. The monomers are assumed to have the same monomer density, ρ_0, which is defined as monomers per unit volume, or the hardcore volume per monomer is ρ_0^{-1}. We will use the convention that all lengths are scaled by the Gaussian radius of gyration, $R_g = b\sqrt{N/6}$. The chain arc length is scaled by the degree of polymerization N. The conformation of the blocks is denoted by a space curve $\vec{R}_i^\alpha(s)$, which specifies the s-th monomer at α-block of the i-th chain. For a given chain configuration $\{\vec{R}_i^\alpha(s)\}$, the concentrations of A and B monomers at a given spatial position $\vec{r}$ are

$$\hat{\phi}_\alpha(\vec{r}) = \frac{N}{\rho_0} \sum_{i=1}^{n_c} \int_0^{f_\alpha} ds\, \delta\left(\vec{r} - \vec{R}_i^\alpha(s)\right), \tag{8.1}$$

where the hat on ϕ_α denotes that these concentrations are a functional of the chain conformations.

For simplicity, we assume that the polymer chains are flexible Gaussian coils, therefore the probability distribution $p_0(\vec{R}_i^\alpha(s))$ for a given block has the standard Wiener form,

$$p_0(\vec{R}_i^\alpha(s)) = A \exp\left[-\frac{3}{2Nb_\alpha^2} \int_0^{f_\alpha} ds \left(\frac{d\vec{R}_i^\alpha(s)}{ds}\right)^2\right], \tag{8.2}$$

where A is a normalization constant. The probability $P_0(\{\vec{R}(s)\})$ of a given chain configuration $\{\vec{R}_i^\alpha(s)\}$ is given by

$$P_0(\{\vec{R}_i^\alpha(s)\}) = \prod_{i=1}^{n_c} \left\{ p_0(\vec{R}_i^A(s)) p_0(\vec{R}_i^B(s)) \delta\left[\vec{R}_i^A(f_A) - \vec{R}_i^B(f_B)\right] \right\}, \tag{8.3}$$

where the delta functions ensure that the two blocks are connected at the end to form a diblock copolymer chain. Extension to more complex block copolymer architectures can be easily done by modifying the above probability distribution.

The partition function of the diblock copolymer melt can be written in terms of the functional integral over all the chain conformations,

$$Z = \frac{z_c^{n_c}}{n_c!} \int D\{\vec{R}(s)\} P_0(\{\vec{R}(s)\}) \prod_{\vec{r}} \delta\left(\hat{\phi}_A(\vec{r}) + \hat{\phi}_B(\vec{r}) - 1\right) e^{-W(\{\hat{\phi}\})}, \tag{8.4}$$

where z_c is the partition function of a copolymer chain due to the kinetic energy, and $W(\{\hat{\phi}\}) = V(\{\hat{\phi}\})/k_B T$ is the intermolecular interaction potential. For simplicity, we assume short-range interactions such that the interaction potential has the Flory–Huggins form,

$$W(\{\hat{\phi}\}) = \rho_0 \chi \int d\vec{r} \hat{\phi}_A(\vec{r})\hat{\phi}_B(\vec{r}), \tag{8.5}$$

where χ is the so-called Flory–Huggins parameter, which varies with temperature. Furthermore, the melt is assumed to be incompressible, mimicking the hardcore monomer–monomer interactions, and a delta function is introduced in the partition function to enforce the incompressibility.

The partition function Z contains all the information about the thermodynamics of the system. However, the evaluation of Z is not possible since the integrand depends on the chain conformation through the concentration variables. A standard algebraic trick is to insert the identity,

$$1 = \int D\{\phi_\alpha\}\delta(\phi_\alpha(\vec{r}) - \hat{\phi}_\alpha(\vec{r})), \tag{8.6}$$

into the expression of the partition function. Furthermore, auxiliary fields can be introduced by converting the delta function to its integral definition,

$$1 = \int D\{\phi_\alpha\}D\{\omega_\alpha\} e^{\int d\vec{r}\omega_\alpha(\vec{r})(\phi_\alpha(\vec{r}) - \hat{\phi}_\alpha(\vec{r}))}, \tag{8.7}$$

where the range of the ω_α integral is along a line in the complex plane from $-i\infty$ to $i\infty$. Substituting this into the expression for Z and rearranging the order of integrations, we can rewrite the partition function of a diblock copolymer melt as a functional integral over the volume fractions and auxiliary fields,

$$Z = \int \prod_\alpha [D\{\phi_\alpha\}D\{\omega_\alpha\}] \prod_r \delta(\phi_A(\vec{r}) + \phi_B(\vec{r}) - 1) e^{-F(\{\phi\},\{\omega\})}, \tag{8.8}$$

where the free-energy functional, or, more precisely, the "Hamiltonian", of the system, $F(\{\phi\},\{\omega\})$, has the form,

$$F(\{\phi\},\{\omega\}) = \frac{\rho_0 R_g^3}{N}\left\{\int d\vec{r}\left[\chi N \phi_A(\vec{r})\phi_B(\vec{r}) - \sum_\alpha \omega_\alpha(\vec{r})\phi_\alpha(\vec{r})\right] - V\ln Q_c(\{\omega_\alpha\})\right\}. \tag{8.9}$$

The quantity $Q_c(\{\omega_\alpha\})$ in the above expression is the single-chain partition function in the external field ω_α,

$$Q_c(\{\omega_\alpha\}) = \frac{1}{V}\int D\{\vec{R}(s)\}P_0(\{\vec{R}(s)\})e^{-\left\{\sum_\alpha \int_0^{f_\alpha} ds\,\omega_\alpha(\vec{R}_\alpha(s))\right\}}, \tag{8.10}$$

which contains the chain-conformation contribution to the total partition function. It is obvious that the single-chain partition function $Q_c(\{\omega_\alpha\})$ is a functional of the fields $\omega_\alpha(\vec{r})$. It is convenient to express the single-chain partition function in terms of the chain propagators $Q_\alpha(\vec{r}, s|\vec{r}')$,

$$Q_c(\{\omega_\alpha\}) = \frac{1}{V}\int d\vec{r}_1 d\vec{r}_2 d\vec{r}_3 Q_A(\vec{r}_1, f_A|\vec{r}_2) Q_B(\vec{r}_2, f_B|\vec{r}_3), \tag{8.11}$$

where the chain propagators are defined by

$$Q_\alpha(\vec{r}, s|\vec{r}') = \int_{\vec{R}(0)=\vec{r}'}^{\vec{R}(s)=\vec{r}} D\vec{R}(s) e^{-\int_0^{f_\alpha} ds\left(\frac{3}{2Nb_\alpha^2}\left[\frac{d\vec{R}(s)}{ds}\right]^2 + \omega_\alpha(\vec{R}(s))\right)}, \tag{8.12}$$

The physical meaning of these propagators is that $Q_\alpha(\vec{r}, s|\vec{r}')$ represents the conditional probability distribution of monomer s at $\vec{r}$, given that monomer 0 is at $\vec{r}'$, in the presence of an external field $\omega_\alpha(\vec{r})$. Alternatively, and more conveniently, it is easy to prove that the propagators can be obtained from the following differential equations,

$$\frac{\partial}{\partial s} Q_\alpha(\vec{r}, s|\vec{r}') = \sigma_\alpha^2 \nabla^2 Q_\alpha(\vec{r}, s|\vec{r}') - \omega_\alpha(\vec{r}) Q_\alpha(\vec{r}, s|\vec{r}'), \tag{8.13}$$

with the initial conditions, $Q_\alpha(\vec{r}, 0|\vec{r}') = \delta(\vec{r}-\vec{r}')$. In later applications, it is convenient to introduce two end-integrated propagators, $q_\alpha(\vec{r}, s)$ and $q_\alpha^+(\vec{r}, s)$, defined by

$$\begin{aligned} q_\alpha(\vec{r}, s) &= \int d\vec{r}' Q_\alpha(\vec{r}, s|\vec{r}'), \\ q_\alpha^+(\vec{r}, s) &= \int d\vec{r}' d\vec{r}'' Q_\alpha(\vec{r}, s|\vec{r}') Q_\beta(\vec{r}', f_\beta|\vec{r}''), \end{aligned} \tag{8.14}$$

where $\beta = B$ if $\alpha = A$ and vice versa. These end-integrated propagators satisfy the same differential equation as $Q_\alpha(\vec{r}, s|\vec{r}')$, with different initial conditions,

$$\begin{aligned} q_\alpha(\vec{r}, 0) &= 1, \\ q_\alpha^+(\vec{r}, 0) &= \int d\vec{r}' Q_\beta(\vec{r}, f_\beta|\vec{r}') = q_\beta(\vec{r}, f_\beta). \end{aligned} \tag{8.15}$$

Henceforth this notation is implicit throughout this chapter. In terms of the end-integrated propagators, the single-chain partition function is given by

$$Q_c(\{\omega_\alpha\}) = \frac{1}{V}\int d\vec{r} q_\alpha^+(\vec{r}, f_\alpha). \tag{8.16}$$

Because the single-chain partition function $Q_c(\{\omega\})$ depends on the field $\omega_\alpha(\vec{r})$ in an implicit form, it is useful to express $Q_c(\{\omega\})$ in terms of a series of $\omega_\alpha(\vec{r})$, leading to an explicit expression. This can be achieved by writing the fields $\omega_\alpha(\vec{r})$ in the form,

$$\omega_\alpha(\vec{r}) = \omega_\alpha^{(0)}(\vec{r}) + \delta\omega_\alpha(\vec{r}), \tag{8.17}$$

where $w_\alpha^{(0)}(\vec{r})$ are some known functions. For many cases, it is convenient to chose $w_\alpha^{(0)}(\vec{r})$ as the mean-field solutions of the system. In terms of $w_\alpha^{(0)}(\vec{r})$ and $\delta w_\alpha(\vec{r})$, the propagator $Q_\alpha(\vec{r}, s|\vec{r}')$ can be obtained as a perturbation series,

$$Q_\alpha(\vec{r}, s|\vec{r}') = Q_\alpha^{(0)}(\vec{r}, s|\vec{r}') + \sum_{n=1}^{\infty} (-)^n \int d\vec{r}_1 \cdots d\vec{r}_n$$

$$\int ds_1 \cdots ds_n \delta w_\alpha(\vec{r}_1) \cdots \delta w_\alpha(\vec{r}_1) \cdots \delta w_\alpha(\vec{r}_n) G_\alpha(\vec{r}, s|\vec{r}_n, s_n)$$

$$G_\alpha(\vec{r}_n, s_n|\vec{r}_{n-1}, s_{n-1}) \cdots G_\alpha(\vec{r}_2, s_2|r_1, s_1) G_\alpha(\vec{r}_1, s_1|\vec{r}', 0), \quad (8.18)$$

where $Q_\alpha^{(0)}(\vec{r}, s|\vec{r}')$ is the propagator solution with the zeroth-order field $w_\alpha^{(0)}(\vec{r})$,

$$\frac{\partial}{\partial s} Q_\alpha^{(0)}(\vec{r}, s|\vec{r}') = \sigma_\alpha^2 \nabla^2 Q_\alpha^{(0)}(\vec{r}, s|\vec{r}') - w_\alpha^{(0)}(\vec{r}) Q_\alpha^{(0)}(\vec{r}, s|\vec{r}'). \quad (8.19)$$

The Green functions $G_\alpha(\vec{r}, s|\vec{r}', s')$ are solutions of the differential equations,

$$\left[\frac{\partial}{\partial s} - \sigma_\alpha^2 \nabla^2 + w_\alpha^{(0)}(\vec{r})\right] G_\alpha(\vec{r}, s|\vec{r}', s') = \delta(s-s')\delta(\vec{r}-\vec{r}'). \quad (8.20)$$

It is easy to show that the $G_\alpha(\vec{r}, s|\vec{r}', s')$ are related to the propagators,

$$G_\alpha(\vec{r}, s|\vec{r}', s') = \theta(s-s') Q_\alpha^{(0)}(\vec{r}, s-s'|\vec{r}'), \quad (8.21)$$

where $\theta(s-s')$ is the Heaviside step function.

Using the perturbation solutions of the propagators, $Q_\alpha(\vec{r}, s|\vec{r}')$, the single-chain partition function Q_c can be obtained as a series expansion of the $\delta w_\alpha(\vec{r})$ terms,

$$Q_c = Q_c^{(0)} + Q_c^{(1)} + Q_c^{(2)} + \cdots$$
$$= Q_c^{(0)} + \frac{1}{V} \sum_{n=1}^{\infty} \sum_{\alpha_1, \cdots, \alpha_n} (-)^n \int d\vec{r}_1 \cdots d\vec{r}_n C_{\alpha_1, \cdots, \alpha_n}^{(n)}(\vec{r}_1, \cdots, \vec{r}_n) \delta w_{\alpha_n}(\vec{r}_n), \quad (8.22)$$

where the expansion coefficients $C_{\alpha_1, \cdots, \alpha_n}^{(n)}(\vec{r}_1, \cdots, \vec{r}_n)$ are appropriate combinations of the single-chain propagators $Q_\alpha^{(0)}(\vec{r}, s|\vec{r}')$. Using the above expression for the single-chain partition function, the single-chain partition function term in the free energy can be written as a cumulant expansion,

$$V \ln Q_c = V \ln Q_c^{(0)} + \sum_{n=1}^{\infty} \frac{(-)^n}{n!} \sum_{\alpha_1, \cdots, \alpha_n} \int d\vec{r}_1 \cdots d\vec{r}_n C_{\alpha_1, \cdots, \alpha_n}(\vec{r}_1, \cdots, \vec{r}_n)$$
$$\delta w_{\alpha_1}(r_1) \cdots \delta w_{\alpha_n}(\vec{r}_n), \quad (8.23)$$

where the coefficients $C_{\alpha_1, \cdots, \alpha_n}(\vec{r}_1, \cdots, \vec{r}_n)$ are given in terms of the single-chain propagators. The physical meaning of $C_{\alpha_1, \cdots, \alpha_n}(\vec{r}_1, \cdots, \vec{r}_n)$ is that it is the nth-order cumulant correlation function of a noninteracting diblock copolymer chain in an external field $w_\alpha^{(0)}(\vec{r})$.

In order to proceed, we expand the concentration variables around some known zeroth-order solution,

$$\phi_\alpha(\vec{r}) = \phi_\alpha^{(0)}(\vec{r}) + \delta\phi_\alpha(\vec{r}). \tag{8.24}$$

The precise choice of $\phi_\alpha^{(0)}(\vec{r})$ will be determined later. The free-energy functional can now be expanded in the form,

$$F = F^{(0)} + F^{(1)} + F^{(2)} + \cdots, \tag{8.25}$$

where the zeroth-order term is given by

$$F^{(0)} = \frac{\rho R_g^3}{N} \left\{ \int d\vec{r} \left[\chi N \phi_A^{(0)}(\vec{r}) \phi_B^{(0)}(\vec{r}) - \sum_\alpha \omega_\alpha^{(0)}(\vec{r}) \phi_\alpha^{(0)}(\vec{r}) \right] - V \ln Q_c^{(0)}(\{\omega_\alpha^{(0)}\}) \right\}. \tag{8.26}$$

The first-order contribution has the form,

$$F^{(1)} = \frac{\rho_0 R_g^3}{N} \int d\vec{r} \sum_\alpha \left\{ \left[\chi N \phi_\beta^{(0)}(\vec{r}) - \omega_\alpha^{(0)}(\vec{r}) \right] \delta\phi_\alpha(\vec{r}) - \left[\phi_\alpha^{(0)}(\vec{r}) - C_\alpha(\vec{r}) \right] \delta\omega_\alpha(\vec{r}) \right\}, \tag{8.27}$$

where $\beta = B$ if $\alpha = A$ and vice versa. The second-order contribution to the free-energy functional is

$$F^{(2)} = \frac{\rho_0 R_g^3}{N} \left\{ \int d\vec{r} \left[\chi N \delta\phi_A(\vec{r}) \delta\phi_B(\vec{r}) - \sum_\alpha \delta\omega_\alpha(\vec{r}) \delta\phi_\alpha(\vec{r}) \right] - \frac{1}{2} \sum_{\alpha\beta} \int d\vec{r} d\vec{r}' C_{\alpha\beta}(\vec{r}, \vec{r}') \delta\omega_\alpha(\vec{r}) \delta\omega_\beta(\vec{r}') \right\}. \tag{8.28}$$

The higher-order terms depend on the field variables $\delta\omega_\alpha(\vec{r})$ only, and they have the generic form,

$$F^{(n)}(\{\delta\omega_\alpha\}) = -\frac{\rho_0 R_g^3}{N} \frac{(-)^n}{n!} \sum_{\alpha_1,\cdots,\alpha_n} \int d\vec{r}_1 \cdots d\vec{r}_n$$
$$C_{\alpha_1,\cdots,\alpha_n}(\vec{r}_1,\cdots,\vec{r}_n) \delta\omega_{\alpha_1}(\vec{r}_1) \cdots \delta\omega_{\alpha_n}(\vec{r}_n). \tag{8.29}$$

It should be noted that, although there is no explicit χN dependence in $F^{(n)}(\{\delta\omega_\alpha\})$ for $n > 2$, there is an implicit χN dependence if the correlation functions are obtained in terms of the mean fields $\omega_\alpha^{(0)}(\vec{r})$.

The cumulant correlation functions are completely determined by the propagators in the zeroth-order solution $\omega_\alpha^{(0)}(\vec{r})$. Explicitly, the first- and second-order cumulant correlation functions are given by [15],

$$C_\alpha(\vec{r}) = \frac{1}{Q_c^{(0)}} \int_0^{f_\alpha} ds\, q_\alpha(\vec{r},s) q_\alpha^+(\vec{r},f_\alpha-s),$$

$$C_{\alpha\alpha}(\vec{r},\vec{r}') = \frac{1}{Q_c^{(0)}} \int_0^{f_\alpha} ds \int_0^{s} ds'\, q_\alpha(\vec{r},f_\alpha-s) Q_\alpha(\vec{r},s-s'|\vec{r}') q_\alpha^+(\vec{r}',s') \quad (8.30)$$

$$+ \frac{1}{Q_c^{(0)}} \int_0^{f_\alpha} ds \int_0^{s} ds'\, q_\alpha(\vec{r}', f_\alpha-s) Q_\alpha(\vec{r}',s-s'|\vec{r}) q_\alpha^+(\vec{r},s'),$$

$$C_{\alpha\beta}(\vec{r},\vec{r}') = \frac{1}{Q_c^{(0)}} \int_0^{f_\alpha} ds \int_0^{f_\beta} ds' \int d\vec{r}_1\, q_\alpha(\vec{r},f_\alpha-s) Q_\alpha(\vec{r},s|\vec{r}_1) Q_\beta(\vec{r}_1,f_\beta-s'|\vec{r}') q_\beta(\vec{r}',s').$$

For simplicity, the superscript in the propagators $Q_\alpha^{(0)}(\vec{r},s|\vec{r}')$ has been omitted in the above expressions.

The free-energy functional obtained above forms the basis for further development. This expression is exact since there has been no approximation in the theory. In principle, it is possible to carry out the integration over the $\delta\omega_\alpha(\vec{r})$ fields, leading to an *exact* free-energy functional $F(\{\delta\phi_\alpha\})$ depending on the concentrations only. In practice, it is not possible to carry out the integration over the $\delta\omega_\alpha(\vec{r})$ fields exactly. Instead, the $\delta\omega_\alpha(\vec{r})$ integrals can be carried out approximately. One popular method is the random-phase approximation (RPA), which corresponds to a saddle-point evaluation of the $\delta\omega_\alpha(\vec{r})$ integrals, leading to a free-energy functional that depends on the density variables $\delta\phi_\alpha(\vec{r})$. For the cases where the zeroth-order solution is the homogeneous phase, the RPA treatment of the $\delta\omega_\alpha(r)$ leads to Leibler's weak-segregation theory [7].

8.3 SELF-CONSISTENT MEAN-FIELD THEORY (SCMFT)

Because exact evaluation of the partition function is in general not possible, a variety of approximate methods have been developed. The most fruitful method is the mean-field approximation, which amounts to evaluating the functional integral using a saddle-point technique. The mean-field theory can be derived by noting that the free-energy functional of the system can be written in the form,

$$F(\{\phi\},\{\omega\}) = \frac{1}{h}\tilde{F}(\{\phi\},\{\omega\}),$$

$$\tilde{F}(\{\phi\},\{\omega\}) = \int d\vec{r}\left[\chi N \phi_A(\vec{r})\phi_B(\vec{r}) - \sum_\alpha \omega_\alpha(\vec{r})\phi_\alpha(\vec{r})\right] - V\ln Q_c(\{\omega_\alpha\}), \quad (8.31)$$

where the parameter h is given by $h = N/\rho_0 R_g^3 \propto N^{-1/2}$. For long polymers, N is large and h is a small number. In the limit $h \to 0$ the functional integral (Eq. 8.8) can be approximated by the largest integrand $e^{-F(\{\phi_\alpha^{(0)}\},\{\omega_\alpha^{(0)}\})}$, where $\phi_\alpha^{(\vec{r})}$ and $\omega_\alpha^{(0)}(\vec{r})$ minimize the free-energy functional. Therefore we expect that the mean-field approximation should give reliable results for high molecular weight block copolymers.

Technically, the mean-field approximation requires that the first-order variation of the free-energy functional be zero, leading to a set of coupled equations determining the mean concentrations $\phi_\alpha^{(0)}(\vec{r})$ and fields $\omega_\alpha^{(0)}(\vec{r})$,

$$\phi_\alpha^{(0)}(\vec{r}) = C_\alpha(\vec{r}) = \frac{1}{Q_c^{(0)}} \int_0^{f_\alpha} ds\, q_\alpha(\vec{r},s) q_\alpha^+(\vec{r}, f_\alpha - s),$$

$$\omega_\alpha^{(0)}(\vec{r}) = \chi N \phi_\beta^{(0)}(\vec{r}) + \eta^{(0)}(\vec{r}),$$

(8.32)

where $\eta^{(0)}(\vec{r})$ is a Lagrangian multipler that is introduced to ensure that the incompressibility condition, $\phi_A^{(0)}(\vec{r}) + \phi_B^{(0)}(\vec{r}) = 1$ is satisfied. The single-chain partition function is given by

$$Q_c^{(0)} = \frac{1}{V} \int d\vec{r}\, q_A^+(\vec{r}, f_A) = \frac{1}{V} \int d\vec{r}\, q_B^+(\vec{r}, f_B).$$

(8.33)

The end-integrated propagators, $q_\alpha(\vec{r},s)$ and $q_\alpha^+(\vec{r},s)$ are solutions of the modified diffusion equations in the mean fields $\omega_\alpha^{(0)}(\vec{r})$,

$$\frac{\partial}{\partial s} q_\alpha(\vec{r},s) = \sigma_\alpha^2 \nabla^2 q_\alpha(\vec{r},s) - \omega_\alpha^{(0)}(\vec{r}) q_\alpha(\vec{r},s),$$

(8.34)

with the initial conditions, $q_\alpha(\vec{r},0) = 1$, $q_\alpha^+(\vec{r},0) = q_\beta(\vec{r},f_\beta)$. Because both $\phi_\alpha^{(0)}(\vec{r})$ and $\omega_\alpha^{(0)}(\vec{r})$ are determined self-consistently from the above equations, the mean-field approximation is often referred as the self-consistent mean-field theory (SCMFT). In the literature, the SCMFT is often referred to simply as the self-consistent field theory (SCFT). In this chapter we try to distinguish the mean-field theory by using the term SCMFT.

Within the mean-field approximation, the free energy per chain of the system is obtained by inserting the mean-field solution into the free-energy expression,

$$f^{(0)} = \frac{N}{\rho_0 R_g^3 V} F^{(0)}$$

$$= \frac{1}{V} \int d\vec{r} \left[\chi N \phi_A^{(0)}(\vec{r}) \phi_B^{(0)}(\vec{r}) - \sum_\alpha \omega_\alpha^{(0)}(\vec{r}) \phi_\alpha^{(0)}(\vec{r}) \right] - \ln Q_c^{(0)}(\{\omega_\alpha^{(0)}\}).$$

(8.35)

It is interesting to examine the structure of the SCMFT for a diblock copolymer melt. Within the mean-field approximation, the parameters entering the theory are the combination χN, the block volume fraction $f_A = 1 - f_B$, and the effective Kuhn lengths σ_α. The parameter χN characterizes the degree of segregation and it is controlled by temperature and the molecular weight. On the other hand, the polymer structure is characterized by the block composition f_A and the block Kuhn lengths $\sigma_\alpha b$. Within the mean-field approximation, the thermodynamic properties of a diblock copolymer melt are completely specified by χN, f_A and σ_α.

It is useful to note that the mean-field density profiles $\phi_\alpha^{(0)}(\vec{r})$ and the free energy are invariant under the transformation, $\omega_\alpha(\vec{r}) \to \omega_\alpha(\vec{r}) + \bar{\omega}_\alpha$, where $\bar{\omega}_\alpha$ are constants. Using this property we can choose the constants such that the mean-field potential $\omega_\alpha(\vec{r})$ satisfies the condition, $\int d\vec{r} \omega_\alpha(\vec{r}) = 0$. In this case the mean-field equations become,

$$\phi_\alpha^{(0)}(\vec{r}) = \frac{1}{Q_c^{(0)}} \int_0^{f_\alpha} ds\, q_\alpha(\vec{r}, s) q_\alpha^+(\vec{r}, f_\alpha - s), \qquad (8.36)$$

$$\omega_\alpha^{(0)}(\vec{r}) = \chi N \left[\phi_\beta^{(0)}(\vec{r}) - f_\beta\right] + \eta^{(0)}(r).$$

The method to solve the SCMFT equations is conceptually straightforward. The first step is to make an initial guess of the mean fields $\omega_\alpha^{(0)}(\vec{r})$, which bears the symmetry of the ordered phase under investigation. The modified diffusion equations with the appropriate initial and boundary conditions are then solved to obtain the propagators, $Q_\alpha(\vec{r}, s | \vec{r}')$. These propagators can then be used to compute the mean-field concentrations, $\phi_\alpha^{(0)}(\vec{r})$. The next step is to adjust the mean fields $\omega_\alpha^{(0)}(\vec{r})$ according to an iterative procedure so as to satisfy the self-consistent equations and the incompressibility condition. For a given set of controlling parameters such as $\{f_A, \chi N, \sigma_\alpha\}$ for a diblock copolymer melt, there are many solutions to the mean-field equations, corresponding to different morphologies. The phase diagram is constructed by finding the structures with the lowest free energy density.

The simplest solution of the mean-field equations is obtained for a homogeneous phase, in which the polymer concentrations and the mean-field potentials are constants, $\phi_\alpha^{(0)} = f_\alpha$, $\omega_\alpha^{(0)} = 0$, leading to the trivial solution $q_\alpha(\vec{r}, s) = q_\alpha^+(\vec{r}, s) = 1$. The free energy per chain of a homogeneous phase is therefore given by

$$f_H^{(0)} = \frac{N}{\rho_0 R_g^3 V} F_H^{(0)} = \chi N \phi_A^{(0)} \phi_B^{(0)} = \chi N f_A (1 - f_A). \qquad (8.37)$$

Because of the complexity of the theory even within the mean-field approximation, analytical solutions of the mean-field theory can only be obtained under special conditions, such as for the homogeneous phase. In order to obtain analytical solutions of the SCMFT equations, approximation methods are needed. In particular, approximate methods have been developed in the weak-segregation limit (WSL) [7] and in the strong-segregation limit (SSL) [9]. The connection of the SCMFT and the WSL and SSL theories has been reviewed by Matsen in [3]. To obtain an exact solution, numerical techniques are required.

Since the formulation of the SCFT for block copolymers by Helfand in 1975 [5], great effort has been devoted to the solution of the SCFT equations. To date, the most efficient and accurate method to solve the self-consistent mean-field equations is the reciprocal-space method developed by Matsen and Schick [14], which is based on the expansion in terms of plane wave-like basis functions. Recently, with the availability of increasing computing power and new numerical techniques, real-space methods have been developed to the level that they can be used to explore the possible phases for a given block copolymer architecture [17–19].

8.3.1 RECIPROCAL-SPACE METHOD

The key aspect of the reciprocal space method is that the functions of interest within the SCMFT, i.e., the mean-field concentrations, the mean-field potentials, as well as the end-integrated propagators are all periodic functions. For an ordered phase, the reciprocal lattice vectors are completely specified by the space group of that structure. For the case of diblock copolymers, the ordered structures are the one-dimensional lamellar phase, the two-dimensional hexagonal phase with space group p6m, the BCC spherical phase with space group Im3m, and the double-gyroid phase with space group Ia3d. For a given ordered phase, the reciprocal lattice vectors $\vec{G}$ can be identified. These reciprocal lattice vectors are ordered according to their magnitudes $|\vec{G}^{(n)}|$. The plane waves $e^{i\vec{G}\cdot\vec{r}}$ corresponding to these reciprocal lattice vectors are used as the basis functions so that the densities and fields can be expanded in the form,

$$\phi_\alpha^{(0)}(\vec{r}) = \sum_{\vec{G}} \phi_\alpha(\vec{G}) e^{i\vec{G}\cdot\vec{r}},$$
$$\omega_\alpha^{(0)}(\vec{r}) = \sum_{\vec{G}} \omega_\alpha(\vec{G}) e^{i\vec{G}\cdot\vec{r}}. \tag{8.38}$$

The ordered structure is then completely specified by the Fourier components $\phi_\alpha(\vec{G})$ and $\omega_\alpha(\vec{G})$. In terms of these Fourier components, the SCMFT equations in Fourier space become

$$\phi_\alpha(\vec{G}) = \frac{1}{Q_c^{(0)}} \sum_{\vec{G}'} \int_0^{f_\alpha} dt\, q_\alpha(\vec{G}', t) q_\alpha^+(\vec{G}-\vec{G}', f_\alpha-t), \tag{8.39}$$

where the single-chain partition function Q_c is given by $Q_c = q_A^+(\vec{G}=0, f_A)$. The fields $\omega_\alpha(\vec{G})$ are determined self-consistently from,

$$\omega_\alpha(\vec{G}) = \chi N \left[\phi_\beta(\vec{G}) - f_\beta \delta_{\vec{G},\vec{0}} \right] + \eta(\vec{G}), \tag{8.40}$$

where $\eta(\vec{G})$ is introduced to ensure the incompressibility condition $\phi_A(\vec{G}) + \phi_A(\vec{G}) = \delta_{\vec{G},\vec{0}}$. In Fourier space, the modified diffusion equations for the end-integrated propagators become,

$$\frac{\partial}{\partial t}q_\alpha(\vec{G},t) = -\sum_{\vec{G}'} H_\alpha(\vec{G},\vec{G}')q_\alpha(\vec{G}',t),$$
$$\frac{\partial}{\partial t}q_\alpha^+(\vec{G},t) = -\sum_{\vec{G}'} H_\alpha(\vec{G},\vec{G}')q_\alpha^+(\vec{G}',t),$$
(8.41)

where the "Hamiltonians" $H_\alpha(\vec{G},\vec{G}')$ are defined by $H_\alpha(\vec{G},\vec{G}') \equiv \sigma_\alpha^2 G^2 \delta_{\vec{G},\vec{G}'} + \omega_\alpha(\vec{G}-\vec{G}')$. The initial conditions for the propagators in Fourier space are,

$$q_A(\vec{G},0) = q_B(\vec{G},0) = \delta_{\vec{G},\vec{0}},$$
$$q_A^+(\vec{G},0) = q_B(\vec{G},f_B),$$
$$q_B^+(\vec{G},0) = q_A(\vec{G},f_A).$$
(8.42)

The solution of the modified diffusion equation can be obtained by making an analogy with the Schrödinger equation in quantum mechanics. In order to do this, we construct the eigenvalues and eigenfunctions of the Hamiltonians $H_\alpha(\vec{G},\vec{G}') \equiv \sigma_\alpha^2 G^2 \delta_{\vec{G},\vec{G}'} + \omega_\alpha(\vec{G}-\vec{G}')$,

$$\sum_{\vec{G}'} H_\alpha(\vec{G},\vec{G}')\psi_n^\alpha(\vec{G}') = \varepsilon_n^\alpha \psi_n^\alpha(\vec{G}).$$
(8.43)

Because these operators are Hermitian operators, all the eigenvalues of them are real and the eigenfunctions form a complete orthonormal set,

$$\sum_{\vec{G}} \psi_n^{\alpha*}(\vec{G})\psi_m^\alpha(\vec{G}) = \delta_{nm},$$
$$\sum_n \psi_n^{\alpha*}(\vec{G})\psi_n^\alpha(\vec{G}') = \delta_{\vec{G},\vec{G}'}.$$
(8.43)

Therefore the propagators can be expanded in terms of the eigenfunctions, leading to solutions of the propagators in the form,

$$q_\alpha(\vec{G},t) = \sum_n e^{-\varepsilon_n^\alpha t} q_n^\alpha(0)\psi_n^\alpha(\vec{G}),$$
$$q_\alpha^+(\vec{G},t) = \sum_n e^{-\varepsilon_n^\alpha t} q_n^{\alpha+}(0)\psi_n^\alpha(\vec{G}),$$
(8.44)

where the coefficients $q_n^\alpha(0)$ and $q_n^{\alpha+}(0)$ are determined by the appropriate initial conditions,

$$q_n^\alpha(0) = \psi_n^{\alpha*}(\vec{0}),$$
$$q_n^{\alpha+}(0) = \sum_m e^{-\varepsilon_m^\beta f_\beta} \psi_m^{\beta*}(\vec{0}) \left[\sum_{\vec{G}'} \psi_m^\beta(\vec{G}')\psi_n^{\alpha*}(\vec{G}') \right].$$
(8.45)

Using these expressions, the single-chain partition function are given by

$$Q_c^{(0)} = \sum_{n,m} e^{-\varepsilon_n^A f_A} \psi_n^A(\vec{0}) \left[\sum_{\vec{G}'} \psi_n^{A*}(\vec{G}') \psi_m^B(\vec{G}') \right] \psi_m^{B*}(\vec{0}) e^{-\varepsilon_m^B f_B}. \quad (8.46)$$

The density profiles are now specified by

$$\phi_\alpha(\vec{G}) = \frac{f_\alpha}{Q_c} \sum_{n,m} \frac{1 - e^{-(\varepsilon_m^\alpha - \varepsilon_n^\alpha) f_\alpha}}{(\varepsilon_m^\alpha - \varepsilon_n^\alpha) f_\alpha} e^{-\varepsilon_n^\alpha f_\alpha} \psi_n^{\alpha*}(\vec{0}) \left[\sum_{\vec{G}'} \psi_n^\alpha(\vec{G}') \psi_m^\alpha(\vec{G} - \vec{G}') \right] q_m^{\alpha+}(0). \quad (8.47)$$

The plane-wave expansion method described above is a general method that applies to any periodic structure. However, using the plane waves as basis functions leads to a large number of Fourier coefficients, and many of these coefficients are related by the symmetry of the system. For an ordered structure, it is possible to reduce the number of independent coefficients by exploiting the point group symmetry of the structure. The basic idea is that, due to the point group symmetry, the Fourier coefficients for the reciprocal lattice vectors within one star are related [26]. A set of new basis functions, which are linear combinations of the plane waves with wave vectors within one star, can be constructed using this observation. Each of these new basis functions is a linear combination of the form,

$$f_n(\vec{r}) = \frac{1}{\sqrt{N_n}} \sum_{i \in n} S_i^n e^{i\vec{G}_i^n \cdot \vec{r}}, \quad (8.48)$$

where the wave vectors $\vec{G}_i^n$ are related by the point group symmetry operation and satisfy the relation $|\vec{G}_i^n|^2 = \lambda_n$, the factor S_i^n assumes the values ± 1 according to the space group and N_n is the number of reciprocal lattice vectors belonging to the n-th star. The values of S_i^n and N_n can be found from the International Table of Crystallography [26]. A periodic function can be expanded in terms of the basis functions $f_n(\vec{r})$. In particular, we have,

$$\begin{aligned} \phi_\alpha^{(0)}(\vec{r}) &= \sum_n \phi_n^\alpha f_n(\vec{r}), \\ \omega_\alpha^{(0)}(\vec{r}) &= \sum_n \omega_n^\alpha f_n(\vec{r}) \end{aligned} \quad (8.49)$$

The original Fourier coefficients are related to the new coefficients by $\phi_\alpha(\vec{G}_i^n) = \phi_n^\alpha S_i^n / \sqrt{N_n}$ and $\omega_\alpha(\vec{G}_i^n) = \omega_n^\alpha S_i^n / \sqrt{N_n}$.

Expanding all periodic functions in terms of $f_n(\vec{r})$, the SCMFT equations can be cast in terms of the coefficients, just as the Fourier representation described above. In particular, the chain-conformation Hamiltonians became symmetric matrices with matrix elements,

$$H_{nm}^\alpha = \sigma_\alpha^2 \lambda_n \delta_{nm} + \sum_l \Gamma_{nml} \omega_l^\alpha, \quad (8.50)$$

where the coefficients Γ_{nml}, or the Γ symbol, are defined by,

$$\Gamma_{nml} = \frac{1}{V}\int d\vec{r} f_n(\vec{r}) f_m(\vec{r}) f_l(\vec{r})$$
$$= \frac{1}{\sqrt{N_n N_m N_l}} \sum_{i \in n} \sum_{j \in m} \sum_{k \in l} S_i^n S_j^m S_k^l \delta_{\vec{G}_i^n + \vec{G}_j^m + \vec{G}_k^l, \vec{0}}. \quad (8.51)$$

The eigenvalues and eigenfunctions of the matrices H_{nm}^α are found by solving the equations

$$\sum_m H_{nm}^\alpha \psi_{mi}^\alpha = \varepsilon_i^\alpha \psi_{ni}^\alpha, \quad (8.52)$$

where i labels the eigenvalues and eigenfucntions. The eigenfunctions are orthonormal and form a complete set,

$$\sum_n \psi_{ni}^\alpha \psi_{nj}^\alpha = \delta_{ij}, \quad \sum_i \psi_{ni}^\alpha \psi_{mi}^\alpha = \delta_{nm}. \quad (8.53)$$

The SCMFT equations can now be cast in terms of the eigenvalues and eigenfunctions. For diblock copolymers, it is useful to construct the following quantities using the eigenvalues and eigenfunctions of the operators H_{nm}^α,

$$L_{ij}^{AB} = \sum_n \psi_{ni}^A \psi_{nj}^B, \quad L_{ij}^{BA} = \sum_n \psi_{ni}^B \psi_{nj}^A;$$
$$S_{n,ij}^A = \sum_{m,l} \Gamma_{nml} \psi_{mi}^A \psi_{lj}^A, \quad S_{n,ij}^B = \sum_{m,l} \Gamma_{nml} \psi_{mi}^B \psi_{lj}^B, \quad \psi_{lj}^B \quad (8.54)$$
$$F_i^A = e^{-\varepsilon_i^A f_A} \psi_{1i}^A, \quad F_i^B = e^{-\varepsilon_i^B f_B} \psi_{1i}^B.$$

Using these quantities, the single-chain partition function and the density coefficients are given by

$$Q_c^{(0)} = \sum_{ij} F_i^A L_{ij}^{AB} F_j^B;$$
$$\phi_n^A = \frac{f_A}{Q_c^{(0)}} \sum_{ijk} g\left(\varepsilon_j^A f_A - \varepsilon_i^A f_A\right) F_i^A S_{n,ij}^A L_{jk}^{AB} F_k^B, \quad (8.55)$$
$$\phi_n^B = \frac{f_B}{Q_c^{(0)}} \sum_{ijk} g\left(\varepsilon_j^B f_B - \varepsilon_i^B f_B\right) F_i^B S_{n,ij}^B L_{jk}^{BA} F_k^A,$$

where the Debye function $g(x)$, $x = (\varepsilon_j^\alpha - \varepsilon_i^\alpha) f_\alpha$, is defined by $g(x) = (1-e^{-x})/x$. The coefficients of the fields ω_n^α are determined from the self-consistent conditions,

$$\omega_n^A = \chi N \left(\phi_n^B - f_B \delta_{n,1}\right) + \eta_n,$$
$$\omega_n^B = \chi N \left(\phi_n^A - f_A \delta_{n,1}\right) + \eta_n, \quad (8.56)$$

where the coefficients η_n are to be adjusted so that the system satisfies the incompressibility condition, $\phi_n^A + \phi_n^B = \delta_{n,1}$.

For a given set of control parameters $\{\chi N, f_A, \sigma_\alpha\}$ and a specific space group, the SCMFT equations (Eqs. (8.50), (8.52), (8.55) and (8.56)) can be solved using an iterative process starting from an initial guess for the fields ω_n^α. Details of the numerical implementation of the reciprocal-space method can be found in [27]. The reciprocal method of SCMFT has been applied to a large number of block copolymer systems [28], leading to a large body of literature on the study of equilibrium phase behavior. These studies have established a quantitative relation between molecular architecture, composition, and equilibrium phase behavior. A good understanding of the block copolymer phase behavior has emerged from these studies. The understanding gained from the SCMFT studies has been very successful in helping to explain the complex mesophases experimentally observed in block-copolymer systems. Figure 8.1 shows a schematic of the ordered phase formed by diblock-copolymers, and the SCMFT phase diagram for diblock copolymer melts is given in Figure 8.2. Further details of the progress can be found in the review of Matsen [3].

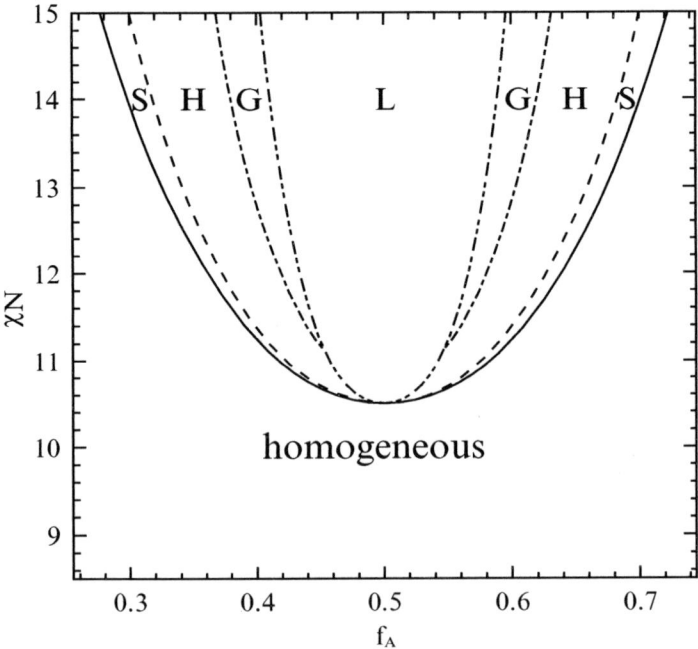

Figure 8.2 SCMFT phase diagram of diblock copolymer melt. The phases are indicated by the symbols (L) lamellar, (G) gyroid, (H) cylindrical and (S) spherical.

8.3.2 REAL-SPACE METHOD: RECENT PROGRESS

The reciprocal-space method described above is numerically efficient for a precise computation of free energies and phase diagrams. However, this method requires the space group of the ordered phase as an input. It is therefore desirable to develop methods that do not require prior knowledge of the symmetry of the phase. One possibility of achieving this is to solve the SCMFT equations in real space.

Since the formulation of SCMFT of block copolymers, a variety of techniques have been developed to solve the SCMFT equations in real space. The procedure of solving the SCMFT equations starts with an initial guess for the mean fields $\omega_\alpha^{(0)}(\vec{r})$. The modified diffusion equations are solved to obtain the end-integrated propagators, and the results are used to compute the mean-field densities according to Eq. (8.36). The fields for the next iterations are obtained using a linear mixture of new (present iteration) and old (previous iteration) fields [27]. The iteration is repeated until the solution becomes self-consistent. The main computational effort in the real-space method resides in solving the modified diffusion equations. One popular method is the Crank–Nicholson scheme [17].

One of the earliest numerical solutions of the SCMFT for diblock copolymers was obtained by Helfand and Wasserman [10]. In more recent years, numerical solutions using real-space methods have been obtained by, among others, Vavasour and Whimore [12]. Due to limited computing power, for these early techniques the generally anisotropic Wigner–Seitz unit cells was approximated by a spherical one, so that the SCMFT equations could be reduced to one-dimensional form. The spherical-unit-cell approximation has been quite successful in the studies of diblock-copolymer melts [12] and blends [29].

Recently there has been a renaissance in the development of real-space methods for SCMFT. This is motivated mainly by the desire to develop methods that do not require a structural input. Towards this goal, Drolet and Fredrickson [17] have suggested a real-space approach that solves the SCMFT equations in a large box using a nonconserved relaxation dynamics. Exploring the effect of different initial configurations, it has been demonstrated that this method can be used to carry out combinatorial screening of block copolymer phases. This approach has been applied to a number of block copolymer systems [30]. Another method of searching for ordered phases of block copolymer systems has been proposed by Bohbot-Raviv and Wang [18], in which the free-energy functional in an arbitrary unit cell is minimized with respect to the composition profiles *and* the dimension of the unit cell. More recently, an efficient algorithm to solve the SCMFT equations has been proposed by Tzeremes *et al.* [19]. Although these new methods have only been applied to a limited number of systems, promising results have been achieved. It is hoped that a combination of these methods and powerful computing facilities will lead to a truly predictive tool based on SCMFT that can be used to explore materials science of block copolymers.

8.4 STABILITY OF ORDERED PHASES: GAUSSIAN FLUCTUATIONS

The SCMFT presented above has been very successful in explaining, sometimes quantitatively, the equilibrium phase behavior of block copolymer systems. This success stems from the fact that the SCMFT is asymptotically exact when the parameter $h = N/\rho_0 R_g^3 \propto N^{-1/2} \to 0$, as discussed above. Therefore for block copolymers with high molecular weight, we expect the SCMFT to give an accurate description of the thermodynamic properties of the system.

The equilibrium structures predicted by SCMFT correspond to the solutions obtained at the extrema of the free-energy functional of the system. These solutions do not necessarily ensure the minimization of the free-energy functional. The mean-field solution may, for example, correspond to a saddle point. In order to investigate the stability of the ordered phases, we have to consider the effect of the higher-order contributions to the free-energy functional. In particular, the Gaussian fluctuation contributions derived above can be used to predict the stability of any ordered structure. In what follows we formulate the theory of Gaussian fluctuations in ordered phases [15,27,31,32].

8.4.1 GAUSSIAN FLUCTUATIONS AND RPA CORRELATION FUNCTIONS

In order to examine the Gaussian fluctuations, it is convenient to chose the zeroth-order solution such that it corresponds to the self-consistent mean field solution, that is, $\phi_\alpha^{(0)}(\vec{r})$ and $\omega_\alpha^{(0)}(\vec{r})$ are chosen such that the first-order free-energy functional vanishes, $F^{(1)} = 0$. With this choice of the zeroth-order solution, the partition function of the system becomes [15]

$$Z = e^{-F^{(0)}(\{\phi^{(0)}\},\{\omega^{(0)}\})} \int \prod_\alpha [D\{\delta\phi_\alpha\}D\delta\omega_\alpha]$$

$$\prod_{\vec{r}} \delta\left(\sum_\alpha \delta\phi_\alpha(\vec{r})\right) e^{-F^{(2)}(\{\delta\phi\},\{\delta\omega\}) - \sum_{n=3}^\infty F^{(n)}(\{\delta\omega\})} . \qquad (8.57)$$

In order to examine the effect of fluctuations on the mean-field solution, we keep only the lowest (second) order. Thus we are dealing with Gaussian fluctuations around the mean-field solution. To this order of approximation, the partition function is

$$Z \approx e^{-F^{(0)}} \int \prod_\alpha [D\{\delta\phi_\alpha\}D\{\delta\omega_\alpha\}] \prod_{\vec{r}} \delta\left(\sum_\alpha \delta\phi_\alpha(\vec{r})\right) e^{-F^{(2)}}, \qquad (8.58)$$

where the Gaussian fluctuation free-energy functional is given by Eq. (8.28).

Because of the incompressibility condition, the fluctuations in the concentrations are not independent. It is convenient to introduce three new collective variables,

$$\delta\phi(\vec{r}) = \delta\phi_A(\vec{r}) - \delta\phi_B(\vec{r}),$$
$$\delta\omega(\vec{r}) = \frac{1}{2}[\delta\omega_A(\vec{r}) - \delta\omega_B(\vec{r})], \quad (8.59)$$
$$\delta\xi(\vec{r}) = \frac{1}{2}[\delta\omega_A(\vec{r}) + \delta\omega_B(\vec{r})].$$

In terms of these new variables, the partition function becomes

$$Z \approx e^{-F^{(0)}} \int D\{\delta\phi\} D\{\delta\omega\} D\{\delta\xi\} e^{-F^{(2)}(\{\delta\phi\}, \{\delta\omega\}, \{\delta\xi\})}, \quad (8.60)$$

where the Gaussian fluctuation free energy is given by

$$F^{(2)} = \frac{\rho_0 R_g^3}{N} \frac{1}{2} \int d\vec{r} d\vec{r}' \{ -(\chi N/2)\delta(\vec{r}-\vec{r}')\delta\phi(\vec{r})\delta\phi(\vec{r}')$$
$$- 2\delta(\vec{r}-\vec{r}')\delta\phi(\vec{r})\delta\omega(\vec{r}')' - C(\vec{r},\vec{r}')\delta\omega(\vec{r})\delta\omega(\vec{r}') - \Delta_1(\vec{r},\vec{r}')\delta\omega(\vec{r})\delta\xi(\vec{r}') \quad (8.61)$$
$$- \Delta_2(\vec{r},\vec{r}')\delta\xi(\vec{r})\delta\omega(\vec{r}') - \Sigma(\vec{r},\vec{r}')\delta\xi(\vec{r})\delta\xi(\vec{r}')\}.$$

In the above expression for the free-energy functional, we have introduced four new functions, $C(\vec{r},\vec{r}')$, $\Delta_1(\vec{r},\vec{r}')$, $\Delta_2(\vec{r},\vec{r}')$ and $\Sigma(\vec{r},\vec{r}')$, which are combinations of the second-order copolymer correlation functions $C_{\alpha\beta}(\vec{r},\vec{r}')$,

$$C(\vec{r},\vec{r}') = C_{AA}(\vec{r},\vec{r}') - C_{AB}(\vec{r},\vec{r}') - C_{BA}(\vec{r},\vec{r}') + C_{BB}(\vec{r},\vec{r}'),$$
$$\Delta_1(\vec{r},\vec{r}') = C_{AA}(\vec{r},\vec{r}') + C_{AB}(\vec{r},\vec{r}') - C_{BA}(\vec{r},\vec{r}') - C_{BB}(\vec{r},\vec{r}'),$$
$$\Delta_2(\vec{r},\vec{r}') = C_{AA}(\vec{r},\vec{r}') - C_{AB}(\vec{r},\vec{r}') + C_{BA}(\vec{r},\vec{r}') - C_{BB}(\vec{r},\vec{r}'), \quad (8.62)$$
$$\Sigma(\vec{r},\vec{r}') = C_{AA}(\vec{r},\vec{r}') + C_{AB}(\vec{r},\vec{r}') + C_{BA}(\vec{r},\vec{r}') + C_{BB}(\vec{r},\vec{r}').$$

For many applications, it is convenient to obtain an effective free-energy functional in terms of the concentration order-parameter $\delta\phi(\vec{r})$ only. This effective free-energy functional can be obtained by performing the functional integral over $\delta\omega(\vec{r})$ and $\delta\xi(\vec{r})$. In general, the integration over $\delta\omega(\vec{r})$ and $\delta\xi(\vec{r})$ cannot be carried out exactly, and a random-phase approximation is usually used, corresponding to a saddle-point approximation of the $\delta\omega(\vec{r})$ and $\delta\xi(\vec{r})$ integrals, while keeping $\delta\phi(\vec{r})$ fixed. At the Gaussian fluctuation level, such an integral can be carried out exactly, leading to an effective free-energy functional $F^{(2)}(\{\delta\phi\})$,

$$Z \approx e^{-F^{(0)}} \left\{ \det\left[\left(\frac{\rho_0 R_g^3}{2\pi N}\right)^2 \Sigma \cdot \tilde{C}\right] \right\}^{-1/2} \int D\{\delta\phi\} e^{-F^{(2)}(\{\delta\phi\})}, \quad (8.63)$$

where $\tilde{C} = C - \Delta_1 \cdot \Sigma^{-1} \cdot \Delta_2$. In these expressions, the inverse operators are defined through the relation $\int d\vec{r}'' O^{-1}(\vec{r}, \vec{r}'') O(\vec{r}'', \vec{r}') = \delta(\vec{r} - \vec{r}')$. The effective Gaussian free-energy functional is given by

$$F^{(2)} \approx \frac{\rho_0 R_g^3}{N} \frac{1}{2} \int d\vec{r} d\vec{r}' \, \delta\phi(\vec{r}) \left[C^{RPA} \right]^{-1}(\vec{r}, \vec{r}') \delta\phi(\vec{r}'), \tag{8.64}$$

where the RPA density-density correlation function can be written in the following matrix form,

$$\left[C^{RPA} \right]^{-1}(\vec{r}, \vec{r}') = \left[\tilde{C}^{-1} - \frac{\chi N}{2} I \right](\vec{r}, \vec{r}'),$$

$$C^{RPA}(\vec{r}, \vec{r}') = \left\{ \left[I - \frac{\chi N}{2} \tilde{C} \right]^{-1} \tilde{C} \right\}(\vec{r}, \vec{r}') \tag{8.65}$$

where I is the identity matrix. Formally, the RPA correlation function obtained above is identical to that obtained by Leibler [7]. The essential difference is that the independent chain correlation functions $C_{\alpha\beta}(\vec{r}, \vec{r}')$ arise from the full mean-field solution, and hence include the anisotropic nature of the system.

The density-density correlation function $C^{RPA}(\vec{r}, \vec{r}')$ can be used to obtain information about several important thermodynamic properties: (1) the mean-field solution is linearly stable when all the eigenvalues of $C^{RPA}(\vec{r}, \vec{r}')$ are positive. The boundary in the $(f_A, \chi N)$ plane where the smallest eigenvalue of $C^{RPA}(\vec{r}, \vec{r}')$ becomes zero defines the spinodal line of that phase. (2) The experimentally observed scattering intensities correspond to the Fourier transform of $C^{RPA}(\vec{r}, \vec{r}')$. The Gaussian fluctuation theory accounts for anisotropic fluctuations and therefore enables the calculation of the scattering function for the periodic-ordered phases. (3) The elastic moduli of the ordered phases can be extracted from the density-density correlation function $C^{RPA}(\vec{r}, \vec{r}')$ [33].

Since the integration over the composition fluctuations is Gaussian, it can be carried out leading to an expression of the partition function in the form,

$$Z \approx e^{-F^{(0)}} \left\{ \det \left[\left(\frac{\rho_0 R_g^3}{2\pi N} \right)^3 \Sigma \cdot \left(I - \frac{\chi N}{2} \tilde{C} \right) \right] \right\}^{-1/2}. \tag{8.66}$$

Therefore including Gaussian fluctuations leads to a correction to the SCMFT free energy,

$$F \approx F^{(0)} + \frac{1}{2} \ln \left\{ \det \left[\left(\frac{\rho_0 R_g^3}{2\pi N} \right)^3 \Sigma \cdot \left(I - \frac{\chi N}{2} \tilde{C} \right) \right] \right\}. \tag{8.67}$$

In principle, the inclusion of the Gaussian fluctuations will lead to the modification of the mean-field phase boundaries, as indicated by the above expression. The computation of this correction involves the calculation of all eigenvalues of the operators Σ and $\tilde{C}$, which requires large-scale computational efforts and has not been carried out.

8.4.2 METHOD OF SOLUTION: RECIPROCAL-SPACE FORMULISM

The above theoretical framework was developed in real space. However, computing the correlation functions in real space can be carried out only for simple geometries, such as the lamellar phase [31]. In order to apply the theory to more complex structures, efficient methods other than the direct real-space computation have to be developed. One particularly useful method is the reciprocal-space technique, which utilizes the symmetries of the ordered phases. The key observation is that the mean-field solution $\omega_\alpha(r) = \sum_{\vec{G}} \omega^\alpha(\vec{G}) e^{i\vec{G}\cdot\vec{r}}$ is a periodic function with reciprocal lattice vectors $\{\vec{G}\}$. Therefore, the modified diffusion equation for the chain propagators is equivalent to the Schrödinger equation of an electron in a crystalline solid. The eigenvalues and eigenfunctions of the Hamiltonian are determined from

$$H_\alpha \psi^\alpha(\vec{r}) = \varepsilon^\alpha \psi^\alpha(\vec{r}), \qquad (8.68)$$

where the Hamiltonian is defined by $H_\alpha = -\sigma_\alpha^2 \nabla^2 + \omega_\alpha(\vec{r})$. It is well known from solid-state theory that the electronic energy in a crystal forms bands according to Bloch's theorem, i.e., the eigenmodes of H_α can be labeled by a band index n and a reduced wave vector $\vec{k}$ within the first Brillouin zone [34]. The normalized eigenfunctions of H_α are Bloch functions of the form

$$\psi^\alpha_{n\vec{k}}(\vec{r}) = \frac{1}{\sqrt{V}} \sum_{\vec{G}} u^\alpha_{n\vec{k}}(\vec{G}) e^{i(\vec{k}+\vec{G})\cdot\vec{r}}, \qquad (8.69)$$

where the set $\{\vec{G}\}$ constitutes the reciprocal lattice vectors of the ordered structure under investigation. The coefficients $u^\alpha_{n\vec{k}}(\vec{G})$ are solutions of the following set of linear eigenvalue equations,

$$\sigma_\alpha^2 [(\vec{k}+\vec{G})^2 - \varepsilon^\alpha_n(\vec{k})] u^\alpha_{n\vec{k}}(\vec{G}) + \sum_{\vec{G}'} \omega^\alpha(\vec{G}-\vec{G}') u^\alpha_{n\vec{k}}(\vec{G}') = 0, \qquad (8.70)$$

where $\varepsilon^\alpha_n(\vec{k})$ are the eigenvalues and $\omega^\alpha(\vec{G})$ are the Fourier components of the auxiliary mean fields.

Because the eigenfunctions $\psi^\alpha_{n\vec{k}}(\vec{r})$ of the Hamiltonian (which is Hermitian) form a complete orthonormal basis, we can use them as basis functions to formulate the theory. This will allow us to take advantage of the symmetry of the ordered phase. In particular, the mean-field propagator can be written in the form

$$Q_\alpha(\vec{r},t|\vec{r}') = \sum_{n\vec{k}} e^{-\varepsilon^\alpha_n(\vec{k})t} \psi^\alpha_{n\vec{k}}(\vec{r}) \psi^{\alpha*}_{n\vec{k}}(\vec{r}'). \qquad (8.71)$$

Similarly, the cumulant correlation functions can be expressed in terms of the Bloch functions. Using the properties of the eigenfunctions, it can be shown that the cumulant correlation functions are diagonal in the lattice vector $\vec{k}$ space,

$$C_\alpha(\vec{r}) = \sum_{n\vec{k}} C_n^\alpha \psi_{n\vec{0}}^\alpha(\vec{r}),$$

$$C_{\alpha\beta}(\vec{r},\vec{r}') = \sum_{n\vec{k}, n'\vec{k}'} C_{nn'}^{\alpha\beta}(\vec{k}) \psi_{n\vec{k}}^\alpha(\vec{r}) \psi_{n'\vec{k}'}^{\beta*}(\vec{r}').$$
(8.72)

In principle the dimension of this eigenvalue problem is infinite. In practice, since the coefficients $\omega^\alpha(\vec{G})$ become smaller for large reciprocal lattice vectors, it can be truncated at a finite number N, leading to an $N \times N$ eigenvalue problem. The solution of this problem can be carried out using standard linear algebra packages. A generic feature of the eigenvalues in a periodic potential is the presence of band gaps at the zone boundaries.

The Gaussian fluctuation contributions derived above can be used to predict the stability of any ordered structure. Specifically, the inverse RPA density-density correlation function $[C^{\text{RPA}}]^{-1}$ can be used to obtain information about the stability, spinodal point and the most unstable mode of the ordered phase. The mean-field solution is linearly stable when all the eigenvalues of C^{RPA} are positive. The boundary in the $(f_A, \chi N)$ plane where the smallest eigenvalue of C^{RPA} becomes zero defines the spinodal line of that phase.

Specifically, the stability analysis starts with the eigenmodes of Gaussian fluctuations [27,32],

$$\int d\vec{r}' [C^{\text{RPA}}]^{-1}(\vec{r},\vec{r}') \Psi_{n\vec{k}}(\vec{r}') = \lambda_n(\vec{k}) \Psi_{n\vec{k}}(\vec{r}),$$
(8.73)

where the eigenfunctions $\Psi_{n\vec{k}}(\vec{r})$ form a complete set. Expanding the fluctuations $\delta\phi(\vec{r})$ in terms of these eigenmodes, $\delta\phi(\vec{r}) = \sum_{n\vec{k}} \delta\phi_{n\vec{k}} \Psi_{n\vec{k}}(\vec{r})$ we can write the effective Gaussian fluctuation energy functional in the form,

$$F^{(2)} = \frac{1}{2} \sum_\lambda \lambda_n(\vec{k}) |\delta\phi_{n\vec{k}}|^2.$$
(8.74)

The eigenvalues $\lambda_n(\vec{k})$ form a band structure. A typical eigenvalue spectrum is shown in Figure 8.3. The anisotropic fluctuations are therefore quantified by the eigenvalue band $\lambda_n(\vec{k})$. The smallest eigenvalue $\lambda_0(\vec{k}_0)$, which occurs at a specific reciprocal vector $\vec{k}_0$, determines the thermodynamic stability of the ordered phase. The ordered structure is stable if $\lambda_0(\vec{k}_0) > 0$ and it becomes thermodynamically unstable when $\lambda_0(\vec{k}_0) \leq 0$. The condition that $\lambda_0(\vec{k}_0) = 0$ defines the spinodal point of the ordered phase being considered. Furthermore, the profiles of the fluctuations are characterized by the eigenfunctions $\Psi_{n\vec{k}}(\vec{r})$. In particular, the most unstable mode $\Psi_{0\vec{k}_0}(\vec{r})$, corresponding to the smallest eigenvalues $\lambda_0(\vec{k}_0)$,

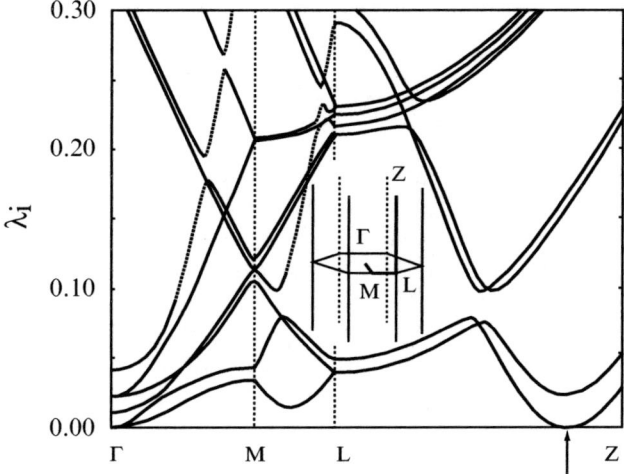

Figure 8.3 The eigenvalues of the operator $[C^{RPA}]^{-1} \cdot \tilde{C} = [I - \frac{\chi N}{2} \tilde{C}]$ for the cylindrical phase at $\chi N = 10.9$, $f = 0.428$, showing the band structure of the eigenvalue spectrum. The insert illustrates the first Brillouin zone of the system. The smallest eigenvalue occurs at the zone boundary, as indicated by the arrow. The smallest eigenvalue is negative in this case, indicating that the cylindrical structure is unstable at this point of the phase space. (Reproduced from M. Laradji et al. Phys. Rev. Lett **78**, 2577 (1997) Copyright (1997) with permission from the American Physical Society).

characterizes the initial kinetics of the transitions away from the ordered phase. The most unstable mode can therefore be used to predict the kinetic pathways of the order–order phase transitions since an unstable (or marginally metastable) structure will initially follow this mode towards a stable structure.

8.4.3 SCATTERING FUNCTIONS OF ORDERED PHASES

The above theoretical framework provides a technique to compute the scattering functions of ordered phases. The experimentally observed scattering intensities correspond to the Fourier transform of the density-density correlation function [35],

$$I(\vec{q}) = \frac{1}{V} \int d\vec{r} d\vec{r}' \, \langle \phi(\vec{r}) \phi(\vec{r}') \rangle e^{-\vec{q} \cdot (\vec{r} - \vec{r}')}, \tag{8.75}$$

where $\phi(\vec{r}) = \phi_A(\vec{r}) - \phi_B(\vec{r})$. For an ordered phase the density profiles can be written in the form, $\phi(\vec{r}) = \sum_{\vec{G}} \phi^{(0)}(\vec{G}) e^{i\vec{G} \cdot \vec{r}} + \delta\phi(\vec{r})$. Inserting this form into the expression of $I(\vec{q})$, we have

$$I(\vec{q}) = \sum_{\vec{G}} |\phi^{(0)}(\vec{G})|^2 \delta_{\vec{q}, \vec{G}} + S(\vec{q}), \tag{8.76}$$

where $S(\vec{q})$ is the scattering function due to composition fluctuations and is given by

$$S(\vec{q}) = \frac{1}{V} \int d\vec{r} d\vec{r}' \langle \delta\phi(\vec{r})\delta\phi(\vec{r}') \rangle e^{-\vec{q}\cdot(\vec{r}-\vec{r}')}. \tag{8.77}$$

Therefore the scattering intensity is composed of two contributions. The first one corresponds to a set of Bragg peaks located at the reciprocal lattice vectors, whose amplitudes are determined by the mean-field solution. The second contribution is due to composition fluctuations and it is given by the density-density correlation function $\langle \delta\phi(\vec{r})\delta\phi(\vec{r}') \rangle$. Within RPA, this correlation function is given by $\langle \delta\phi(\vec{r})\delta\phi(\vec{r}') \rangle = C^{RPA}(\vec{r}, \vec{r}')$. Therefore the anisotropic Gaussian fluctuation theory enables the computation of the scattering functions

$$S(\vec{q}) = \frac{1}{V} \int d\vec{r} d\vec{r}' C^{RPA}(\vec{r}, \vec{r}') e^{-\vec{q}\cdot(\vec{r}-\vec{r}')}$$
$$= \sum_{\vec{G}} \sum_{n,n'} C^{RPA}_{nn'}(\vec{q}-\vec{G}) u^{A*}_{n,\vec{q}-\vec{G}}(\vec{G}) u^{A}_{n',\vec{q}-\vec{G}}(\vec{G}). \tag{8.78}$$

8.4.4 APPLICATION TO DIBLOCK COPOLYMERS

Using this exact formulation of the anisotropic Gaussian fluctuation theory, the stability of the ordered diblock copolymer phases has been examined [27,31,32]. The spinodal lines of the ordered phases were computed and the kinetic pathways between the various phases have been investigated. Furthermore, the scattering functions of the ordered phase were calculated. Due to the limited computing power, the analysis was performed in the weak-segregation regime. In this region of the phase diagram the gyroid phase has higher free energy and was not included in the analysis.

For diblock copolymers in the weak-segregation regime, it was found that the lamellar, cylindrical, and spherical one-phase regions are encapsulated by their spinodal lines, indicating the robustness of these three "classical" phases that are observed in diblock copolymer melts. Because details of the investigations have been published previously [27,32], we will use the lamellar phase as an example to illustrate the information that can be extracted from the study of Gaussian fluctuations.

For a lamellar structure, a typical scattering function is presented in Figure 8.4 for $\chi N = 10.8$ and $f = 0.462$. At this point of the phase diagram the lamellar phase is metastable, and the equilibrium phase is the cylindrical phase. The scattering function has two strong Bragg peaks at $q_z = \pm k_0$, $q_x = 0$ (where $k_0 = 2\pi/D$ and D is the period of the layers), and two higher-order Bragg peaks at $q_z = \pm 2k_0$, $q_x = 0$. These peaks indicate a relatively strong lamellar ordering. Besides these Bragg peaks, four additional scattering peaks located

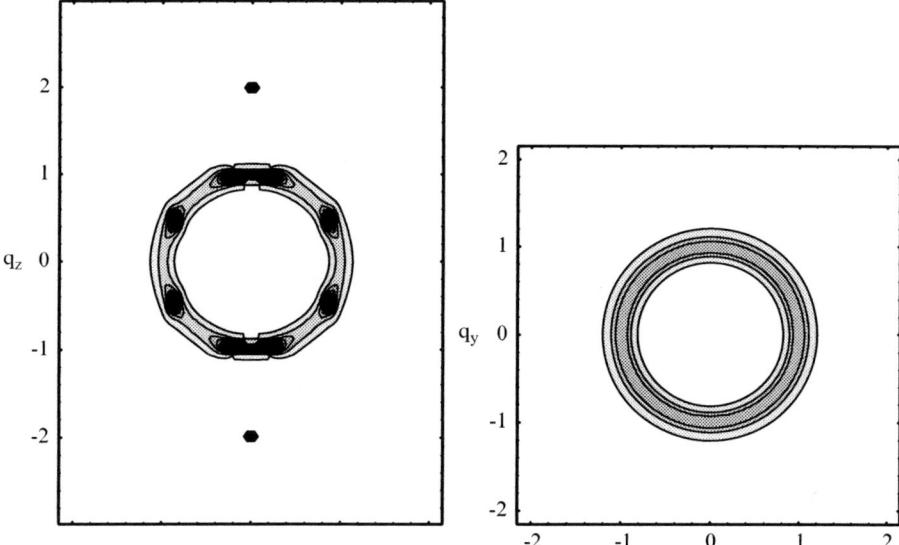

Figure 8.4 Calculated scattering function of the lamellar phase at $\chi N = 10.8$, $f = 0.462$. The left and right figures show the x–z and x–y cuts, respectively. (Reproduced from M. Laradji et al. Macromolecules **30**, 3242 (1997) Copyright (1997) with permission from the American Chemical Society).

$q_z = \pm k_0/2$, $q_x = \pm\sqrt{3}k_0/2$ are observed. These four peaks correspond to the smallest eigenvalues and therefore, the most unstable mode of the structure. It can be shown that these modes are in-plane fluctuation modes with hexagonal ordering, as can be seen from the six-fold symmetry of the scattering peaks at $q_z = \pm k_0$, $q_x = 0$ and $q_z = \pm k_0/2$, $q_x = \pm\sqrt{3}k_0/2$. These calculated scattering functions are in agreement with experiments [36]. The fact that the most unstable mode leads to six-fold cylindrical structures can also be seen from the real-space contour plots of the density profiles (Figure 8.5).

The fact that the most unstable mode of a lamellar phase is infinitely degenerate in the x–y plane can be used to show that, when the lamellar phase is driven into other ordered phases, the kinetics of the transition proceeds through a long-lived intermediate modulated-layered state that may correspond to the experimentally observed perforated layered structures. If one direction of the most unstable mode is excited, it will lead to the undulation of the layers that eventually form cylindrical structures (Figure 8.5). However, because of the degeneracy of the fluctuation modes, it is most likely that more than one direction of the fluctuation modes will be excited, leading to the formation of a perforated layered structure (Figure 8.6).

The stability analysis has been carried out for other phases and information about the kinetic pathways between the ordered structures has been obtained. Many of the predictions, especially the transition pathways and epitaxial rela-

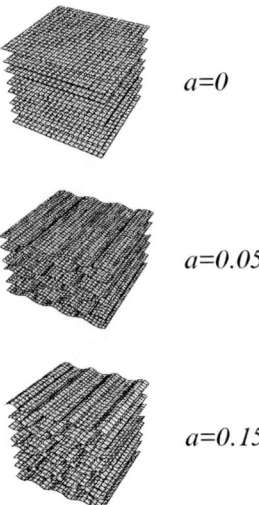

Figure 8.5 Three-dimensional contour plots of the lamellar structure showing the effects of the most unstable mode. The contour plots are defined by $\phi_A(\vec{r}) = \phi_A^{(0)}(\vec{r}) + a\Psi_{0\vec{k}_0}(\vec{r})$., where a is the amplitude of the most unstable mode. It is interesting to note that, for relatively large amplitude, the fluctuation mode drives the lamellae into a hexagonally packed cylindrical structure. (Reproduced from M. Laradji *et al*. Macromolecules **30**, 3242 (1997) Copyright (1997) with permission from the American Chemical Society).

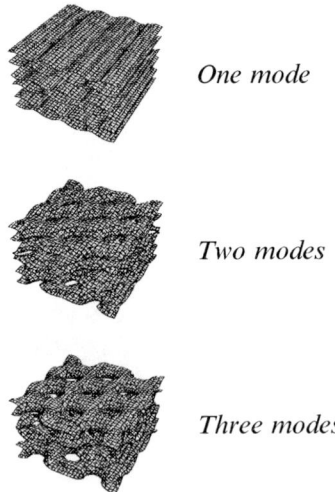

Figure 8.6 Three-dimensional contour plots showing the effects of the degeneracy of the most unstable modes. The parameters are the same as in Figure 8.4. It is obvious that the simultaneous excitation of the most unstable modes leads to a perforated layered structure. (Reproduced from M. Laradji *et al*. Macromolecules **30**, 3242 (1997) Copyright (1997) with permission from the American Chemical Society).

tions between the cylindrical and spherical structures, are in good agreement with experiments [37,38] and results from time-dependent Ginzburg–Landau theory [39].

Besides studying block copolymer melts, the Gaussian fluctuation theory can be used to study mixtures and other mesoscopic systems. Using the theory of anisotropic Gaussian fluctuations and modeling the lipids as diblock copolymers, Li and Schick [40] have studied the role of nonlamellar-forming lipids in biological membranes. The model system is a mixture of two diblock copolymers, one of which forms lamellar phases, while the other forms inverted hexagonal phases. The fluctuation modes were used to examine the effect of the nonlamellar-forming lipids on the formation of nonlamellar structures in the mixtures.

8.5 CONCLUSIONS AND DISCUSSION

In this chapter the self-consistent field theory for block copolymers has been reviewed. Although a diblock copolymer melt is used as a model system, it should be emphasized that the theoretical methods of deriving the SCFT and techniques of solving the SCMFT equations are very flexible and versatile. Extension of the theory to more complex block copolymers is straightforward. The current formulation allows a systematic analysis of the thermodynamic properties of block-copolymer systems. The equilibrium phases and phase diagrams can be obtained by solving the SCMFT equations, while the stability of the ordered phases can be analyzed using the theory of Gaussian fluctuations.

Within the mean-field approximation, the reciprocal-space method of Matsen and Schick [14] provides an efficient and accurate numerical technique to solve the SCMFT equations for given ordered structures. This has led to a comprehensive understanding of the equilibrium phase behavior of simple block copolymer systems. Valuable insights into the physics of the self-assembly in block copolymer systems have been obtained from the numerical solutions. In particular, the formation of different structures can be explained using the concepts of spontaneous interfacial curvature and packing frustration [3].

The reciprocal method of SCMFT requires *a priori* knowledge of the structure of the phases. The recently introduced real-space methods have the potential to predict the phases without prior assumptions about the structures [17–19]. It is feasible that a combination of the real-space and reciprocal-space methods will lead to a numerical platform that is capable of predicting the phases and phase diagrams for complex block copolymers. In this scheme the real-space method can be used to carry out a combinatorial search for the possible candidate structures, and the reciprocal-space method can then be used to obtain accurate free energies of these candidate structures.

In order to understand the thermodynamic stability of the ordered phases, particular attention has been paid to the theory of anisotropic Gaussian fluctuations. Our progress stems from the observation that any ordered structure has the symmetry of its space group. Using general symmetry arguments, it has been shown that for an ordered structure, the anisotropic fluctuation modes can be classified by a wave vector $\vec{k}$ within the first irreducible Brillouin zone and a band index n. The eigenvalues of the Gaussian fluctuation operator form a band structure, similar to the electronic energy band structure in a crystalline solid. The fluctuation modes are described by Bloch functions, which are plane waves modulated by periodic functions. What emerges from this observation is a powerful technique to compute the eigenvalues and eigenmodes of Gaussian fluctuations in ordered phases. This formulation of the fluctuation modes enables the calculation of the stability (spinodal) lines and the scattering functions of the ordered phases [27,32]. It should be emphasized that the symmetry argument are not restricted to block copolymer systems. The same general statement on the nature of anisotropic fluctuation modes can be applied to any ordered systems [16]. The application of the theory to diblock copolymer melts [27,32] has led to the determination of the spinodal lines of the ordered structures. The corresponding most unstable modes are obtained and used to predict the pathways and epitaxial relations between ordered structures. Furthermore, the scattering functions of the ordered phases are calculated. Many of these predictions are in good agreement with theoretical studies based on time-dependent Ginzburg–Landau theory [39] and experimental observations [37,38].

Despite the success of the theory, our example stability analysis was carried out in the weak segregation regime. This was dictated by the computing power available to us at that time. It is desirable to extend the theory to intermediate to strong segregation regimes. This extension will, in particular, resolve the region of stability of the gyroid phase [41]. With the availability of high-performance computing facilities, such calculations will become feasible.

Another future project using the Gaussian fluctuation theory is the calculation of phase diagrams. In principle, including Gaussian fluctuations will lead to corrections to the mean-field free energy for the ordered phases (Eq. 8.50). This correction will lead to shifts of the phase boundaries for block copolymers. Carrying out a calculation of this correction to the mean-field phase diagram will give us insights into the significance of the fluctuations effects.

ACKNOWLEDGEMENTS

I would like to thank Dr I. W. Hamley for the invitation to contribute this chapter. The Gaussian fluctuation theory resulted from a fruitful collaboration with Drs C. Yeung, M. Laradji, R. C. Desai, and especially Dr J. Noolandi. I would like to acknowledge many useful discussions with Drs M. Matsen, M. Schick, Z.-G. Wang, and M. D. Whitmore. The research at McMaster is support by NSERC and the Research Corporation.

REFERENCES

1. F. S. Bates and G. H. Fredrickson, *Physics Today* **52**, 32 (1999)
2. I. W. Hamley, *The Physics of Block Copolymers*, Oxford University Press, Oxford (1998)
3. M. W. Matsen, *J. Phys.: Condens. Matter* **14**, R21 (2002)
4. S. Edwards, Proc. Phys. Soc. **85**, 613 (1965)
5. E. Helfand, *J. Chem. Phys.* **62**, 999 (1975); *Macromolecules* **8**, 552 (1975)
6. K. M. Hong and J. Noolandi, *Macromolecules* **14**, 727 (1981)
7. L. Leibler, *Macromolecules* **13**, 1602 (1980)
8. T. Ohta and K. Kawasaki, *Macromolecules* **19**, 2621 (1986)
9. A. N. Semenov, *Sov. Phys. - JETP* **61**, 733 (1985)
10. E. Helfand and Z. R. Wasserman, *Macromolecules* **9**, 879 (1976)
11. K. R. Shull, *Macromolecules* **25**, 2122 (1992)
12. J. D. Vavasour and M. W. Whitmore, *Macromolecules* **25**, 5477 (1992)
13. M. Banaszak and M. W. Whitmore, *Macromolecules* **25**, 3406 (1992)
14. M. W. Matsen and M. Schick, *Phys. Rev. Lett.* **72**, 2041 (1994)
15. A.-C. Shi, J. Noolandi and R. C. Desai, *Macromolecules* **29**, 6487 (1996)
16. A.-C. Shi, *J. Phys.: Condens. Matter* **11**, 10183 (1999)
17. F. Drolet and G. H. Fredrickson, *Phys. Rev. Lett.* **83**, 4317 (1999)
18. Y. Bohbot-Raviv and Wang, *Phys. Rev. Lett.* **85**, 3428 (2000)
19. G. Tzeremes, K. O. Rasmussen, T. Lookman and A. Saxena, *Phys. Rev. E* **65**, 041806 (2002)
20. V. Ganesan and G. H. Fredrickson, *Europhys. Lett.* **55**, 814 (2001)
21. G. H. Fredrickson, V. Ganesan and F. Drolet, *Macromolecules* **35**, 16 (2002)
22. F. Schmid, *J. Phys.: Condens. Matter* **10**, 8105 (1998)
23. M. W. Whitmore and J. D. Vavasour, *Acta. Polymerica* **46**, 341 (1990)
24. E. F. David and K. S. Schweizer, *J. Chem. Phys.* **100**, 7784 (1994); M. Guenza and K. S. Schweizer, *J. Chem. Phys.* **106**, 7391 (1997)
25. For a recent review of polymer simulations, see S. C. Glotzer and W. Paul, *Annu. Rev. Mater. Res.* **32**, 401 (2002)
26. *International Tables of X-Ray Crystallography*; N. F. M. Henry and K. Lonsdale, Eds., Kynoch Press, Birmingham, UK (1965)
27. M. Laradji, A.-C. Shi, J. Noolandi and R. C. Desai, *Macromolecules* **30**, 3242 (1997)
28. M. W. Matsen and F. S. Bates, *Macromolecules* **29**, 1091 (1996)
29. A.-C. Shi and J. Noolandi, *Macromolecules* **27**, 2936 (1994); **28**, 3103 (1995)
30. F. Drolet and G. H. Fredrickson, *Macromolecules* **34**, 5317 (2001)
31. C. Yeung, A.-C. Shi, J. Noolandi and R. C. Desai, *Macromol. Theory Simul.*, **5**, 291 (1996)
32. M. Laradji, A.-C. Shi, R. C. Desai and J. Noolandi, *Phys. Rev. Lett.* **78**, 2577 (1997)
33. A.-C. Shi and Z.-G. Wang, unpublished.
34. N. W. Ashcroft and N. D. Mermin, *Solid State Physics* (Saunders, 1976)
35. P. M. Chaikin and T. C. Lubensky, *Principles of Condensed Matter Physics* (Cambridge University Press, Cambridge, 1995)
36. I. W. Hamley, K. A. Koppi, J. H. Rosedale, F. S. Bates, K. Almdal and K. Mortensen., *Macromolecules* **26**, 5959 (1993)
37. C. Y. Ryu, M. E. Vigild and T. P. Lodge, *Phys. Rev. Lett.* **81**, 5354 (1998)
38. C. Y. Ryu and T. P. Lodge *Macromolecules* **32**, 7190 (1999)
39. S. Qi and Z-G Wang, *Phys. Rev. Lett.* 76, 1679 (1996); *Phys. Rev.* **E55, 1682 (1997)**
40. X-J. Li and M. Schick, *J. Chem. Phys.* 112, 10599 (2000)
41. M. W. Matsen, *Phys. Rev. Lett.* **80**, 201 (1998); M. Laradji, R. C. Desai, A.-C. Shi and J. Noolandi *Phys. Rev. Lett.* **80**, 202 (1998)

9 Lithography with Self-Assembled Block Copolymer Microdomains

CHRISTOPHER HARRISON
Schulumberger-Doll Research, 36 Old Quarry Road, Ridgefield CT 06877, USA

JOHN A. DAGATA
National Institute of Standards and Technology, Gaithersburg MD 20899, USA

DOUGLAS H. ADAMSON
Princeton Materials Institute, Princeton University, Princeton NJ 08540, USA

9.1 INTRODUCTION

9.1.1 MOTIVATION

The quest for faster, cheaper, and more powerful electronics has driven the semiconductor industry to ever smaller feature sizes, ca. 130 nm at the time of writing. The ingenuity and success of this industry are breathtaking [1]. As the expense of lithographic technologies has increased and the importance of computational power has grown, alternatives to conventional photolithography have been put forward by academic and industrial researchers. One such alternative, the focus of this review, has been the use of block-copolymer microdomains as a lithographic template. Rather than using photolithography or electron beam lithography as a means of pattern formation, a block copolymer is allowed to self-assemble into the desired structure. Some degree of intelligent guidance may be utilized, depending upon the application's needs. While self-assembly is generally limited to a few periodic forms of high symmetry (e.g. spheres or cylinders), it turns out that there is great use for such structures in lithography, particularly that related to information storage.

There are many research groups working on block-copolymer lithography; many more have expressed interest in joining the field. Researchers who are new to this area will find the broad review of block copolymer thin films by Fasolka and Mayes [2] to be a highly valuable resource. However, to this researcher's knowledge, no comprehensive review of block copolymer lithography currently exists. Therefore, we undertake one such review here that will enable lithography researchers to quickly come up to speed.

Developments in Block Copolymer Science and Technology. Edited by I. W. Hamley
© 2004 John Wiley & Sons, Ltd ISBN: 0 470 84335 7

9.1.2 ANECDOTAL ORIGIN OF BLOCK COPOLYMER LITHOGRAPHY

According to the often-told story, sometime in the late 1980s, P. M. Chaikin was in the office of L. J. Fetters at Exxon Research and Engineering where he noticed an electron micrograph of a hexagonal array of dots [3]. The array was produced by a microphase-separated block copolymer system with a lattice spacing of 30 nm. Realizing that this length scale was perfect for electron transport measurements of the so-called Hofstadter butterfly pattern [4], he began research to harness these patterns as lithographic templates, first putting P. Mansky on the task [5]. With the help of N. Thomas's students, they were able to show that polystyrene-polyisoprene block copolymers were largely compatible with semiconductor-based lithographic techniques. The addition of R. A. Register brought further copolymer expertise to the project, especially concerning the use of ozone for templating nanostructures, the focus of one of us (C. H.) during his first years in graduate school. Progress was further facilitated with the synthesis of a wide range of block copolymers by one (D. H. A.) of us. The project accelerated with the addition of Miri Park whose nanolithography skills had been honed at the Cornell Nanofabrication Facility. Indeed, the majority of the initial publications resulted from numerous trips from Princeton to Ithaca, a perilous journey during the winter. Subsequent group members continued to develop the technology such that the polymer pattern could be transferred to semiconductor and metallic films. Since then many research groups have joined the field and prototypes of products based on this technology are being evaluated.

9.1.3 EMERGING TECHNOLOGY

A further motivation to writing this chapter has been the use of copolymer lithography by researchers at the Corporate Research and Development Center of the Toshiba Corporation. Section 9.7 [6] details their clever use of copolymer lithography for information storage. These researchers demonstrate that CoCrPt films can be patterned without difficulty on hard-drive platters that have been embossed for microdomain templating. While applications such as this were one of the motivations for academic work over the past decade, until recently there has been a paucity of projects with a truly applied focus.

9.1.4 OVERVIEW OF CHAPTER

We present an overview of this chapter's organizational layout. In Section 9.2 we describe polymer synthesis and the resulting self-assembled structures, both in bulk and thin films. In Section 9.3 we describe the imaging technologies necessary for working in this field. Building on this, we describe the means to control microdomain orientation in Section 9.4. Section 9.5 discusses the chemical and metallic modifications that are possible to optimize a copolymer pattern for lithographic use. Section 9.6 describes the progress of researchers applying conventional lithographic tools to copolymer templates. Finally, Section 9.7 discusses currently emerging applications and possibilities for the future.

9.2 SELF-ASSEMBLED STRUCTURES

Self-assembly is a smart means of using chemistry and thermodynamics to select a desired pattern. While self-assembly can be used in a variety of systems ranging from surfactant-templated silicates [7] (nanometers) to colloidal dispersions (microns) [8], the focus here will be on block copolymers (tens of nanometers). Block copolymers with $\chi N \gg 10$ microphase separate above their glass transition temperature, where χ is the Flory–Huggins interaction parameter and N is the degree of polymerization [9,10]. The resulting morphology depends largely on the relative volume fraction of the components. Some of the more commonly seen morphologies are lamellae, cylinders, and spheres (Figure 9.1). The length scale of microdomains is determined by the length of the polymer

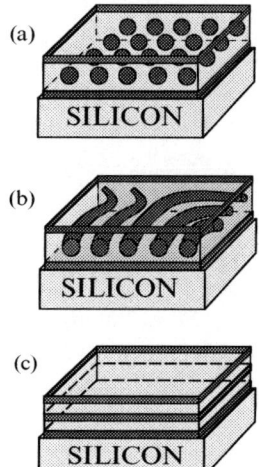

Figure 9.1 Schematic of block-copolymer microdomains in thin films. Panels a and b show one layer of spheres and cylinders, respectively (darker component). Note the additional wetting layers above and below the microdomains that serve to satisfy the interfacial constraints; the specific configuration varies depending upon the copolymer system. Panel c shows lamellar sheets oriented parallel to the substrate by the strong alignment influence of the substrate and vacuum/polymer interfaces. (Reproduced with permission from the American Association for the Advancement of Science)

chains. The thermodynamics of these structures have been extensively examined in bulk, and more recently, in thin films [2].

9.2.1 CHEMICAL SYNTHESIS

The narrow polydispersity of block copolymers (necessary for self-assembly with a good degree of long-range order) is greatly facilitated by the use of living anionic polymerization techniques [11–13]. Anionic polymerization affords a fairly large selection of monomers and yields materials with well-defined composition and well-controlled molecular mass. Due to these advantages, anionic polymerization is the typical method used in the synthesis of polymers for self-assembly. Recently, however, some evidence has emerged that certain polydisperse systems microphase separate to produce periodic structures with good order [14].

Nonpolar hydrocarbon monomers such as styrene, isoprene, and butadiene are polymerized in hydrocarbon solvents such as benzene or cyclohexane. Initiation is achieved with the use of alkyllithiums such as sec-butyllithium and molecular mass is controlled by the ratio of initiator to monomer. The living nature of anionic polymerization allows the syntheses of block copolymers by sequential addition of the monomers. After one monomer is exhausted, the chain remains reactive, or "living." The addition of the second monomer then continues the polymerization to form a block copolymer. Such techniques are used to synthesize polystyrene-polyisoprene or polystyrene-polybutadiene copolymers (PS-PI or PS-PB, respectively).

Polar monomers such as 2-vinylpyridine and methyl methacrylate are normally polymerized in polar solvents such as tetrahydrofuran and at low temperature ($-78\,^\circ$C). In addition, additives such as LiCl are often added to help lower the rates of termination reactions to levels insignificant in the time frame of the reaction. Block copolymers made with nonpolar and polar monomers start with the nonpolar monomer because of its greater reactivity. These active centers are then typically capped with 1,1-diphenylethylene to lower their reactivity before the addition of the polar monomer. This helps eliminate side reactions resulting from addition of the active center to electrophilic sites in the polar monomers. The two polar polymers, polystyrene-2-vinylpyridine (PS-P2VP) and polystyrene-poly methyl methacrylate (PS-PMMA) have been extensively studied in thin films.

Ring-opening anionic polymerization is used in the synthesis of polyferrocenyldimethylsilane (PFS) (see section 9.6.3). This mechanism involves nucleophilic addition of a polymer anion to a cyclic monomer. The monomer then ring opens, leading to incorporation of the monomer into the growing chain and generation of a new anion. This method is not as common as those previously mentioned, but can result in well-defined polymers with unique chemical composition.

Experimentally, there are two general methods for anionic polymerization. One is high-vacuum polymerization, and the other is inert-atmosphere polymerization. High-vacuum polymerization has the advantage of higher levels of purity over longer periods of time. This can be very important for high molecular mass polymers or polymers containing coupling agents that have long reaction times (months) and must be kept very clean for a long time. The disadvantage is the added effort and time needed to run a reaction under high vacuum. The reactor must be made with glass-blowing techniques and the reactants added by break seals.

Inert-atmosphere techniques, on the other hand, are complementary to high-vacuum techniques. Rather than using a vacuum, an inert gas such as nitrogen or argon is used to maintain the absence of moisture, oxygen or carbon dioxide. Less effort is required, and for simple diblock copolymers, the loss of some purity may not be detrimental to the self-assembly of the polymers.

A relatively new technique used for the random PI-PS brushes mentioned later in this review (Section 9.2.2) is controlled-radical polymerization [15–17]. This technique reduces the rate of radical recombination by lowering the effective concentration of radicals. This new method opens the possibility of random polymers such as those used in brushes. These polymers have a relatively narrow polydispersity (especially for radically produced polymers).

Hydrogenation is often used to improve the chemical and thermal stability of polymers. Hydrogenation of isoprene, for instance, saturates the double bonds to form poly(ethylene propylene) or PEP [18]. This polymer is chemically distinct from PI and has different properties. One advantage is that PEP is much less sensitive to oxidation than PI, and so can be heated in the presence of oxygen with no significant degradation. Hydrogenation is done under hydrogen pressure, with either soluble or insoluble catalysts. It is possible to hydrogenate a diene in the presence of styrene, or to hydrogenate the styrene as well.

9.2.2 SELF-ASSEMBLY IN THIN FILMS

While bulk systems are easier to analyze, effective and rapid implementation of copolymer lithography is contingent upon the fabrication and control of microdomains in resist-like thin films. Use of such films can take advantage of well-developed resist technologies (spin coating and film characterization) thereby speeding their adaption into fabrication environments. However, the structure formed by copolymers in bulk or melt state may differ from that of thin films. Such films can easily be fabricated by spin coating from dilute (ca. 1% wt) solutions onto smooth silicon wafers where the thickness can be easily tuned by the usual parameters of spin speed and solution concentration. For films of thickness comparable to the microdomain spacing (10 nm to 100 nm thick), the influence of the interface dominates, leading to structures different from that in bulk. A large fraction of polymer in thin films is devoted to satisfying the wetting constraints, e.g. PMMA preferably wets silicon wafers in a PS-PMMA system [2]. Microdomains are often submerged inside the film, though this depth depends upon the molecular mass and chemistry of the copolymer system [18]. Films thinner than a critical thickness often exhibit no microdomains as all polymeric material is used to wet the interfaces. Microdomain structure varies as well – many researchers have noted that copolymers that form cylinders in bulk form spheres in sufficiently thin films [19]. Lastly, thin films often suffer from kinetic or pinning influences from the surface that lead to short microdomain correlation lengths or grain sizes. Surface modifications have been employed to try to ameliorate surface pinning [20].

9.3 IMAGING MICRODOMAINS IN THIN FILMS

Perhaps the first challenge faced by researchers working on copolymer lithography was finding a fast, reliable, and robust imaging technique. Rigorous diagnosis of the success or failure of each lithography step requires analysis via an imaging method; since there can be dozens of steps in multilayer lithography a rapid technique is essential. There are two dominant imaging techniques that have emerged as strong research tools: the scanning electron microscope (SEM) and the atomic force microscope (AFM). The instrument choice depends mainly upon which technique generates higher contrast for each particular copolymer system, and of course, availability and researcher preference.

9.3.1 ELECTRON MICROSCOPY

The earliest work on imaging block copolymer microdomains relied heavily upon transmission electron microscopy (TEM), and it still proves to be a useful tool to this day [19]. Samples are either microtomed or solvent cast to produce thin (ca. 100 nm) sections. PS-PI or PS-PB samples can be stained with osmium tetroxide to increase contrast. Osmium tetroxide reacts selectively with unsaturated double bonds such as found in PI or PB microdomains so as to provide mass contrast [21]. Unfortunately, TEM requires that the samples be freestanding or transferred to a transparent support (e.g. carbon), a cumbersome and time-consuming process that is largely incompatible with silicon or GaAs wafers. While silicon nitride membranes can be employed for TEM, these expensive and delicate structures are not easily accessible to all researchers [22].

Compared to TEM, scanning electron microscopy (SEM) suffers from lower resolution but its ease of use and ability to image surface features have made it the workhorse of microlithography. Indeed, easy access to SEMs in clean-room environments plays a large role in its choice by many researchers. Furthermore, developments in the past decade have made low-voltage, high-resolution

SEM more commonplace in clean rooms, narrowing the resolution gap between SEM and TEM [23]. Low-voltage SEM (around 1 kV) is advantageous for the insulating nature of copolymer masks increases its susceptibility to charging effects. In some cases, a thin metal coating is used to decrease sample charging, but researchers often find this unnecessary when using the latest generation of low-voltage scanning electron microscopes–an advantage not be overlooked, as coating a sample with a nonuniform metal film can obscure important features. Examples of SEM micrographs of spherical and cylindrical microdomains are shown in Figure 9.2

Microdomains that present surface topography, such as those formed by PS-PMMA copolymers, are readily imaged by SEM with standard topography enhancing operating procedures. However, microdomains of other copolymer systems such as PS-PI are often submerged beneath a surface wetting layer, requiring additional steps to enhance contrast. First, conventional OsO_4 staining is used to produce contrast in a manner consistent with TEM sample preparation. In some cases, the operating voltage can be used to image slightly submerged microdomains. Specifically, low-voltage (1 kV) SEM enhances features within the top 10 nm, but by increasing the operating voltage to 5 kV, features as deep as 25 nm have been successfully imaged [24]. Secondly, for those features further submerged beneath the surface, reactive ion etching has been shown to provide an even higher level of contrast; optimal choice of the etching gas to minimize (or maximize) the selectivity is paramount [18]. Plasma etching has the added advantage of providing depth information, important in multilayer systems [25].

9.3.2 ATOMIC FORCE MICROSCOPY

Atomic force microscopy has become a powerful tool for examining the surface of block copolymer films. Researchers have produced a set of publications on imaging copolymer microdomains and proved its utility in many, though not all, copolymer systems [26–29]. Unfortunately, AFMs are often not readily available in the typical clean-room setting, though their presence is growing. Additionally, an inexperienced researcher can easily damage the AFM tip, the recognition of which is a learned skill and leads to a very shallow learning curve.

An AFM generates contrast by sensing either topographic features or variation in mechanical properties, the latter by measuring the phase of an oscillating tip. Tapping mode appears to be the preferred imaging method as indicated by the frequency of publications that describe its use in imaging polymer films. For PS-PMMA copolymer systems, Morkved and coworkers [22] have argued that contrast originates from purely topographic effects, which is plausible as the moduli of PS and PMMA are virtually identical over a wide range of temperatures. Alternatively, a modulus difference between microdomains and

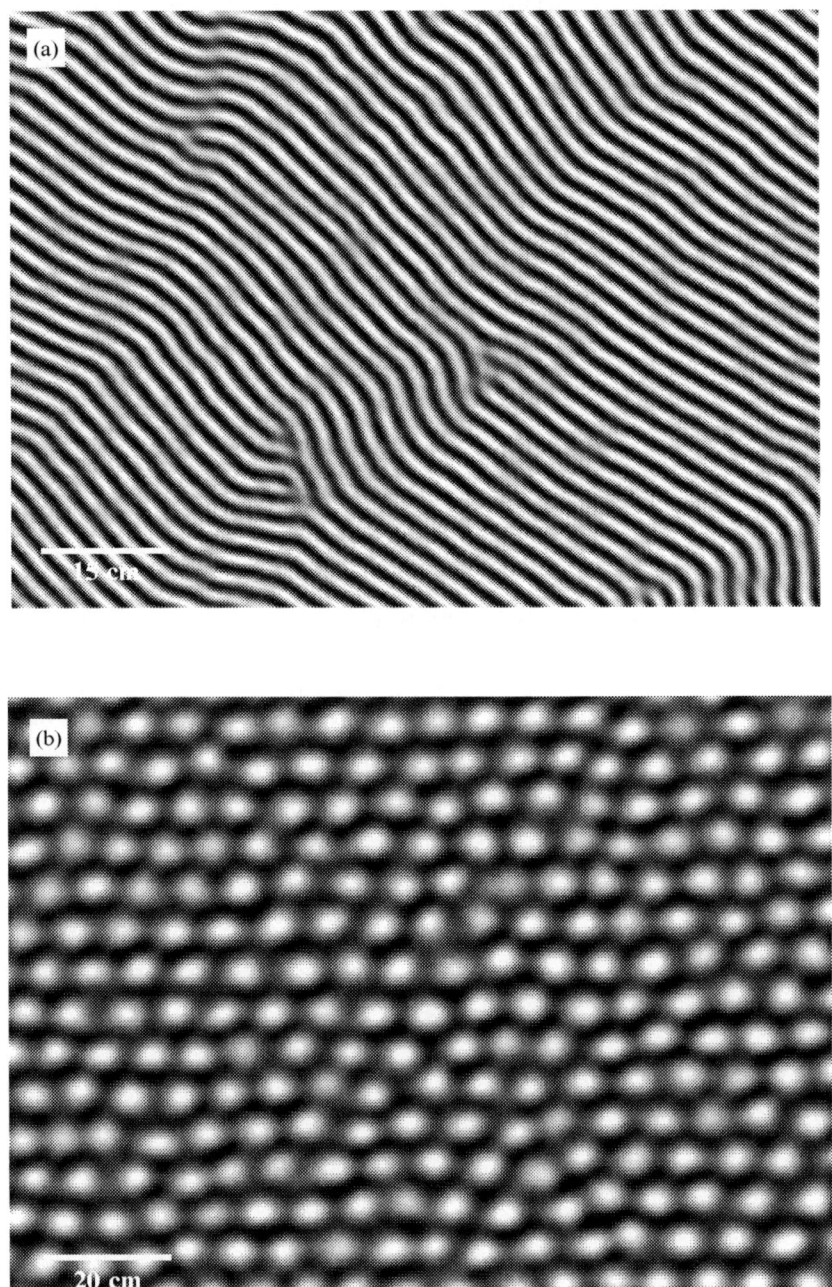

Figure 9.2 Panels (a), (b) show representative SEM images of cylinders and spheres, respectively. These images can be interpreted as plan views of the corresponding panels of Figure 9.1. Bar = 100 nm.

the matrix (hard and soft) is sufficient to give rise to contrast in PS-PEP systems [18]. These researchers have argued that there is virtually no topography in their samples and that the tip taps through a wetting layer of the softer material to sense the harder microdomains. However, contrast may originate from an entirely different mechanism: polymer–tip interactions that strongly depend upon the polymer chemistry. Lastly, it should be pointed out that it is always difficult to completely decouple any one of these effects from the others without extremely careful experiments [28,30].

9.4 MECHANISMS TO CONTROL ORIENTATION

Control over the alignment of microdomains in block copolymer films greatly increases their utility for lithography. For example, if spherical microdomains were arranged onto a lattice and modified to form metal nanodots, in theory each one could function as an addressable memory bit at a density of 10^{11} per square centimeter. Similar arguments for cylindrical microdomains suitably modified to form wires suggest that circuitry could be fabricated if only orientational control could be imposed. With these goals in mind, much effort has been made to control the orientation of films cast from copolymer solutions. The surface imposed by a flat interface (i.e. silicon wafer) does strongly affect the microdomains orientation, but the resulting orientation is not typically desirable. For example, symmetric copolymers that form lamellae usually orient with their planes parallel to the wafer interfaces, thereby reducing their utility as lithographic templates (Figure 9.1c). In an analogous fashion, cylindrical microdomains in thin films typically orient parallel to the wafer plane, but their in-plane orientation varies throughout the film. In what follows, we discuss techniques that have been exploited to control microdomain orientation with an eye towards optimizing a lithographic template.

9.4.1 ORIENTATION CONTROL THROUGH MICROFABRICATION OF TEMPLATES

Perhaps the simplest manner in which to control the orientation of microdomains is to impose physical or chemical topography. Until the last year or so, most efforts in this field had been largely restricted to influencing the features of polymer islands on topographically or chemically patterned substrates. Heier and coworkers [31,32] demonstrated that periodic chemical patterning influenced the local thickness of polymer films subsequently applied. They also demonstrated that a moderate degree of lamellar orientation can be achieved with chemically patterned substrates. Nealey and coworkers [33–35] have used an approach where advanced lithography techniques are used to alter the chemistry of the surface layer in a periodic fashion, thereby influencing the microdomain orientation of subsequently applied layers. Additionally, it has been shown that patterned substrate topography can be used to manipulate the morphology of block copolymer films, such as islands that adopt an anti-conformal arrangement with respect to surface topography [36–38]. While this has led to interesting scientific questions, control of cylindrical or spherical microdomain orientation is the most challenging and most rewarding goal.

Segalman et al. [39] have recently demonstrated that arrays of microfabricated mesas can be used to template PS-P2VP microdomains. Micrometer-scale structures were fabricated on silicon wafers and copolymers were subsequently applied via spin coating. These researchers found that the step edge could be used to template the edge of a single layer of microdomains, and hence the microdomain lattice itself (Figure 9.3). Interestingly enough, there is a slight difference in angle between the step and the microdomain lattice. Sibener and coworkers [40] have also begun to demonstrate control in a similar fashion with a PS-PMMA system. Such work is very exciting because it forwards a means of microdomain control, a crucial enabler of copolymer lithography.

The authors have demonstrated an analogous, though surprisingly different alignment mechanism (Figure 9.4). Silicon wafers were photolithographically patterned with a wide variety of mesas and trenches (typically $4 \times 50\,\mu$m) and a copolymer was applied via spin coating at an average thickness of 30 nm. Highly elongated islands consisting of one layer of cylinders (PS microdomains in a PEP matrix) were observed on the mesas. Surprisingly enough, the islands

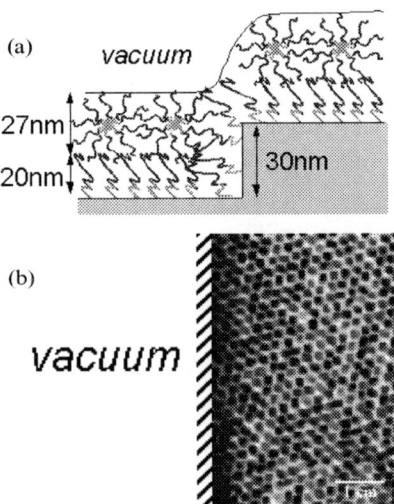

Figure 9.3 (a) Schematic of copolymer chains in the vincinity of a 30 nm step. (b) AFM image of microdomain lattice with lattice orientation templated by mesa edge. Note that microdomains extend to the edge of the step and that the lattice orientation is slightly askew from the step edge. Bar = 150 nm. (Reproduced from R. A. Seyalman et al. Adr. Mater. **12**, 1152 (2001), Copyright (2001) with permission from Witey – VCH).

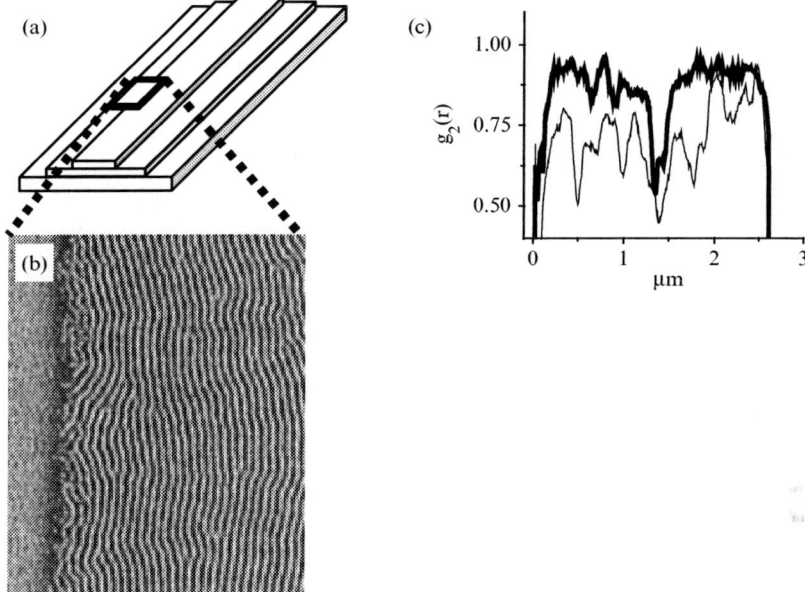

Figure 9.4 (a) Copolymer island (top structure) on silicon mesa (middle structure) fabricated on silicon wafer (bottom). Mesas are typically fifty micrometers long by four micrometers wide, copolymer film is close to three micrometers wide. (b) After spin coating the copolymer film retracts from the mesa edge during annealing and the microdomains orient along the longer dimension of the island. Bar = 200 nm. (c) Plot of microdomain alignment with mesa edge as a function of distance from mesa edge. $g_2(r)$ is the cross-correlation function of the microdomain alignment with the mesa edge where a value of 1.0 indicates perfect alignment. Note that the plot range almost spans the three-micrometer width of the copolymer island of panel (a). The middle of the mesa corresponds to 1.5 micrometers on the graph. At early times (thin line) $g_2(r)$ varies between 0.5 to 0.75, reflecting the poor alignment of the microdomains. At later times (thick line) the microdomains become more aligned and $g_2(r)$ is closer to 1.0 except for a misalignment region in the middle of the mesa. This region (dip on the graph at 1.5 micrometers) disappears with further annealing.

extended along the length of the mesas but the film retracted from the mesa edge. These elongated islands were observed to effectively orient the cylinders over dozens of micrometers, though presumably the alignment length is potentially as long as the wafer.

Cheng *et al.* [41] have recently demonstrated that the spherical microdomains of a polystyrene-polyferrocenyldimethylsilane (PS-PFS) block copolymer can be templated by microfabricated grooves. In this case the topographic relief structures were microfabricated by interference lithography, though presumably conventional lithographic techniques would work as well if high resolution was obtained. These authors demonstrated that microdomains subsequently applied aligned with grooves and that the ordering proceeded from the walls inward. Furthermore, they were able to use these microdomains to fabricate silica posts by reactive ion etching. Presumably this technique could be extended to fabricate an array of W-Co dots as well.

Hot-stage AFM is being used to investigate the kinetics of microdomain alignment with respect to a microfabricated step edge [26,42]. There has been some evidence that the alignment near a step edge proceeds faster than the coarsening process in the absence of a step. The results here are especially pertinent to the use of microfabricated steps to align microdomains. Defects may be repelled or absorbed by the step; one may be able to analyze this motion in an analogous fashion to the motion of image charges. Cheng and coworkers [41] have confirmed these observations by pointing out that PS-PFS microdomains in grooves align first at the outside and the alignment proceeds towards the middle.

Finally, it has been shown that the trenches formed by embossing – a much cheaper process than microlithography – can be used to control the lattice orientation of spherical microdomains for the purposes of information storage (see Section 9.7) [43,44]. A thorough understanding of the alignment process would provide insight into the use of any one of the techniques mentioned in this section.

9.4.2 ELECTRIC FIELDS

It has been shown that electric fields can be used to align polymeric microdomains, typically by taking advantage of the mismatch of the dielectric constants of two or more blocks. Amundson and coworkers [45–47] demonstrated the effectiveness of such techniques in a bulk PS-PMMA system where the respective dielectric constants are listed as 2.55 and 3.78, evidently sufficient for macroscopic alignment. Birefringence analysis was performed to show that the order parameter, a measure of the degree of field alignment, grew with time upon application of fields in the range of 15 kV per cm until saturation. Additionally, defect analysis was performed to correlate the birefringence data with electron microscopy images.

There have been two main techniques for electric-field alignment in thin films: in-plane and out-of-plane electrodes. The potential of microfabricated in-plane electrodes has been elegantly demonstrated by Morkved *et al.* [48] who showed that sufficiently strong electric field fields could be generated to align cylindrical microdomains (Figure 9.5). Though these researchers worked with a kinetically hindered PS-PMMA system, impressive alignment was demonstrated by annealing above the glass transition temperature in the presence of an electric field. TEM analysis of such samples was facilitated by sample fabrication on silicon nitride membranes. Unfortunately, the region of microdomain alignment extended only a micrometer or so beyond the metal electrodes. These researchers estimate that 30 kV/cm is necessary to orient the microdomains. While the technique demonstration was beautiful, its ultimate

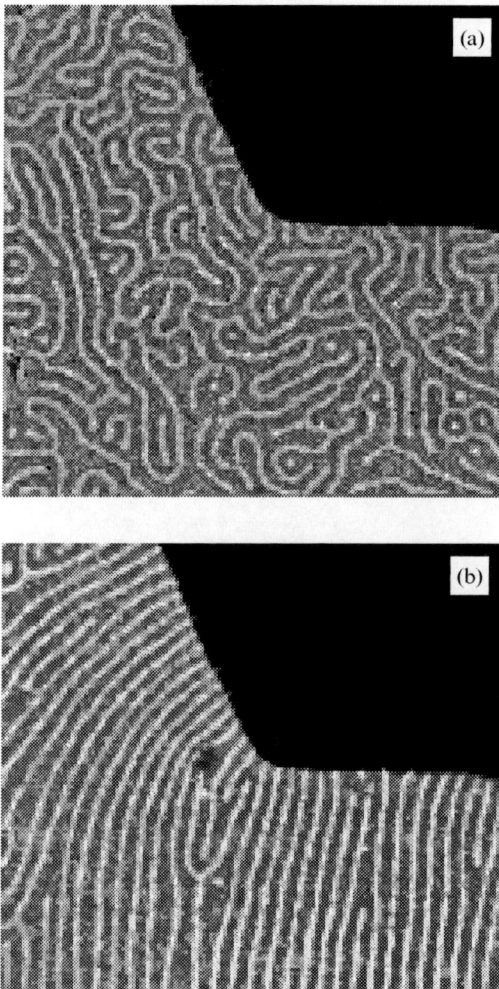

Figure 9.5 (a) AFM image of cylindrical microdomains in the vicinity of an electrode; field has not yet been applied and the average orientation is random. (b) After application of the electric field the microdomains have oriented parallel to the field lines emerging normal to the electrode edge. Bar = 150 nm. (Reproduced from T. L. Morkved *et al. Science* **273**, 931 (1996), copyright (1996) with permission from the American Association for the Advancement of Science).

use as a mean of generating macroscopic microdomain control would be challenging as the entire wafer would need to be patterned with electrodes, a technical challenge.

Thurn-Albrecht *et al.* [49] have shown that out-of-plane electrodes can force cylindrical PS microdomains to orient perpendicular to the substrate in a

PS-PMMA system (Figure 9.6). The alignment field (30 V/μm), produced by metal electrodes above and below the polymer film, acts like a parallel-plate capacitor with a micrometer-scale separation. While this technique is very successful in aligning cylinders perpendicular to the substrate, the field has little or no effect on the packing of the cylinders. A plan view of these structures reveals a polycrystalline lattice-like structure where each grain adopts a random orientation. Furthermore, these researchers have used lithography techniques to erode away PMMA cylinders and back fill with various metals, such as Co

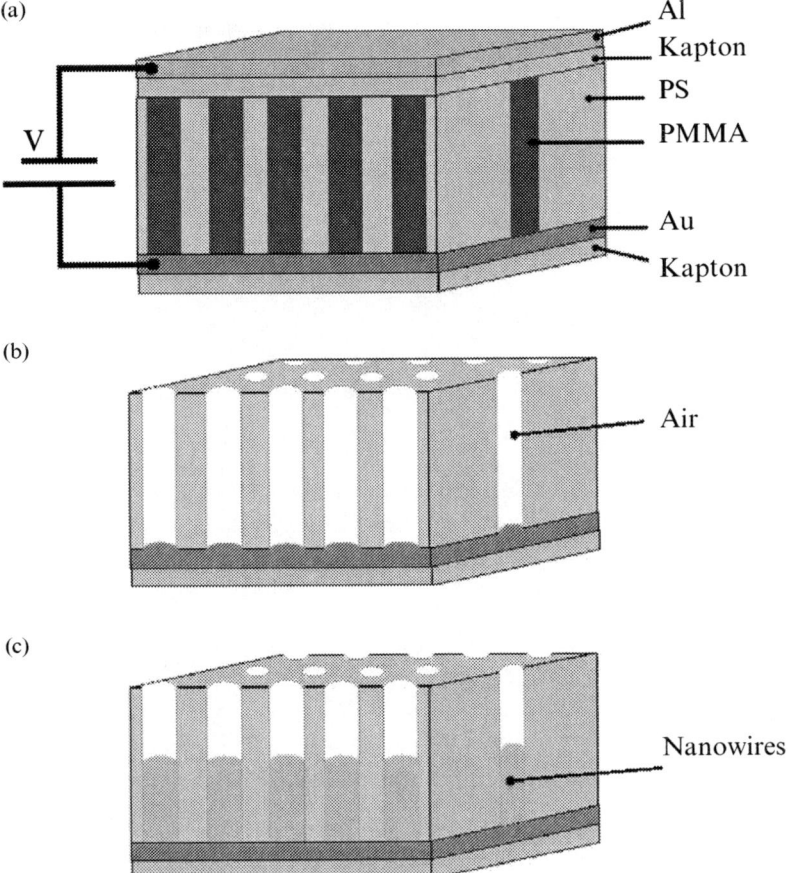

Figure 9.6 Fabrication steps to make an array of nanowires oriented perpendicular to the substrate. (a) The PS-PMMA copolymer forms cylinders oriented perpendicular to the substrate when poled with an electric field above the glass temperature. (b) The oriented PMMA cylinders are removed by exposure to deep ultraviolet light. (c) Co or Cu wires are grown in cylindrical holes by electrodeposition. (Reproduced from T. Thurn-Albrecht *et al. Science*, **290**, 2126 (2000), copyright (2000) with permission from the American Association for the Advancement of Science).

and Cu. Coercivity measurements were performed and found to be rather large, $H_c = 800$ Oe at 300 K. Here, the wire diameter is much smaller than the theoretical critical single-domain behavior (around 50 nm) so that single-domain behavior is possible. These authors suggest that coercivity can be tailored by the aspect ratio of the wires and their packing density. This method shows great promise and is an impressive combination of science and technology.

9.4.3 SHEAR ALIGNMENT

Shear alignment of bulk polymeric samples has been a standard technique for decades and its applicability to block copolymers for the purposes of controlling order has been soundly demonstrated by both Kornfield and Winey [50–54]. Such shearing techniques have been extended to thick polymer films by Albalak and Thomas [55] using counter-rotating cylinders (roll casting). The simplicity of shear alignment is appealing but challenging to utilize on films with thicknesses comparable to that of a microdomain. However, some progress has been made. Typically, a film is cast on a silicon wafer and a second wafer – treated with adhesion-preventing polymers – is pressed on top and held under pressure, usually at an elevated temperature. Chou and coworkers [56] are beginning to investigate the utility of such techniques to align single layers of microdomains for the purposes of microfabrication. This technique necessitates clean-room conditions as any dust particles between the two wafers will prevent adequate contact. One alternative route is to replace the treated wafer with a robust strip of conformal silicone (poly(dimethyl siloxane)). Pressure could be applied to the polymer-coated wafer through the PDMS at moderately high temperatures. A few randomly distributed dust particles would not prevent a large degree of contact between the surfaces. Such techniques may make this technique accessible to a wider range of researchers.

9.4.4 ALIGNING MICRODOMAINS VERTICALLY THROUGH INTERFACE CONTROL

Perhaps the earliest contribution to the control of microdomain orientation via interface control was the experiments of Mansky *et al.* [5], who demonstrated that cylindrical microdomains could be oriented perpendicular to a film by solvent casting on an aqueous surface. It was also demonstrated here that ozone could be used to selectively remove PB microdomains from a PS matrix, thereby enabling subsequent research in lithography. However, the uniformity of this sample preparation method was highly lacking – only certain randomly distributed regions of samples prepared in this manner yielded perpendicularly orinted cylinders.

Building on this work, Huang et al. [57,58] have demonstrated that clever tuning of the interfacial energies of thin films can be used to control microdomain orientation in PS-PMMA copolymers. Lamellae were induced to orient perpendicular to a substrate through the use of random copolymers, as demonstrated by both small-angle neutron scattering and electron microscopy. PS-PMMA films were typically cast on surfaces tailored to be neutral to wetting by either block, leading to the perpendicular orientation. Challenges remain, however, in assuring the uniformity of alignment throughout the sample, and increasing the correlation lengths of the microdomains. Perhaps a combination of this technique with alignment-inducing physical topography would be a winning combination.

9.4.5 DIRECTIONAL CRYSTALLIZATION

Temperature gradients are extensively used to grow single-crystal silicon ingots, and their application to polymer microdomains is a natural and intriguing extension. One typically starts with a seed crystal with a well-determined lattice orientation and immerses it into a melt to trigger crystallization. Zone refining (repeated movement of a melting and crystallizing zone) can be exploited to exclude impurities and defects from the region of interest. Bodycomb and Hashimoto [59,60] have applied this methodology to bulk lamellar systems to form well-oriented samples without shear. Further application of this technique to thin films – facilitated by temperature gradient stages – may provide an additional pathway to macroscopic sample orientation [61].

Alternatively, one can directionally solidify a crystallizable material for the purposes of templating the orientation of microdomains. Researchers have shown that benzoic acid can be suitably crystallized and polystyrene-polyethylene diblock copolymers that are dissolved into benzoic acid will form microdomains with an orientation perpendicular to the substrates [62]. While this is very powerful, it not clear that suitable solvents can be found for a wide variety of copolymer systems such that this can be generalized. Additionally, the grain size of the cylinders is very small, so that addressability of the cylinders is rendered difficult.

9.4.6 CONTROL VIA FLUIDICS

Thomas and coworkers [62] have used a novel application of microfluidic technology to begin to control microdomain array orientation. Standard soft lithography techniques were used to fabricate microfluidic channels out of poly (dimethyl siloxane) (PDMS) that were placed onto a semiconductor wafer [63]. Polymer solutions were allowed to flow into the channels and the solvent evaporated, resulting in microdomains in controlled areas of the wafer [64]. At present, little control of the microdomain patterns has been demonstrated, but this technique has great potential to place polymer films in specific areas of a wafer, a useful addition for high-throughput measurements.

9.4.7 EMBOSSING

Perhaps the simplest, yet most effective method for templating microdomain orientation is with physical embossing, such as in the manufacturing of compact discs. For this process a hard master with relief structures is manufactured. The platter to be embossed is coated with a relatively deformable material, such as a resist. The resist to be embossed is coated with a "nonstick" surface, such as chemical species with fluorine groups. Copolymer films can be spin coated onto such grooved surfaces to template the location of copolymer microdomains [56]. Koji and coworkers have successfully used this templating methodology for information technology (Section 9.7).

9.5 METALLIC INCOPORATION OR MODIFICATIONS TO THE COPOLYMER TEMPLATE

For the purposes of patterning media such as silicon wafers, differentiation of the microdomains is often desirable. For many systems, both copolymer blocks are carbonaceous, leading to an etch resistance between the microdomains and the matrix that is less than desirable (see Section 9.6). However, this can be differentiated by for example, incorporating metal clusters into one block. In what follows we describe suitable modification schemes.

9.5.1 METAL INCORPORATION VIA EVAPORATION

The simplest means to template the placement of metal is to evaporate it onto the proper choice of a copolymer film. For those copolymer blocks with sufficiently disparate metal–polymer interactions, interfacial energies can be used to tailor the ultimate location of metallic particles after coalescence. Segregation of metals into microdomains have been demonstrated in this manner by Jaeger and coworkers [65,66]. While only a limited set of metals satisfy the constraints, metal films so fabricated would act as robust masks (Figure 9.7). Most interestingly, chains of such segregated particles lead to fascinating transport properties.

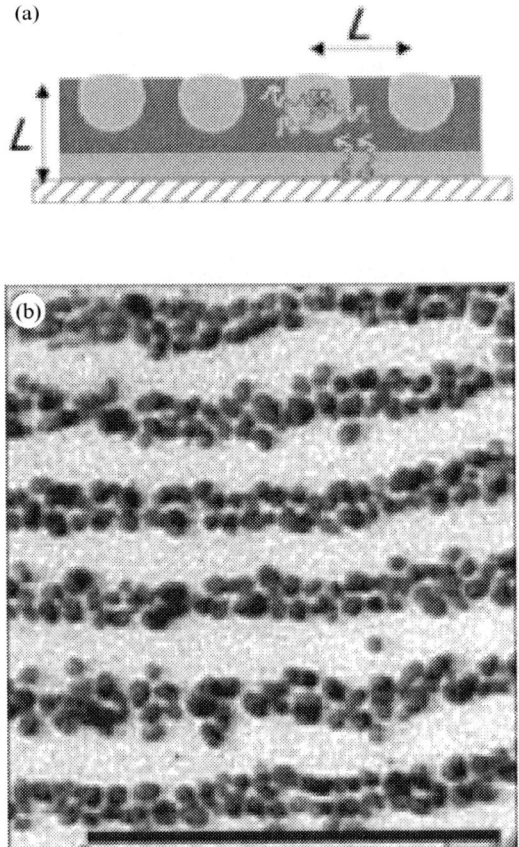

Figure 9.7 (a) Schematic of PS-PMMA in thin film, where both PS and PMMA component are exposed to free surface. (b) Evaporated gold on copolymer template, followed by annealing leads to segregation with cylinder loading fractions up to 30 %. Bar = 250 nm. (Reproduced from W. A. Lopes and H. M. Jaeger, *Nature*, **414**, 735 (2001), copyright (2001) with permission from Nature Publishing Group).

9.5.2 METAL INCORPORATION VIA DIRECT SYNTHESIS

Cohen and coworkers [67–69] have made strong contributions to the field of block copolymers by directly synthesizing metal nanoclusters in copolymer materials. Starting with organometallic monomers, a variety of processing means, such as heating or reduction with hydrogen have been employed to convert the metal atoms to clusters. These researchers have demonstrated the fabrication of silver, platinum, copper, nickel, and gold nanoclusters. The metal clusters follow the block copolymer microdomain templates to varying degrees, with perhaps the greatest success with silver. These techniques have the possibility of fabricating microdomains with ultimate etch resistance as many metals are not attacked by reactive ion etching. One disadvantage, however, is that the metal clusters rarely occupy more than half the microdomain volume, minimizing the effectiveness of a microdomain mask. Generating continuous metal lines in the form of cylinders remains a challenge.

Spatz et al. [70] have made large contributions to the field of nanoparticle patterning through the development of successful gold particle fabrication schemes and extensive characterization. Ultrathin films of PS-P2VP are typically applied to substrates by dip coating mica substrates into a micellar solution. Films are created with an average thickness less than 10 nm (Figure 9.8). Favorable interactions between PS and titanium resulted in preferential deposition of evaporated titanium onto PS domains, thereby dramatically increasing their etch resistance. Control of the spacing, though not the resulting microdomain order, was demonstrated. Argon ion milling was then used to transfer the pattern into the gallium arsenide substrate. An alternative approach was demonstrated as well, whereby PS-P2VP micelles were formed in solution and treated to bind $AuCl_4$ ions to the micelle cores. Reduced gold particles inside micelles that are transferred to a substrate act as the lithographic mask into gallium arsenide [71]. Whether any of these processes lead to patterned GaAs with quantum-dot-like characteristics has yet to be shown. Such processes, unfortunately, lead to many crystal defects, preventing the fabrication of quantum dots.

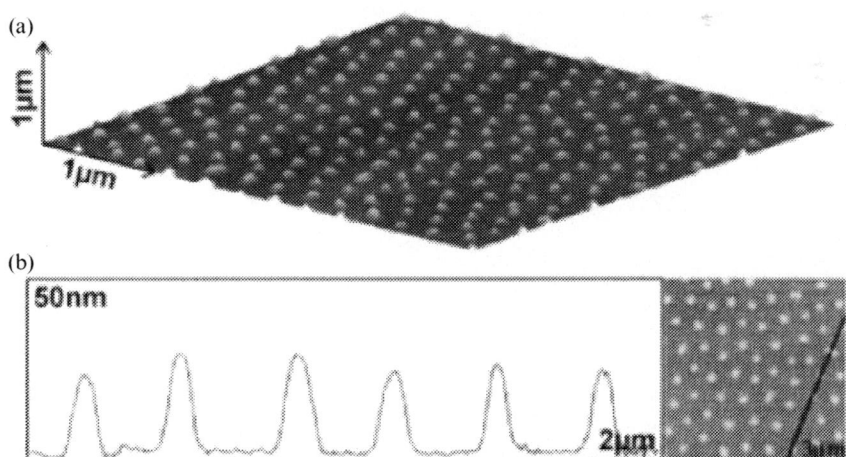

Figure 9.8 (a) AFM image of a layer of micelles formed in a solution of PS-P2VP and applied to mica. A layer of Ti/TiO_2 was subsequently applied and the micelle height increased, indicating a segregation of the evaporated coating. (b) Height profile of AFM image indicating 27 nm corrugation height. (Reproduced from J. P. Spatz et al. Adv. Mater., **10**, 849 (1998), copyright (1998) with permission from Wiley–VCH).

9.6 LITHOGRAPHIC USE OF TEMPLATES

In this section we focus on the patterning of materials by copolymer templates. Building on the film preparations of Section 9.5 we discuss chemical-based wet-etching processes and plasma-based dry-patterning techniques. Ensuring the compatibility of copolymer etching steps with commonly used processes is advantageous as it should speed industrial adoption of this technology. We highlight two successful copolymer templating processes – ozonated PS-PI films and polyisoprene-polyferrocenyldimethylsilane.

9.6.1 PATTERN TRANSFER METHODOLOGIES: WET AND DRY

Perhaps the simplest and oldest pattern transfer technique is wet etching, which is surprisingly effective even for copolymer templates. Typical use of wet etching involves photolithography or electron beam lithography to selectively expose substrate areas for dissolution. An aggressive liquid etchant is then used to remove exposed areas of a wafer. One historical drawback has been the isotropic nature of many etchants so that high aspect features are difficult to fabricate. However, as will be discussed in Section 9.6, some success has been reported for copolymer templates.

Dry-etching processes such as reactive ion etching (RIE) and plasma etching are the dominant tools for pattern transfer of submicrometer features. A pattern is generated on a resist-coated wafer and this pattern is transferred by a directed plasma of high-energy ions. Plasmas typically consist of gases of CF_4, O_2, SF_6, Cl_2, or argon. Etching takes place by a combination of physical bombardment (sputtering) and chemical reactivity to make volatile compounds. Copolymer lithography was initially demonstrated with a low-power, low-pressure CF_4 plasma using a PS-PI system. Low-pressure etching conditions were used to maximize the average path length of the ions so as to facilitate anisotropic etching [72,73].

9.6.2 PATTERNING VIA AN OZONATED COPOLYMER TEMPLATE

The earliest patterning techniques took advantage of ozonation to dissolve microdomains away from the matrix, leaving a single array of pores in a thin film. This porous film then acted as a standard etching mask, where the pores presented less etch resistance than the matrix. CF_4-based RIE was shown to be the most effective tool in pattern transfer from the thin film to the substrate. It was first demonstrated that standard semiconductor materials – such as silicon or silicon nitride – could be easily and uniformly patterned by this technique [5,72,73]. Starting with a copolymer-coated silicon wafer, an hexagonal array of holes with a 20 nm depth and a 40 nm lattice constant was uniformly etched into the sample (Figure 9.9). These researchers were

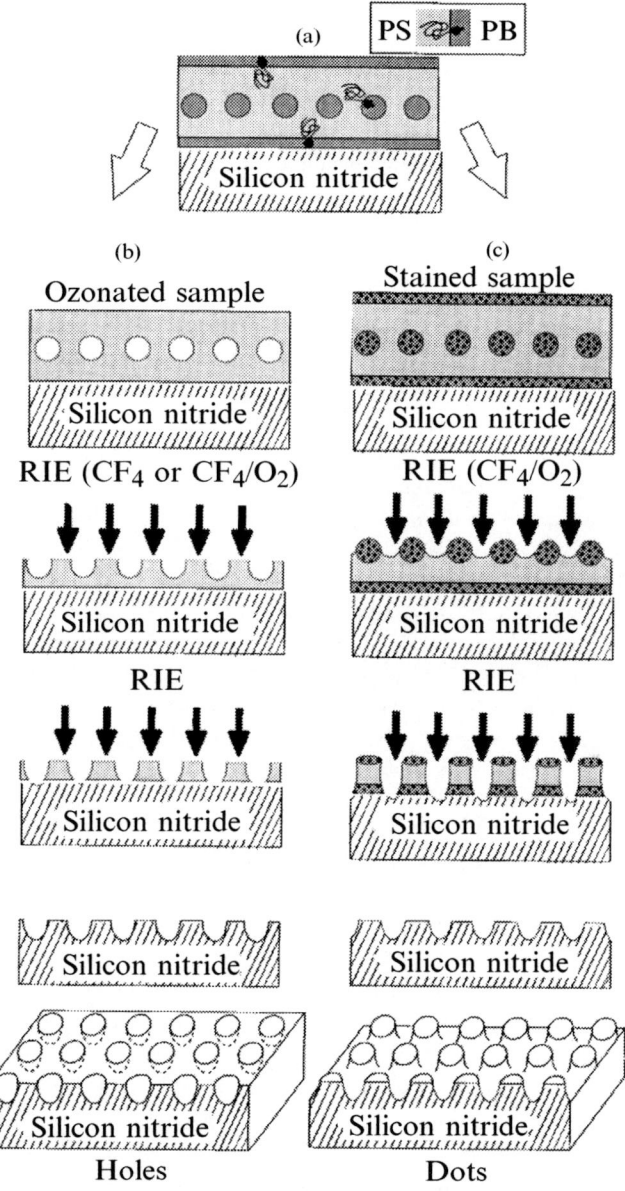

Figure 9.9 Two complementary fabrication strategies used by Chaikin and coworkers to pattern substrates. The left side shows the removal of microdomains by ozonation, thereby acting as a positive resist. The right side shows the crosslinking of microdomains by staining, thereby acting as a negative resist. In both cases reactive ion etching is used to transfer the template to the substrate. The two processes are used to fabricate holes or dots, respectively. (Reproduced from M. Park *et al. Science*, **278**, 1401 (1997), copyright (1997) with permission from the American Association for the Advancement of Science).

challenged to fabricate deeper holes as the polymer film matrix presented little etch resistance. This technique was further extended to fabricate an array of gold dots via a boot-strapping trilevel procedure. Wafers were coated with a thin layer of polyimide and then coated with an even thinner film of silicon nitride via plasma-enhanced chemical vapor deposition. A single layer of copolymer microdomains was applied via spin coating. After appropriate annealing and ozonation, the copolymer film was used to pattern through the silicon nitride via CF_4 RIE. Using the silicon nitride film as a mask, O_2 RIE was used to pattern cylindrical holes down to the silicon wafer, onto which gold was evaporated. The polyimide was dissolved away, leaving an array of gold dots on the wafer [74]. The ultimate goal of these researchers is two-fold; primarily they are interested in the electron-transport properties of patterned metal films, but additionally the possibilities for information storage (one bit per dot) are lucrative [75]. Further extension of this technique includes the fabrication of metal wires using a template of cylindrical microdomains.

Li *et al.* [76] were able to employ ozonation technologies to pattern GaAs substrates using a copolymer template and a combination of wet- and dry-etching technologies. The ultimate goal of this project is to generate quantum dots with a tighter size distribution than currently possible via Stranski–Krastanow growth or metal-organic chemical vapor deposition [77,78]. Starting with a copolymer-coated GaAs wafer, the microdomains were dissolved away and the films were plasma etched, generating topographical contrast. Wet etching (mixture of ammonium hydroxide and hydrogen peroxide) was used to transfer the pattern into GaAs with both the (100) and (311)B orientation. As wet etching is often more gentle than direct reactive ion etching, this technique may play a role where disruption of the crystal lattice is not acceptable.

9.6.3 PATTERNING VIA PI-PFS COPOLYMER TEMPLATE

Thomas and coworkers [41,79–82] have demonstrated that a polyisoprene-polyferrocenyldimethylsilane (PI-PFS) copolymer is a more robust mask as compared to purely organic copolymers. The ferrous component – which formed the microdomains – was demonstrated to be strongly resistant to an oxygen-based plasma. Metal is incorporated directly during polymerization rather than subsequent modifications to the template. Etch-resistance ratios as high as 50 were reported for the organic vs ferrous material for an oxygen plasma. In one demonstration, a boot-strapping process was used where PI-PFS was used to pattern silicon oxide via oxygen-based RIE, the patterned silion oxide was used to mask tungsten for etching via CHF_3, and finally the patterned tungsten film was used to pattern a cobalt film. Optimal choice of the etching gas and conditions used at each step led to optimal patterning (Figure 9.10). These researchers showed that the coercivity increased as a function of etch depth; arrays of individual dots had a higher coercivity than a continuous

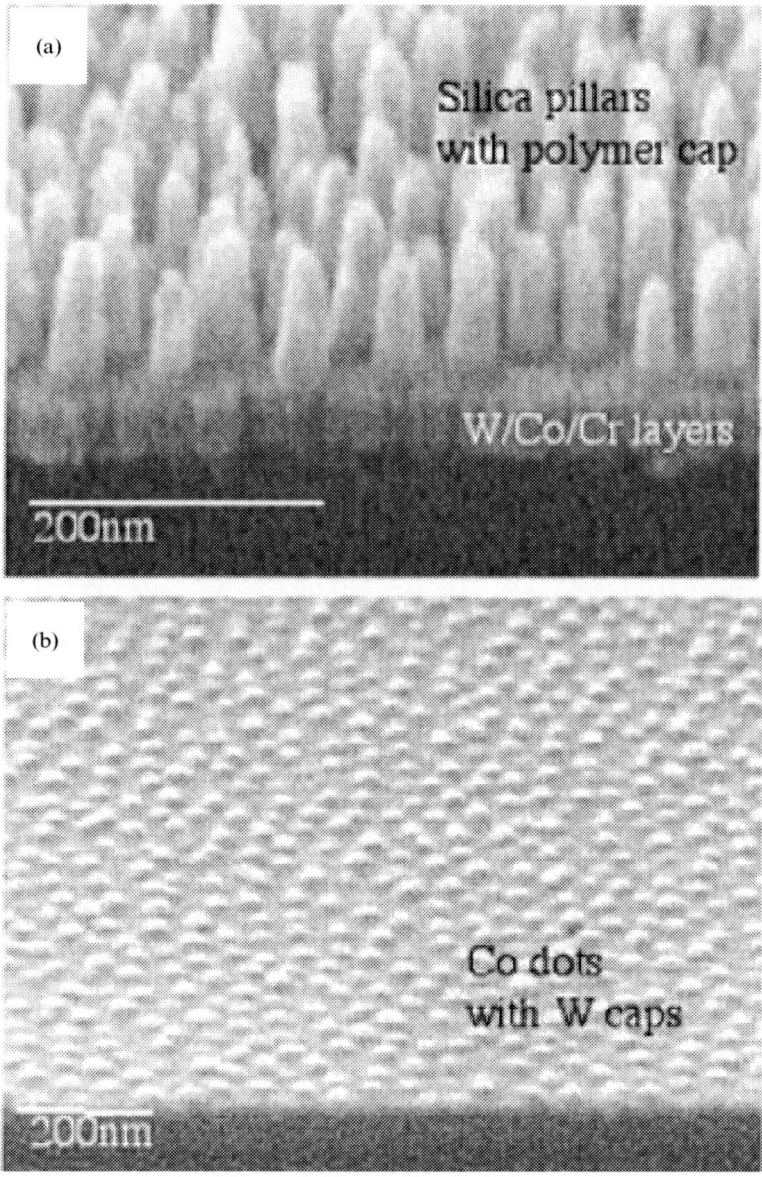

Figure 9.10 Fabrication strategy used by Thomas and coworkers. (a) Pillars of silicon oxide topped with oxidized PFS after etching with CHF_3 based RIE. (b) Tungsten-topped cobalt dots produced as final product. Bar = 200 nm. (Reproduced from J. Y. Cheng *et al. Adv. Mater.* **13**, 1174 (2001) copyright (2001) with permission from Wiley-VCH).

film. The authors argued that domain-wall motion, necessary to switch magnetization direction, was impeded by defects introduced by etching.

9.6.4 PATTERNING VIA PS-PMMA COPOLYMER TEMPLATE

PS-PMMA systems can be used as lithographic templates in a similar manner to PS-PI systems but with a different degradation mechanism. PMMA degrades quickly when exposed to ultraviolet light and can then be washed away. Therefore, after fabrication of a thin film of PS-PMMA microdomains, the PMMA can be degraded and removed, producing a matrix of voids. These can then be used as a lithographic template for the purposes of masking a wafer or filling with metal [49]. Alternatively, PMMA is less resistant to dry etching with an oxygen plasma, providing an additional route to patterning (further discussed in Section 9.7).

9.7 CURRENT APPLICATIONS OF BLOCK COPOLYMER LITHOGRAPHY

Asakawa and coworkers [43] have demonstrated that PS-PMMA copolymer films can be used for patterning magnetic media for information storage. Copolymers consisting of PS and PMMA have the advantage of being both controllable by an electric field and also having etch-rate ratios of at least a factor of two under CF_4-based dry etching. Taking advantage of the latter, these researchers developed a three-step etching process whereby PS-PMMA patterns could be directly transferred to 6.26 cm (2.5 inch) glass disks designed for hard-drive applications. A metal film was sputtered onto the glass disk consisting of Ti (adhesion promoter) followed by $Co_{74}Pt_{26}$, the magnetic layer for information storage. A novolac-based resist was then spin coated onto this for embossing. A nickel master with spiral relief structures was used to imprint the novalac resist disk where the spiral widths ranged from 60 to 250 nm (Figure 9.11) [43]. The spiral pattern was transferred to the disk at a pressure of 1000 bar resulting in spiral grooves. A PS-PMMA diblock copolymer solution was spin coated onto this patterned disk and annealed, inducing microdomains to cluster in the lines. While the ordering of the dots along the grooves is not extremely regular (as is often the case with PS-PMMA systems), they are well defined and appear undamaged from this process. A modification of the above process was undertaken as well – after oxygen etching to remove the PMMA microdomains, spin-on glass (SOG) was applied that selectively filled the holes. Ion milling through the SOG was then used to pattern the underlying metal film. Measurements revealed that the coercivity of such films increased as compared to the continuous film. These researchers used magnetic force microscopy to demonstrate that the media could be erased by DC magnetic fields.

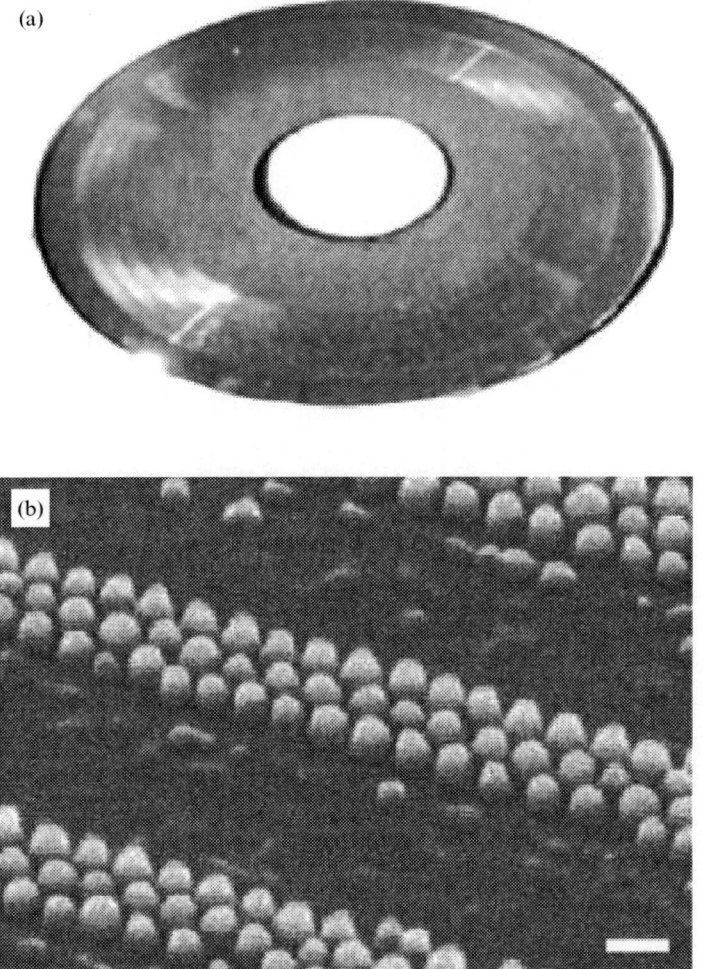

Figure 9.11 Fabrication strategy employed by researchers at Toshiba. (a) 6.25 cm (2.5 inch) HDD glass plate used for information storage. Circular lines originate from interference colors of embossed lines. (b) SEM micrograph of CoCrPt dots formed from copolymer template in substrate grooves. Bar = 50 nm. (Reproduced from K. Asakawa *et al. J. Photopolym. Sci. Technol*, **15**, 465 (2002) copyright (2002) with permission from J. Photopolym. Sci. Technol).

9.8 SUMMARY AND FUTURE DIRECTIONS

During the past decade, research on copolymer lithography began as a trickle and has now increased to a torrent. The elementary patterning techniques first demonstrated by Chaikin and Register have been further developed by the techniques of the research groups of Thomas, Kramer, and Russell. It is soundly demonstrated that there are multiple paths to taking a self-assembled copolymer pattern with a 20 nm feature size and transferring it to a semiconductor or metallic substrate. Unfortunately, it has yet to be demonstrated that an entire hexagonal array of metal dots can be generated with lattice-like registry. However, as Asawaka and coworkers have shown, there may be alternative paths to information storage that do not require such registry. A combination of the emerging orientation control techniques and the patterning technologies of copolymer systems such as PI-PFS should prove to be a winning combination.

Lastly, researchers who are able to take the long view realize that the techniques and insight developed here may best come to fruition with a self-assembling system that is not based on copolymers. For example, applications may emerge more quickly with silicates templated by surfactants than with self-assembling copolymers. Even so, it is doubtless that the techniques and processes developed for copolymer lithography will assist researchers in these fields as well.

ACKNOWLEDGEMENTS

CH thanks Ian Hamley for the opportunity to write this review and cheerfully acknowledges fruitful discussions with Michael Fasolka. CH has unlimited gratitude to his advisors at Princeton – Paul, Rick, and David – whose amazing degree of intelligence and patience never ceases to astound him. JAD acknowledges support by the NIST Office of Microelectronics Programs for portions of this work.

REFERENCES

1. C. Y. Chang and S. M. Sze, *ULSI Technology*. McGraw-Hill Higher Education, New York, 2 edition, 1996.
2. M. J. Fasolka and A. M. Mayes, *Ann. Rev. Mater.*, **31**:323, 2001.
3. P. M. Chaikin, personal communication.
4. D. Hofstadter, *Phys. Rev. B*, **14**:2239, 1976.
5. P. Mansky, P. M. Chaikin, and E. L. Thomas, *J. Mater. Sci.*, **30**:1987, 1995.
6. Certain commercial equipment, instruments, or materials are identified in this chapter in order to specify the experimental procedure adequately. Such identification is not intended to imply recommendation or endorsement by the National Institute of Standards and Technology, nor is it intended to imply that the materials or equipment identified are necessarily the best available for the purpose.
7. C. T. Kresge, M. E. Leonowicz, W. J. Roth, J. C. Vartuli, and J. S. Beck, *Nature*, **359**:710, 1992.
8. W. B. Russel, D. A. Saville, and W. R. Schowalter. *Colloidal Dispersions*. Cambridge University Press, New York, 1992.

9. F. S. Bates, *Science*, **251**:898, 1991.
10. F. S. Bates and G. H. Fredrickson, *Annu. Rev. Phys. Chem.*, **41**:525, 1990.
11. L. J. Fetters, D. J. Lohse, D. Richter, T. A. Witten, and A. Zirkel, *Macromolecules*, **27**:4639, 1994.
12. R. P. Quirk and L. J. Fetters, *Comprehensive Poymer Science*, volume 7, chapter 1, page 1. Pergamon Press, 1989.
13. K. A. Davis and K. Matyjaszewski, *Adv. Polym. Sci.*, **159**:1, 2002.
14. D. Bendejacq, V. Ponsinet, M. Joanicot, Y. L. Loo, and R. A. Register, *Macromolecules*, **35**:6645, 2002.
15. K. Matyjaszewski, *Controlled Living Radical Polymerization: Progress in ATRP, NMP, and RAFT*, volume 768. American Chemical Society, Washington D.C., 2 edition, 2000.
16. K. Matyjaszewski and J. Xia, *Chem. Rev.*, **101**:2921–2990, 2001.
17. C. J. Hawker, A. W. Bosman, and E. Harth, *Chem. Rev.*, **101**:3661–3688, 2001.
18. C. Harrison, Z. D. Cheng, S. Sethuraman, D. A. Huse, P. M. Chaikin, D. A. Vega, J. M. Sebastian, R. A. Register, and D. H. Adamson *Phys. Rev. E*, **66**:011706, 2002.
19. C. S. Henkee, E. L. Thomas, and L. J. Fetters. *J. Mater. Sci.*, **23**:1685–1694, 1988.
20. C. Harrison, P. M. Chaikin, D. A. Huse, R. A. Register, D. H. Adamson, A. Daniel, E. Huang, P. Mansky, T. P. Russell, C. J. Hawker, D. A. Egolf, I. V. Melnikov, and E. Bodenschatz, *Macromolecules*, **33**:857, 2000.
21. A. E. Ribbe, J. Bodycomb, and T. Hashimoto. *Macromolecules*, **32**:3154, 1999.
22. T. L. Morkved, W. A. Lopes, J. Hahm, S. J. Sibener, and H. M. Jaeger *Polymer*, **39**:3871, 1998.
23. D. W. Schwark, D. L. Vezie, J. R. Reffner, and E. L. Thomas, *J. Mater. Sci. Lett.*, **11**:352, 1992.
24. C. Harrison, M. Park, P. M. Chaikin, R. A. Register, D. H. Adamson, and N. Yao, *Polymer*, **30**:2733, 1998.
25. C. K. Harrison, M. Park, P. M. Chaikin, R. A. Register, and D. H. Adamson, *Macromolecules*, **31**:2185, 1998.
26. M. L. Trawick, D. E. Angelescu, P. M. Chaikin, M. J. Valenti, and R. A. Register, *Rev. Sci. Instrum.*, submitted.
27. B. K. Annis, D. W. Schwark, J. R. Reffner, E. L. Thomas, and B. Wunderlich, *Makromol. Chem.*, **193**:2589, 1992.
28. R. G. Winkler, J. P. Spatz, S. Sheiko, M. Moller, P. Reineker, and O. Marti, *Phys. Rev. B*. **54**:8908, 1996.
29. A. Knoll, R. Magerle, and G. Krausch. *Macromolecules*, **34**:4159, 2001.
30. J. P. Spatz, S. Sheiko, M. Moller, R. G. WInkler, P. Reineker, and O. Marti, *Nanotechnology*, **6**:40, 1995.
31. J. Heier, J. Genzer, E. J. Kramer, F. S. Bates, S. Walheim, and G. Krausch, *J. Chem. Phys.*, **111**:11101, 1999.
32. J. Heier, E. J. Kramer, S. Walheim, and G. Krausch, *Macromolecules*, **30**:6610, 1997.
33. R. D. Peters, X. M. Yang, Q. Wang, J. J. de Pablo, and P. F. Nealey, *J. Vac. Sci. Technol. B*, **18**: 3530, 2000.
34. R. D. Peters, X. M. Yang, and P. F. Nealey. *Macromolecules*, **35**:1822, 2002.
35. X. M. Yang, R. D. Peters, P. F. Nealey, H. H. Solak, and F. Cerrina, *Macromolecules*, **33**:9575, 200.
36. Z. Li, S. Qu, M. H. Rafailovich, J. Sokolov, M. Tolan, M. S. Turner, J. Wang, S. A. Scharz, H. Lorenz, and J. P. Kotthaus, *Macromolecules*, **30**:8410, 1997.
37. M. J. Fasolka, T. A. Germer, and A. Karim. *Langmuir*. in preparation.
38. M. J. Fasolka, P. Banerjee, A. M. Mayes, G. Pickett, and A. C. Balazs, *Macromolecules*, **33**:5702, 2000.

39. R. A. Segalman, H. Yokoyama, and E. J. Kramer, *Adv. Mater.*, **12**:1152, 2001.
40. D. Sundrani and S. J. Sibener, *Macromolecules*, **35**:8531, 2002.
41. J. Y. Cheng, C. A. Ross, E. L. Thomas, H. I. Smith, and G. J. Vancso, *Appl. Phys. Lett.* **81**:3657, 2002.
42. M. L. Trawick, M. Megens, C. Harrison, D. E. Angelescu, D. A. Vega, P. M. Chaikin, R. A. Register, and D. H. Adamson, *Scanning*, submitted.
43. K. Asakawa, T. Hiraoka, H. Hieda, M. Sakurai, Y. Kamata, and K. Naito, *J. Photopolym. Sci. Technol.*, **15**:465, 2002.
44. K. Naito, H. Hieda, M. Sakurai, Y. Kamata, and K. Asakawa, *IEEE Trans. Magn.* **38**:1949, 2002.
45. K. Amundson, E. Helfand, Xina Quan, S. Hudson, and S. D. Smith, *Macromolecules*, **27**:6559–6570, 1994.
46. K. Amundson, E. Helfand, Xina Quan, and S. D. Smith, *Macromolecules*, **26**:2698–2703, 1993.
47. K. Amundson and E. Helfand, *Macromolecules*, **26**:1324–1332, 1993.
48. T. L. Morkved, M. Lu, A. M. Urbas, E. E. Ehrichs, H. M. Jaeger, P. Mansky, and T. P. Russell, *Science*, **273**:931, 1996.
49. T. Thurn-Albrecht, J. Schotter, C. A. Kastle, N. Emley, T. Shibauchi, L. Krusin-Elbaum, K. Guarini, C. T. Black, and M. T. Tuominen, *Science*, **290**:2126, 2000.
50. Z. R. Chen, J. A. Kornfield, S. D. Smith, J. T. Grothaus, and M. M. Satkowski, *Science*, **277**:1248, 1997.
51. D. L. Polis, K. I. Winey, A. J. Ryan, and S. D. Smith *Phys. Rev. Lett.*, **83**:2861, 1999.
52. I. W. Hamley, *J. Phys. Condens. Matter*, **13**:R643, 2001.
53. R. H. Colby, *Curr. Opin. Colloid Interfure Sci.*, **1**:454, 1996.
54. G. H. Fredrickson and F. S. Bates, *Annu. Rev. Mater. Sci.*, **26**:501, 1996.
55. R. J. Albalak and E. L. Thomas, *J. Polym. Sci. Part B: Polym. Phys.*, **31**:31, 1993.
56. L. Zhuang, "Controlled Self-Assembly in Homopolymer and Diblock Copolymer", Ph. D. Thesis, Princeton University (2002).
57. E. Huang, T. P. Russell, C. Harrison, P. M. Chaikin, R. A. Register, and C. Hawker, *Macromolecules*, **31**:7641, 1998.
58. E. Huang, P. Mansky, T. P. Russell, C. Harrison, P. M. Chaikin, R. A. Register, C. Hawker, and J. Mays. *Macromolecules*, **33**:80, 2000.
59. T. Hashimoto, J. Bodycomb, Y. Funaki, and K. Kimishima, *Macromolecules*, **32**:952, 1999.
60. J. Bodycomb, Y. Funaki, K. Kimishima, and T. Hashimoto, *Macromolecules*, **32**:2075, 1999.
61. J. Carson Meredith, A. Karim, and E. J. Amis. *Macromolecules*, **33**: 5760, 2000.
62. C. Park, C. De Rosa, and E. L. Thomas. *Macromolecules*, **34**:2602, 2001.
63. E. Kim, Y. N. Xia, and G. M. Whitesides, *Phys. Rev. B*, **376**:581, 1995.
64. T. Deng, Y. Ha, J. Y. Cheng, C. A. Ross, and E. L. Thomas, *Langmuir*, **18**:6719, 2002.
65. T. L. Morkved, P. Wiltzius, H. M. Jaeger, D. G. Grier, and T. A. Witten, *Appl. Phys. Lett.*, **64**:422, 1994.
66. W. A. Lopes, and H. M. Jaeger, *Nature*, **414**:735, 2001.
67. Y. Ng, C. Chan, R. R. Schrock, and R. E. Cohen, *Chem. Mater.*, **4**:24, 1992.
68. B. H. Sohn and R. E. Cohen, *Acta Polymer*, **47**:340, 1996.
69. R. T. Clay and R. E. Cohen, *Supramolec. Sci.*, **2**:183, 1995.
70. J. P. Spatz, P. Eibeck, S. Mosmer, M. Moller, T. Herzog, and P. Ziemann, *Adv. Mater.*, **10**:849, 1998.
71. J. P. Spatz, T. Herzog, P. Eibeck, S. Mosmer, P. Ziemann, and M. Moller, *Adv. Mater.*, **11**:149, 1999.

72. M. Park, C. Harrison, P. M. Chaikin, R. A. Register, and D. H. Adamson, *Science*, **276**:1401, 1997.
73. C. Harrison, M. Park, P. M. Chaikin, R. A. Register, and D. H. Adamson, *J. Vac. Sci. Technol. B.*, **16**:544, 1998.
74. M. Park, P. M. Chaikin, R. A. Register, and D. H. Adamson, *Appl. Phys. Lett.*, **79**:257, 2001.
75. C. Harrison, M. Park, R. A. Register, D. H. Adamson, P. Mansky, and P. M. Chaikin. Method of nanoscale patterning and products made thereby, *United States Patent 5948470*, 1999.
76. R. R. Li, P. D. Dapkus, M. E. Thompson, W. G. Jeong, C. K. Harrison, P. M. Chaikin, R. A. Register, and D. H. Adamson, *Appl. Phys. Lett.*, **76**:1689, 2000.
77. T. R. Ramachandran, A. Madhukar, I. Mukhametzhanov, R. Heitz, A. Kalburge, Q. Xie, and P. Chen, *J. Vac. Sci. Technol. B*, **16**:1330, 1998.
78. F. Heinrichsdorff, A. Krost, N. Kirstaedter, M. H. Mao, M. Grundmann, D. Bimberg, A. O. Kosogov, and P. Werner. *Jpn. J. Appl. Phys. Part 1-Regular Papers and Short Notes and Review Papers*, **36**:4129, 1997.
79. J. Y. Cheng, C. A. Ross, V. Z. H. Chan, E. L. Thomas, R. G. H. Lammertink, and G. J. Vancso, *Adv. Mater*, **13**:1174, 2001.
80. R. G. H. Lammertink, M. A. Hempenius, V. Z. H. Chan, E. L. Thomas, and G. J. Vancso, *Chem. Mater.*, **13**:429, 2001.
81. R. G. H. Lammertink, M. A. Hempenius, J. E. van den Enk, Chan V. Z. H., E. L. Thomas, and G. J. Vancso, *Adv. Mater.*, **12**:98, 2000.
82. J. Y. Cheng, C. A. Ross, E. L. Thomas, H. I. Smith, R. G. H. Lammertink, and G. J. Vancso, *IEEE Trans. Magn.*, **38**:2541, 2002.

10 Applications of Block Copolymer Surfactants

MICHAEL W. EDENS and ROBERT H. WHITMARSH
Polyglycols Research and Development, The Dow Chemical Company, Freeport, TX 77541, USA

10.1 INTRODUCTION

The chemical literature abounds with references to the class of compounds known as polyoxyalkylene block copolymers. The majority of these versatile polymers are made by polymerizing propylene oxide (PO) onto an initiator, usually propylene glycol, followed by "capping" or "tipping" with ethylene oxide (EO). Copolymers made using 1,2-butylene oxide (BO) as the hydrophobe are included in this category of products, however commercial uses of these materials are not common. The reverse arrangement, an internal EO block capped with PO or BO, also exists. These synthetic pathways result in a potentially infinite number of EO and PO (or BO) combinations that possess a wide variety of physical properties. Viscosity, cloud point, HLB, physical state (liquid, paste, or solid), and gelling temperature are all variables that can be adjusted by manipulation of the EO/PO quantity and ratio. These polymers, which were developed and commercialized by Wyandotte Chemicals Corporation over 50 years ago [1,2] are still being used by today's scientists to improve cosmetics, medicines, cleaners, lubricants, and other formulated products. There seems to be no end to the widespread use of EO/PO block copolymers to improve products that improve lives.

The applications of EO/PO (or BO) block copolymers have been reviewed on many occasions [3–7]. This chapter will focus on the most recent developments reported for this class of block copolymers, while giving a brief historical perspective on each major application area. As one becomes familiar with the literature covering these products, it is clear that the long history of the PLURONIC® and TETRONIC® trademarks (BASF Corporation) has caused many authors to generally refer to block copolymers initiated with propylene glycol by their Pluronic designation and products initiated with ethylenediamine by the Tetronic designation, rather than by their chemical names. To assist the reader in identifying the products discussed, Table 10.1 gives the Pluronic designation as well as the chemical composition for many of the

Developments in Block Copolymer Science and Technology. Edited by I. W. Hamley
© 2004 John Wiley & Sons, Ltd   ISBN: 0 470 84335 7

Table 10.1 Polyoxyalklene block copolymer nomenclature (Reproduced from M. W. Edens, in Nonionic Surfactants. Polyoxyalkylene Block Polymers (V. M. Nace, ed.) Surfactant Science Series No. 60 (1996), copyright (1996) with permission from Marcel Dekker Inc.).

		Poloxamer structure $(EO)_x(PO)_y(EO)_x$		
Pluronic	Poloxamer[a]	Hydrophope (y)	% EO (x)	Polymer MW
F-68	188	1750 (30)	80 (80)	8750
F-77	217	2050 (35)	70 (54)	6835
F-87	237	2250 (39)	70 (60)	7500
F-88	238	2250 (39)	80 (102)	11 250
F-98	288	2750 (47)	80 (125)	13 750
F-108	338	3250 (56)	80 (148)	16 250
F-127	407	4000 (69)	70 (106)	13 335
L-35	105	5950 (103)	50 (68)	11 900
L-43	123	1200 (21)	30 (6)	1715
L-44	124	1200 (21)	40 (9)	2000
L-61	181	1750 (30)	10 (2)	1945
L-62	182	1750 (30)	20 (5)	2190
L-63	183	1750 (30)	30 (9)	2500
L-64	184	1750 (30)	40 (13)	2915
L-72	212	2050 (35)	20 (6)	2565
L-81	231	2250 (39)	10 (3)	2500
L-92	282	2750 (47)	20 (8)	3440
L-101	331	3250 (56)	10 (4)	3610
L-121	401	4000 (69)	10 (5)	4445
L-122	402	4000 (69)	20 (11)	5000
P-65	185	1750 (30)	50 (20)	3500
P-103	333	3250 (56)	30 (16)	4645
P-104	334	3250 (56)	40 (25)	5415
P-105	335	3250 (56)	50 (37)	6500
P-123	403	4000 (69)	30 (19)	5715
17R1[b]	171[c]	1410	10	1565
25R2[b]	252[c]	2100	20	2625
25R8[b]	258[c]	2100	80	10 500
31R1[b]	311[c]	2450	10	2720

[a] Nomenclature as follows – first two digits are the M_w of the hydrophobe divided by 100 and the last digit is the percentage of ethylene oxide in the molecule divided by ten. For example, Poloxomer 188 has a hydrophobe $M_w = 1800$ approximately, and contains 80 % EO.
[b] Reverse block copolymers (PO-EO-PO)
[c] Meroxapol

common products. It is also important to point out these types of products are offered by many chemical companies, both global and regional producers. Among the larger companies offering EO/PO block copolymers in their product line are Huntsman Corporation (Surfonic® POA series), Uniqema (Synperonic™ series) and The Dow Chemical Company (Tergitol™ L series).

10.2 MEDICAL APPLICATIONS

Medical applications for block copolymers of ethylene oxide and propylene oxide are many and diverse [6]. They are utilized for improved drug delivery/stability in efficient patient-acceptable formulations, for coatings to reduce protein adhesion or clotting, and for structural gels and wound coverings, as a few examples. While no completely new applications have been found since the last review [7], the following paragraphs provide a sense of how the use of the alkylene oxide block copolymers continues to provide improvements in many medical areas.

It is axiomatic that, no matter how effective the drug, it will be ineffective if the dosage form is unacceptable to the patient. Improved patient comfort and compliance with drug delivery by suppository takes advantage of the gelling properties of formulations based on ethylene oxide/propylene oxide block copolymers. Delivered as liquids, the formulations gel *in situ*. Kim and coworkers [8–10] reported that a blend of two ethylene oxide/propylene oxide block copolymers, together with a bioadhesive polymer, provided the necessary gel strength and bioadhesive force for acceptable delivery, retention and drug release. Improved absorption of insulin-loaded suppository gels based on F-127 was reported by Barichello *et al.* [11]. The improvement resulted from adding unsaturated fatty acids to the formulations.

Alkylene oxide block copolymers are not yet the cure for cancer but they have been successfully employed by a number of groups to help in the treatment of this disease. Formulations with ethylene oxide/propylene oxide block copolymers improve targeted delivery of the therapeutic agents. March [12] patented improvements in the introduction of genetic material into living cells using compositions of the genetic material containing at least 15% of an EO/PO copolymer such as Poloxamer 407.

Cancer is only one of the diseases susceptible to such treatments. Kabanov and Alakov [13] found that a number of problems associated with antineoplastic treatment for cancer, such as low stability in the bloodstream, low solubility and poor transport across cell membranes, can be overcome with the use of ethylene oxide/propylene oxide block copolymers in the delivery system. In a later patent [14] these workers showed that it is important to form micelles of the alkylene oxide copolymer in the drug delivery formulation. Each of the above applications takes advantage of the block structure to improve drug compatibility within the bloodstream. Gladysheva and coworkers [15] went a step further and actually chemically bonded the anticarcinogenic agent to the block copolymer. As with many PEGylated drugs, this conjugate circulated for a longer period at higher concentrations that did the underivatized drug, resulting in higher concentrations throughout the body.

Drug delivery is impacted in many ways. Block copolymers of EO and PO are employed in nearly all of the delivery methods. Suppositories with better patient acceptance are discussed above. Ethylene oxide/propylene oxide block copolymer formulations can also be found in orally administered drugs, recently patented by Kabanov et al. [14]. Hoang and Khan [16] described a skin-preparation composition containing iodine. An ethylene oxide/propylene oxide block copolymer was an important component of this composition that prevented degradation of the iodine. Long-lasting, single-dose injectables are described by Paavola et al. [17]. In their paper, the gel-forming properties of the EO/PO block copolymers are used to advantage for controlled drug release. The structural differences of the gels, more than the macroviscosity, seem to regulate drug release. A recent paper by Bohner et al. [18] reported the ways in which ethylene oxide/propylene oxide block copolymers, such as Pluronic F-108, influenced the circulation lifetimes of polystyene latex particles in the blood. Delivery of drugs to specific target cells and organs of the body via intravenous colloidal carriers was greatly improved by covering the particles with hydrophilic, nonionic polymers. Nonionic, water compatible, flexible and well-hydrated polymers were preferred, and the alkylene oxide block copolymers were a nearly perfect choice. This approach was reviewed in a paper by Storm et al. [19]. A more recent review by Allen et al. [20] summarized the use of ethylene oxide/propylene oxide block copolymers for drug delivery. Their emphasis was on the micelle properties of the block copolymers in aqueous systems that determine the effectiveness of delivering hydrophobic drugs.

The usefulness of EO/PO block copolymers in many biomedical applications is derived from their influence on the adsorption of plasma proteins. In a paper addressing the adsorption of proteins onto films prepared with a wide range of Pluronic materials, Retzinger and coworkers [21] developed a model for this phenomenon. Espadas-Torre and Meyerhoff [22] took advantage of this effect by incorporating high molecular weight ethylene oxide/propylene oxide block copolymers within ion-selective membranes. The incorporation reduced platelet adhesion, which was a key factor in blood clotting. Armstrong et al. [23] demonstrated that the addition of Pluronic F-68 to whole blood reduced the amount of aggregation. In a follow-up study, Edwards et al. [24] showed that while aggregation did initially occur in the presence of Pluronic F-68, the rate of disaggregation was greatly enhanced. Hunter and Duncan [25] described a similar clot-dissolving enhancement. A method for applying and retaining the Pluronic polymers on a biological surface was described in a paper by Messersmith et al. [26]. The approach taken was to chemically bond the block copolymer to a strongly bioadhesive material. By incorporating Pluronic F-68 into a process for filtering blood components, Green and Goodrich [27] were awarded a patent for an improved method of separating particles. The EO/PO block copolymer functioned in this application by reducing particle adherence to one another and thereby improving separation.

The controlled delivery of drugs from a liquid formulation that thermally gels at body temperature has been described for improved suppositories and long-lasting single-dose injectables [8,9]. Other workers have taken advantage of solution gelling using the body's own water. Scherlund *et al.* [28] reported on a local anaesthetic formulation based on Lutrol F-68 (identical to Pluronic F-68) that displayed gelation *in situ*. As formulated with a eutectic mixture of lidocaine and prilocaine, the system was an isotropic low viscosity liquid. When injected within the mouth, the presence of saliva caused a transition into a rigid hexagonal phase that remained at the application site.

In a series of patents Viegas *et al.* [29–31] showed how in-situ-formed gels based on ethylene oxide and propylene oxide copolymers can be used for wound care and surgical dressings. They were able to take advantage of the drug delivery ability of the gels as well as the structural strength of the gel.

10.3 COAL AND PETROLEUM APPLICATIONS

Block copolymers have been reported for many years as surfactants for the demulsification of crude oil and tars [32–36].

Retter *et al.* [37] reported the use of EO/PO block copolymer surfactants to optimize the recovery of mineral oil from water. They outlined a procedure involving emulsification of the oils by adding amphiphiles, breaking the emulsion, followed by extraction of the oils. The emulsification step was dependent on the HLB of the amphiphiles. Optimum performance was obtained at HLB 29 using Pluronic F-68. Zaki *et al.* [38,39] recently studied PO/EO reverse block copolymers as demulsifiers for water-in-oil emulsions. A series of reverse copolymers were made by synthesizing a core of 4000 and 6000 molecular weight (MW) EO and adding various amounts of PO. Model emulsions were prepared by stabilizing water in benzene emulsions with asphaltenes. The authors determined that demulsification efficiency is directly proportional to HLB and molecular weight. Maximum efficiency was found using a demulsifier of 6000 MW EO and 26 moles PO. These same authors also report that raising the temperature led to increased demulsification efficiency for PO/EO block copolymers. They also reported that increased salinity decreased efficiency and noted that the pH of the aqueous phase should be approximately 7.

Tadros *et al.* [40] reported on the effect of EO/PO block copolymers on the rheology of coal/water slurries. In order for the rheology of concentrated slurries (> 65% solids) to be balanced, the slurry must be stabilized against flocculation and must also be protected against settling under the influence of gravity. To accomplish this, a mild flocculant must be added. The authors, investigation used Synperonic polyglycols from Uniqema with a constant PO block (55 units), and from 4 to 147 units of EO. The block copolymers evaluated were L-101, P-104, P-105, and F-108. F-108 was found to be an excellent flocculant at low concentrations, but to be a mild flocculant and actually a restabilization agent at concentrations from 0.3 to 1.0 % by weight of the formulation.

Polat and Chander [41] reported on the effect of EO/PO block copolymers on the wetting of coal. They used contact-angle measurements to determine that

Pluronic L-64 and P-104 gave the highest contact angles and best wetting performance. Paterson et al. [42] studied the extraction of polycyclic aromatic hydrocarbons from coal-tar-contaminated soil using nonionic surfactants. For this the EO/PO block copolymers contained a 56-unit PO core and varying amounts of EO. The block copolymers tested included Pluronic P-103, P-105, and F-108. The highest desorption efficiency for phenanthrene and anthracene was obtained with Pluronic P-103. The authors reported that increasing EO block size decreased efficiency.

Crawford [43] reported the development of high-temperature and high-pressure (HTHP) fluid-loss-control aids that function under drilling conditions of 204 °C and up to 138 MPa. These aids are primarily composed of Gilsonite in combination with solubilized lignite and carbon black. Pluronic L-101 was used at 1.5–5.0% by weight as an emulsifying surfactant.

10.4 AGRICULTURAL APPLICATIONS

Polyoxyalkylene block copolymer surfactants have found wide use in agricultural formulations as emulsifiers and dispersing agents [44]. In many instances emulsifiable concentrates are the preferred method of applying chemicals, thus avoiding the use of solvents.

Hester [45,46] reported on the use of EO/PO block copolymers as compatibility agents for agricultural pesticide compositions. Specifically, he used them for tank mixes where the various components being blended have cationic and anionic surfactants present and thus could be incompatible. Under certain circumstances, these agents caused the formation of a precipitate or sediment. The author reported using nonionic surfactants with HLB of 18–23, which worked surprisingly well to compatibilize these mixtures. Formulations using Pluronic P-103, P-104, P-105, F-108, and F-127 gave systems with no sediment and no flocculation. Blends of P-105 and L-61 worked especially well at less than 1% concentration.

Haas et al. [47] reported the use of polymeric surfactants to create pesticidal suspensions. In many instances the active agents must be milled and then dispersed before they can be delivered to the desired target. The authors investigated several types of polymeric surfactants as processing aids. Pluronic surfactants were found to be useful for increasing the milling efficiency of certain active ingredients. Higher MW and higher EO content improved the efficiency.

10.5 LATEX AND EMULSION APPLICATIONS

An interesting new application for alkylene oxide block copolymers has been their use in the formation of latexes from solutions of polymers in supercritical carbon dioxide. Examples, using both the common EO/PO as well as EO/BO block copolymers, can be found in these applications.

The research group led by Johnston [48] was the first to demonstrate that using EO/PO or EO/BO block copolymers allowed the precipitation of uniform microspheres of amorphous polymers from CO_2 solution without flocculation or agglomeration. Heater and Tomasko [49] followed up this work using reversed Pluronic 17R1 as a surfactant stabilizer to reduce agglomeration during epoxy-resin processing with carbon dioxide as an antisolvent. Johnston's group continues to explore this area. In a recent paper [50] they reported on the synthesis of water-dispersible polymer particles in carbon dioxide. Block copolymer surfactants made with ethylene oxide and propylene oxide or butylene oxide were used in the synthesis since they are hydrophilic and soluble in both water and CO_2.

Cameron and Sherrington [51] provided an interesting application of block copolymers of ethylene oxide and propylene oxide. They reported on the formation of stable, high internal phase, nonaqueous emulsions. The volume fraction of the nonpolar internal phase was as high as 0.9. The low-volume, polar-continuous phase solublized the polyethylene oxide (PEO) blocks resulting in highly stable emulsions.

10.6 PAPER (DEINKING) APPLICATIONS

Deinking of office waste paper, a mixture containing xerographic and laser-printed paper, is a difficult task. Moon and Nagarajan [52] reported that it can be accomplished using a simple formulation and process based on Pluronic surfactants. Careful selection of the proper Pluronic grade allowed for a much simpler formulation than is typically used. A theoretical paper by Gandini *et al.* [53], used model surfactant systems based on ethylene oxide/propylene oxide block copolymers to elucidate the complex interface features related to deinking in paper recycling.

10.7 CLEANING AND DETERGENT APPLICATIONS

Polyoxyalkylene block copolymers have long been used in all aspects of cleaning from hard surface cleaners to laundry rinse aids [54–56]. In recent times their utility has been hampered by their relatively slow biodegradability [57]. However, these compounds still find use today due to their stability and foaming characteristics.

Pancheri and Mao [58] reported the use of block copolymers for making high-suds liquid detergent compositions. The products were useful for washing hard surfaces and demonstrated a superior ability to handle grease. In these compositions, the polymeric surfactant was used from 0.1% to 10% by weight, with the most preferred range being 0.5% to 2%. The authors used Pluronic F-38, L-41, L-42, F-47, F-68, L-81, L-82, and others. Cheung et al. [59,60] described the development of hard surface cleaning and disinfecting compositions. The use of a fluorosurfactant along with nonionic surfactants such as the Pluronic L series allowed higher water content and lower amounts of organic solvents than with traditional cleaners.

Hsu et al. [61] reported the use of polyoxyalkylene copolymer surfactants to develop a clear, heavy-duty liquid laundry detergent that can suspend encapsulated particles from 300–5000 micrometers in size. The inventors used special thickening polymers to suspend the particles, as well as 10–40% by weight of nonionic surfactants like Pluronic L-65. Johnson and Franklin [62] reported using Pluronic surfactants along with alkoxylated quaternary ammonium compounds to formulate low foaming compositions for autodishwashing and other clean-in-place applications. Murphy [63] described novel dry-cleaning compositions that use nonionic surfactants to minimize the forces that bind dirt and stains to fabric. Surfactants, such as Pluronic L-62 used in low levels, also increased the solubility of the contaminants in the stain-removal solvents.

Takashima [64] reported formulating cleaning solutions for electronic parts using polyoxyalkylene block copolymer surfactants. These parts must be cleaned with high-pH cleaners to remove fines and organic substrates. However, traditional cleaners eroded silicon and other metals. The author formulated a system with various organic cleaners and used Pluronic polyols (L-31, L-61, L-44, L-64, and L-68) to prevent metal erosion. Richter et al. [65] investigated drain-treatment products suitable for killing micro-organisms that cause food-borne diseases, such as listeriosis and salmonellosis. Pluronic surfactants were added to the formulation to control the solubility of components in the sanitizing formulation. Stewart et al. [66] reported using Pluronic polyols and Pluronic derivatives as metal-chelating surfactants, which were used for coupling to a protein with an amino acid sequence having affinity for metal ions.

10.8 PERSONAL CARE APPLICATIONS

The physical and dermatological properties of ethylene oxide/propylene oxide block copolymers cause them to find widespread use in personal care applications. Many of the early commercial uses of EO/PO block copolymers were for cosmetic formulations [67]. There have been many review articles written by Schmolka and others [68,69] about the use of polyoxyalkylene block copolymers in personal care applications.

Ramirez and Vishnupad [70] reported that they were able to produce post-foaming, clear gels using Pluronic and Tetronic polyols in conjunction with volatile hydrocarbons in anionic or amphoteric detergent formulations. Unlike earlier formulations, these products were not self-foaming when enclosed in containers and could easily withstand transportation and storage. Pluronic polyols used include L-43, F-61, F-68, 31R1, and 17R2. Generally, the use level was 5% with pentane as the foaming agent.

Assini [71] described a method and system for delivering hair dyes. In this system gels made with Carbopol® were the preferred delivery form of the dye. When peroxide dyes were used, a nonionic surfactant was required and Pluronic F-127 was used at 1–3% of the formulation.

Brewster et al. [72, 73] reported the development of clear cosmetic sticks using Pluronic polyols. These sticks were used to deliver deodorant, antiperspirants, lipstick, etc. The authors indicated current products could be harsh to the skin when formulated with soaps, but attempts to reduce harshness by replacing the soap with alkoxylated fatty alcohols reduced the desired clarity of the sticks. The use of Pluronic F-127 and F-108 at approximately 4% of the formulation produced a stick with both clarity and rigidity, and was also mild to the skin.

Polyoxyalkylene polyols have found widespread use in oral care compositions. Trom and Oxman [74] reported the use of Pluronic polyols to control the viscosity of tooth-etching solutions. Earlier solutions had the problem of being either too thin (would run when applied to the tooth) or too thick (could not be easily applied). The present invention used Pluronic F-127 or F-108 at 17–26 wt% of the formulation. This mixture was thin at room temperature but will gel and stay in place when applied to the tooth as it experiences a subsequent rise in temperature. These same researchers reported [75] a method of whitening teeth that has viscosity control provided by polyoxyalkylene block copolymers. Once again, current systems may be too thick or thin to provide effective whitening. When Pluronic polyols are present, the systems gelled on contact with the teeth and remained in place. Hoic et al. [76] developed a two-component tooth-whitening composition that effervesced when mixed by brushing. This formulation used Pluronic F-127 at 15–25% by weight as a thickener. Curtis et al. [77] reported a two-part system to enhance peroxide whitening of teeth that called for brushing with an aqueous alkaline rinse, then brushing with a peroxide-containing dentifrice. This system used Pluronic F-127 and F-108 at 1% each to solubilize flavor components and provide foaming action.

Lee et al. [78] reported the development of an antitartar dental product containing calcium phosphate salts. The preferred form for these products was a gel that remained in place on the teeth when applied. Pluronic F-127 was used at 18–25% by weight to provide gelling properties. Pluronic F-88, F-99 and F-108 were also mentioned. Gaffar et al. [79] reported the discovery of anticalculus and antiplaque compositions that used Pluronic F-68 and L-44 as surfactants.

10.9 METAL-CLEANING APPLICATIONS

Polyoxyalkylene block copolymers have long been used as components of cleaners for ferrous and nonferrous metals [80].

Das [81] reported the use of solutions containing Pluronic surfactants as rinse aids for metal surfaces, primarily aluminum, used as food and beverage containers. In the course of metal forming, the aluminum was washed with acidic cleaners to remove metal fines. The metal was then washed with tap water followed by deionized water and dried in a hot-air oven. Retention of water on the surface of the metal resulted in long drying times at high temperatures, resulting in decreased production and spotting (from water) on the metal. The rinse aid reported by the author used Pluronic polyols along with alcohol alkoxylates to give increased drainage of the rinse water so drying can be accomplished at lower temperatures and shorter oven times. An added benefit was that the aluminum surface had a lower coefficient of friction after rinsing, which resulted in improved molding of the metal. Pluronic surfactants used were L-31, L-61, L-64, L-63, and L-43. Reverse Pluronics 10R5, and 10R8 were also reported. In practice, concentrated rinsing solutions were prepared with the Pluronic surfactants present at 15–20% by weight, and then diluted for use in the rinsing process. Hnatin and Reichgott [82] reported a similar rinse aid for aluminum using Pluronic L-61. These authors refered to the Pluronic polyol as a defoamer for the rinsing process.

10.10 LUBRICATING APPLICATIONS

The physical properties of EO/PO block copolymers make them excellent additives for various types of lubrication systems.

Khan and Khan [83] reported the use of Pluronic polyols as lubricants for intravenous catheters. Specifically they developed a lubricant that does not require the use of CFCs as solvent or carrier of the lubricant. Catheters are typically lubricated to minimize drag between the catheter and the patient's skin. Also, a lubricant can be applied to minimize adhesion between the catheter and the needle to facilitate removal. Historically, a silicone-based lubricant was used for this purpose and was applied to the catheter by dissolving in CFC, the CFC then being allowed to evaporate. Today CFCs are no longer used and flammable organic solvents such as alcohols or hydrocarbons have taken their place. The authors reported using Pluronic polyols in combination with phospholipids to form a water-soluble lubricant system that can be applied by dipping, brushing, or spraying. After the water evaporates, the Pluronic polyol and the phospholipid provide lubricity. The authors reported using Pluronic P-123 at 4–8% concentration.

Winicov et al. [84] reported the use of Pluronic polyols as lubricants for conveyor systems that handle thermoplastic articles. Many aqueous lubricants can adversely affect certain types of materials such as polyethylene terephthalate (PET), polybutylene terephthalate (PBT), polycarbonate, etc. The authors reported that lubricants made with Pluronic F-108 showed marked superiority over other commercially available lubricants with regard to lubricity and crazing of the polymers. They observed positive effects for lubricity and crazing with increasing MW and EO content of the block copolymer.

10.11 CAPILLARY ELECTROPHORESIS APPLICATIONS

Chu and coworkers [85,86] have demonstrated a new application for ethylene oxide/butylene oxide block copolymers. Blending the triblock polymers $BO_6EO_{46}BO_6$ and $BO_{10}EO_{27}BO_{10}$ formed a medium useful for the separation of double-stranded DNA. While neither block copolymer worked alone, various combinations performed well as the gel media. This use of block copolymers of ethylene oxide and butylene oxide extended and improved upon the work done by Rill et al. [87–89] and Chu et al. [90–93] with corresponding ethylene oxide/propylene oxide block copolymers as the separation medium for capillary electrophoresis.

10.12 SURFACE-TREATMENT APPLICATIONS

Inhibiting microbial colonization of surfaces in contact with aqueous systems is a common concern. This is a considerable problem where large volumes of water must be used and subsequently discarded. Examples include cooling water, pulping and papermaking, and metal-working systems. A formulation based on an ethylene oxide/propylene oxide block copolymer in combination with an anionic alkylsulfosuccinate or amphoteric surfactant was effective in inhibiting microbial fouling of such hydrophilic surfaces as acrylics, ceramics and metals. This formulation was patented by Wright and Johnson [94]. Wright [95] also reported the use of Pluronic polyols in the development of compositions for inhibiting microbial adhesion on surfaces in contact with an aqueous system such as cooling water, or water streams in pulping or papermaking. Often when nonionic surfactants are used in these systems, the concentration needed to kill bacteria is substantially higher than for cationic surfactants. A system formulated with EO/PO block copolymer surfactants was found to substantially inhibit bacterial colonization of surfaces without exhibiting toxicity toward the target population. Use of Pluronic L-101 at a 1:2 ratio with the active ingredient dioctylsulfosuccinate allowed for a substantial reduction in the amount of active ingredient needed.

10.13 CORROSION-INHIBITION APPLICATIONS

It is known that alkylene oxide block copolymers and their esters function as corrosion inhibitors [96]. Rangelov and Mircheva [97] reported on the mechanism by which the block copolymers of EO and PO protect steel surfaces from corrosion. Their study included the triblock copolymer $EO_{24}PO_{15}EO_{24}$ and the pentablock $EO_8PO_{15}EO_{10}PO_{15}EO_8$. The protection mechanism proposed was consistent with the block structures and loss of corrosion protection with increasing temperature.

10.14 MISCELLANEOUS APPLICATIONS

For over 50 years EO/PO block copolymer surfactants have found widespread use in numerous commercial applications. Industrial and academic researchers continue to use these adaptable polymers to enhance systems and formulations for the benefit of society.

Hill and Brown [98] reported the formation of water-free, cosolvent-free ultramulsions by dispersing polydimethylsiloxanes in polyoxyalkylene surfactants. These ultramulsions combined certain characteristics of emulsions with those of microemulsions and were used in coatings for products such as antigas/antacids, plaque-control agents, medical lubricants, polishes, etc. These ultramulsions were stable and water and solvent free. Copolymers used include Pluronic F-68, F-88, F-108, and F-127. Concentrations used were 25–90% of the formulation.

Allen et al. [99–102] took advantage of the gel properties of Pluronic surfactants to create gel compositions for shoe construction. This shoe had a liquid-filled bladder that conformed to the shape of the foot as it was inserted into the shoe, and then the liquid gelled when subjected to heat from the foot. This gel-filled bladder now held the shape of the foot. Solutions made from EO/PO block copolymers and water had the desirable property of gelling at a higher temperature than the liquid state. Prior art described filling compositions that were not water based and had gelling temperatures below that of the liquid state. The primary gelling agent used was Pluronic F-127 at 22–28% by weight of the formulation. Ersfeld et al. [103] reported the development of a polyurethane custom shoe insert that would not only custom-form to the foot, but also provided a cooling effect. This insert was made using 4% Pluronic F-38 as a surfactant to facilitate the reaction between the isocyanate and the polyol.

Turcotte and Lockwood [104] reported the development of a heavy-duty automobile antifreeze composition that used Pluronic L-61 at 0.05% by weight as a defoamer. Schubert et al. [105] described a method for cleaning that used automobile antifreeze via an ion exchange/carbon bed and then reintroducing necessary performance enhancing additives. The authors reported using Pluronic L-61 as the primary antifoam agent in the reconstituted material.

Berke et al. [106] were able to increase the amount of air entrained in cementitious compositions by use of triblock surfactants. Increasing the amount of air in cement is advantageous, since it leads to increased resistance to frost attack and deterioration due to repeated freezing and thawing. Many shrinkage-reducing agents added to cement reduce the amount of entrained air, but this was overcome by using Pluronic F-68 at less than 1 % on a solids basis. When these surfactants were used, entrained air increased 3-fold.

Smith and Kellett [107] reported the use of Pluronic L-61 to assist in the formation of a uniform liquid dispersion of a fabric-softener composition on dryer sheets. Novich et al. [108] described a new composition for impregnating glass-fiber strands for use in reinforcing thermoplastic resins. This formulation used Pluronic F-108 as an emulsifying agent.

REFERENCES

1. T. H. Vaughn, H. R. Suter, L. G. Lundsted, and M. K. Kramer, *J. Am. Oil Chemists' Soc.* **28**:294 (1951).
2. D. R. Jackson and L. G. Lundsted, U.S. Patent 2,677,700 to Wyandotte Chemicals Corporation (1954).
3. I. R. Schmolka in *Nonionic Surfactants* (M. J. Schick, ed.) Surfactant Science Series 1, Marcel Dekker, New York, 1996, p. 300.
4. L. G. Lundsted and I. R. Schmolka, in *Block and Graft Copolymerization*, Vol. 2 (R. J. Ceresa, ed.), John Wiley & Sons, Ltd., London, 1976, pp.113–272.
5. I. R. Schmolka, *J. Am. Oil Chem. Soc.* **54**:110 (1977).
6. M. W. Edens, in *Nonionic Surfactants. Polyoxyalkylene Block Copolymers* (V. M. Nace, ed.) Surfactant Science Series 60, Marcel Dekker, New York, 1996, pp.185–210.
7. *Amphiphilic Block Copolymers* (P. Alexandridis and B. Lindman, eds.) Elsevier, Amsterdam, 2000, pp. 305–408.
8. H. Choi, Y. Oh, and C. Kim, *Int. J. Pharm.* **165**:23 (1998).
9. H. Choi, J. Jung, J. Ryu, S. Yoon, Y. Oh, and C. Kim, *Int. J. Pharm.* **165**:33 (1998).
10. H. Choi, M. Lee, M. Kim, and C. Kim, *Int. J. Pharm.* **190**:13 (1999).
11. J. M. Barichello, M. Morishita, K. Takayama, Y. Chiba, S. Tokiwa, and Nagai, T., *Int. J. Pharm.*, **183**:125 (1999).
12. K. L. March, U. S. Patent 5,552,309 to Indiana University Foundation (1996).
13. A. Y. Kabanov and V. Y. Alakhov, U. S. Patent 6,060,518 to Supratek Pharma Inc. (2000).
14. A. V. Kabanov, V. Y. Alakhov and E. V. Batrakova, U. S. Patent 6,277,410 to Supratek Pharma Inc. (2001).
15. I. P. Gladysheva, O. V. Polekhing, T. A. Karmakova, E. R. Nemtsova, R. I. Yakubovskaya, W.-C. Shen, A. R. Kennedy and N. I. Larionova, *J. Control. Release*, **74**:303 (2001).
16. M. Q. Hoang and M. A. Khan, U. S. Patent 5,922,314 to Becton, Dickinson and Company (1999).
17. A. Paavola, J. Yliruusi and R. Rosenburg *J. Control. Release* **52**:169 (1998).
18. M. Bohner, T. A. Ring and K. D. Caldwell *Marcromolecules* **35**:6724 (2002).
19. G. Storm, S. O. Belliot, T. Daemen and D. D. Lasic, *Adv. Drug Delivery* **17**:31 (1995).
20. C. Allen, D. Maysinger and A. Eisenberg, *Colloids Surfact. B. Biointerf.* **16**:3 (1999).
21. S. M. O'Conner, A. P. DeAnglis, S. H. Gehrke and G. S. Retzinger, *Biotechnol. Appl. Biochem.* **31**:185 (2000).
22. C. Espadas-Torre and M. E. Meyerhoff, *Anal. Chem.* **67**:3108 (1995).
23. J. K. Armstrong, H. J. Meiselman and T. C. Fischer, *Thrombosis Res.* **79**:437 (1995).

24. C. M. Edwards, J. A. May, S. Heptinstall and K. C. Love, *Thrombosis Res.* **81**:511 (1996).
25. R. L. Hunter and A. Duncan, U. S. Patent 5,648,071 to Emory University (1997).
26. K. Huang, B. P. Lee, D. R. Ingram and P. B. Messersmith, *Biomacromolecules* **3**:397 (2002).
27. T. C. Green and R. P. Goodrich U. S. Patent 6,051,146 to COBE Laboratories, Inc. (2000).
28. M. Scherlund, K. Weber-Berger, A. Brodin and M. Malmsten, *Eur. J. Pharm. Sci.* **14**:53 (2001).
29. T. X. Viegas, L. E. Reeve and R. L. Henry, U. S. Patent 5,587,175 to MDV Technologies Inc. (1996).
30. T. X. Viegas, L. E. Reeve and R. L. Henry, U. S. Patent 5,593,683 to MDV Technologies Inc. (1997).
31. T. X. Viegas, L. E. Reeve and R. L. Henry, U. S. Patent 6,346,272 to MDV Technologies Inc. (2002).
32. G. P. Canevari and R. J. Fiocco, U.S. Patent 3,331,765 to Esso Research and Engineering Co. (1967).
33. H-F. Fink, G. Koerner, and G. Rossmy, U.S. Patent 4,029,596 to Th. Goldschmidt AG (1977).
34. F. Staiss, R. Böhm, and R. Kupfer, *Fossil Fuels, Deriv. Relat. Prod.* **6**:334 (1991).
35. S. E. Taylor, *Chem. Ind. (Lond.)* **20**:770 (1992).
36. M. Amaravathi and B. P. Pandey, *Res. Ind.* **36**:198 (1991).
37. U. Retter, M. Koch, and I. Nehls, *Fresenius J. Anal. Chem.* **364**:777 (1999).
38. N. N. Zaki, M. E. Abdel-Raouf, and A-A. A. Abdel-Azim, *Monatshefte für Chemie* **127**:621 (1996).
39. N. N. Zaki, M. E. Abdel-Raouf, and A-A. A. Abdel-Azim, *Monatshefte für Chemie* **127**:1239 (1996).
40. Th. F. Tadros, P. Taylor, and G. Bognolo, *Langmuir* **11**:4678 (1995).
41. H. Polat and S. Chander, *Colloids Surf. A: Physicochem. Eng. Aspects* **146**:199 (1999).
42. I. F. Paterson, B. Z. Chowdhry, and S. A. Leharne, *Chemosphere* **38**:3095 (1999).
43. D. Crawford, U.S. Patent 6,395,686 B2, Unassigned (2002).
44. L. G. Lundsted and I. R. Schmolka, in *Block and Graft Copolymerization*, Vol. 2 (R.J. Ceresa, Ed.), John Wiley & Sons, Ltd., London, 1976, pp. 114–117.
45. D. M. Hester, U.S. Patent 5,939,358 to Zeneca, Limited (1999).
46. D. M. Hester, U.S. Patent 5,686,384 to Zeneca, Limited (1997).
47. S. Haas, H-W. Hässlin, and C. Schlatter, *Colloids Surf. A: Physicochem. Eng. Aspects* **183–185**:785 (2001).
48. S. Mawson, M. Z. Yates, M. L. O'Neill and K. P. Johnston, *Langmuir* **13**:1519 (1997).
49. K. J. Heater and D. L. Tomasko, *J. Supercrit. Fluids* **14**:55 (1998).
50. J.-J. Shim, M. Z. Yates and K. P. Johnston, *Ind. Eng. Chem. Res.* **40**:536 (2001).
51. N. R. Cameron and D. C. Sherrington, *J. Chem. Soc. Faraday Trans.* **92**:1543 (1996).
52. T. Moon and R. Nagarajan, *Colloids Surf. A: Physiochem. Eng. Aspects* **132**:275 (1998).
53. D. Beneventi, B. Carre and A. Gandini, *Langmuir* **18**:618 (2002).
54. K. M. Hellsten and S. O. Santamaki, U.S. Patent 3,850,831 to Mo Och Domsjo Aktiebolag (1974).
55. L. A. Gilbert, U.S. Patent 4,284,524 to The Proctor & Gamble Company (1981).
56. L. G. Lundsted and I. R. Schmolka, in *Block and Graft Copolymerization*, Vol. 2 (R.J. Ceresa, Ed.), John Wiley & Sons, Ltd., London, 1976, pp. 131–159.

57. A. Hettche and E. Klahr, *Tenside* **19**:127 (1982).
58. E.J. Pancheri and M. H. K. Mao, U.S. Patent 5,167,872 to The Procter & Gamble Company (1992).
59. T. W. Cheung, D. T. Smialowicz, and M. H. Mehta, U.S. Patent 6,440,916 B1 to Reckitt & Coleman Inc. (2002).
60. T. W. Cheung, D. T. Smialowicz, and M. H. Mehta, U.S. Patent 6,306,810 B1 to Reckitt Benckiser Inc. (2001).
61. F-L. G. Hsu, D. S. Murphy, K. M. Neuser, M. Bae-Lee, D. J. Kuzmenka, K. Garufi, P. G. Lazare, C. J. Buytenhek, P. W. Van Reeven, C. C. Verburg, and F. T. Van De Scheur, U.S. Patent 6,362,156 B1 (2002).
62. A. K. Johnson and R. Franklin, U.S. Patent 6,426,014 B1 to Akzo Nobel N.V. (2002).
63. D. S. Murphy, U.S. Patent 6,309,425 B1 to Unilever Home & Personal Care, USA (2001).
64. M. Takashima, U.S. Patent 6,472,357 B2 to Sumitomo Chemical Company, Limited (2002).
65. F. L. Richter, J. Wilson, and D. E. Pedersen, U.S. Patent 6,197,321 B1 to Ecolab, Inc. (2001).
66. R. Stewart, K.D. Caldwell, C. Ho, and L. Limberis, U.S. Patent 6,087,452 to University of Utah (2000).
67. C. Pacifico, J. G. Kramer, and R. M. Abbott, *J. Soc. Cosmetic Chemists* **3**:303 (1952).
68. I. R. Schmolka, *Am. Perfumer Cosmetics* **82**:25 (1967).
69. I. R. Schmolka, *Cosmetics Perfumery* **89**:63 (1974).
70. J. E. Ramirez and M. Vishnupad, U.S. Patent 6,096,702 to Imaginative Research Associates, Inc. (2000).
71. A. Assini, U.S. Patent 5,554,197 to Anthony Bernard Incorporated (1996).
72. D. A. Brewster, M. Kuznitz, and J. R. Faryniarz, U.S. Patent 5,128,123 to Chesebrough-Pond's USA Co. (1992).
73. M. Kuznitz, D. A. Brewster, J. R. Faryniarz, and L. Cancro, U.S. Patent 5,114,717 to Chesebrough-Pond's USA Co. (1992).
74. M. C. Trom and J. D. Oxman, U.S. Patent 6,312,667 B1 to 3M Innovative Properties Company (2001).
75. J. D. Oxman and M. C. Trom, U.S. Patent 6,312,666 B1 to 3M Innovative Properties (2001).
76. D. A. Hoic, D. B. Viscio, and J. G. Masters, U.S. Patent 6,254,857 B1 to Colgate-Palmolive Company (2001).
77. J. P. Curtis, L. C. Beck, B. G. Reinhard, and K. N. Rustogi, U.S. Patent 6,174,516 B1 to Colgate-Palmolive Company (2001).
78. G. J. Lee, A. G. Ziemkiewicz, D. R. Williams, and S. R. Barrow, U.S. Patent 6,207,139 B1 to Unilever Home & Personal Care USA (2001).
79. A. Gaffar, J. J. Affilitto, N. Nabi, and M. T. Joziak, U.S. Patent 6,214,320 B1 to Colgate-Palmolive Company (2001).
80. R. J. Lipinski, U.S. Patent 3,239,467 to Lord Corporation (1966).
81. N. Das, U.S. Patent 5,279,677 to Coral International, Inc. (1994).
82. M. D. Hnatin and D. W. Reichgott, U.S. Patent 5,286,300 to Man-Gill Chemical Company (1994).
83. M. A. Khan and A. J. Khan, U.S. Patent 6,066,602 to Becton Dickinson and Company (2000).
84. E.Winicov, C. Foret, C. Palmer, M. W. Griffith, T. C. Hemling, U.S. Patent 5,663,131 to West Agro, Inc. (1997).
85. D. Liang, T. Liu, L. Song and B. Chu, *J. Chromatogr. A.* **909**:271 (2001).

86. T. Liu, D. Liang, L. Song, V. M. Nace and B. Chu, *Electrophoresis* **22**:449 (2001).
87. R. L. Rill, B. R. Locke, Y. Liu and D. H. Van Winkle, *Proc. Natl. Acad. Sci. USA* **95**:1534 (1998).
88. R. L. Rill, Y. Liu, D. H, Van Winkle and B. R. Locke, *J. Chromatogr. A*. **817**:287 (1998).
89. Y. Liu, B. R. Locke, D. H. Van Winkle and R. L. Rill, *J. Chromatogr. A*. **817**:367 (1998).
90. C. Wu, T. Liu, B. Chu, D. K. Schneider and V. Graziano, *Marcomolecules* **30**:4574 (1997).
91. C. Wu, T. Liu and B. Chu, *Electrophoresis* **19**:231 (1998).
92. C. Wu, T. Liu and B. Chu , *J. Non-Cryst. Solids* **235–237**:605 (1998).
93. D. Liang and B. Chu, *Electrophoresis* **19**:2447 (1998).
94. J. B. Wright and C. Johnson, U. S. patent 6,241,898 to BetzDearborn, Inc. (2001)
95. J. B. Wright, U.S. Patent 6,110,381 to BetzDearborn, Inc. (2000).
96. M. W. Edens, in *Nonionic Surfactants. Polyoxyalkylene Block Copoly*mers (V. M. Nace, ed.) Surfactant Science Series 60, Marcel Dekker, New York, 1996, 201.
97. S. Rangelov and V. Mircheva, *J. Mater. Sci. Lett*. **16**:209 (1997).
98. I. D. Hill and D.G. Brown, U.S. Patent 5,538,667 to WhiteHill Oral Technologies, Inc. (1996).
99. B. Allen, Z. Petrovic, I. Javni, and N. M. Goldman, U.S. Patent 5,985,383 to Acushnet Company (1999).
100. B. Allen and N. M. Goldman, U.S. Patent 5,939,157 to Acushnet Company (1999).
101. B. Allen, Z. Petrovic, I. Javni, and N. M. Goldman, U.S. Patent 5,955,159 to Acushnet Company (1999).
102. B. Allen, Z. Petrovic, and I. Javni, U.S. Patent 5,766,704 to Acushnet Company (1998).
103. D. A. Ersfeld, R. E. Anderson, M. L. Ruegsegger, K. R. McGurran, and R. A. Mallo, U.S. Patent 6,280,815 B1 to 3M Innovative Properties Company (2001).
104. D. E. Turcotte and F. E. Lockwood, U.S. Patent 6,228,283 B1 to Ashland, Inc. (2001).
105. D. C. Shubert, G. R. Myers, and R. C. Richardson, U.S. Patent 5,174,902 to BG Products, Inc. (1992).
106. N. S. Berke, M. C. Hicks, and J. J. Malone, U.S. Patent 6,358,310 B1 to W. R. Grace & Co.-Conn. (2002).
107. J. A. Smith and G. W. Kellett, U.S. Patent 6,254,932 B1 to Custom Cleaner, Inc. (2001).
108. B. E. Novich, K. Lammon-Hilinski, W. J. Robertson, X. Wu, V. Velpari, E. L. Lawton, and W. B. Rice, U.S. Patent 6,419,981 B1 to PPG Industries Ohio, Inc. (2002).

11 Development of Elastomers Based on Fully Hydrogenated Styrene–Diene Block Copolymers

CALVIN P. ESNEAULT, STEPHEN F. HAHN, GREGORY F. MEYERS

The Dow Chemical Company Corporate Research and Development, Midland MI 48667, USA

ABSTRACT

Thermoplastic elastomers have been prepared by performing catalytic hydrogenation of both the polystyrene and the polybutadiene segments in polystyrene-*b*-polybutadiene-*b*-polystyrene (SBS) triblocks to give poly(cyclohexylethylene)-*b*-poly(ethylene-*co*-1-butene)-*b*-poly(cyclohexylethylene) (CEBC) triblocks. These fully saturated materials were observed to be microphase separated at compositions and molecular weights that have commonly been used in the unsaturated starting materials. The development of these materials focused on the preparation of thermoplastic elastomers having good melt processability and mechanical strength, while exhibiting the lowest room temperature modulus. The modulus was minimized both by lowering the prehydrogenation styrene content and by modifying the polybutadiene microstructure to increase the 1-butene level of the ethylene/1-butene elastomeric segment in the final polymer. Increasing the 1-butene content also had the effect of lowering the melt viscosity in these polymers by decreasing the level of phase segregation. The properties of these block copolymers were further modified by the incorporation of a variety of additives. The addition of mineral oil lowered the order–disorder transition (T_{ODT}) of the CEBC block copolymers, but additive levels up to 24% by weight retained good elastomeric properties at room temperature. The mineral oil additive effectively decreased the bulk modulus of the elastomer below levels that could be obtained with the neat polymer, and the oil dramatically lowered the melt viscosity. The addition of low molecular weight poly(ethylene-*co*-1-octene) samples also lead to a decreased melt viscosity, but the T_{ODT} was not significantly influenced, suggesting that these polymers form a separate macrophase that coexists with the microphase-separated CEBC block copolymer.

Developments in Block Copolymer Science and Technology. Edited by I. W. Hamley
© 2004 John Wiley & Sons, Ltd   ISBN: 0 470 84335 7

11.1 INTRODUCTION

Thermoplastic elastomers (TPEs) are materials that display elastic deformation and recovery similar to that of crosslinked rubber but that can be melt processed in the manner of thermoplastic polymers [1]. Commonly, TPE materials consist of polymeric segments or blocks that are well above their glass transition temperature during use, allowing for response to external stress by large-scale deformation, combined with other blocks or segments that are below their glass transition temperature [2]. Living anionic polymerization techniques permit the synthesis of block copolymers based on the higher T_g block from styrene monomer and a lower T_g elastic segment from conjugated diene monomers such as butadiene and isoprene. Modern anionic polymerization techniques provide excellent control over block structure and molecular weight and allow for the synthesis of polymers with quantitative block connections at compositional crossover points and little or no homopolymer contamination [3]. These styrene/diene block copolymers and their derivatives have become the predominant commercial type of thermoplastic elastomer in terms of volume. This is due to the low cost and availability of the monomers and the development of effective large-scale polymerization processes.

11.1.1 MICROPHASE SEPARATION

The preparation of block copolymers in which the two polymeric components are immiscible gives rise to materials with characteristic patterns of compositional heterogeneity [4–6]. This phenomenon, commonly termed *microphase separation*, is derived from the balance between component immiscibility and chain stretching at the interfacial boundary. Periodic structures on a size scale of roughly 10–50 nm are commonly observed. It is this type of morphology that provides the styrene/diene-type block copolymers with their characteristic TPE properties. Discrete rubbery (diene) and glassy (styrene) phases form, with the glassy phase serving to limit permanent deformation of the elastomeric diene upon distortion. Phase separation is driven by segregation strength described by χN, where χ is the Flory–Huggins interaction parameter and N the degree of polymerization, as discussed in Chapter 1. Overall morphological behavior of block copolymers is also influenced by the relative volume fraction of the two block components.

Two-component block copolymers commonly display upper critical solution temperature (UCST) behavior [7]. They form ordered, microphase-separated morphologies at lower temperatures but can be heated to temperatures where the discrete heterogeneity is lost. The transition point from a heterogeneous microstructure to a compositionally homogeneous state is termed the order-to-disorder transition (T_{ODT}). At any given degree of polymerization N, the highest T_{ODT} exists in systems with equal volumes of the two components, at least if the two components are conformationally symmetric. Changing to uneven block compositions tends to decrease the T_{ODT}.

11.1.2 MATERIALS DESIGN

The practical materials design of microphase-separated polymers focuses first on the synthesis of structures that are microphase separated at use temperature to give access to the useful mechanical properties brought about by phase heterogeneity [8]. Additionally, materials can also be designed such that they also can be processed above the T_{ODT} since phase disordering is often accompanied by a precipitous decrease in the melt viscosity of the system. Control over the position of T_{ODT} is brought about by manipulation of the degree of polymerization (changing N), changes in the identity of one of the components (changing χ), or by changing the relative volume fractions of the copolymer components.

11.1.3 HYDROGENATION

Anionic polymerization typically requires that the carbon–carbon double bond that is incorporated into the growing chain be conjugated with another π bonding sequence that can stabilize the propagating carbanionic chain end during polymerization. For the polymerization of styrene with butadiene or isoprene, this means that the final polymer structure retains unsaturated species in the form of alkenes and phenyl rings. These functional structures are prone to further reaction during processing and in use, giving rise to temperature limitations in processing, discoloration, and eventual degradation of material properties. In order to improve the resistance to oxidation of these materials, chemical modification has been employed to transform the unsaturated species into less-reactive chemical structures [9]. Of these chemical modifications, hydrogenation has been the most commonly performed. However, the reactivities of the diene and styrene blocks towards hydrogenation are quite different. Hydrogenation of the diene block, which requires less stringent conditions, has been reported using hydrogen with either homogeneous or heterogeneous catalysts, as well as with chemically induced hydrogenation chemistry [10]. Hydrogenation transforms polybutadiene into the equivalent of a copolymer of ethylene and 1-butene, (EB), whose composition depends on the microstructure of the diene in the starting materials (Figure 11.1) [11]. Polyisoprene is hydrogenated to give the equivalent of an alternating copolymer of ethylene and propylene, (EP), with a small amount of pendant isopropyl groups due to 3,4 monomer insertion (Figure 11.1) [12].

Although more difficult than hydrogenation of the diene segments, conditions have also been discovered that will lead to complete saturation of both the diene-based and styrenic portions of the block copolymer [8,13], and the most

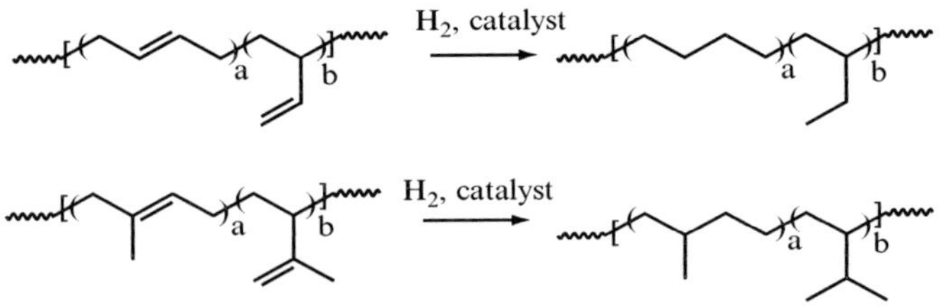

Figure 11.1 Repeat unit structures of polybutadiene and its hydrogenation product poly(ethylene-*co*-1-butene) and polyisoprene and its hydrogenation product poly(ethylene-*alt*-propylene).

effective of these reactions utilize heterogeneous catalysts. Typically, higher temperatures and higher hydrogen pressures are required for complete reaction in comparison with the diene-only hydrogenation. Recently, porous heterogeneous supports have been discovered that are more efficient than traditional catalysts in the hydrogenation of styrene and styrene block copolymers [13].

If one begins with a base styrene/diene block copolymer, hydrogenating either the polydiene segments alone, or hydrogenating both polydiene and polystyrene segments, changes the nature of the block units and necessarily changes the respective Flory–Huggins χ parameters. The χ parameter represents the nonideal part of the free energy of mixing, and χ is positive for most polymer pairs (there is overall net repulsion). While it has not been possible to accurately predict χ for a given polymer pair, trends in the magnitude of χ have been related to differences in the solubility parameters of the two polymeric components;

$$\chi = [v_{\text{ref}}/k_B T](\delta_2 - \delta_1)^2, \tag{11.1}$$

where v_{ref} is an arbitrary reference volume, k_B is the Boltzmann constant, T is temperature, and δ_1 and δ_2 are the solubility parameters of the polymeric species [14]. Hydrogenation of the central polybutadiene block in a polystyrene-*b*-polybutadiene-*b*-polystyrene block copolymer (denoted as SBS) leads to a polystyrene-*b*-poly(ethylene-*co*-1-butene)-*b*-polystyrene material (denoted as SEBS). There is a larger difference in solubility parameters in the hydrogenated polymer SEBS than the starting SBS, and an increase in the level of phase segregation has been observed [15–18]. Full saturation of SBS (Figure 11.2) to poly(cyclohexylethylene)-*b*-poly(ethylene-*co*-1-butene)-*b*-poly(cyclohexylethylene) (denoted as CEBC) would be predicted to lead to smaller solubility parameter differences than for the starting SBS [15,16,19], the reverse of the situation with only polydiene saturation. When these materials have been prepared, however, the relative strength of the phase segregation has been found to be less than that of SEBS but similar to that of the SBS starting material [18]. The implications of this stronger-than-expected phase separation for CEBC will be explored in later sections.

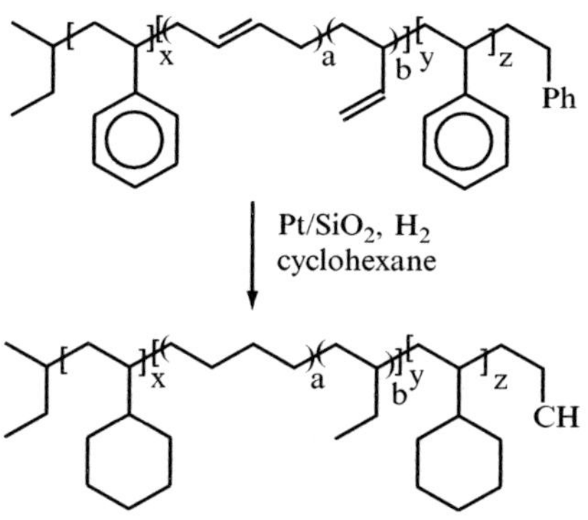

Development of Elastomers

Figure 11.2 Chemical structures of poly(styrene-*b*-butadiene-*b*-styrene) and its hydrogenation product poly(cyclohexylethylene-*b*-[ethylene-*co*-1-butene]-*b*-cyclohexylethylene).

A complication in hydrogenated styrene/diene block copolymer systems arises from the fact that the level of phase segregation is further influenced by the microstructure of the polydiene block. For blends of polystyrene (PS) and hydrogenated polybutadiene (polyethylene-*co*-1-butene, or PEB), studies of cloud-point behavior showed that the level of phase segregation decreases with increasing 1-butene content [17]. This trend has also been observed for PCHE/PEB blends [20], where PCHE denotes fully hydrogenated PS, which we designate poly(cyclohexylethylene). The fundamental reasons for these differences in the phase segregation with respect to microstructure for components with similar solubility parameters have not been fully explained.

Similar to hydrogenation of SBS, the polydiene-only hydrogenation of polystyrene-*b*-polyisoprene-*b*-polystyrene block copolymer (denoted as SIS) leads to a polystyrene-*b*-poly(ethylene-alt-propylene)-*b*-polystyrene material (denoted as SEPS), and complete hydrogenation of SIS leads to poly(cyclohexylethylene)-*b*-poly(ethylene-*alt*-propylene)-*b*-poly(cyclohexylethylene) material (denoted as CEPC). The syntheses and characterizations of CEBC and CEPC copolymers have been reported previously [21,22]. A principle motivation for using hydrogenation to make these block copolymers is the potential to dramatically improve the sensitivity of these materials towards oxidative, thermal and radiation-induced degradation. Complete hydrogenation also provides materials with a lower Flory–Huggins χ parameter than the corresponding SEBS or SEPS block copolymers, thus allowing for the preparation of materials that have the potential to be processed in a disordered state without requiring processing aids.

11.1.4 PRIOR ART IN CEBC AND CEPC BLOCK COPOLYMERS

CEPC triblock polymers were prepared by a group at the Shell Oil Company [21]. The SIS precursor polymers were prepared by sequential anionic polymerization, followed by hydrogenation using a Ni/kieselguhr catalyst. For the most part, these materials were intended to have elastomeric properties, with prehydrogenation styrene content ranging from 2 to 54% by weight. The use of isoprene in these polymers means that hydrogenation produced an amorphous poly(ethylene-*alt*-propylene) center block upon hydrogenation [12]. Although the microstructure of the polyisoprene block was not reported, the conditions used for the polymerization are known to provide a predominantly 1,4 repeat unit [3]. The hydrogenated polymers displayed improved tensile strength compared to the unsaturated starting material, and they also had higher tensile modulus, elongation at break, and permanent set. The weatherability of the hydrogenated polymers was studied, and an analysis of mechanical properties showed the expected improvement over those of the starting polymers. Subsequent work [22] extended this approach to block copolymers in which butadiene was employed in the center block to give an EB central block. The synthetic approach was identical to that used for the hydrogenated SIS copolymers with one key difference – the use of butadiene requires that the microstructure of the butadiene block be manipulated in order to vary the proportions of 1,2- and 1,4- microstructures. This was performed by the addition of tetrahydrofuran during the anionic polymerization process. This work showed clearly the influence of the diene microstructure on the elastomeric properties of the hydrogenated block copolymers. (Figure 11.3). Saturation of polybutadiene with high levels of the 1,4 microstructural repeat unit produces polyethylene repeat units, which have been shown to crystallize within the PE domains of the microphase-separated material when the matrix is glassy [23]. The incorporation of 1,2 (vinyl) repeat units provides the equivalent of a 1-butene monomer unit in the saturated polymer, reducing or eliminating crystallinity in the saturated diene block [11]. At 40% 1,2 repeat unit incorporation, the hydrogenated block copolymers showed the best combination of elastomer mechanical properties (the lowest glass transition temperature, highest % rebound, and highest tensile strength at 75 °C and 100 °C). This chemistry was further explored by workers at Borg-Warner as part of a larger study on the synthesis and characterization of hydrogenated styrene-diene (CEBC and CEPC) block copolymers [24,25]. These workers also recognized the influence of the hydrogenated polybutadiene microstructure on elastomer properties, and modified the polybutadiene microstructure by adding THF during the polymerization step. Retention of properties at elevated temperatures and after weather exposure differentiated these materials from the unhydrogenated block copolymers. More recently the preparation and characterization of elastic fibers from fully hydrogenated SBS block copolymers and their blends with polyolefin copolymers prepared using single-site metallocene catalysts has been described [26]. These materials showed improved melt spinnability (due to lower melt viscosity) compared to partially saturated SEBS block copolymers. Electron microscopy showed the CEBC/polyolefin blends to be macrophase separated into distinct polyolefin regions and microphase-separated block copolymer regions.

Development of Elastomers

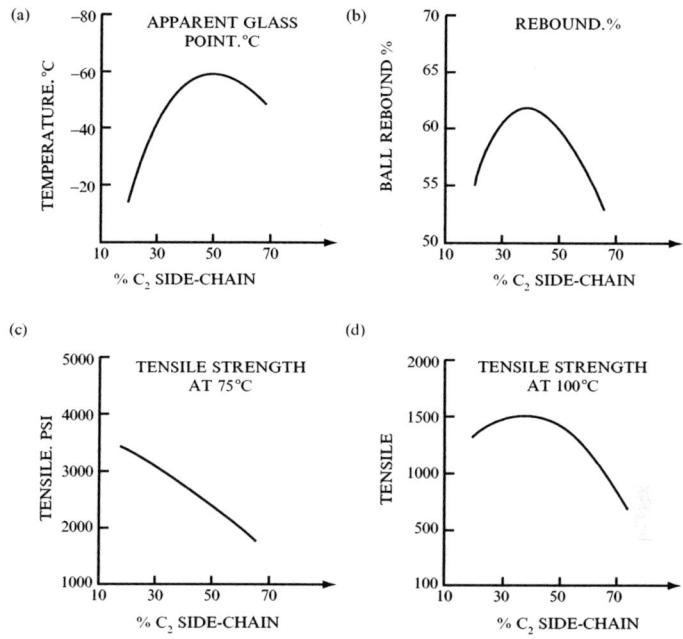

Figure 11.3 Cover page of US Patent 3,431,323, with plots of apparent glass point and ball rebound with respect to % C_2 side chain (parts a and b) and tensile strength with respect to % C_2 side chain at 75 °C (part c) and 100 °C (part d). The % C_2 side chain is equivalent to the 1-butene content in the hydrogenated polybutadiene block.

11.2 ELASTOMER DEVELOPMENT

Thermoplastic elastomers with the general structure CEBC are of interest for a variety of applications where rapid elastomeric response (recoverable low-stress deformation), good UV and oxidative stability, and exceptional processability are required. Creating such elastomers that have low tensile set (residual elongation after tensile stretching) with the lowest possible modulus at ambient temperatures is of special interest. Soft elastomers have found utility in a variety of applications, ranging from soft-touch applications (handles and grips made by molding a soft elastomeric skin over a supporting hard core) to fibers and surgical gloves. A major focus of development in this work was to understand the design factors that influence the modulus of CEBC block copolymers and to construct materials that had low moduli while retaining necessary mechanical properties.

11.2.1 EXPERIMENTAL

The block copolymers studied in this effort were prepared by sequential anionic polymerization of styrene and butadiene in cyclohexane solvent with initiation by *sec*-butyl lithium. All block copolymers had monomodal molecular weight distributions with polydispersity indices (M_w/M_n) less than 1.1. Hydrogenation was performed on solutions of the block copolymer in cyclohexane in the presence of a catalyst consisting of Pt on a porous silica support. Hydrogenation levels were typically greater than 99% for the polystyrene blocks and quantitative for the polybutadiene blocks. Polydispersity did not significantly change as a result of the hydrogenation process. Polybutadiene segment microstructure was modified by the addition of dry tetrahydrofuran (THF) at varying concentrations. Polymerization in cyclohexane provided polybutadiene with about 8% vinyl repeat units, and samples with up to 60% vinyl content could be prepared using the cosolvent THF [3,17]. The melt viscosity of the final polymer was assessed by performing melt flow rate measurements according to a standard test method protocol [27] with a 5 kg weight at 235 °C; values are given in units of grams polymer per ten minute increment. Mechanical properties were determined using compression-molded samples that were roll milled prior to testing. Tensile testing was performed using a type L die [28] with an Instron instrument with a 1000 N load cell at a rate of 254 cm/min. Tensile set was determined by elongating tensile specimens to 200%, holding for 30 s, returning to the starting position for a 60 s recovery, then measuring the amount of residual elongation. Dynamic mechanical analysis was performed using a Rheometrics Dynamic Shear Rheometer using a 25-mm diameter plate-plate sample geometry. Morphology was assessed by tapping mode AFM (TMAFM) characterization of spun-cast and annealed thin films. Polymer solutions (2 wt% polymer in decalin) were spun cast onto Si wafers at room temperature and were annealed in a nitrogen-purged oven at 200 °C for twelve hours. AFM images were obtained on a Nanoscope III instrument using a Dimension 3000 Large Sample AFM (Digital Instruments, Inc., Santa Barbara, CA) and "G" scanner head. The system is also equipped with extender electronics for phase imaging. Integral silicon tips (TESP, 225 µm long) were used and imaging was obtained with a free amplitude, A_o, of about 1.0 V and a set-point voltage, A_{sp}, of about 0.5 V.

11.2.2 SAMPLE-COMPOSITION NOMENCLATURE

The protocol for identifying individual samples of CEBC copolymers has been developed using a numbering system to describe key attributes. A series of three numbers separated by periods designates the total molecular weight (in thousands), the weight percentage styrene prior to hydrogenation, and the percentage of 1,2 butadiene microstructure in the polybutadiene block, respectively. For example, a hydrogenated triblock copolymer that is based on a copolymer with $M_n = 60$ kg/mol, 32% polystyrene by weight prior to hydrogenation, and 40% 1,2-polybutadiene microstructure is designated 60.32.40. All of the polymers examined in this project had linear triblock architectures.

11.2.3 ELASTOMER DEVELOPMENT

Styrene-based thermoplastic elastomers are typically linear triblock or multi-arm radial structures prepared with terminal PS blocks [3], and the polystyrene content ranges from roughly 15 to 40% by weight. In this composition range, the polystyrene phase usually adopts a cylindrical morphology (18 to 40 vol% PS) or a spherical morphology (below 18 vol% PS) [5,6]. Prior work [11,22] has established that a midblock microstructure comprising 40 wt% 1-butene or more is essentially free of crystallinity. Based on the preceding considerations, this work focused on completely hydrogenated styrene/butadiene linear triblock polymers having < 50% styrene by weight. Our goal was to determine the relative influence of the amount of PCHE hard segment and the effects of butadiene microstructure of the unsaturated precursor on final properties – ultimate tensile strength, tensile modulus, and melt viscosity.

In general, the block copolymers prepared in this study behaved in a manner consistent with materials that were microphase separated. Dynamic mechanical analysis of a 40% 1-butene CEBC showed two distinct glass transition events, one at about $-53\,°C$ due to the ethylene-co-1-butene block and another at about $125\,°C$ due to the PCHE block, with a relatively flat plateau between these transitions. A characteristic DMA spectrum is given in Figure 11.4.

The morphology was examined for some of these samples to verify that a microphase-separated structure existed. Figure 11.5 shows AFM images for a

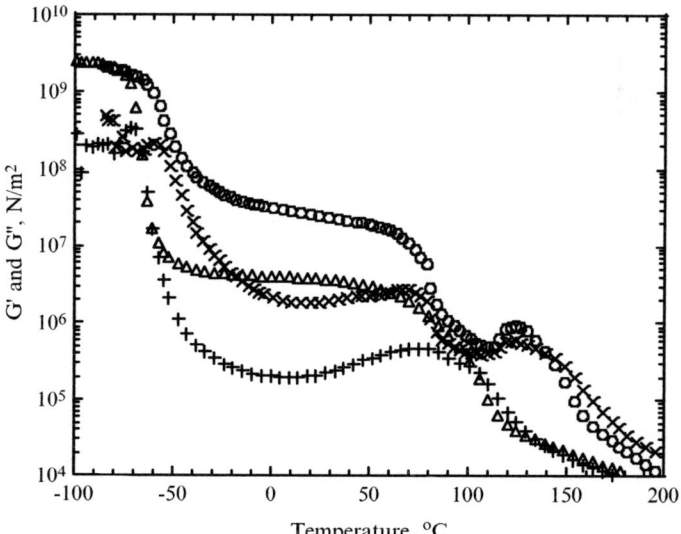

Figure 11.4 Dynamic mechanical analysis of SBS 66.32.40 ($G'\triangle$; $G''+$) and its hydrogenation product CEBC ($G'\bigcirc$, $G''\times$) with respect to temperature at a frequency of 1 rad/s.

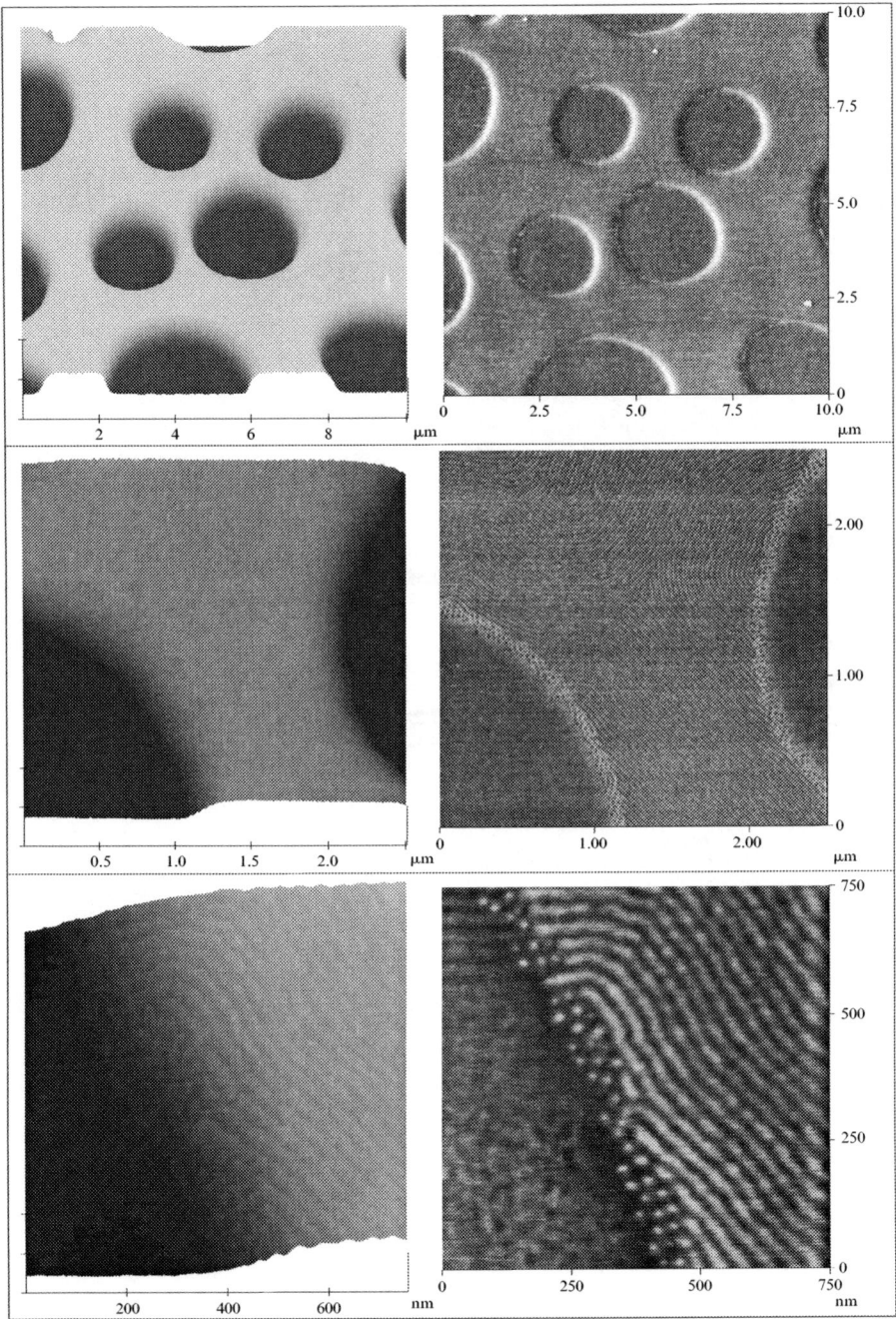

Figure 11.5 Tapping mode AFM images of an annealed 81.25.40 thin film. Left column shows height (perspective) images ($z = 50$ nm/div) and right column shows corresponding phase images (z range = 20 degrees).

hydrogenated block copolymer of the structure 81.25.40, in the center of the composition range that was examined in this development project. Following annealing, this film is 37 nm thick. There is not enough material in this film to support a continuous film layer, hence the formation of "holes" is observed. Based on TMAFM analysis of an area that had been scratched to reveal the substrate, the 37 nm film consists of an 11 nm brush layer and a 26 nm thick triblock layer. Within the holes the brush layer alone coats the wafer surface. Near the edge of a hole it is possible to observe both the long axis of the cylinders (lying in the plane of the film) and the ends of hexagonally packed cylinders where the film is discontinuous at the hole boundary. The cylinder diameters are measured to be 26 nm, consistent within a single layer. This analysis is consistent with a film in which a layer about one morphological domain in thickness is supported on a brush layer that is half of one domain in thickness. The expected cylindrical morphology is observed for the uppermost layer, while no distinct morphology is observed in the brush layer.

11.2.4 THE INFLUENCE OF PCHE CONTENT

Two series of samples were synthesized in which the total block copolymer molecular weight and diene microstructure were maintained at similar levels but in which the hard-segment content in the final polymer was systematically changed. The characteristic properties of these polymers are summarized in Table 11.1.

The first series of CEBC block copolymers had a total molecular weight of about 60 kg/mol and a poly-1-butene microstructure level of about 40%, while the PCHE content was varied from 20 to 45%. The influence of the styrene content on these materials was first evaluated by tensile testing. The modulus of styrenic block copolymers has been studied previously for a variety of SBS and

Table 11.1 Mechanical properties of fully hydrogenated triblock elastomers with variable PCHE block content.

Polymer	MFR	Shore A	Tensile Strength (MPa)	% Elong.	Force to 100% elong. (MPa)	Force to 300% elong. (MPa)	% tensile set
60.20.40	65	56.9	36.3	530	1.9	4.4	9.2
58.25.40	28.8	74.7	38.5	470	2.9	6.9	7.5
61.32.40	34.0	83.6	42.5	440	3.3	8.0	8.4
65.45.40	6.2	88.2	39.8	460	8.0	19.9	12.9
115.10.34	7.9	57.7	16.2	590	2.0	3.9	9.0
116.14.32	4.7	62.5	21.1	460	2.5	5.2	8.6
120.16.40	0.44	70.6	31.4	460	2.8	6.1	13.9
120.18.39	3.26	56.3	23.3	510	1.9	3.9	6.7
148.19.40	0.47	58.6	24.8	550	1.7	3.6	4.5

SIS materials [29–31] with much of the emphasis placed on materials that exhibit a cylindrical morphology. The modulus can be modeled by a rule-of-mixtures approach [30], such that in the composition range commonly observed in thermoplastic elastomers the modulus is profoundly influenced by the polystyrene content [31]. It must also be noted that the mechanical properties of these materials are known to be anisotropic [32–34]. The melt processing of block copolymers is known to orient the cylinders in the plane of melt flow [31,32] and it is anticipated that the cylinders in these block copolymer samples are aligned in the plane of the tensile elongation, but no attempt was made to interrogate the level of orientation. These materials exhibited high ultimate tensile strength and high elongation at break. These results are typical for microphase-separated block copolymers where there is negligible interference from any residual crystallinity. Modulus values increased in a systematic fashion as the PCHE content was increased, as evidenced by Shore A hardness and tensile stress at 100 and 300% extension. The melt index decreased with increasing styrene content, a result of increasing phase incompatibility as the volumes of the two phases become more similar. The CEBC block copolymer prepared by hydrogenating an SBS with 45% styrene showed a step change in modulus compared with the other samples suggesting a discontinuity in morphology (for example, from cylindrical to a gyroid or lamellar structure).

A second series of polymers was prepared at low PCHE content (PS weight % between 10 and 20 before hydrogenation). Molecular weights in this series of polymers were higher than the first set since the hard-segment content of these polymers were in the compositional range where the block copolymers were expected to be disordered at 60 kg/mol. These materials showed roughly the same range for Shore A hardness and tensile moduli as the lower molecular weight, 20% hard-segment sample from the first series. Also note that the lowest PCHE sample (from a 10% PS SBS) had very low tensile strength, suggesting that this sample was not microphase separated. These results suggest that, for a material containing approximately 40% 1-butene, there is a lower limit of modulus that can be obtained simply by lowering the PCHE content. For the CEBC elastomers of this study, Shore A hardness reached a plateau of approximately 60 at less than 20% hard-segment content. This study was expanded to include a larger number of samples of hydrogenated triblock copolymers with 40 wt% 1-butene and various polystyrene contents. A plot of the tensile stress at 100% elongation of these materials with respect to PS content is shown in Figure 11.6. These data further confirm the initial findings concerning the influence of PCHE content on CEBC modulus.

11.2.5 INFLUENCE OF POLYBUTADIENE MICROSTRUCTURE

The modulus of hydrogenated polybutadiene homopolymers has been shown to be directly related to the 1-butene content in the polymer [11]. In the absence of

Development of Elastomers

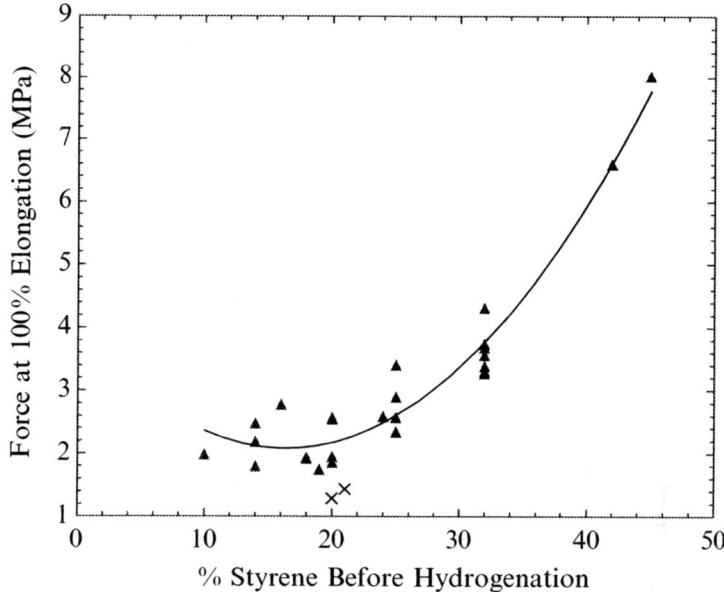

Figure 11.6 The influence of the hard-segment content on force required to bring about 100% elongation for CEBC elastomers with 40% 1,2 butadiene microstructure (▲) and 60% 1,2 butadiene microstructure (x). The line is a polynomial fit for the data obtained from the 40% 1,2 butadiene based polymers ($y = 0.00698x^2 - 0.23x + 3.98$, $R^2 = 0.936$).

crystallinity, as the 1-butene content increases, the entanglement density of the polymer decreases, and the plateau modulus decreases. Thus, both the PCHE hard-segment level and the hydrogenated polybutadiene microstructure should contribute to the overall CEBC modulus [30]. In order to determine the influence of the diene microstructure on the modulus of fully saturated block copolymer elastomers, and in order to determine how mechanical properties vary with 1-butene content, block copolymers were prepared with polybutadiene blocks that had higher 1,2-butadiene microstructure content. The properties of these polymers are listed in Table 11.2.

Here, the CEBC block copolymers for the reference case (20% hard-segment, 40% 1-butene) gave slightly higher Shore A values than the previous data. In comparison, block copolymers with 20% hard-segment but with higher 1-butene contents (51% to 61%) gave significantly lower Shore A hardness values and lower tensile moduli. Tensile testing showed that these more highly branched materials had lower tensile strength and lower moduli, but higher elongation at break. Figure 11.7 shows stress–strain curves for a variety of block copolymers with varying 1,2 butadiene contents. This clearly shows the differentiation in high strain tensile response between materials with differing 1-butene content. The melt flow values of the higher 1,2 butadiene samples were

Table 11.2 Mechanical properties of fully hydrogenated triblock elastomers with variable 1-butene content.

Polymer	MFR	Shore A	Tensile strength (MPa)	% elong.	Force at 100% elong. (MPa)	Force at 300% elong. (MPa)
90.20.40	5.7	69.8	35.4	480	2.6	5.8
100.20.40	4.8	67.8	32.1	490	2.6	5.4
76.20.51	68.5	56.2	26.5	590	1.9	3.6
80.20.60	212	48.5	10.3	960	1.2	1.8
87.21.61	73.5	50.5	23.8	780	1.4	2.4
80.20.1	13.9	62.1	29.5	575	2.0	4.1

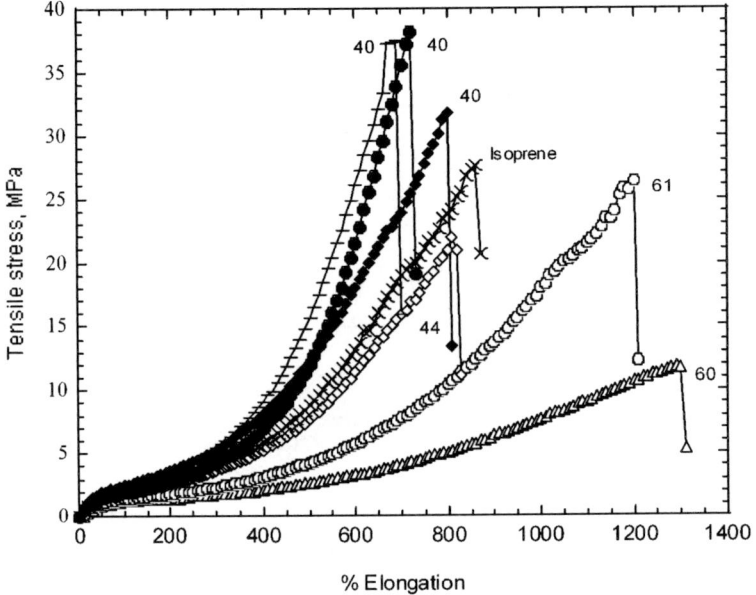

Figure 11.7 Stress–strain tensile curves for CEBC and CEPC block copolymers 60.20.40 (●), 90.20.40 (−), 100.20.40 (◆), 144.18.44 (◇), 80.20.60 (△), 87.21.61(○), and 80.20.EP (×). Branching levels are also given on the plot next to the associated tensile curve.

also significantly higher than those with lower 1,2 butadiene segments. The extent of phase segregation between PCHE and hydrogenated polybutadiene has been shown to be influenced by the butadiene microstructure [20]. PCHE is more incompatible with polyethylene than it is with poly-1-butene. This supports the observation that increasing the level of 1-butene in this series of elastomeric block copolymers provides materials with lower melt viscosity. Practical considerations limit the level of 1,2-butadiene to about 60% using THF as a polar modifier in the polymerization process, since higher THF levels lead to detectable levels of chain-end deactivation during the polymerization.

Development of Elastomers

The stress–strain curve of an elastomeric CEPC copolymer (prepared by hydrogenating an SIS) is also illustrated. This copolymer shows a tensile modulus that is intermediate between the 40% and 60% 1,2 EB-containing copolymers, with ultimate tensile properties that are very similar to the 40% EB samples.

11.2.6 ADDITIVES

As demonstrated above, the properties of block copolymers can be varied substantially by careful control of block structure, composition, and microstructure. Further differentiation in properties is commonly brought about by the use of additives. The addition of low molecular weight compounds that are preferentially sequestered in the low T_g block can effectively extend the properties of the microphase-separated copolymers, thus lowering the modulus of the soft block while maintaining a generic elastomeric response [35,36]. Compounding can also improve processability in block copolymers. This effect is particularly dramatic for the polystyrene-b-poly(ethylene-co-1-butene)-b-polystyrene (SEBS) block copolymers. The hydrogenation leads to a pronounced increase in the level of phase segregation compared to the SBS starting material. The increased phase segregation renders compounding more selective (by providing a larger difference in solubility parameters between the phases) and, on the other hand, makes it necessary to compound in order to render the polymers processable.

The addition of compounds to CEBC block copolymers was interrogated by selecting a single block copolymer with a representative molecular weight and PCHE content and by then adding a series of hydrocarbon compounds to it. Block copolymer CEBC 66.32.40 was selected, since it has an accessible order–disorder transition (T_{ODT}) of 230 °C. The temperature T_{ODT} was determined by rheological characterization. Frequency sweeps were performed at a series of temperatures, and discontinuities in plots of the shear storage modulus (G') with respect to the shear loss modulus (G'') were used to identify T_{ODT}. This analysis is shown in Figure 11.8. A discontinuity in the melt viscosity at low frequencies is also evident from the melt viscosity (Figure 11.9).

Low molecular weight mineral oils are commonly added to block copolymers to improve processability. For this study the hydrocarbon oil WITCO 200 Mineral Oil (molecular weight of 414 g/mol) or squalane (molecular weight 422 g/mol) were added to block copolymer 66.32.40 by solution blending in cylcohexane solvent with subsequent solvent devolatilization. The influence of the additive on melt viscosity and T_{ODT} were determined by dynamic mechanical analysis in a manner identical to that used for the neat copolymer. As anticipated, the CEBC/mineral oil blends had distinctly lower melt viscosities and lower moduli compared to the neat polymer. These blends retain useful tensile strength up to surprisingly high levels of added oil. The elongation at break does not drop significantly until 36% oil has been added (Figure 11.10).

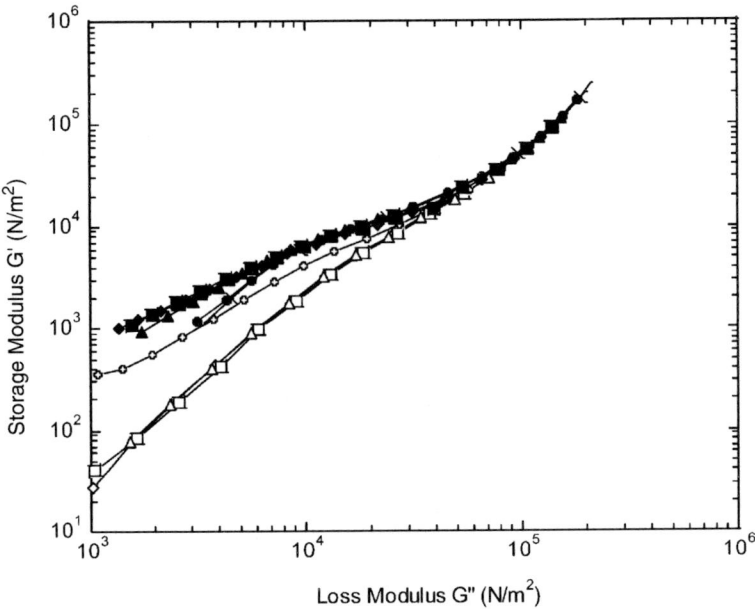

Figure 11.8 Dynamic mechanical analysis, plot of shear storage modulus G' with respect to shear loss modulus G'' for CEBC 66.32.40 as obtained from frequency sweeps at various temperatures above and below the order–disorder transition (240 ◇; 235 □; 230 △; 225 ○; 220 ◆; 215 ■; 210 ▲; 200 ●; 190°C ×)

For comparison, blends were also prepared with a commercially available SEBS triblock copolymer, Kraton G1652 [37]. Figure 11.11 shows stress–strain plots of Kraton G1652 and CEBC 66.32.40 with identical levels of oil. Despite the greater similarity of the solubility parameters in the CEBC polymers compared to the SEBS polymer, these materials show very similar blend elastomeric properties even up to 24% mineral oil addition.

The addition of mineral oil systematically softens the CEBC block copolymers and lowers the observed T_{ODT}. Another approach to modify the melt viscosity of the CEBC block copolymers was to incorporate a polymeric additive with a chemical structure similar to that of the center block, which might be relatively miscible with the center EB block but less so with the terminal PCHE blocks. A series of low molecular weight poly(ethylene-1-octene) copolymers were blended with CEBC 66.32.40 to assess the efficacy of this approach. These copolymers have a density of 0.87 g/cc and were differentiated by their melt-flow index values, which range from 1000 for the lowest molecular weight to 5 for the highest [38]. These materials were melt blended with the CEBC block copolymer and were characterized by dynamic mechanical spectroscopy. While all of these blends showed substantially lower melt viscosity compared to the CEBC starting material, only the lowest molecular weight poly(ethylene-*co*-1-octene) blend showed any measurable decrease in T_{ODT} and only at the highest additive level (36%). The observed T_{ODT} for the higher melt index PE/CEBC

Development of Elastomers

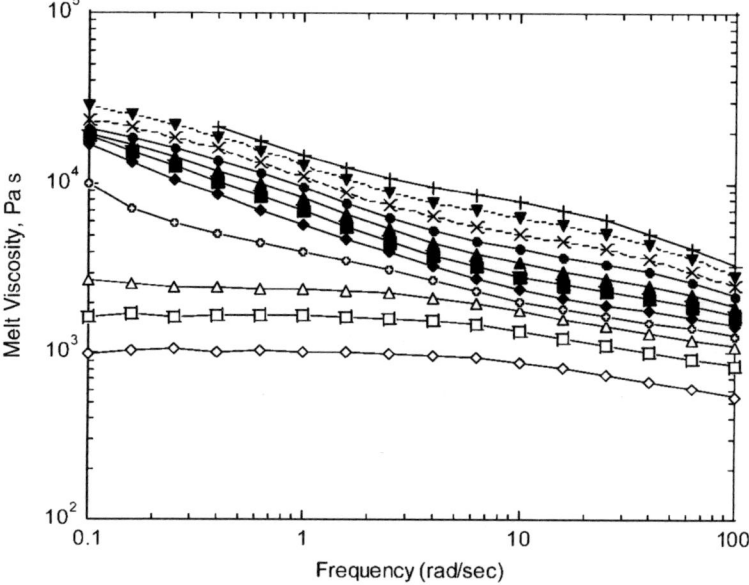

Figure 11.9 Dynamic mechanical analysis of CEBC block copolymer 66.32.40 showing the melt viscosity with respect to frequency for a series of temperatures above and below the order-to-disorder temperature (190, +; 195, ▼; 200, ×; 205, ●; 210, ▲; 215, ■; 220, ♦; 225, ○; 230, △; 235, □; 240°C ◇).

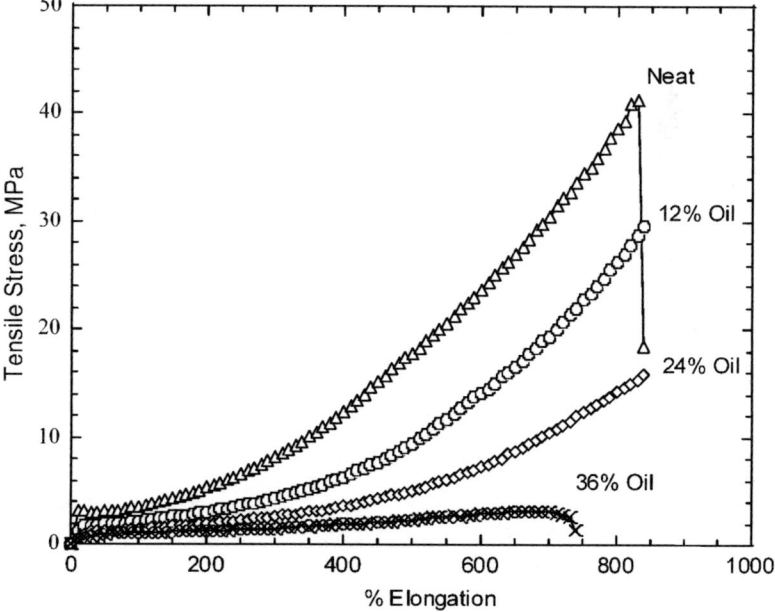

Figure 11.10 Stress–strain tensile curves for CEBC 66.32.40 neat (△) and with varying levels of added mineral oil (12% ○, 24% ◇, 36% ×)

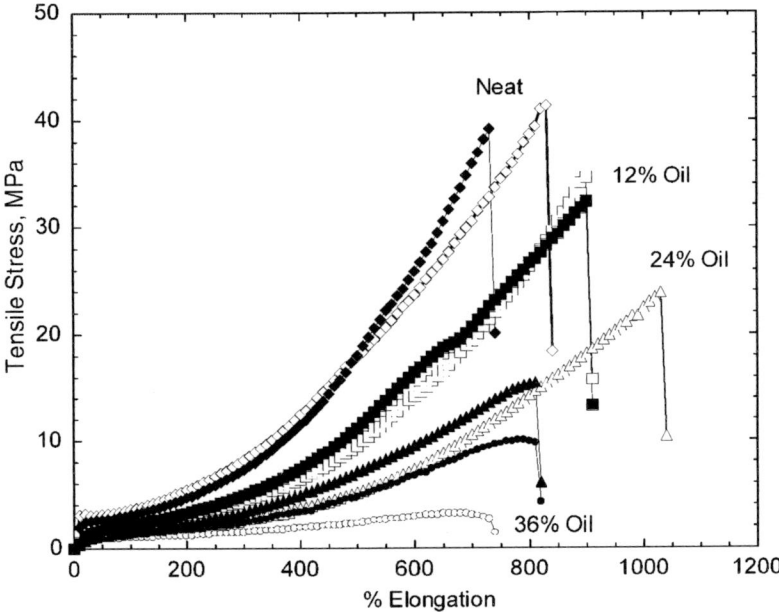

Figure 11.11 Stress–strain tensile curves for neat and oil-extended CEBC (neat ◇, 12 % □, 24 % △, 36 % ○) 66.32.40 and SEBS Kraton G1652 (neat ◆, 12 % ■, 24 % ▲, 36 % ●) block copolymers.

blends did not differ significantly. This behavior is consistent with the added copolymer either mixing selectively with the EB center block or forming a macrophase that is separate from either block copolymer microphase, as observed previously for similar systems [26]. The influence of a variety of the additives on CEBC melt viscosity is summarized in Figure 11.12.

11.3 SUMMARY AND CONCLUSIONS

The development of CEBC block copolymers for elastomeric applications has been pursued. Characterization of triblock copolymers showed these systems to be microphase separated at molecular weights and compositions where the starting SBS copolymers are microphase separated. The CEBC block copolymers behave as thermoplastic elastomers at PCHE contents ranging from 14 % to about 40 % based on the polystyrene content in the starting material, and the PCHE block provides higher heat performance compared to equivalent SBS or partially hydrogenated SEBS block copolymer. Efforts to prepare elastomers with lower room temperature moduli showed that lowering the PCHE content decreased the modulus until about 20 % PCHE, where the modulus became dominated by the hydrogenated polybutadiene phase. Further reduction in

Development of Elastomers 359

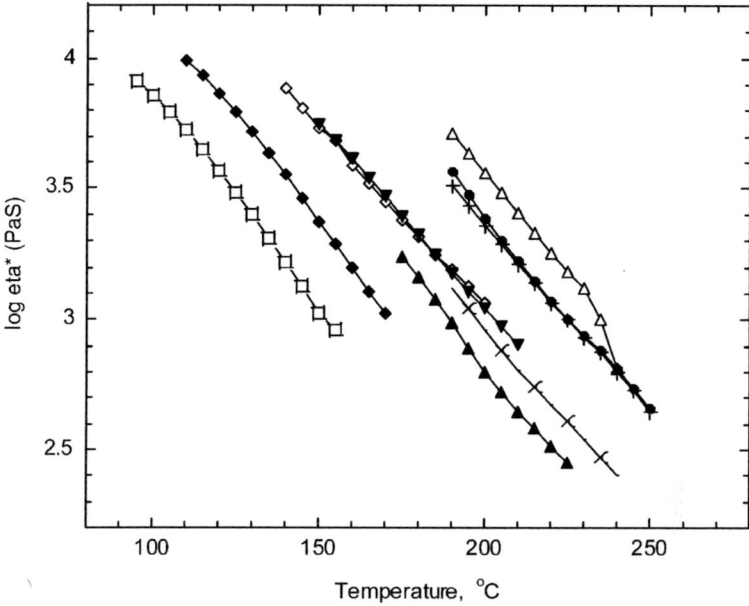

Figure 11.12 Melt viscosity of CEBC 66.32.40 (△) and blends with squalene (12%, ◇; 24% □) mineral oil (12%, ○; 21% ◆) and low molecular weight polyethylenes (12% PE 500, ●; 36% PE 500, ×; 12% PE 1000, +; 36% PE 1000, ▲) with respect to temperature at a frequency of 40 rad/s.

modulus was achieved by increasing the 1-butene content of the poly(ethylene-co-1-butene) center block in low PCHE content elastomers.

Compounding of CEBC block copolymers was performed using both mineral oil and low molecular weight poly(ethylene-co-1-octene). Mineral-oil compounding led to a decrease in modulus and in melt viscosity, and was accompanied by a decrease in the T_{ODT} with increasing mineral oil content. Blends with low molecular weight poly(ethylene-co-1-octene) also exhibited decreased melt viscosity, but did so without significantly influencing T_{ODT}.

Table 11.3 Influence of additives on fully hydrogenated triblock elastomer 66.32.40.

Additive	Wt%	T_{ODT}	MFR[a]	Tensile str. (MPa)	% elong.	Force at 100% elong. (MPa)	Shore A	% tensile set
None	–	230	4.5	41.2	450	3.6	81.5	9.1
Min. oil	12	175	26.3	35.5	580	2.1	70.8	7.2
Min. oil	24	125	124	25.3	700	1.45	52.8	9.0
Min. oil	36	–	512	3.2	880	1.0	37.2	16.4

[a] 200 °C, 5 kg weight

ACKNOWLEDGEMENTS

The authors would like to thank Molly T. Reinhardt for her work on the synthesis of the base block copolymers, and Matt Larive, Guy Mazlowski, and Avani Patel for the hydrogenation of these samples. Faye Brown and Lisa Lopez-Soto performed a majority of physical property testing of these materials.

REFERENCES

1. *Thermoplastic Elastomers*, 2nd Edn, Holden, G., Legge, N. R., Quirk, R. P., Schroeder, H. E., Eds., Hanser, New York, 1996.
2. Holden, G., Legge, N. R., in ref. 1, Chapter 3.
3. *Anionic Polymerization Principles and Practical Applications*, Hsieh, H., Quirk, R. P., Marcel Dekker: New York, 1996.
4. Bates, F. S., Fredrickson, G. H. *Physics Today*, 1999, **52**, 32.
5. Bates, F. S., Fredrickson, G. H. *Annu. Rev. Phys. Chem.* 1990, **41**, 525
6. *The Physics of Block Copolymers*, Hamley, I. W., Oxford University Press, New York, 1998.
7. Balsara, N. P. *Thermodynamics of Polymer Blends, pp 257–268, in Physical Properties of Polymers Handbook*, J. E. Mark ed., AIP Press, New York, 1996.
8. Bates, F. S., Fredrickson, G. H., Hucul, D. A., Hahn, S. F. *AIChE J.* 2001, **47**, 762–765.
9. McGrath, M. P., Sall, E. D., Tremont, S. J. *Chem. Rev.* 1995, **95**, 381–398.
10. Hahn, S. F. *J. Polym. Sci. Part A: Polym. Chem.* 1992, **30**, 397–408.
11. Carella, J. M., Graessley, W. W., Fetters, L. J. *Macromolecules*, 1984, **17**, 2775–2786.
12. Gotro, J. T., Graessley, W. W., *Macromolecules* 1984, **17**, 2767–2775.
13. Hucul, D. A., Hahn, S. F. *Adv. Mater.* 2000, **12**, 1855–1858.
14. Hildebrand, J. H., Scott, R. L. *The Solubility of Non-Electrolytes*, Dover, New York 1964.
15. Gergen, W. P., Lutz, R. G., Davison, S. in Ref. 1, p. 297.
16. Gehlsen, M. D., Bates, F. S. *Macromolecules*, 1993, **26**, 4122–4127.
17. Han, C. D., Chun, S. B., Hahn, S. F., Harper, S. Q., Savickas, P. J., Meunier, D. M., Li, L., Yalcin, T. *Macromolecules*, 1998, **31**, 394–402.
18. Adams, J. L., Quiram, D. J., Graessley, W. W., Register, R. A., Marchand, G. R. *Macromolecules*, 1998, **31**, 201–204.
19. Cochran, E. W., Bates, F. S. *Macromolecules* 2002, **35**, 7368–7374.
20. Han, C. D., Lee, K. M., Choi, S., Hahn, S. F. *Macromolecules*, 2002, **35**, 8045–8055.
21. Haefele, W. R., Dallas, C. A., Deisz, M. A. U.S. Patent 3,333,024, 1967.
22. Jones, R. C. U.S. Patent 3,432,323 1969.
23. Hamley, I. W., Fairclough, J. P. A., Terrill, N. J., Ryan, A. J., Lipic, P. M., Bates, F. S., Towns-Andrews, E. *Macromolecules*, 1996, **29**, 8835–8843.
24. Pendleton, J. F., Hoeg, D. F., Goldberg, E. P. US Patent 3,598,886, 1971.
25. Pendleton, J. F., Hoeg, D. F., Goldberg, E. P. *Adv. Chem. Ser.* 1973, **129**, 27–38.
26. Patel, R. M., Hahn, S. F., Esneault, C., Bensason, S. *Adv. Mater.* 2000, **12**, 1813–1817.
27. Melt-flow rate measurements were obtained using ASTM D 1238 test method.
28. ASTM D 1822 Type L die was used to prepared tensile samples.

29. Odell, J. A., Keller, A. *Polym. Eng. Sci.*, 1977, **17**, 544–559.
30. Arridge, R. G. C., Folkes, M. J. Block Copolymers and Blends as Composite Materials, in *Processing, Structure and Properties of Block Copolymers*, Elsevier Applied Science Publishers, New York, 1985.
31. Honeker, C. C., Thomas, E. L. *Chem. Mater.* 1996, **1702**, 1714.
32. Daniel, C., Hamley, I. W., Mortensen, K. *Polymer,* 2000, **41**, 9239–9247.
33. Honeker, C. C., Thomas, E. L., Albalak, R. J., Hajduk, D. A., Gruner, S. M., Capel, M. C. *Macromolecules*, 2000, **33**, 9395–9406.
34. Honeker, C. C., Thomas, E. L. *Macromolecules*, 2000, **33**, 9407–9417.
35. Mischenko, N., Reynders, K., Scherrenberg, R., Mortensen, K., Fontaine, F. Reynears, H. *Macromolecules*, 1994, **27**, 2345–2347.
36. Mischenko, N., Reynders, K., Koch, M. H. J., Mortensen, K., Pederson, J. S., Fontaine, F., Graulus, R. Reynaers, H. *Macromolecules*, 1995, **28**, 2054–2062.
37. Kraton G1652 is a product of Kraton Polymers. This polymer is an SEBS with 30% polystyrene and a Shore A hardness of 75 (further information is available at *www.kraton.com*).
38. The sample designated as PE 30 is commercially available under the tradename Affinity EG 8200, and PE 5 is available as ENGAGE 8400, both products of the Dow Chemical Company.

Index

ABC triblock 7, 16, 72, 82, 113, 129, 161–162, 165, 177, 179–181, 183, 188, 192–195
Additives, effect of 355–358, 359
Adhesion 163, 165, 327
Adhesion promotor 91
Amphiphilic copolymers 1, 14, 202
Anionic polymerization 2, 16, 32–33, 40–59, 95, 100–102, 113, 214, 216, 298–299, 342, 343, 345–346, 347
Antinucleation 16
Architecture 1–2, 32, 50–56, 72
Artificial skin 246
Asphalt, 198
Asymmetric wetting 9
Atomic force microscopy 11, 18, 217, 234, 235, 300, 301, 303, 313, 348, 349–351
Atom Transfer Radical Polymerization (ATRP) 3, 53, 73, 75–78, 85, 86, 89–90, 93, 96–97, 105–106, 11, 112
Avrami kinetics 131–133, 230–232, 235, 238–239

Bead model 251, 257–258
Bicontinuous structure 202, 245, 255–256. See also Gyroid structure, Double Diamond structure
Bioadhesive 327–328
Biodegradability 331
Blend miscibility 19, 22–23, 162–165
Blends of block copolymers 22–23, 183–197
Block copolymer/homopolymer binary blends 19–20, 165–172, 177–183
Block copolymer/homopolymer ternary blends, 21–22, 172–177
Body-Centred Cubic (BCC) structure 4, 128, 129, 131, 135–137, 141, 142, 149, 159, 168, 169, 201, 265, 276, 280, 288, 291, 302. See also Spherical structure
Breakout crystallization 227–228, 236–237, 239
Bridging 177–178, 191–192, 194, 199
Brownian motion 254
Brush, see Polymer brush

Cancer treatment 327
Capillary electrophoresis 335
Carnahan-Starling equation 13
Cationic polymerization 2, 3, 48–49, 53, 57, 59, 95, 96–98, 102, 112, 113, 214
Ceramic 202
Chain folding 15, 16–17, 215, 228–229
Chain localization 20, 164–165, 178, 191–192
Chemical patterning 303
Chi parameter, see Flory-Huggins interaction parameter
Cleaning 331–332
Close-Packed Sphere structure 128, 129, 152, 169
Coating 237
Comb-shaped copolymer 54, 88, 105, 115–118
Compatibilization 19, 21, 91, 163–165, 172, 195
Composition fluctuations 5, 21
Confined crystallization 16, 143, 178, 218–228, 236–239
Conformational asymmetry 5, 129
CONTIN 13
Continuum model 246, 252
Controlled radical polymerization 3, 31, 73–79, 90, 96–97, 99, 105, 111, 112, 214, 299
Convection-diffusion model 253–254
Core-shell cylinder structure 195

Core-shell gyroid structure 195
Corrosion inhibition 335–336
Coupling reaction 50, 54–55, 90–91, 114, 117
Critical micelle concentration (cmc) 12
Crosslink 1, 129
Crystal thickening 215
Crystallinity 16, 145, 150, 215, 216, 219
Crystallite orientation, see stem orientation
Crystallization kinetics 16, 18, 141, 170, 227, 230–239
Cubic phase (micellar) 12, 14
Cyclic copolymers 2–4, 55–56
Cylindrical micelles 12, 14, 198
Cylindrical structure. See Hexagonal structure
Cylindrical structure (thin films) 261–262, 300, 303–309, 313, 351–352

Degree of crystallinity, see Crystallinity
Demicellization 177
Demulsification 329
Density fluctuations, during crystallization 151, 152
Depolarized light scattering 130
Detergency 331
Dewetting 9
Dielectric relaxation 177
Difunctional initiator 41, 54–55, 57, 98, 106, 112
Differential scanning calorimetry (DSC) 141, 150, 169, 216, 234, 236
Directional crystallization 310
Domain spacing scaling 18, 170–171, 184–185
Double diamond structure 20, 131
Double gyroid structure, see Gyroid structure
Drug delivery 246, 247, 255, 327–328
Dynamic light scattering 13
Dynamic mechanical analysis 348, 349, 356
Dynamical equation 253–254

Elastomeric property 342, 346, 347–359
Electric field alignment 306–309, 318
Elongation at break 346, 355–356

Embossing 306, 311, 318–319
Emulsification 128, 289, 292
Epitaxial transition 128, 289, 292
Epoxy 198
Etching 301, 306, 309, 311, 313, 314–318
Extensional flow 219. See also Roll casting

Face-Centred Cubic (FCC) structure, see Close-Packed Sphere structure
Flocculation 329–331
Flory-Huggins interaction parameter 4, 159, 171, 194, 246, 253, 255, 257–258, 261, 268–269, 297, 342, 344, 345
Flory-Huggins theory 245–251
Flow-induced orientation. See Shear-induced orientation
Flower micelle 199
Fluropolymer 43–45, 49, 75
Freezing point depression 236

Gaussian curvature 181
Gel 199–201, 246, 247, 327–329, 333, 336
Ginzburg-Landau model 246, 292
Gradient copolymer 117
Graft copolymer 1, 114–118, 181
Grain boundary 129, 183
Gyroid structure 4, 128, 129, 130, 131, 133–135, 139, 167–168, 187–188, 195, 213, 265, 276, 292
Gyroid structure (crystallization in) 225, 238

Hard spheres 13–14
Helfand model 3, 250, 266, 273, 276, 281
Herringbone structure 130, 134
Heterogeneous nucleation 230
Hexagonal perforated layer structure. See Perforated layer structure
Hexagonal structure (bulk) 4, 128, 129, 131–137, 141–142, 159, 167–168, 180, 186, 188, 213, 265, 276, 280, 288–289, 291, 302
Hexagonal structure (crystallization) 16–18, 219–220, 222, 224, 234–239
Hexagonal structure (film) 11, 261–262, 297

High-impact materials 246
High Internal Phase Emulsion 165
Holes 9, 11, 18, 351
Homogeneous nucleation 230, 236, 238
Hydrogenation 16, 41, 54, 145, 188–189, 214–215, 299, 341, 343–346, 348
Hydrolysis 41, 107
Hydrosilylation 50, 52, 93

Incommensurability 9, 18
Interfacial curvature 165–166, 180–181, 182, 186, 194–195, 200
Interfacial tension 21, 165, 172, 203
Invariant 150
Ionomer 204
Islands 9, 11, 18, 303–305

Kinetic nucleation theory 18
Knitting pattern structure 196–197
Kraton 7, 92

L_3 phase 197
Lamellar structure (bulk) 4, 7, 128–133, 138–141, 145–146, 150–151, 159, 166, 167, 174, 177, 181, 183, 186, 188–189, 193–194, 196, 213, 265, 276, 280, 288–289
Lamellar structure (crystal) 16–18, 144–148, 150–151, 218, 221–224, 228–230, 238, 239
Lamellar structure (film) 9, 10, 11–12, 171, 297, 310
Lamellar structure (solution) 12, 197–198, 201, 245, 255–256
Lattice model 246, 250–251
Leibler theory 3, 266, 273
Lifshitz point 21, 174, 188, 191
Light scattering 4, 12, 148, 150–152
Liquid chromatography 40
Linking reaction 54–55. See also Coupling reaction
Lithography 295–297, 303–306, 308–310, 313, 314–320
Living radical polymerization. See Controlled radical polymerization
Looping 178, 199
Lubricant 334–336
Lyotropic phase 12, 14–15, 255

Macroinitiator 3, 32, 51–54, 55, 56–59, 74–78, 85–86, 88–94, 99–103, 106–107, 111, 112–114
Macrophase separation 19–20, 21, 22, 163, 172–174, 178, 184–185, 188, 190, 245, 297, 358
Magnetic media 309, 318
Magnetic field alignment 152
MALDI-TOF 38, 40
Mechanism transformation 95
Mean-field theory 246–263, 265–293. See also Self-consistent field theory
Melt viscosity 341, 343, 346, 348, 349, 354–359
Mesodyn 246–264
Metal evaporation 311–312
Methacrylate polymers 40–43, 49, 74–75, 77–79, 85–86, 88, 97, 99–100, 102, 105–106, 110
Micelle 1, 12, 128, 164, 245, 255–256, 313, 327
Micelle dimensions 12, 13, 14
Micellization 177
Microemulsion 21, 172–176
Microfluidics 310–311
Microphase separation 9, 20–22, 171–174, 185, 190, 245, 342–343, 346
Miktoarm star. See Mixed arm star
Mixed arm star 72, 106, 112–114, 129, 181, 204
Modulated lamellar structure 289
Modulus 199–200, 347, 351–353, 355, 358–359
Molar mass distribution. See polydispersity
Molecular simulation 246
Molecular weight distribution. See Polydispersity
Multiblock copolymers 1, 2, 71, 72, 177
Multifunctional initiator 51–55, 107, 112, 115, 117

Nanocluster 312
Nanolithography 296
Nanoparticle 14
Nanoreactor 14
Nanowire 308
Neutron reflectivity 11, 19

Nitroxide-mediated polymerization 83–86, 88, 99, 100, 106, 110
NMR 4
Non-centrosymmetric lamellar structure 193–194

One component approximation. See Single component approximation.
Onsager coefficient 253–254
Order-disorder transition (ODT) 4, 5, 7, 9, 12, 127, 141, 148, 152, 167–168, 172, 174, 177, 196, 226, 341–343, 355–357, 359
Order-disorder transition, effect of shear on 148, 152
Order-order transition (OOT) 127, 128, 131–137, 141, 167–168
Ordering pathway 131–141, 246, 287, 289
Ozone etching 309, 314–316

Pattern dynamics 152, 246
Pentablock copolymer 40, 129
Perforated lamellar structure 4, 20, 131, 133, 159, 167, 262–262, 289, 290
Perforated lamellar structure (crystallization) 225
Personal care applications 332–333
Photoinduced controlled radical polymerization 86–88, 105
Photolithography 295, 304, 314
Physical gel 199–201
Plasma etching 301, 314, 316
Polarized optical microscopy 135, 137, 151, 223, 230
Polymersome 198
Pluronic 2, 12, 15, 248, 255, 325, 327, 329–334, 336–337
Polydispersity 33, 35–40, 298
Polyelectrolyte 2
Polymer brush 14, 163, 165, 166, 170, 177, 180, 184
Polymer brush theory 18
Polyurethane 1, 171
PRISM theory 267
Propagator 252, 270–274, 276–278

Radical polymerization 3, 31. See also Controlled radical polymerization
Random copolymer 172, 181, 183, 195–196
Random phase approximation 250, 273, 282–288
Reactive ion etching 301, 306, 313, 316
Resist 299
Reversible Addition-Fragmentation Transfer (RAFT) 73, 79–83, 88, 90, 92, 94, 96
Rheology 5, 14, 142, 172, 199–200, 348
Ring-opening polymerization 4, 32, 56–59, 96–98, 99, 112, 214–215, 298
Rod-coil copolymer 9
Roll casting 129, 309
Rouse dynamics 254

Scanning electron microscopy (SEM) 300–310
Scattering function 287–289
Schulz-Zimm distribution 36–38
Scheutjens-Fleer model 250–251
Self-consistent field theory 5–7, 12, 14, 15, 21, 22, 127–128, 131, 176–178, 184, 186, 200, 250, 261, 265–292
Semiconductor patterning 295–296, 303–306, 311–320
Shear alignment 7, 14, 17, 129–130, 149, 221, 258–260, 309
Shore A hardness 351–354
Silica 202
Siloxane 47–48
Single component approximation 22, 186
Size-exclusion chromatography (SEC) 38
Small-angle light scattering. See Light scattering
Small-Angle Neutron Scattering (SANS) 12, 13, 145, 148, 310
Small-Angle X-ray Scattering (SAXS) 4, 12, 13, 17, 19, 129–132, 138, 144–145, 148–152, 168, 172, 184–185, 216–217, 221–224, 228, 231–233, 235, 258
Spherical structure (films) 11, 297, 300, 304–306
Spherical structure (crystallization) 219, 226–227, 231–237, 239
Spherulite 16, 130, 148, 151, 152, 217, 218, 223, 226, 231, 239
Spin coating 9, 299, 304, 316

Spinodal 172, 286
Sponge structure 181–182, 197
Stable free radical polymerization 83, 96, 105–106
Star block copolymer 50–55, 107–111, 181
Star homopolymer 50, 58–59
Stem orientation 17–18, 143–145, 152, 219–226, 228–230, 238–239
Stepwise synthesis 32, 40
Stokes-Einstein equation 13
Stripe pattern 11
Strong segregation limit 5, 12, 21, 184, 237, 266, 275
Supercritical CO_2 76, 330–331
Suppository 327–328
Surface-induced order 9
Surface tensiometry 12
Swelling 171, 178, 181, 202
Symmetric wetting 9

Templated crystallization 236–238
Templating (of inorganic material) 15, 201–202, 297
Tensile strength 346–349, 351–358
Terracing 18, 261
Tetrafunctional initiator. See Multifunctional initiator
Thermoset 198
Time-dependent Ginzburg-Landau model 292
Tooth whitening 333
Topography 303–306

Transfer agent 95, 114
Transformation polymerization 94–106
Transmission electron microscopy (TEM) 4, 8, 11, 19, 20, 131, 139–140, 164, 168, 172, 173, 176, 181–182, 184, 189, 195, 219–221, 227–238, 235–236, 300–301
Trifunctional initiator. See Multifunctional initiator

Undulation mode 134, 137, 141
Upper Critical Solution Temperature (UCST)
UV degradation 318

Vesicle 14, 198, 204, 245

Weak segregation limit 5, 12, 148, 166, 184, 237, 266, 273, 275, 288
Wide-Angle X-ray Scattering (WAXS) 17, 19, 130, 144–152, 216–217, 219, 221–224, 231–233
Wigner-Seitz cell 128, 131, 180, 281
Wound covering 327

X-ray reflectivity 11, 230
X-ray scattering. See Small-Angle X-ray Scattering.

Zimm plot 13
Zone refining 310
Zwitterionic end group 31